Principles of Thermodynamics

JUI SHENG HSIEH, Ph.D.
Professor of Mechanical Engineering
New Jersey Institute of Technology

SCRIPTA BOOK COMPANY
Washington, D.C.

McGRAW–HILL BOOK COMPANY

New York   St. Louis   San Francisco   Düsseldorf   Johannesburg
Kuala Lumpur   London   Mexico   Montreal   New Delhi   Panama
Paris   São Paulo   Singapore   Sydney   Tokyo   Toronto

Library of Congress Cataloging in Publication Data

Hsieh, Jui Sheng, date.
  Principles of thermodynamics.

  1. Thermodynamics.   I. Title.
QC311.H78   536'.7   73–21592
ISBN 0–07–030630–3

PRINCIPLES OF THERMODYNAMICS

1 2 3 4 5 6 7 8 9 0   K P K P   7 9 8 7 6 5

This book was set in Press Roman type by Scripta Graphica.
The supervising editor was Glenda Hightower;
the designer was Victor Enfield; and the production
supervisor was Keith Wilkinson. The compositor was Bernie
Doenhoefer.
It was printed and bound by The Kingsport Press.

To my wife
  Mary Cheng-Lian Wang Hsieh
and our children
  Lawrence Szeche
  Esther Szeyu
  Vivian Szelun

CONTENTS

PREFACE

Thermodynamics is relevant to many branches of science and engineering and its basic laws have a wide range of applicability. This book aims at providing a clear and unified treatment of various thermodynamic systems. The diversity of topical coverage will help the student to understand the power of thermodynamic arguments in solving practical problems for present and future applications.

This book is intended primarily for a second course in classical thermodynamics at the senior or first-year graduate level. For readers who have no prior knowledge of thermodynamics, a comprehensive review of the fundamentals of the first and second laws is given in Chap. 1. This is followed in Chap. 2 with a detailed review of the general thermodynamic relationships pertaining to simple compressible systems. These general relations are then applied in the next three chapters to three types of compressible systems. Chapter 3 is devoted to the equations of state and the evaluation of thermodynamic properties of single-component gas systems. Chapter 4 introduces the fundamentals of multicomponent systems, including gaseous mixtures and liquid solutions. Chapter 5 deals with stability and multicomponent phase equilibrium. Beginning in Chap. 6, the book turns to systems other than compressible. Elastic systems and systems with interfacial tension are treated in Chap. 6. Systems involving external force fields, including magnetic, electric, and gravitational, are treated in Chap. 7. The next three chapters relate to the thermodynamics of low temperatures. Chapter 8 deals with the production of cryogenic temperatures, including gas-liquefaction, and magnetic cooling. Chapter 9 introduces the subject of superconductivity and superfluidity. Chapter 10 is devoted to the third law of thermodynamics and negative Kelvin temperatures. A study of reaction equilibrium in Chap. 11 concludes the book.

Throughout the preparation of this book the student has been foremost in the author's mind. Sufficient detail is given in the presentation of the subject matter. Important derivations are not left to the student but are contained in the main body of the text. A large number of completely solved examples are provided to illustrate the theories and applications. In addition, there are homework problems at the end of each chapter. Answers to most of these problems are provided at the end of the book.

The amount of background in mathematics assumed for the reader is a knowledge of calculus. A few mathematic procedures that require special attention are described in the book.

All major countries except the United States use the metric system of units. Now the United States has committed herself to convert to metric units. However, since at the present time it is essential for the students to have the ability to work with a variety of units, various systems of units are used in the examples and problems. Nevertheless, all basic relations in this book have been developed in a general way and are valid for any system of units.

Tables and diagrams of thermodynamic properties for a number of substances are included in the Appendix. A table of conversion factors is also included.

This book is designed to be a textbook for a one-semester course. However, sufficient material has been provided to allow the instructor considerable latitude in the choice of topics in order to meet the needs of his students. The following list of chapters and sections may be considered as a balanced package for mechanical engineering students:

Chaps. 2 and 3
Secs. 4-1 through 4-4
Secs. 5-1 through 5-4, 5-6 and 5-9
Secs. 6-1 through 6-5
Secs. 7-1 through 7-3, 7-5 and 7-6
Secs. 8-1 through 8-6
Secs. 9-1, 9-2, and 9-6
Secs. 10-1, 10-2, 10-5, and 10-6
Secs. 11-1 through 11-4

If chemical engineering aspects are to be emphasized, Chaps. 4, 5, 10, and 11 should be covered without omission.

It is a great pleasure to acknowledge my indebtedness to Dr. W. B. Kay of the Ohio State University who has kindly read through the entire original manuscript and made many valuable suggestions. I am also indebted to Drs. R. P. Kirchner and C. E. Wilson of New Jersey Institute of Technology who also read the entire manuscript and made helpful suggestions. I would like to express my appreciation to Dr. J. P. Holman of Southern Methodist University, Dr. W. R. Schowalter of Princeton University, and Dr. J. H. Potter of Stevens Institute of Technology for the valuable comments that resulted from their reading of the manuscript. A special debt of gratitude is due to Mr. B. J. Clark of McGraw-Hill

Book Company for his editorial suggestions and comments. Finally, I wish to express my thanks and appreciation to my wife Mary for her diligence and patience in typing the manuscript.

J. S. Hsieh

LIST OF SYMBOLS

Latin letters

a	Activity
a, A	Specific Helmholtz function and total Helmholtz function
A	Area
A_c	As defined in Eq. (8-9)
a, b, c, A, B, C	Constants in equations of state
$\mathbf{B}$	Magnetic induction
B, C, D	Second, third, and fourth virial coefficients
B_J	Brillouin function
c	Specific heat
$\mathbf{C}$	Capacitance
C_c	Curie constant
c_p, C_p	Specific heat and heat capacity at constant pressure
c_v, C_v	Specific heat and heat capacity at constant volume
c_σ, C_σ	Specific heat and heat capacity at constant stress
c_ϵ, C_ϵ	Specific heat and heat capacity at constant strain
C_L	Heat capacity at constant length
$C_{\mathbf{H}}$	Heat capacity at constant magnetic field
$C_{\mathbf{M}}$	Heat capacity at constant magnetic moment
$c_{\sigma, \mathbf{E}}$	Specific heat at constant stress and constant electric field
$c_{\sigma, \mathbf{P}}$	Specific heat at constant stress and constant polarization
$c_{\epsilon, \mathbf{E}}$	Specific heat at constant strain and constant electric field
$c_{\epsilon, \mathbf{P}}$	Specific heat at constant strain and constant polarization
$\mathbf{D}$	Electric displacement
D_m	Demagnetizing factor
e	Electronic charge
e	Naperian logarithm base
e, E	Specific stored energy and total stored energy
E	Electric potential
$\mathbf{E}$	Electric field or intensity
f	Fugacity
F	Degree of freedom
F	Force

g, G	Specific Gibbs function and total Gibbs function
g	Gravitational acceleration
g_L	Landé splitting factor
ΔG_R	Gibbs function change of reaction
h, H	Specific enthalpy and total enthalpy
ΔH_F	Enthalpy of formation
ΔH_R	Enthalpy of reaction
$\mathbf{H}$	Magnetic field or intensity
$\mathbf{H}_c$	Critical or threshold field of a superconductor
$\mathbf{H}_0$	Critical field at $0°\,\mathrm{K}$
i	Electric current
$\mathbf{I}$	Total magnetic moment
J	Total angular momentum quantum number
k	Ratio of specific heats c_p/c_v
k_B	Boltzmann constant
k_2	Proportional constant in Henry's law
$K(T)$	Equilibrium constant
l, L	Length
L	Latent heat
m	Mass
$\dot{m}$	Mass rate of flow
m	Gram-ionic mass
M	Molecular weight
$\mathbf{M}$	Magnetization or magnetic moment per unit volume
n	Number of moles
N	Number of paramagnetic ions
N	Number of turns of winding
N_A	Avogadro's number
p	Pressure
p_c	Critical pressure
p_r	Reduced pressure p/p_c
p_r	Relative pressure as used in gas tables
$\mathbf{P}$	Electric polarization or electric dipole moment per unit volume
$\mathbf{P'}$	Total electric moment
q, Q	Heat transfer per unit mass and total heat transfer
$\dot{Q}$	Rate of heat transfer
r	Number of components
r, R	Radius
R	Universal gas constant
R_i	Individual gas constant

s, S	Specific entropy and total entropy
ΔS_R	Entropy change of reaction
t, T	Temperature and absolute temperature
T_c	Critical temperature
T_0	A standard reference temperature
T_0	Transition temperature of a superconductor
T_r	Reduced temperature T/T_c
T^*	Magnetic temperature
u, U	Specific internal energy and total internal energy
ΔU_R	Internal energy of reaction
v, V	Specific volume and total volume
v_c	Critical volume
v_r	Reduced volume v/v_c
v_r	Relative volume as used in gas tables
v, V	Velocity
v_2	Velocity of second sound
w, W	Work per unit mass and total work
$\dot{W}$	Rate of work or power
x	Mole fraction
x	Quality
y, Y	A molal property and corresponding extensive property
Y	Young's isothermal modulus of elasticity
Y_s, Y_T	Adiabatic and isothermal elastic stiffness coefficients
z	Elevation
Z	Compressibility factor
Z_c	Critical compressibility factor
Z	Electric charge

Greek letters

α	An equation of state constant
α	Coefficient of thermal expansion
α	Coefficient of thermal strain
β	Coefficient of thermal stress
γ	Activity coefficient
γ	An equation of state constant
γ	Surface tension
Γ	As defined in Eq. (6-55)
Δ	Finite difference
ϵ	Strain
ϵ_0	Permittivity of free space

ζ	As defined in Eq. (8-4)
η	Thermal efficiency
θ	Angle
Θ	A characteristic constant
κ	Dielectric constant
κ	Isothermal compressibility
κ_{G-L}	Ginzburg-Landau constant
κ_s	Adiabatic compressibility
κ_s, κ_T	Adiabatic and isothermal elastic compliance coefficients
λ	Penetration depth
μ	Chemical potential
μ_B	Bohr magneton
μ_J	Joule-Thomson coefficient
μ_0	Permeability of free space
ν	Stoichiometric coefficient
ξ	Coherence length
π	Osmotic pressure
π	$\pi = 3.14159$
ρ	Density
ρ_n	Density of normal fluids
ρ_s	Density of superfluids
σ	Stress
τ	Thickness of surface layer
τ	Time
φ	Number of phases
ϕ	$\phi = \displaystyle\int_{T_0}^{T} c_p \frac{dT}{T}$ as defined in gas tables
χ	Magnetic susceptibility
Ψ	Centrifugal potential
Ψ	Gravitational potential
ω	Angular velocity

Subscripts

f	Saturated liquid
fg	Difference in property for saturated vapor and saturated liquid
g	Saturated vapor
i	Saturated solid
i	ith component in a multicomponent system

1	Component 1
1	State 1

Superscripts

—	Bar over symbol (such as $\bar{V}_i$, $\bar{U}_i$, $\bar{H}_i$, $\bar{S}_i$, $\bar{G}_i$, and $\bar{A}_i$) denotes partial molal property
$\circ$	Pure component
*	Ideal gas state
$\oplus$	Standard state
(α)	α phase
(β)	β phase
(n)	Normal conducting phase
(s)	Superconducting phase
(Δ)	Interfacial layer

Special notations

d	Differential change in a property
$đ$	Differential change in a path function
δ	Virtual variation away from equilibrium state
Δ	Δ = final − initial
$(\partial y/\partial x)_z$	The partial derivative of y with respect to x, keeping z constant
$\displaystyle\sum_{i=1}^{r} y_i$	The sum $y_1 + y_2 + \cdots + y_r$

Symbols for units

amp	Ampere
atm	Atmosphere
bar	Bar
Btu	British thermal unit
c	Centi, a prefix with a multiple of 10^{-2}
coul	coulomb
°C	Degree Celsius
cal	Calorie
d	Deci, a prefix with a multiple of 10^{-1}
ft	Foot
°F	Degree Fahrenheit
g	Gram
hp	Horsepower

hr	Hour
in.	Inch
k	Kilo, a prefix with a multiple of 10^3
kJ	Kilojoule
kN	Kilonewton
°K	Degree Kelvin
lbf	Pound force
lbm	Pound mass
m	Meter
min	Minute
mm Hg	mm of mercury
psia	Pound force per square inch absolute
rpm	Revolution per minute
°R	Degree Rankine
sec	Second

Principles of Thermodynamics

Chapter 1
FUNDAMENTALS

1-1 DEFINITIONS

Thermodynamics is the basic science which deals with heat and other forms of energy, their transformations, and their relationships with properties of substances. The science of *classical thermodynamics* has been developed without an inquiry into the structure of matter. It is concerned only with the average characteristics of large aggregations of molecules, not with the characteristics of individual molecules. In other words, classical thermodynamics takes the macroscopic point of view and deals with macroscopic phenomena. On the other hand, *statistical thermodynamics* considers the microscopic structure of matter and adopts the laws of mechanics on the statistical analysis of the individual particles. This text treats selected topics on classical thermodynamics.

In a thermodynamic analysis a collection of matter within certain prescribed boundaries is specified in order to focus study. The boundaries may be either movable or fixed. The collection of matter thus specified is called a thermodynamic *system*, and everything external to the system which has relation with the system is called the *surroundings* or the *environment*. A system of fixed content will be referred to as *closed*, and one whose content can be varied by passage of matters across its boundaries will be referred to as *open*.

The *state* of a system at any instant is its condition of existence at that instant. A *property* of a system is any quantity or characteristic that depends upon the state of the system. Thermodynamic properties may be classified as intensive properties and extensive properties. An *intensive property* of a system is independent of the extent of mass of the system. Pressure, temperature, and specific volume are familiar examples of intensive properties. An *extensive property* of a system is one whose value for the entire system is equal to the sum of its values for all parts of the system. Volume, energy, and mass are examples of extensive properties.

When any property of a system changes there is a change of state, and the system is said to undergo a *process*. A process whose initial and final states are identical is called a *cycle*.

A property of a system depends only on the state of the system and not on how that state was attained. The uniqueness in the value of a property at a state introduces the name *state* or *point function* for a property. In contrast, quantities which depend on the path of the process by which a system changes between two states are called *path functions*. Because a property is a point function, its differential must be an exact or perfect differential in mathematical term. The line integral of the differential of a property is independent of the path or curve connecting the end states, and in the special case of a complete cycle this integral vanishes.

A quantity of matter which is homogeneous in chemical composition and in physical structure is called a *phase*. All substances can exist in solid, liquid, and gaseous phases; solids can exist in different forms. A system consisting of a single phase is called a *homogeneous* system; a system consisting of more than one phase is called a *heterogeneous* system.

1-2 THERMODYNAMIC EQUILIBRIUM

A system is in a state of equilibrium if a change of state cannot occur while the system is not subject to interactions with the environment. When a system is in equilibrium there must be no unbalanced potential which tends to promote a change of state. The unbalanced potential may be mechanical, thermal, chemical, or electrical, or any combination of them. When there is no unbalanced force within a system, it is said to be in a state of *mechanical equilibrium*. If the change in pressure with elevation due to gravitational force is neglected, a system of fluid in a state of mechanical equilibrium should have uniform pressure. When there is no temperature gradient within a system, it is said to be in a state of *thermal equilibrium*. When a system has no tendency to undergo either a chemical reaction or a process such as diffusion or solution, the system is said to be in a state of *chemical equilibrium*. When there is no electrical potential gradient within a system, it is said to be in a state of *electrical equilibrium*. If all the conditions for mechanical, thermal, chemical, and electrical equilibrium are satisfied, the system is said to be in a state of *thermodynamic equilibrium*, or simply in an *equilibrium state*. When any of the conditions for a complete thermodynamic equilibrium is not satisfied, the system is said to be in a *nonequilibrium state*. Classical thermodynamics deals only with systems in equilibrium states.

When a homogeneous system is in thermodynamic equilibrium its state can be represented by a single point on a set of thermodynamic coordinates. If it is not in thermodynamic equilibrium, some of its macroscopic coordinates vary from one part to another and are not single-valued and unique so that it is impossible to locate a single point on a state diagram.

1-3 REVERSIBLE AND IRREVERSIBLE PROCESSES

During a thermodynamic process some unbalanced potential must exist either within the system or between it and the surroundings to promote the change of state. If the unbalanced potential is infinitesimal so that the system is infinitesimally close to a state of equilibrium at all times, such a process is called *quasistatic*. A quasistatic process may be considered practically as a series of equilibrium states and its path can be represented graphically as a continuous line on a state diagram.

In contrast, any process taking place due to finite unbalanced potentials is *nonquasistatic*. Such a process cannot be described by means of macroscopic coordinates that are characteristic qualities of the entire system. Thus the complete path of a nonquasistatic process is indeterminate and cannot be represented on a state diagram, although it is possible to indicate the general direction of change by locating the terminal equilibrium states.

A system is said to have undergone a *reversible* process, if at the conclusion of the process, the initial states of the system and the surroundings can be restored without leaving any net change at all elsewhere. If when the system and the surroundings are restored to their respective initial states there are net changes left elsewhere, the process is said to be *irreversible*.

A reversible process must be quasistatic, so that the process can be made to traverse in the reverse order the series of equilibrium states passed through during the original process, with no change in magnitude of any energy transfer but only a change in direction. Besides being quasistatic, a reversible process must not involve any effects associated with such phenomena as solid or fluid friction, electric resistance, inelastic deformation, and hysteresis in magnetization or polarization.

Reversible processes are idealized processes which can never be realized in the laboratory but can be approximated as closely as one wishes. For example, a gas confined in a cylinder with a well-lubricated piston can be made to undergo an approximately reversible process by pushing or pulling the piston in slow motion or by dividing the process into very small steps. This is true because in the limit in any stage of the process it could be turned into the opposite direction by an infinitesimal change of the external conditions.

1-4 TEMPERATURE AND THERMOMETRY

When two systems which initially are each in thermal equilibrium are brought in thermal contact through a rigid metal plate, while isolated from all other surroundings, each of the two systems will in general undergo a certain change of state until mutual thermal equilibrium between the two is reached. Two systems in mutual thermal equilibrium have a common characteristic, namely that of being in thermal equilibrium with each other. This characteristic is solely a function of the states of the systems, and therefore is a property of each system. This property is the temperature.

When two systems are each in thermal equilibrium with a third system, they are in thermal equilibrium with each other. This statement is known as the *zeroth law* of thermodynamics. In terms of temperature the zeroth law states that when two systems are each equal in temperature to a third system, they are equal in temperature to each other. This obvious fact forms the basis of temperature measurement.

In order to establish the means of temperature measurement we choose a reference system for which there is some physical property that changes with temperature. This reference system is called a thermometer and the selected physical property whose changes are used as indication of changes in temperature is called the *thermometric property*. A small quantity of mercury enclosed in a glass capillary is called a mercury-in-glass thermometer, in which the expansion of the mercury is used as the thermometric property. When two wires of different materials are joined at their ends with different temperatures existing at the two junctions, an electromotive force is generated which may be used as a measure of the temperature of one junction if the temperature at the other junction is known; such a thermometer is known as a *thermocouple*. A gas, such as oxygen, nitrogen, or helium, at very low pressures may be considered as an ideal gas. The temperature of such a gas is a unique function of volume when its pressure is held constant, or a unique function of pressure when its volume is held constant. Thus, if the pressure of the gas is held constant while its volume is used as the thermometric property, a constant-pressure gas thermometer is formed. On the other hand, if the volume of the gas is held constant while its pressure is used as the thermometric property, a constant-volume gas thermometer is formed.

To establish quantitative measurements of temperature in terms of numbers, easily reproducible states (called fixed points) of an arbitrary standard system are assigned arbitrary numbers. Formerly, two standard fixed points, the ice point and the steam point, were used. The ice point is defined as the equilibrium temperature of pure ice and air-saturated water, and the steam point as the equilibrium temperature of pure water and water vapor, both at one atmosphere pressure. However, in practice it is difficult to realize the condition of equilibrium between pure ice and air-saturated water as required by the definition of the ice point. As ice melts it forms a layer of pure water around itself which prevents the direct contact of pure ice and air-saturated water. Due to this difficulty in obtaining the ice point with accuracy, the two-fixed-point method was abandoned in 1954 by international agreement. The new method adopts a single standard fixed point, namely, the triple point of water, which is the equilibrium state in which ice, liquid water, and water vapor coexist. A value of 273.16 degrees Kelvin ($^{\circ}$K) on the Kelvin scale is assigned for the temperature at this fixed point. The zero of the Kelvin scale is the absolute zero of temperature.

A temperature scale based on the single-fixed-point method mentioned above may be established through the use of an ideal gas thermometer. Figure 1-1 shows schematically a constant-volume gas thermometer. It consists of a quantity of gas contained in a bulb with a capillary tube leading to a mercury monometer. The volume of the gas is kept constant by raising or lowering the mercury reservoir until the mercury in the left-hand side of the U-tube stands at the level indicated by a fixed mark. The height z of the mercury column gives the gas pressure, which is an indication of the gas temperature.

Let p_{TP} be the absolute pressure of the gas when the bulb of a constant-volume gas thermometer is immersed in water at its triple-point temperature, which has an assigned value of 273.16°K; and p be the absolute pressure of the gas when the bulb is immersed in any system whose temperature T is to be measured. Experiment shows that the value of the ratio p/p_{TP} depends on the kind and the amount of gas in the bulb. However, when the procedure is repeated with less and less gas in the bulb, the limiting value of the

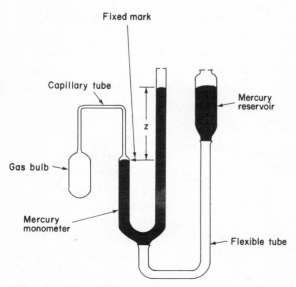

FIG. 1-1. Schematic diagram of a constant-volume gas thermometer.

ratio p/p_{TP}, and therefore the value of $T/273.16$, when extrapolated to $p_{TP} = 0$, become fixed irrespective of the kind of gas used. This is illustrated in Fig. 1-2 which shows the uniqueness of the indicated temperature when $p_{TP} \rightarrow 0$ as given by a gas thermometer using different gases for different series of measurements. Thus, in the limiting condition a unique value of temperature T is defined by the equation

$$T = 273.16 \lim_{p_{TP} \to 0} \left(\frac{p}{p_{TP}} \right) \tag{1-1}$$

The temperature so defined is called the *ideal gas temperature* in degrees Kelvin.

A second absolute temperature scale in use is the Rankine scale, defined by setting

$$T(^\circ R) = 1.8\, T(^\circ K)$$

which corresponds to the assignment of 491.69 degrees Rankine ($^\circ R$) to the triple point of water.

Associated with the Kelvin scale is the Celsius scale on which the triple point of water is, by definition, 0.01 degrees Celsius ($^\circ C$). The relation between the Kelvin and Celsius scales is

$$T(^\circ K) = t(^\circ C) + 273.15$$

Associated with the Rankine scale is the Fahrenheit scale. The relation between

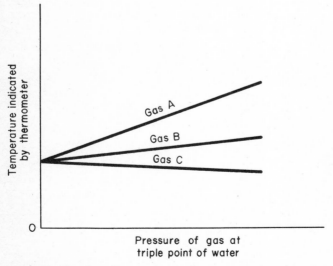

FIG. 1-2. Results of measurement from a constant-volume gas thermometer.

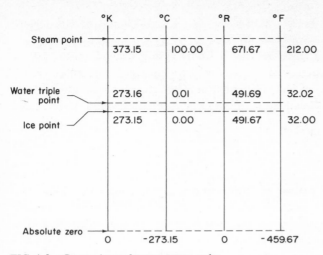

FIG. 1-3. Comparison of temperature scales.

the Rankine and Fahrenheit scales is

$$T(^{\circ}R) = t(^{\circ}F) + 459.67$$

and that between the Fahrenheit and Celsius scales is

$$t(^{\circ}F) = 1.8 \, (^{\circ}C) + 32$$

These four temperature scales are compared in Fig. 1-3.

1-5 HEAT

When two systems at different temperatures are placed in thermal contact with each other and isolated from all other systems, a change in the temperature or other thermodynamic properties in both systems will always occur due to energy transfer. The energy being transferred from one system to another solely by reason of a temperature difference between the two systems is called *heat*. According to this definition, heat can be identified only as it crosses the boundary of a system, and therefore is a form of energy in transit. It is not an energy contained in a system and not a property of a system. Upon entering a system, heat is converted into potential or kinetic energy of the molecules, the atoms, or the subatomic particles that form the system.

As a convention of sign, heat transferred to a system is considered positive and heat transferred from a system is considered negative. The symbol Q is used to represent heat transfer to a system, and q is used to represent heat transfer per unit mass or per mole of the system.

A process which involves no heat interaction between the system and its surroundings is called an *adiabatic* process.

The amount of heat transferred to or from a system depends on the path of the process by which the system changes between two given end states. Heat is therefore a path function. The differential of a path function is inexact and will be denoted by the symbol $\bar{d}$ to distinguish from the symbol d for exact differentials.

1-6 WORK

In mechanics, work is defined as the product of a force and the displacement when both are measured in the same direction. In thermodynamics, work is an interaction between a system and its surroundings by reasons other than a temperature difference. The system under consideration may be, for instance, a compressible fluid, a paramagnetic solid, or a dielectric material. However, a work interaction between a system and its surroundings may not always involve a recognizable force and a displacement. We therefore adopt the following general definition: *Work* is done by a system if the sole effect on the surroundings can be reduced to the lifting of a weight. The magnitude of work is the product of the weight and the distance it could be lifted. It is always possible to reduce the external effect of any work interaction to the lifting of a weight through some mechanism whether it is real or by reasonable imagination.

Like heat, work is an interaction across the boundary of a system and therefore is also a form of energy in transit. Unlike heat, however, work is not an energy in transit because of a difference in temperature. Neither heat nor work is an energy contained in a system and neither are properties of a system.

As a convention of sign, work done by a system is considered positive and work done on a system is considered negative. The symbol W is used to represent work done by a system, and w is used to represent work per unit mass or per mole of the system.

In this section we seek to derive the expression for work in relation to volume changes. Consider as a system a quantity of a compressible fluid contained in a cylinder provided with a movable piston, as shown in Fig. 1-4. As the piston moves to the right a small distance dl under the action of fluid pressure, work is done by the fluid of the amount

$$\bar{d}W = F\,dl$$

where F is the total force on the piston due to fluid pressure. The force F will probably vary as the piston moves, but for any position of the piston it may be written as

$$F = p \, A$$

where p is the fluid pressure against the piston and A the piston area. If the process undergone by the fluid is quasistatic, p is then the uniform pressure of the fluid that is unique throughout the fluid system. Thus we have

$$dW = p \, A \, dl$$

But the product $A \, dl$ is a differential volume dV, so finally,

$$dW = p \, dV \tag{1-2}$$

It is often convenient to write equations for a unit mass of the system. Thus

$$dw = p \, dv \tag{1-2a}$$

where v is the specific volume.

These equations represent the work done by the fluid during an infinitesimal quasistatic process. However, if there are dissipative effects, as when the piston moves with friction, part of the work done by the fluid would become energies of the fluid and the piston-cylinder machine parts. Thus less work would be delivered out through the piston. In other words, Eq. (1-2) applies in general to reversible processes only.

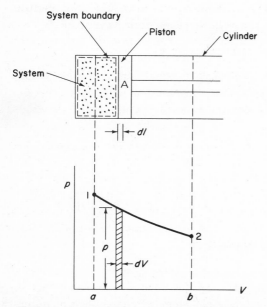

FIG. 1-4. Work done on the moving boundary of a closed system in a reversible process.

The total work done by the fluid as the fluid expands reversibly from state 1 to state 2 is

$$W_{12} = \int_1^2 p \, dV \tag{1-3}$$

This work is represented by the area $12ba1$ under the curve 12 on a p-V diagram as shown in Fig. 1-4. If the process had proceeded from 2 to 1, work of the amount

$$W_{21} = \int_2^1 p \, dV = \text{area } 21ab2$$

would be done on the fluid.

It is possible to connect two states of a system by many different reversible processes. The work done by or on the system depends on the path of the process. Work is therefore a path function.

Although the equation $dW = p \, dV$ was derived for the case of fluid expansion against a piston, it is correct for any closed system with a moving boundary of any shape under a uniform hydrostatic pressure.

In later chapters we will study a few different systems which can perform work reversibly other than $p \, dV$. The expressions for work always take the form

$d(\text{work}) = (\text{intensive property}) \cdot d(\text{extensive property})$

Even though the factors within a given work expression may not bring to mind recognizable force and displacement, it is convenient to refer to all types of work as a product of a force and a displacement, and to write

$d(\text{work}) = (\text{generalized force}) \cdot d(\text{generalized displacement})$

For a system which can perform work reversibly in different independent modes we can write

$d(\text{work})_{\text{total}} = \Sigma \left[(\text{generalized force}) \cdot d(\text{generalized displacement}) \right]$

1-7 THE FIRST LAW OF THERMODYNAMICS

The first law of thermodynamics is essentially the law of conservation of energy applied to thermodynamic systems. Through his famous experiments in 1843 Joule was led to the postulate that heat and work were equivalent quantities. This postulate is generally known as the *first law of thermodynamics*. When a system is carried through a cycle, the first law is expressed by

$$\oint dW = \oint dQ \qquad (1\text{-}4)$$

where both work and heat are measured in the same units.

Equation (1-4) may be rewritten as

$$\oint (dQ - dW) = 0$$

It follows that since the cyclic integral of the quantity $(dQ - dW)$ is always zero, this quantity $(dQ - dW)$ must be the differential of a property of the system. To define this property more clearly, consider a system which changes from state 1 to state 2 by process A and returns to state 1 by either process B or process C as depicted on the property diagram of Fig. 1-5. For the cycle composed of processes A and B we have

$$\int_{1,A}^{2} (dQ - dW) + \int_{2,B}^{1} (dQ - dW) = 0 \qquad (a)$$

and for the cycle composed of processes A and C we have

$$\int_{1,A}^{2} (dQ - dW) + \int_{2,C}^{1} (dQ - dW) = 0 \qquad (b)$$

Comparing Eqs. (a) and (b) reveals that

$$\int_{2,B}^{1} (dQ - dW) = \int_{2,C}^{1} (dQ - dW)$$

Since B and C are arbitrary processes between states 2 and 1, it follows that the quantity $\int (dQ - dW)$ is the same for all processes between the two states. In other words, $\int (dQ - dW)$ depends only on the end states and not on the path followed between the two end states. Therefore, $\int (dQ - dW)$ is a point function and is a property of the system. This property is called stored energy and is denoted by the symbol E. Thus, we can write

$$\Delta E = \int (dQ - dW) = Q - W$$

or $\qquad Q = \Delta E + W \qquad (1\text{-}5)$

Equation (1-5) is a useful expression of the first law for a system of fixed mass. In words it reads that the heat added to a system during a process is equal to the gain in stored energy of the system plus work done by the system. The symbol Δ will always be used to mean "final-minus-initial." Thus, in Eq. (1-5)

$$\Delta E = E_{\text{final}} - E_{\text{initial}}$$

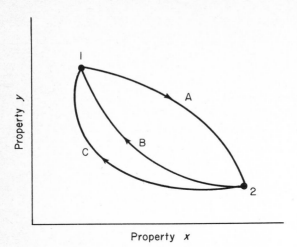

Property x

FIG. 1-5. Stored energy is a point function.

Since E is a property, its differential is exact and shall be denoted as dE. Thus for an infinitesimal process the first law is written as

$$dQ = dE + dW \qquad (1\text{-}5a)$$

The property E represents all the energy of a system in a given state. It includes the kinetic energy KE and potential energy PE of the system as a whole and all other energy of the system which we call the *internal energy*. The symbol U will be used to denote the internal energy of a system, and the symbol u will be used to denote specific or molal internal energy. Thus we write

$$E = U + KE + PE = U + \frac{m V^2}{2} + mgz \qquad (1\text{-}6)$$

For unit mass of a system, this is written

$$e = u + \frac{V^2}{2} + gz \qquad (1\text{-}6a)$$

where V is the velocity of the system, z is the elevation of the center of gravity of the system from an arbitrary datum plane, and g is the acceleration due to gravity.

Internal energy U, as introduced above, is an extensive property of a system and represents all the energy associated with the motion and configuration of the molecules and a number of other microscopic forms. In the absence of motion and gravity effects the first law for a system of fixed mass may be written in integrated form as

$$Q = \Delta U + W \tag{1-7}$$

and in differential form

$$dQ = dU + dW \tag{1-7a}$$

The sum of the internal energy and the pressure-volume product consists entirely of properties and is itself a property. The frequent occurrence of this new property warrants giving it a name and a symbol. We call it *enthalpy*. It is represented by the symbol H. Thus

$$H = U + p V \tag{1-8}$$

For a unit mass or unit mole we write

$$h = u + p v \tag{1-8a}$$

It is interesting to note that although the development of classical thermodynamics is mainly based on the concept of a closed system, a major portion of its application involves bulk flow through an open system. In the analysis of an open system involving bulk flow, the study is no longer focused on a fixed mass, but rather on a region of space called a *control volume*, depicted in Fig. 1-6a. The boundary of a control volume is called a *control surface*. When a fluid flows across a control surface, in addition to the transportation of stored energy, work is required to push the fluid into or out of the control volume. Consider the process by which a fluid element of volume V with cross-sectional area A and length L is pushed across a control surface, as depicted in Fig. 1-6b. The force acting on this element is pA where p is the pressure. This force acts through a distance L and hence the work done in pushing the element across the control surface is $p \, A \, L = p \, V$. This work is called *flow work*. When unit mass of a fluid with specific volume v crosses a control surface, the flow work is pv. Thus, in Fig. 1-6a, the flow work per unit mass at section 1 is $p_1 v_1$, which is the work done on the system by the fluid outside the control volume. Similarly, the flow work per unit mass at section 2 is $p_2 v_2$, which is the work done by the system on the fluid outside the control volume.

Obviously, in addition to flow works, a control volume may involve other forms of work, such as work done on a piston or a rotating shaft. We will call these other forms of work shaft work W_{shaft}. The first law equation for a control volume can then be written as

$$Q - W_{shaft} + \sum_{\substack{\text{streams} \\ \text{entering}}} m(e + p v) - \sum_{\substack{\text{streams} \\ \text{leaving}}} m(e + p v) = E_f - E_i \tag{1-9}$$

where E_f and E_i are respectively the final and initial stored energies of the

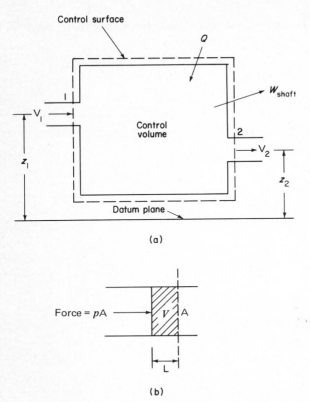

FIG. 1-6. (a) Schematic diagram of a control volume. (b) Development of the expression of flow work.

control volume. In differential form the first law equation can be written as

$$đQ - đW_{shaft} + [(e + p\,v)đm]_{in} - [(e + p\,v)đm]_{out} = dE_{CV} \tag{1-9a}$$

where E_{CV} is the stored energy of the control volume. Note that in the preceding equation we use $đm$ instead of dm, because the amount of mass crossing the control surface is not a property, even though the mass within the control volume is a property.

A simple but important special case of flow problems is that of *steady flow* and *steady state* (or simply, steady flow). The flow is steady when no mass is accumulating within the control volume, and the state is steady when the properties at each point within the control volume is unchanging in time. In other words, in a steady-flow-steady-state system the rate of mass flow into the control volume equals the rate of mass flow out of it, and the energy stored

within the control volume remains constant. Under these conditions, the first
law equation for a control volume can be written as

$$Q - W_{shaft} + \sum_{\substack{streams \\ entering}} m(e + p\,v) - \sum_{\substack{streams \\ leaving}} m(e + p\,v) = 0 \qquad \text{STEADY FLOW} \qquad (1\text{-}10)$$

When a steady-flow control volume has only a single inlet and a single outlet,
as the one shown in Fig. 1-6a, we must have

Rate of mass-inflow = rate of mass-outflow

Substituting Eqs. (1-6a) and (1-8a) into Eq. (1-10), dividing through by the
constant-mass flow rate, and rearranging, we have

$$q + h_1 + \frac{V_1^2}{2} + g\,z_1 = w_{shaft} + h_2 + \frac{V_2^2}{2} + g\,z_2 \qquad \text{STEADY FLOW} \qquad (1\text{-}10a)$$

where q and w_{shaft} are, respectively, heat added to the control volume and shaft
work done by the control volume with per unit mass flowing in and out the
control volume.

1-8 THE SECOND LAW OF THERMODYNAMICS

It is well known that physical processes in nature proceed toward equilibrium
spontaneously. Liquids flow from a region of high elevation to one of low
elevation; gases expand from a region of high pressure to one of low pressure;
heat flows from a region of high temperature to one of low temperature, and
material diffuses from a region of high concentration to one of low
concentration. A spontaneous process can proceed only in a particular direction.
Energy from external source is required to reverse such processes. The second
law of thermodynamics epitomizes our experiences with respect to the
unidirectional nature of thermodynamic processes.

The second law of thermodynamics may be stated in many different forms,
one of which, known as the *Clausius statement of the second law*, is as follows: it
is impossible for any device to operate in a cycle in such a manner that the sole
effect is the transfer of heat from one body to another body at a higher temperature. In the case of a heat pump or refrigerator, heat does flow from a region of
low temperature to one of high temperature, but only when work is added to the
machine from an outside source.

The second law is one of the fundamental laws of nature. It cannot be
derived from any other law, but when one statement of it is accepted as a
postulate, all other statements of it can then be proved. In reference to another
statement of the second law, let us investigate whether heat can be converted

completely into work. Consider two heat reservoirs with a heat pump and a hypothetical heat engine as shown in Fig. 1-7. Assume that the hypothetical heat engine could convert all the heat it received into work. This work could then be used to drive the heat pump. Thus the system comprised of both devices could cause the sole effect of transferring heat from the heat reservoir at a lower temperature to the one at a higher temperature while operating cyclically. This is a violation of the Clausius statement of the second law. We are led to the belief that the hypothetical engine we have postulated cannot exist. Thus we conclude that no process is possible whose sole result is the absorption of heat from a reservoir and the conversion of this heat into work. This is the *Kelvin-Planck statement of the second law.*

The Kelvin-Planck statement as stated above excludes the possibility of producing net work by a cyclic device that exchanges heat only with a single heat reservoir at a fixed temperature. This means that a heat engine needs two or more heat reservoirs at different levels of temperature. Of particular interest is an idealized engine, known as the *Carnot engine*, which employs two heat reservoirs at different temperatures and operates on the following cycle (Fig. 1-8):

Process ab: A reversible isothermal expansion at T_1, absorbing a quantity of heat Q_1.

Process bc: A reversible adiabatic expansion.

Process cd: A reversible isothermal compression at a lower temperature T_2, rejecting a quantity of heat Q_2.

Process da: A reversible adiabatic compression.

For this cycle, according to the first law, the net work output is

$$W = Q_1 - Q_2$$

The thermal efficiency of the cycle, defined as the ratio of the net work output to the heat input, is given by

$$\eta = \frac{W}{Q_1} = \frac{Q_1 - Q_2}{Q_1} = 1 - \frac{Q_2}{Q_1} \qquad \text{CARNOT EFFICIENCY} \qquad (1\text{-}11)$$

Note that the second law imposes the condition that $\eta < 1$ for any heat engine. Moreover the efficiency of the Carnot engine is an upper limit which cannot be exceeded by any engine operating between the same two temperatures. This is stated in the following corollary of the second law, known as the *Carnot principle*: No engine can be more efficient than a reversible engine operating between the same temperature limits, and all reversible engines operating between the same temperature limits have the same efficiency. The proof of this principle is as follows:

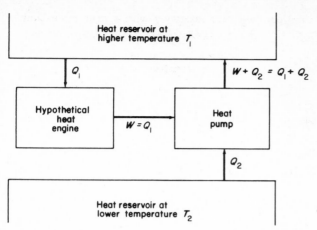

FIG. 1-7. Impossibility of converting heat completely into work in a cycle.

Consider any heat engine A and a reversible heat engine R operating simultaneously between the same temperature limits, as depicted in Fig. 1-9a. Assume at first that $\eta_A > \eta_R$. But $\eta_A = W_A/Q_1$ and $\eta_R = W_R/Q_1$, the amounts of heat supplied to both engines being the same. Therefore we have $W_A > W_R$. Now let the reversible engine R be reversed to operate as a heat pump, as depicted in Fig. 1-9b. Since the work W_R required to drive the heat pump R is less in magnitude than the work output W_A of engine A, engine A can drive heat pump R and still have a net work output of $(W_A - W_R)$. The heat reservoir at

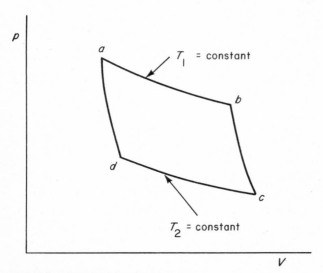

FIG. 1-8. Carnot cycle on p-V plane.

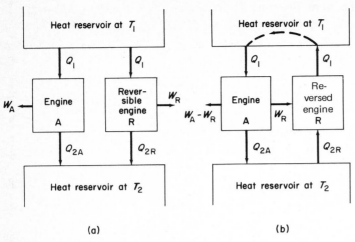

FIG. 1-9. Proof of the Carnot principle.

temperature T_1 could be eliminated by having heat pump R discharge heat directly into engine A. Thus the composite system comprising the heat engine and the heat pump constitutes a device which could perform the sole result of absorbing heat from a single heat reservoir (the one at T_2) and converting the heat into work. This is a violation of the Kelvin-Planck statement of the second law. Consequently our original assumption that $\eta_A > \eta_R$ is false and the first point of the Carnot principle is proved.

To prove the second point of the principle, let both A and R be reversible engines. Assume that these two reversible engines have different efficiencies. By reversing the less efficient engine to operate as a heat pump and following the same procedure used earlier, we again have a situation that violates the Kelvin-Planck statement of the second law. We conclude that our assumption that two reversible engines operating between the same temperature limits can have different efficiencies is false. Thus the second point of the Carnot principle is proved.

1-9 THERMODYNAMIC TEMPERATURE SCALE

Since according to the Carnot principle, the efficiencies of all Carnot engines operating between the same temperatures T_1 and T_2 are the same regardless of the kind of working substance, these efficiencies must be a function of the two temperatures alone. Therefore from Eq. (1-11) we may write

$$\frac{Q_1}{Q_2} = \phi(T_1, T_2) \tag{a}$$

where ϕ denotes some function. To determine the nature of the function ϕ, consider three Carnot engines A, B, and C operating between the pairs of temperatures (T_1, T_2), (T_1, T_m), and (T_m, T_2) respectively, as shown in Fig. 1-10. For these three engines Eq. (*a*) becomes

$$\frac{Q_1}{Q_2} = \phi(T_1, T_2) \qquad \text{for engine A}$$

$$\frac{Q_1}{Q_m} = \phi(T_1, T_m) \qquad \text{for engine B}$$

$$\frac{Q_m}{Q_2} = \phi(T_m, T_2) \qquad \text{for engine C}$$

Since

$$\frac{Q_1}{Q_2} = \frac{Q_1}{Q_m} \frac{Q_m}{Q_2}$$

we have

$$\phi(T_1, T_2) = \phi(T_1, T_m)\, \phi(T_m, T_2)$$

Since T_1, T_m, and T_2 are independent, the above equation can be satisfied only when the function is of the form

$$\phi(T_1, T_2) = \frac{f(T_1)}{f(T_2)} \qquad\qquad (b)$$

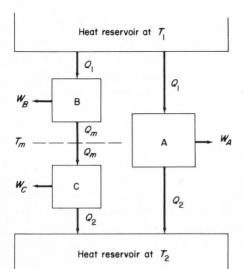

FIG. 1-10. Establishment of thermodynamic temperature scale by means of Carnot engines.

where f denotes another function, so that

$$\frac{f(T_1)}{f(T_2)} = \frac{f(T_1)}{f(T_m)} \frac{f(T_m)}{f(T_2)}$$

Combining Eqs. (a) and (b), we obtain

$$\frac{Q_1}{Q_2} = \frac{f(T_1)}{f(T_2)} \qquad\qquad (c)$$

The mathematical form of $f(T)$ is completely arbitrary. An infinite number of choices could be made, but obviously the simplest choice is to set $f(T) = T$, whereupon Eq. (c) becomes

$$\frac{Q_1}{Q_2} = \frac{T_1}{T_2} \qquad\qquad (1\text{-}12)$$

Equation (1-12) was proposed by Kelvin to define a temperature scale called the *thermodynamic* or *absolute temperature* scale. This scale is based on the amounts of heat transferred to and from a Carnot engine and is independent of the nature of the thermometric substance. Assigning a numerical value of 273.16 to the triple point of water results in the Kelvin scale, whereupon we may write

$$T = 273.16 \frac{Q}{Q_{TP}} \qquad\qquad (1\text{-}13)$$

where Q and Q_{TP} denote the amounts of heat exchange with a heat reservoir at any Kelvin temperature T and a heat reservoir at the Kelvin temperature of the triple point of water respectively. Equation (1-13) introduces the concept of absolute zero, i.e., $T = 0$ when $Q = 0$, a condition which cannot be physically accomplished although it can be approached as a limiting case.

The thermodynamic temperature scale we just defined is equivalent to the ideal gas temperature scale defined in Sec. 1-4. The equivalence of these two temperature scales will be illustrated in an example in Sec. 2-9 in connection with a study of ideal gas relations.

1-10 ENTROPY

Consider a system undergoing a reversible cycle as represented by the continuous curve of Fig. 1-11. It is possible to subdivide this cycle into a number of small Carnot cycles as indicated. The isotherms and part of the adiabats of the small Carnot cycles form a zigzag curve which follows closely the path of the original cycle. The remaining parts of the adiabats of the small Carnot cycles cancel out because each section is traversed once in a forward direction and once in a

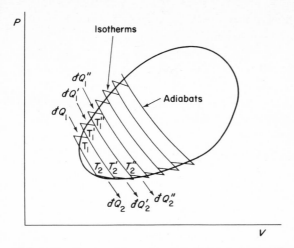

FIG. 1-11. A reversible cycle subdivided into infinitesimal Carnot cycles.

reverse direction. As the number of Carnot cycles is increased, the zigzag curve can be made to approach the original cycle to any desired degree.

Let dQ_1, dQ_1', dQ_1'', ..., dQ_2, dQ_2', dQ_2'', ..., denote the respective algebraic amounts of heat exchanged which are positive when absorbed and negative when given off by the system. Then we may write for the small Carnot cycles,

$$\frac{dQ_1}{T_1} + \frac{dQ_2}{T_2} = 0 \qquad \frac{dQ_1'}{T_1'} + \frac{dQ_2'}{T_2'} = 0 \ldots, \text{etc.}$$

Adding the preceding equations gives

$$\frac{dQ_1}{T_1} + \frac{dQ_2}{T_2} + \frac{dQ_1'}{T_1'} + \frac{dQ_2'}{T_2'} + \cdots = \sum \frac{dQ}{T} = 0$$

In the limit, upon replacement of the summation of finite terms by a cyclic integral, we obtain

$$\oint \frac{dQ_{\text{rev}}}{T} = 0 \tag{1-14}$$

where the subscript "rev" has been used to recall that the original cycle as well as the small Carnot cycles must all be reversible. The foregoing equation states that the integral of dQ/T when carried out over a reversible cycle is equal to zero. It follows that the differential dQ_{rev}/T is a perfect differential and the integral $\int dQ_{rev}/T$ is a property of the system. This property, called *entropy*, is denoted by the symbol S. Thus we may write

$$dS = \frac{dQ_{rev}}{T} \tag{1-15}$$

This is the defining expression for the entropy. Integrating along a reversible path between two equilibrium states 1 and 2 gives

$$\Delta S_{12} = S_2 - S_1 = \int_1^2 \frac{dQ_{rev}}{T} \tag{1-16}$$

Since entropy is a point function, it does not matter what particular reversible path is followed in the integration as long as it is reversible.

According to Eq. (1-15) the small quantity of heat transferred to a system in an infinitesimal reversible process is

$$dQ_{rev} = T\,dS \tag{1-17}$$

The heat transferred to the system in a reversible process between states 1 and 2 is then

$$Q_{12} = \int_1^2 T\,dS \tag{1-18}$$

This last integral is represented by the area under the path 1 to 2 of a reversible process plotted on a temperature vs. entropy diagram as shown in Fig. 1-12. Since heat is a path function, this integral shall be different for different reversible processes.

For a reversible adiabatic process $dQ_{rev} = 0$, then from Eq. (1-15) $dS = 0$.

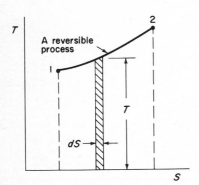

FIG. 1-12. Area under the curve on T-S plane represents heat.

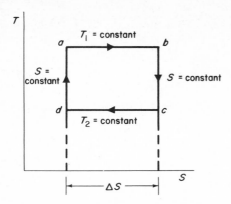

FIG. 1-13. Carnot cycle on T-S plane.

This means that during a reversible adiabatic process there is no change in entropy and the process is isentropic.

It is convenient for many applications to use T and S as the two coordinates to describe thermodynamic processes. Of particular interest is the representation of the Carnot cycle on the T-S plane. Figure 1-13 depicts such a plane. For the Carnot cycle:

Heat added = $Q_1 = T_1 \Delta S$

Heat rejected = $Q_2 = T_2 \Delta S$

Net work = $Q_1 - Q_2 = (T_1 - T_2)\Delta S$

Thermal efficiency $\eta = \dfrac{Q_1 - Q_2}{Q_1} = \dfrac{(T_1 - T_2)\Delta S}{T_1 \Delta S}$

$$= \frac{T_1 - T_2}{T_1} = 1 - \frac{T_2}{T_1} \tag{1-19}$$

We recall that Eq. (1-14) is for a reversible cycle. For an irreversible cycle the analogous expression is

$$\oint \frac{dQ_{irr}}{T} < 0 \tag{1-20}$$

This is the *inequality of Clausius*. The proof is as follows: In virtue of the Carnot principle, the efficiency of an irreversible engine is smaller than that of a reversible engine operating between the same temperature limits, or expressed symbolically,

$\eta_{irr} < \eta_{rev}$

where the subscripts "irr" and "rev" indicate irreversible and reversible respectively. In terms of heat quantities involved in an elementary cycle, this may be written as

$$1 - \frac{dQ_{2\,irr}}{dQ_{1\,irr}} < 1 - \frac{dQ_{2rev}}{dQ_{1rev}}$$

Whence

$$1 - \frac{dQ_{2\,irr}}{dQ_{1\,irr}} < 1 - \frac{T_2}{T_1}$$

or

$$\frac{dQ_{1\,irr}}{T_1} - \frac{dQ_{2\,irr}}{T_2} < 0$$

If we let $dQ_{2\,irr}$ represent the algebraic amount of heat rejection, which is negative, we may then write

$$\frac{dQ_{1\,irr}}{T_1} + \frac{dQ_{2\,irr}}{T_2} < 0$$

Summing up all the equations analogous to the preceding yields the desired expression of the inequality of Clausius.

Upon combining Eqs. (1-14) and (1-20), we conclude that

$$\oint \frac{dQ}{T} \leqslant 0 \tag{1-21}$$

where the sign of equality applies to a reversible cycle, and the sign of inequality applies to an irreversible cycle.

It cannot be overemphasized that the entropy of a system is a state function; it depends only on the state that the system is in, and not on how that state is reached. If a system goes from state 1 to state 2, its entropy changes from S_1 to S_2. However, it is only when the system travels along a reversible path between the two end states that the entropy change $S_2 - S_1$ equals $\int_1^2 (dQ_{rev}/T)$. If the path is irreversible $\int_1^2 (dQ_{irr}/T)$ has a different value although the change in entropy is the same as before. The relation that does exist between the change in entropy and the integral $\int(dQ/T)$ along any arbitrary path can be obtained in the following manner:

Let a system change from state 1 to state 2 by a reversible process A and return to state 1 by either a reversible process B or an irreversible process C as depicted in Fig. 1-14. The cycle made up of the reversible processes A and B is a reversible cycle. Following Eq. (1-14) we write

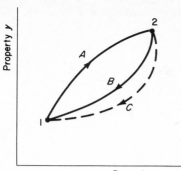

FIG. 1-14. Reversible and irreversible cycles.

$$\int_{1,A}^{2} \frac{dQ}{T} + \int_{2,B}^{1} \frac{dQ}{T} = 0$$

The cycle made up of the reversible process A and the irreversible process C is an irreversible cycle. Equation (1-20) then gives

$$\int_{1,A}^{2} \frac{dQ}{T} + \int_{2,C}^{1} \frac{dQ}{T} < 0$$

Subtracting the first equation from the second gives

$$\int_{2,C}^{1} \frac{dQ}{T} - \int_{2,B}^{1} \frac{dQ}{T} < 0$$

Transposing, we have

$$\int_{2,B}^{1} \frac{dQ}{T} > \int_{2,C}^{1} \frac{dQ}{T}$$

Since process B is reversible, dQ/T may be replaced by dS in the first integral, thus

$$\int_{2,B}^{1} \frac{dQ}{T} = \int_{2,B}^{1} dS$$

But since entropy is a property,

$$\int_{2,B}^{1} dS = \int_{2,C}^{1} dS$$

Therefore,

$$\int_{2,C}^{1} dS > \int_{2,C}^{1} \frac{dQ}{T}$$

For the general case we can then write

$$dS \geqslant \frac{dQ}{T} \tag{1-22}$$

or

$$\Delta S_{12} = S_2 - S_1 \geqslant \int_{1}^{2} \frac{dQ}{T} \tag{1-22a}$$

where the equality holds for a reversible process and the inequality for an irreversible process. This is one of the most important equations of thermodynamics. It expresses the influence of irreversibility on the entropy of a system.

For an isolated system $dQ = 0$; then, according to Eq. (1-22),

$$dS_{isolated} \geqslant 0 \tag{1-23}$$

This is the *principle of increase of entropy.* In accordance with the first law, an isolated system can only assume those states for which the total internal energy remains constant. Now in accordance with the second law as expressed by Eq. (1-23), of the states of equal energy, only those states for which the entropy increases or remains constant can be attained by the system.

There are two combinations of thermodynamic properties involving the entropy that are of great utility in thermodynamics. They are the *Helmholtz function A* and the *Gibbs function G* defined by the following equations:

$$A = U - TS \tag{1-24}$$

$$G = H - TS \tag{1-25}$$

Of course any combination of properties and their derivatives is also a property. Hence the functions A and G are thermodynamic properties which, like the internal energy U and the enthalpy H, are extensive ones. For unit mass or on a molal basis, lower-case letters are used. Thus

$$a = u - Ts \tag{1-24a}$$

$$g = h - Ts \tag{1-25a}$$

1-11 PRINCIPLE OF CARATHÉODORY*

In the last three sections we have studied the second law of thermodynamics and the concept of entropy following the traditional engineering method due to Carnot, Kelvin, and Clausius. In order to establish the entropy as a function of state we had to have recourse to the usual thermodynamic paraphernalia of idealized heat engines and engine efficiencies. Through more rigorous use of mathematics, Carathéodory developed in 1909 a method which gives the same result without recourse to mechanical devices. The *Carathéodory statement of the second law,* often referred to as the *principle of Carathéodory,* reads as follows: There exist arbitrarily close to any given state of a system other states which cannot be reached from it by reversible adiabatic processes.

This postulate is of course more abstract than that of Kelvin, and its development was originally set in purely mathematical terms so that its physical implication became obscured. Recent development on this topic tends to strip it of the greater part of its mathematics. We now give a brief exposition of a modified version[1] of the method.

The procedure of the modified method is first to use the Kelvin-Planck statement of the second law to establish the existence of reversible adiabatic surfaces, and then to deduce the existence of an entropy function, without recourse to Carnot cycles and Carnot principle. For the sake of easy visualization, let us consider a system with two reversible work modes, for which the first law is written as

$$dQ_{rev} = dU + Y_1 dX_1 + Y_2 dX_2 \tag{a}$$

where Y_1 and Y_2 are two generalized forces, and X_1 and X_2 are two generalized displacements. For a reversible adiabatic process we have then

$$dU + Y_1 dX_1 + Y_2 dX_2 = 0 \tag{b}$$

According to the state principle (Sec. 2-1) there are three independent variables. Let us choose the empirical temperature t and the two generalized displacements X_1 and X_2 as the three independent variables. Thus U, Y_1, and Y_2 are then functions of t, X_1, and X_2.

Figure 1-15 is a graph with t, X_1, and X_2 as the three rectangular coordinates. Point a represents an arbitrarily chosen equilibrium state. Point b represents an equilibrium state which can be reached from state a by a reversible

---

*This section may be omitted or delayed in reading without loss of continuity.

[1] M. W. Zemansky, "Heat and Thermodynamics," 5th ed., p. 198, McGraw-Hill Book Company, New York, 1968.

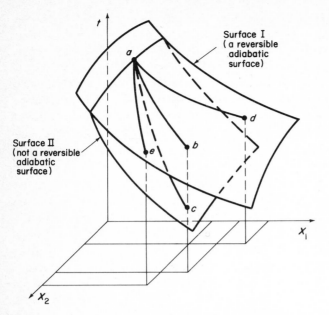

FIG. 1-15. Demonstration of the existence of nonintersecting reversible adiabatic surfaces.

adiabatic process, satisfying Eq. (*b*). Point *c* represents another equilibrium state which has the same values of X_1 and X_2 as point *b* but at a different value of *t*. We will show that state *c* cannot be reached from state *a* by a reversible adiabatic process. Let us assume at first that it is possible to reach state *c* from state *a* by a reversible adiabatic process *ac*. We have then the cycle *acba*, carrying the system from state *a* to state *c* reversibly and adiabatically, then from *c* to *b* by process *cb* at constant X_1 and X_2, and finally back to *a* by the reversible adiabatic process *ba*. When *c* is at a lower temperature than *b* (as the case shown in Fig. 1-15), since there is no reversible work done in process *cb*, we conclude from the first law that heat (or its equivalent due to irreversible effect) must be added to the system during this process. On the other hand, during the two reversible adiabatic processes *ac* and *ba* there is no heat transfer but there is work done. Consequently, the system in going through the cycle *acba* has performed the sole result of absorbing heat from a reservoir and converting this heat into work. This is a violation of the Kelvin-Planck statement of the second law. Therefore we conclude that state *c* cannot be reached from state *a* by a reversible adiabatic process.

From the preceding discussions it is clear that for the given values of X_1 and X_2 corresponding to points *b* and *c*, *b* is the single state which can be reached from state a by a reversible adiabatic process. For different values of X_1 and X_2, there would be different single states (such as d and e) which can be reached from state a by reversible adiabatic processes. All state points that can be reached from the initial state a by reversible adiabatic processes lie on a two-dimensional surface as depicted by surface I in Fig. 1-15. All state points which do not lie on this surface cannot be reached from a by reversible adiabatic processes. Thus this surface is a unique one; in other words, it is the only reversible adiabatic surface corresponding to the initial state a. For other initial states lying outside surface I, there would be other unique reversible adiabatic surfaces.

An interesting characteristic of reversible adiabatic surfaces is that they cannot intersect. For example, as depicted in Fig. 1-15, when surface I containing path ab and surface II containing path ac intersect, then it would be possible to proceed from an initial state, such as state a on the line of intersection, to two different final states b and c having the same values of X_1 and X_2 along two different reversible adiabatic processes. We have just shown that this is impossible. In other words, surface II cannot be a reversible adiabatic surface.

In view of the preceding discussions it is evident that the t-X_1-X_2 three-dimensional thermodynamic space may be considered as being formed by a family of nonintersecting reversible adiabatic surfaces, whose equation can be written symbolically as

$$\phi = \phi(t, X_1, X_2)$$

where for each reversible adiabatic surface there is a constant value of the function ϕ. Thus, ϕ must be a function of state, which has the special characteristic that $\phi = $ constant for a reversible adiabatic process. Any point in the t-X_1-X_2 space can be defined by specifying the values of ϕ, X_1, and X_2.

Consider U as a function of ϕ, X_1, and X_2, then the total differential of U is

$$dU = \left(\frac{\partial U}{\partial \phi}\right)_{X_1,X_2} d\phi + \left(\frac{\partial U}{\partial X_1}\right)_{\phi,X_2} dX_1 + \left(\frac{\partial U}{\partial X_2}\right)_{\phi,X_1} dX_2$$

Substituting this equation into Eq. (a) leads to

$$đQ_{rev} = \left(\frac{\partial U}{\partial \phi}\right)_{X_1,X_2} d\phi + \left[Y_1 + \left(\frac{\partial U}{\partial X_1}\right)_{\phi,X_2}\right] dX_1 + \left[Y_2 + \left(\frac{\partial U}{\partial X_2}\right)_{\phi,X_1}\right] dX_2$$

$$(c)$$

Since ϕ, X_1, and X_2 are independent variables, and $đQ = 0$ whenever $d\phi = 0$, Eq. (c) must reduce to the form

$$\bar{d}Q_{\text{rev}} = \left(\frac{\partial U}{\partial \phi}\right)_{X_1, X_2} d\phi \qquad (d)$$

If now we define

$$\lambda = \left(\frac{\partial U}{\partial \phi}\right)_{X_1, X_2} \qquad (e)$$

then Eq. (d) becomes

$$\bar{d}Q_{\text{rev}} = \lambda \, d\phi \qquad (f)$$

Since as defined in Eq. (e), λ is a function of ϕ, X_1, and X_2, but ϕ is a function of t, X_1, and X_2, it follows that by eliminating X_2 we may express λ as a function of t, ϕ, and X_1. Furthermore, we will now demonstrate that λ is a function of only t and ϕ. For this purpose consider a composite system consisting of one main system and one reference system in thermal equilibrium at a common empirical temperature t as depicted in Fig. 1-16. The independent variables for the various functions as specified in the figure were deduced by the same procedure as that used for a single system.

When heat is transferred reversibly from a heat reservoir to the composite system, we have from Eq. (f),

$$\bar{d}Q_{\text{rev}} = \lambda \, d\phi \qquad \text{for the main system}$$

$$\bar{d}Q'_{\text{rev}} = \lambda' \, d\phi' \qquad \text{for the reference system}$$

$$\bar{d}\hat{Q}_{\text{rev}} = \hat{\lambda} \, d\hat{\phi} \qquad \text{for the composite system}$$

Since

$$\bar{d}\hat{Q}_{\text{rev}} = \bar{d}Q_{\text{rev}} + \bar{d}Q'_{\text{rev}}$$

then

$$\hat{\lambda} \, d\hat{\phi} = \lambda \, d\phi + \lambda' \, d\phi'$$

or

$$d\hat{\phi} = \frac{\lambda}{\hat{\lambda}} \, d\phi + \frac{\lambda'}{\hat{\lambda}} \, d\phi' \qquad (g)$$

Meanwhile, since $\hat{\phi}$ is a function of t, ϕ, ϕ', X_1, and X'_1, we also have

$$d\hat{\phi} = \frac{\partial \hat{\phi}}{\partial t} \, dt + \frac{\partial \hat{\phi}}{\partial \phi} \, d\phi + \frac{\partial \hat{\phi}}{\partial \phi'} \, d\phi' + \frac{\partial \hat{\phi}}{\partial X_1} dX_1 + \frac{\partial \hat{\phi}}{\partial X'_1} \, dX'_1 \qquad (h)$$

Comparing Eqs. (g) and (h) gives

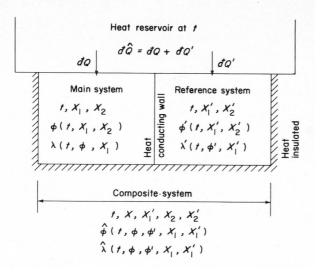

FIG. 1-16. A composite system consisting of two subsystems receives heat from a reservoir.

$$\frac{\partial \hat{\phi}}{\partial t} = 0 \qquad \frac{\partial \hat{\phi}}{\partial X_1} = 0 \qquad \frac{\partial \hat{\phi}}{\partial X_1'} = 0 \tag{i}$$

and also

$$\frac{\lambda}{\hat{\lambda}} = \frac{\partial \hat{\phi}}{\partial \phi} \qquad \frac{\lambda'}{\hat{\lambda}} = \frac{\partial \hat{\phi}}{\partial \phi'} \tag{j}$$

From Eq. (i) it is apparent that $\hat{\phi}$ does not depend on t, X_1, or X_1'. It must depend only on ϕ and ϕ'. Therefore from Eq. (j) it follows that the two ratios $\lambda/\hat{\lambda}$ and $\lambda'/\hat{\lambda}$ must depend only on ϕ and ϕ'. However, each separate λ must depend on something else besides ϕ because if λ depended only on ϕ, then since $\bar{d}Q_{rev} = \lambda d\phi$, $\bar{d}Q_{rev}$ would equal $f(\phi)\, d\phi$, Q would become a point function. Now we see from Eq. (j) that if $\lambda(t, \phi, X_1)/\hat{\lambda}(t, \phi, \phi', X_1, X_1') =$ a function of ϕ and ϕ', then $\hat{\lambda}$ certainly cannot contain X_1'. Similarly, if $\lambda'(t, \phi', X_1')/\hat{\lambda}(t, \phi, \phi', X_1, X_1') =$ a function of ϕ and ϕ', then $\hat{\lambda}$ certainly cannot contain X_1. Therefore we must have $\hat{\lambda} = \hat{\lambda}(t, \phi, \phi')$. Whereupon we conclude that

$$\lambda = \lambda(t, \phi) \qquad \text{and} \qquad \lambda' = \lambda'(t, \phi')$$

Thus we have demonstrated that λ is a function of t and ϕ only.

In order to satisfy the condition that the ratio of two λ's be a function of ϕ and ϕ', and each separate λ be a function of t and ϕ (and/or ϕ'), the λ function must take the form

$$\lambda = T(t)f(\phi) \tag{k}$$

The function $T(t)$ is a universal function of the empirical temperature, meaning that it is the same for all systems at the same empirical temperature. When the ratio of two λ's is taken, this function cancels out, leaving the ratio of the λ's as a function of the ϕ's.

Substitution of Eq. (k) into Eq. (f) results in

$$dQ_{rev} = T(t)f(\phi)d\phi \tag{l}$$

Now we may put

$$dS = f(\phi)d\phi \tag{m}$$

where S will also be a function of state, called the *entropy,* which has the characteristic that $dS = 0$ for a reversible adiabatic process. Thus we finally obtain the desired equation

$$dQ_{rev} = T\,dS \tag{1-26}$$

where $T = T(t)$ is the *absolute temperature,* which is a universal function of the empirical temperature. We have thus arrived at the same result expressed in Eq. (1-17) by an entirely different method. Note that although we employed a system with three independent variables in the foregoing discussions, the results are the same for all systems with more or fewer than three independent variables.

1-12 THERMODYNAMIC SURFACES

A substance which is homogeneous and invariable in chemical composition, irrespective of the phase or phases in which it exists, is called a *pure substance.* In other words, a pure substance is a one-component system. For instance, pure oxygen or pure nitrogen in any phase or any combination of phases is a pure substance. A chemically stable mixture, such as air, which is essentially a mixture of oxygen and nitrogen, may be treated as a pure substance as long as it remains in one phase. A system of gaseous air and liquid air, however, is not a pure substance, because the composition of the two phases are different.

We now proceed to examine in a descriptive way the properties of pure substances and the phases in which a pure substance may exist. We are interested in this section in systems whose volume changes are of most importance. The effects of motion, capillarity, solid distortion, and external force fields (electric, magnetic, and gravitational) are not significant.

The relationships between pressure, specific volume, and temperature can be clearly understood with the aid of a three-dimensional p-v-T surface. Figure 1-17 illustrates a normal type of substance which contracts on freezing and Fig. 1-18 illustrates water which expands on freezing. The projections on the p-T and p-v planes are also shown on these figures. Any point on the p-v-T surface represents an equilibrium state of the substance. Because a reversible process may be considered as a series of equilibrium states, all points along such a process must lie on the p-v-T surface. Several constant-temperature lines and constant-pressure lines are shown in Figs. 1-17 and 18.

There are three regions, labeled solid, liquid, and gas (or vapor) in Figs. 1-17 and 1-18 where the substance exists only in a single phase. Between the single-phase regions lie the two-phase regions, where two phases coexist in equilibrium. They are labeled *S-L, L-V,* and *S-V* for solid-liquid, liquid-vapor, and solid-vapor coexistent regions respectively. In the two-phase regions, lines of constant pressure are also lines of constant temperature. Thus the two-phase portions of the p-v-T surface are made up of straight lines parallel to the volume axis; such surfaces are called ruled surfaces. Therefore, the two-phase surfaces appear as lines when projected on the p-T plane.

There are states of fixed pressure and temperature where solid, liquid, and vapor phases can coexist in equilibrium. These states are called the triple phase, or, more commonly, the *triple point.* On a p-v-T surface the triple point is a line of constant pressure and temperature. On some diagrams, as shall be seen later, it appears as an area of constant pressure and temperature. Triple point appears as a point only on a p-T diagram. Triple-point data for a number of substances are given in Table 1-1.

A line which separates a single-phase region from a two-phase region is called a saturation line. Any state represented by a point on a saturation line is known as a *saturation state.* Thus, the line separating the liquid phase and the liquid-vapor coexisting phase is the saturated liquid line, and the line separating the vapor phase and the liquid-vapor coexisting phase is the saturated vapor line. Any state represented by a point on a saturated liquid line is a saturated liquid state; similarly, any state represented by a point on a saturated vapor line is a saturated vapor state. When a mixture of liquid and vapor coexist in equilibrium, the liquid and vapor are respectively in a saturated liquid state and a saturated vapor state, and they are called *saturated liquid* and *saturated vapor* respectively. Similar definitions apply to saturated solid and saturated solid line.

The saturated liquid line and saturated vapor line meet at a point, called the *critical point*, which marks the termination of any distinction between liquid and vapor phases. In the neighborhood of the critical point, the properties of the liquid phase and the vapor phase approach each other, and at the critical point the properties of both phases become identical. In a pure substance the critical point is the state of highest pressure and temperature at which distinct liquid and vapor phases can coexist. Beyond the critical point the liquid phase and

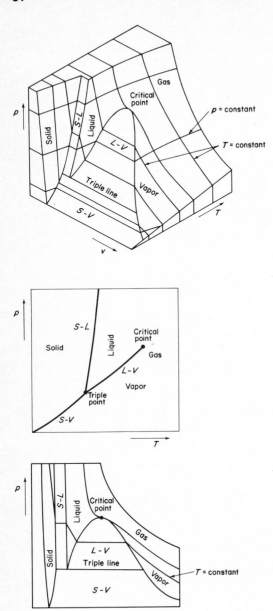

FIG. 1-17. Pressure-volume-temperature surface and projections for a substance which contracts on freezing.

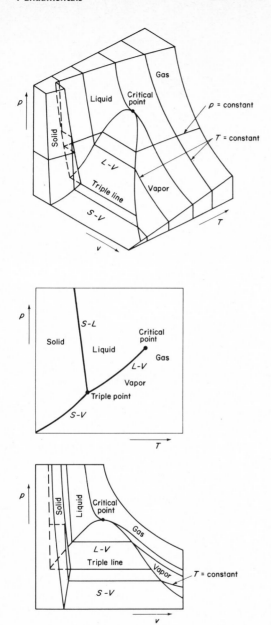

FIG. 1-18. Pressure-volume-temperature surface and projections for water which expands on freezing.

TABLE 1-1. Triple-point Data

Substance	Formula	Temperature °K	Pressure mm Hg
Acetylene	C_2H_2	192.4	962
Ammonia	NH_3	195.42	45.58
Argon	A	83.78	515.7
Carbon dioxide	CO_2	216.55	3885.1
Carbon monoxide	CO	68.14	115.14
Ethane	C_2H_6	89.88	0.006
Ethylene	C_2H_4	104.00	0.9
Hydrogen	H_2	13.84	52.8
Hydrogen sulfide	H_2S	187.66	173.9
Krypton	Kr	115.6	538
Methane	CH_4	90.67	87.7
Neon	Ne	24.57	324
Nitric oxide	NO	109.50	164.4
Nitrogen	N_2	63.15	94.01
Oxygen	O_2	54.35	1.14
Sulfur dioxide	SO_2	197.69	1.256
Water	H_2O	273.16	4.587
Xenon	Xe	161.3	611

vapor phase merge into each other. The pressure, temperature, and volume at the critical point are called the *critical pressure, critical temperature,* and *critical volume* respectively. The critical point data for substances of interest are given in Appendix Table A-5.

It should be mentioned that a pure substance can exist in different solid phases. At extremely high pressures several different crystalline solid phases of water have been observed. These are illustrated in Fig. 1-19. In this figure, different two-phase and triple-phase regions are clearly shown. Notice that besides the usual solid-liquid-vapor triple point, there are other triple points involving two or three solid phases.

In thermodynamics the properties of liquids, vapors, and liquid-vapor two-phase mixtures are of great importance. Figure 1-20 is a *p-v* plane showing in particular the liquid and vapor phases. In this figure, point *f* represents a saturated liquid state and *g* a saturated vapor state. The temperature of these saturation states is called the saturation temperature corresponding to the given pressure; and this pressure is called the saturation pressure (or vapor pressure) corresponding to the given temperature. The region to the right of the saturated vapor line is the *superheated-vapor* region, in which a vapor at a given pressure is at a higher temperature than the saturation temperature at that pressure. The region to the left of the saturated liquid line is the *subcooled-* (*or compressed-*)

liquid region, where a liquid at a given pressure is at a lower temperature than the saturation temperature at that pressure.

A state inside the liquid-vapor region, such as state M in Fig. 1-20 represents an equilibrium state of a mixture of liquid and vapor. The liquid phase is at state f and the vapor phase is at state g, irrespective of the location of state M on line fg. The location of state M on line fg depends on the relative masses of each phase. Let the mass fraction of vapor in the mixture be denoted as the *quality* and symbolized as x. In a mixture of mass m, the mass of vapor is xm and the mass of liquid is $(1 - x)m$. Each extensive property of the mixture is equal to the sum of the values of the same property for the two phases. Thus, the total volume of the mixture at state M is $V_M = V_g + V_f$ or

$$m\,v_M = x\,m\,v_g + (1 - x)m\,v_f$$

where v_M is the specific volume of the mixture at state M, v_g is the specific

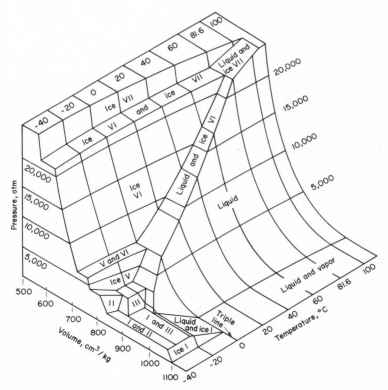

FIG. 1-19. Pressure-volume-temperature surface showing different solid phases of water. (*Based on measurements of Bridgman.*)

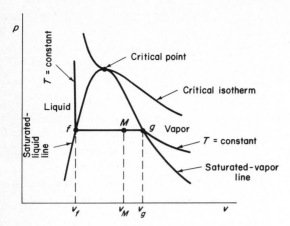

FIG. 1-20. Pressure-volume diagram showing liquid and vapor phases.

volume of saturated vapor at state g, and v_f is the specific volume of saturated liquid at state f. The preceding equation may be arranged to write

$$v_M = x\,v_g + (1 - x)v_f = v_f + x\,v_{fg} = v_g - (1 - x)v_{fg} \tag{1-27}$$

where v_{fg} is defined as $(v_g - v_f)$.

Expressions analogous to the preceding equations can be derived for internal energy, enthalpy, and entropy. These expressions for unit mass of the mixture are

$$u_M = x\,u_g + (1 - x)u_f = u_f + x\,u_{fg} = u_g - (1 - x)u_{fg} \tag{1-28}$$

$$h_M = x\,h_g + (1 - x)h_f = h_f + x\,h_{fg} = h_g - (1 - x)h_{fg} \tag{1-29}$$

$$s_M = x\,s_g + (1 - x)s_f = s_f + x\,s_{fg} = s_g - (1 - x)s_{fg} \tag{1-30}$$

where $u_{fg} = u_g - u_f$, $h_{fg} = h_g - h_f$, and $s_{fg} = s_g - s_f$.

Heat must be added to a substance to cause it to change phase in a reversible process. Application of the first law to a phase change for a pure substance shows that the heat required is equal to the difference in enthalpy between the final and the initial phases at the same pressure and temperature. The amount of heat required to cause a phase change is commonly known as the *latent heat*. A more appropriate name is the *enthalpy of phase change* because it is a property of the system. Thus we may write

Enthalpy (or latent heat) of vaporization = $h_{fg} = h_g - h_f$

Enthalpy (or latent heat) of fusion = $h_{if} = h_f - h_i$

Enthalpy (or latent heat) of sublimation = $h_{ig} = h_g - h_i$

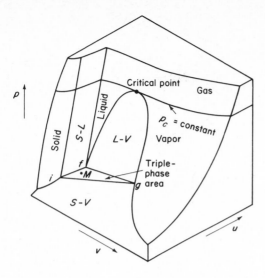

FIG. 1-21. Pressure-volume-internal energy surface for a substance which contracts on freezing.

wherein the subscripts *g, f,* and *i* denote saturated vapor, saturated liquid, and saturated solid phases respectively.

Our discussion so far has been in reference to *p-v-T* surfaces. However, properties other than pressure, volume, or temperature may also be used as coordinates to construct a thermodynamic surface. Figure 1-21 shows a *p-v-u* surface for a normal type of substance which contracts on freezing. Figure 1-22 shows a *u-v-s* surface and Fig. 1-23 an *h-s-p* surface, both for water.

It is to be noted that the triple point becomes a triangular area of constant pressure and temperature on both the *p-v-u* and *u-v-s* surfaces. The three corners

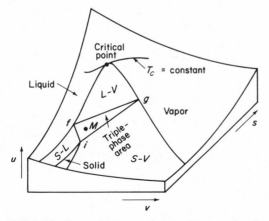

FIG. 1-22. Internal energy-volume-entropy surface for water.

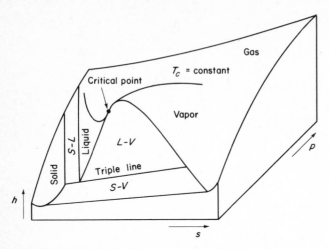

FIG. 1-23. Enthalpy-entropy-pressure surface for water.

of the triangle represent the states of vapor, liquid, and solid phases, and any point in the triangle, such as point M in Figs. 1-21 and 22, represents a mixture of the three phases coexisting at the triple-point pressure and temperature. Let mass fractions of vapor, liquid, and solid in the mixture be denoted by x, y, and z respectively. We have

$$x + y + z = 1$$

Any specific extensive property, such as specific volume, of the three-phase mixture can be determined from an equation of the form

$$v = x\,v_g + y\,v_f + z\,v_i \tag{1-31}$$

where v_g, v_f, and v_i denote the specific volumes of saturated vapor, liquid, and solid respectively at the triple-phase conditions.

    In thermodynamic analyses it is convenient to use two-dimensional property diagrams. The frequently used two-dimensional property diagrams are p-v, T-s, h-s (Mollier diagram), p-h, u-v, and others. Property data of substances of interest are given in the Appendix in the form of these two-dimensional diagrams in addition to tabulations. The properties of helium with certain unique features will be studied later in connection with low temperatures.

PROBLEMS

    1-1. A quantity of a substance in a closed system is made to undergo a reversible process, starting from an initial volume of 1 m³ and an initial pressure of 10^5 newtons/m^2. The final volume is 2 m^3. Compute the work done by the substance if

(a) The pressure times volume remains constant

(b) The pressure is proportional to the square of the volume

1-2. A horizontal cylinder fitted with a sliding piston contains 0.1 m^3 of a gas at a pressure of 1 atm. The piston is restrained by a linear spring. In the initial state, the gas pressure inside the cylinder just balances the atmospheric pressure of 1 atm on the outside of the piston and the spring exerts no force on the piston. The gas is then heated reversibly until its volume and pressure become 0.15 m^3 and 2 atm, respectively.

(a) Write the equation for the relation between the pressure and volume of the gas.

(b) Calculate the work done by the gas.

(c) Of the total work done by the gas, how much is done against the atmosphere? How much against the spring?

1-3. A rectangular chamber of 1 ft width contains water with a free surface. One end of the chamber is fitted with a sliding piston as shown in Fig. P1-3. In the initial condition the chamber length x is 2 ft and the water height y is 1.5 ft. Calculate the work done by the water on the piston when the chamber length is increased slowly from 2 to 2.5 ft. The density of water is 62.4 lbm/ft^3. The atmospheric pressure is 14.7 psia.

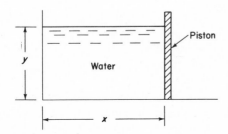

FIG. P1-3.

1-4. An elastic sphere initially has a diameter of 1 m and contains a gas at a pressure of 1 atm. Due to heat transfer the diameter of the sphere increases to 1.1 m. During the heating process the gas pressure inside the sphere is proportional to the sphere diameter. Calculate the work done by the gas.

1-5. Water at 100°F and 50 psia flows through a large pipe at a rate of 100 lbm/min into an insulated tank. Fifteen lbm/min of steam at 300°F and 50 psia flows into the same tank through another pipe which has a cross-sectional area of 0.7 sq in. The steam then condenses and mixes with the water to form a stream of heated water which leaves the tank through a large pipe at a pressure of 50 psia. The enthalpy of water is approximately equal to $(t - 32)$ Btu/lbm, where t is the water temperature in °F. Calculate the temperature of the heated water leaving the tank in steady flow.

1-6. Air flows in a large pipeline at a pressure of 10 atm and a temperature of 30°C. Connected to this pipeline is a vessel which originally contains 1,000 cm^3 of air at 5 atm and 30°C. When the valve on this vessel is opened, air flows from the pipeline into the vessel until the pressure in the vessel becomes 10 atm. If this process occurs adiabatically, what is the final temperature of the air in the vessel? For air, $pv = 2.83T$, $h = 0.24T$, and $u = 0.171T$, where p is in atm, v in cm^3/g, T in °K, and h and u in cal/g.

1-7. Solve the above problem if the vessel is initially evacuated.

1-8. A rigid tank has a volume of 2 ft^3 and contains saturated water vapor at a temperature of 230°F. Because of heat transfer the temperature inside the tank drops to 212°F. Calculate the total mass of steam in the tank, the quality of the steam at the final state, the volume and mass of liquid in the final state, and the volume and mass of vapor in the final state. Calculate also the heat transfer.

1-9. In a constant low-pressure reversible process the rate of heat transfer to one kg mole of oxygen gas per unit temperature rise may be expressed as a function of temperature by the equation

$$\frac{dq}{dT} = 6.0954 + 3.2533 \times 10^{-3}T - 1.0171 \times 10^{-6}T^2$$

where q is in kcal/kg mole, and T is in °K. Calculate the heat added and the increase in entropy for one kg mole of the gas during a reversible heating process from 300 to 400°K at a constant low pressure.

1-10. A block of iron weighs 100 kg and has a temperature of 100°C. When this block of iron is immersed in 50 kg of water at a temperature of 20°C, what will be the change of entropy of the combined system of iron and water? For the iron $dq = 0.11dT$, and for the water $dq = 1.0dT$, wherein q denotes heat transfer in cal/g and T denotes temperature in °K.

1-11. Steam at a pressure of 10 bars and a temperature of 250°C flows into an expander at a rate of 1350 kg/hr. It leaves the expander at 0.25 bars. The process is reversible and adiabatic. Find the power delivered.

1-12. Steam enters a turbine at a pressure of 100 bars and a temperature of 400°C. At the exit of the turbine the pressure is 1 bar and the entropy is 0.6 joule/g°K greater than at the inlet. The process is adiabatic and changes in KE and PE may be neglected. Find the work done by the steam in joules/g. What is the mass rate of flow of steam in kg/sec that is required to produce a power output of 1000 kw?

1-13. A mixture of H_2O at the triple-point condition $(0.01°C)$ containing 20% solid, 30% liquid, and 50% vapor by mass enters a tube at the rate of 5 kg/sec. Heat is added to the mixture at constant pressure and the H_2O emerges out from the tube all saturated vapor. Calculate the rate of heat supplied.

1-14. An insulated tank initially contains 100 ft^3 of water vapor and 500 ft^3 of liquid water at a pressure of 200 psia. For how long can the steam be withdrawn through a pressure-reducing valve at a pressure of 50 psia and at a rate of 2,000 lbm/hr? State and justify any assumption you make.

1-15. A refrigerator uses a reversed Carnot cycle with steam as the working fluid. At the begining of the isothermal expansion, the steam has a pressure of 0.5 psia and a quality of 35%; at the end of the isothermal expansion, the steam is dry and saturated. The heat-rejection process is at $250°F$. For one pound of steam, calculate

(a) the net work input per cycle
(b) the coefficient of performance as defined by the expression

$$\text{Coefficient of performance of a refrigerator} = \frac{\text{heat absorbed}}{\text{net work input}}$$

1-16. One pound of Freon-12 in a piston-and-cylinder machine undergoes a thermodynamic cycle composed of the following reversible processes:

(1-2) Saturated Freon-12 vapor at $40°F$ is expanded isothermally to 5 psia.
(2-3) It is then compressed adiabatically back to its initial volume.
(3-1) It is finally cooled at constant volume back to the initial saturated vapor state.

Sketch the p-h, T-s, and p-v diagrams for the cycle, and calculate the work done, the heat transferred, and the entropy change for each of the three processes. When this machine is used as a heat pump, determine the coefficient of performance as defined by the expression

$$\text{Coefficient of performance of a heat pump} = \frac{\text{heat rejected}}{\text{net work input}}$$

1-17. One kilogram of steam in a closed system undergoes a thermodynamic cycle composed of the following reversible processes:

(1-2) The steam, initially at 10 bars and having a quality of 40%, is heated at constant volume until its pressure rises to 35 bars.
(2-3) It is then expanded isothermally to 10 bars.
(3-1) It is finally cooled at constant pressure back to its initial state.

Sketch the T-s diagram for the cycle, and calculate the work done, the heat transferred, and the change of entropy for each of the three processes. What is the thermal efficiency of the cycle?

Chapter 2
GENERAL THERMODYNAMIC RELATIONSHIPS

2-1 INTRODUCTION

The basic laws of thermodynamics are quite general and may be applied to systems of considerable complexity, involving mechanical, chemical, electrical, magnetic, and other effects. Fortunately the majority of systems encountered in thermodynamic analyses are of restricted nature. For instance, there are a vast number of thermodynamic systems composed of so-called pure substances which are homogeneous and invariable in chemical composition. In addition, in many cases the effects of motion, capillarity, solid distortion, and external force fields (electric, magnetic, and gravitational) are absent or can be neglected. We have studied the properties of such substances in Sec. 1-12. A system composed of such substances can perform work in a reversible process by volume change only ($p\,dV$ work). The term *simple compressible system* will be applied to such a system. The word "simple" is used to indicate that the system has only one reversible work mode. In Chaps. 6 and 7 we will study other simple systems, such as a simple elastic system for which the only reversible work mode is elastic elongation and a simple magnetic system for which the only reversible work mode is magnetization.

Experimental observations have shown that there is a uniqueness about the number of independent properties required to specify an equilibrium state of a system. The number of independent properties required obviously depends on the number of different ways of energy interactions. Besides heat interaction there may be a number of different modes of work interactions. The relation between the number of independent properties and the number of different energy interactions may be summarized in a postulate, known as the *state principle*, which is as follows: The number of independent properties required to specify the equilibrium state of a system is equal to the number of possible reversible work modes plus one. The "plus one" accounts for heat interaction. According to this postulate, two independent properties are required to specify

45

an equilibrium state of a simple system. This is known as the two-property rule. The significance of the term independent property should be emphasized, because only when the properties are independent do they suffice to specify the state. For example, pressure and temperature are not independent properties for a pure substance in the form of a mixture of liquid and vapor because an indefinite number of states could exist at a given pressure and temperature.

In the previous chapter we have defined and used several thermodynamic properties. Some of them are directly measurable, but others cannot be measured and must be calculated from data on other properties and quantities which can be measured. We are now ready to develop some useful general relations between thermodynamic properties that shall facilitate such calculations. We will restrict our attention for the time being to simple systems which require only two independent properties to determine their thermodynamic states. In particular we will devote this chapter to simple compressible systems. Once the thermodynamic relations are developed for such systems it is a simple matter to write the analogous relations for other simple systems.

2-2 FUNDAMENTALS OF PARTIAL DERIVATIVES

Most general thermodynamic relations for simple systems involve the rate of change of one property with respect to another while a third property is held constant. Thus, the methods of partial differential calculus play an important part in the development of thermodynamic relations and a brief review of the rules of partial differentiation is in order.

Let z denote a dependent property of a simple system as defined by the two independent properties x and y. The functional dependence may be stated symbolically as

$$z = z(x, y)$$

According to the calculus, the total differential of the dependent variable z is given by the equation

$$dz = \left(\frac{\partial z}{\partial x}\right)_y dx + \left(\frac{\partial z}{\partial y}\right)_x dy \tag{2-1}$$

Total differential $z = z(x, y)$

The partial derivatives in the above equation are written with the subscript to indicate the independent variable that is held constant in forming the derivative. In thermodynamics such a notation is necessary in order to show clearly the independent variables of which the dependent variable is considered to be a function.

Equation (2-1) may be written as

$$dz = M \, dx + N \, dy$$

where

$$M = \left(\frac{\partial z}{\partial x}\right)_y \quad \text{and} \quad N = \left(\frac{\partial z}{\partial y}\right)_x$$

Then

$$\left(\frac{\partial M}{\partial y}\right)_x = \frac{\partial^2 z}{\partial y\,\partial x} \quad \text{and} \quad \left(\frac{\partial N}{\partial x}\right)_y = \frac{\partial^2 z}{\partial x\,\partial y}$$

Since the order of differentiation is immaterial, we must have

$$\left(\frac{\partial M}{\partial y}\right)_x = \left(\frac{\partial N}{\partial x}\right)_y \tag{2-2}$$

This is the necessary and sufficient condition that $dz = M\,dx + N\,dy$ be an exact differential and that z be a point function, which is, of course, true because z is a thermodynamic property.

Because of the exactness of dz in Eq. (2-1), the line integral $\int dz$ depends only on the two end states and is independent of the path or curve connecting the end states. For a complete cycle or closed curve where the two end states have identical coordinates, it is clear that

$$\oint dz = \oint \left[\left(\frac{\partial z}{\partial x}\right)_y dx + \left(\frac{\partial z}{\partial y}\right)_x dy\right] = 0$$

Usually independent variables of a partial derivative are so chosen as to be particularly convenient for a given case. In thermodynamic analyses there is often a need to interchange the variables of a given partial derivative or to transform it to new sets of variables. We will now review some mathematical relations which are useful in the transformation of state variables.

Let us consider four state variables x, y, z, and α, any two of which might be selected as the independent variables for a given simple system. When x is considered to be a function of y and α, then

$$dx = \left(\frac{\partial x}{\partial y}\right)_\alpha dy + \left(\frac{\partial x}{\partial \alpha}\right)_y d\alpha \tag{2-3}$$

When y is considered to be a function of z and α, then

$$dy = \left(\frac{\partial y}{\partial z}\right)_\alpha dz + \left(\frac{\partial y}{\partial \alpha}\right)_z d\alpha \tag{2-3a}$$

Substitution of Eq. (2-3a) in Eq. (2-3) gives

$$dx = \left[\left(\frac{\partial x}{\partial y}\right)_\alpha\left(\frac{\partial y}{\partial z}\right)_\alpha\right] dz + \left[\left(\frac{\partial x}{\partial \alpha}\right)_y + \left(\frac{\partial x}{\partial y}\right)_\alpha\left(\frac{\partial y}{\partial \alpha}\right)_z\right] d\alpha \tag{2-3b}$$

On the other hand, when x is considered to be a function of z and α, we have

$$dx = \left(\frac{\partial x}{\partial z}\right)_\alpha dz + \left(\frac{\partial x}{\partial \alpha}\right)_\alpha d\alpha \qquad (2\text{-}3c)$$

Since α and z are independent, the coefficients of $d\alpha$ and dz in Eqs. (2-3b) and (2-3c) must be respectively equal. Then

$$\left(\frac{\partial x}{\partial \alpha}\right)_z = \left(\frac{\partial x}{\partial \alpha}\right)_y + \left(\frac{\partial x}{\partial y}\right)_\alpha \left(\frac{\partial y}{\partial \alpha}\right)_z \qquad (2\text{-}4)$$

$$\left(\frac{\partial x}{\partial z}\right)_\alpha = \left(\frac{\partial x}{\partial y}\right)_\alpha \left(\frac{\partial y}{\partial z}\right)_\alpha$$

The last equation may be put in the following more symmetrical form:

$$\left(\frac{\partial x}{\partial y}\right)_\alpha \left(\frac{\partial y}{\partial z}\right)_\alpha \left(\frac{\partial z}{\partial x}\right)_\alpha = 1 \qquad \text{CHAIN RELATION} \qquad (2\text{-}5)$$

This equation is sometimes called the chain relation, each partial derivative in the equation being a link of the chain. This relation may be extended or shortened by adding or subtracting links. Thus, if there are five state variables x, y, z, γ, and α involved in the transformation, the chain relation becomes

$$\left(\frac{\partial x}{\partial y}\right)_\alpha \left(\frac{\partial y}{\partial z}\right)_\alpha \left(\frac{\partial z}{\partial \gamma}\right)_\alpha \left(\frac{\partial \gamma}{\partial x}\right)_\alpha = 1$$

Likewise, for three variables x, y, and α, the chain relation reduces to

$$\left(\frac{\partial x}{\partial y}\right)_\alpha \left(\frac{\partial y}{\partial x}\right)_\alpha = 1 \qquad \text{or} \qquad \left(\frac{\partial x}{\partial y}\right)_\alpha = \frac{1}{(\partial y/\partial x)_\alpha} \qquad (2\text{-}6)$$

which is a simple reciprocal relation.

When there are only three state variables involved in a transformation, there is a very useful partial differential relation which may be obtained as follows: The total differentials of $x = x(y, z)$ and $y = y(x, z)$ are

$$dx = \left(\frac{\partial x}{\partial y}\right)_z dy + \left(\frac{\partial x}{\partial z}\right)_y dz$$

$$dy = \left(\frac{\partial y}{\partial x}\right)_z dx + \left(\frac{\partial y}{\partial z}\right)_x dz$$

Combining the above two equations gives

$$\left[1 - \left(\frac{\partial x}{\partial y}\right)_z \left(\frac{\partial y}{\partial x}\right)_z\right] dx = \left[\left(\frac{\partial x}{\partial y}\right)_z \left(\frac{\partial y}{\partial z}\right)_x + \left(\frac{\partial x}{\partial z}\right)_y\right] dz$$

Since x and z are independent, the above equation must be true for all values of dx and dz. Letting $dz = 0$ and $dx \neq 0$ results in the simple reciprocal relation of

Eq. (2-6). Whereas letting $dx = 0$ and $dz \neq 0$ results in

$$\left(\frac{\partial x}{\partial y}\right)_z \left(\frac{\partial y}{\partial z}\right)_x + \left(\frac{\partial x}{\partial z}\right)_y = 0$$

which may be arranged in the following symmetrical form

$$\left(\frac{\partial x}{\partial y}\right)_z \left(\frac{\partial y}{\partial z}\right)_x \left(\frac{\partial z}{\partial x}\right)_y = -1 \qquad \text{CYCLIC RELATION} \tag{2-7}$$

This equation is sometimes called the cyclic relation because the variables are permuted cyclically. Notice that unlike the chain relation, this equation cannot be extended or shortened.

2-3 SIMPLE COMPRESSIBLE SYSTEMS

In the previous chapter we studied the first and second laws separately. We will combine them now to produce a very important relation for simple compressible systems of fixed mass. The first law for an infinitesimal process is

$$đQ = dU + đW$$

If the process is reversible, then according to the second law

$$đQ = T\,dS$$

For a reversible process of a simple compressible closed system

$$đW = p\,dV$$

whence

$$T\,dS = dU + p\,dV \quad \text{or} \quad dU = T\,dS - p\,dV \tag{2-8}$$

This is the very basic relation which combines the first and second laws as applied to a simple compressible system of fixed mass. Although we used the condition of a reversible process to derive this equation, we see that once it is written it is a relation solely among the properties of a system. If the system passes from one equilibrium state to another, the property changes are the same whether the process taking place is reversible or irreversible. However, if the process is not reversible the term $T\,dS$ in Eq. (2-8) is not the heat added ($đQ$) and the term $p\,dV$ is not the work done ($đW$) for the system. Moreover, it is imperative to notice that in order to evaluate the change in the properties U, S, or V by the use of Eq. (2-8), one must choose a suitable reversible path to perform the integration.

The combined first and second laws as expressed by Eq. (2-8) may be used in connection with the definitions of enthalpy, Helmholtz function, and Gibbs function to form three other important relations. From the definition of enthalpy we have the relation

$$dH = dU + p\,dV + V\,dp$$

Substituting Eq. (2-8) into this relation gives   ENTHALPY

$$dH = T\,dS + V\,dp \tag{2-9}$$

From the definition of Helmholtz function we have the relation

$$dA = dU - T\,dS - S\,dT$$

Substituting Eq. (2-8) into this relation gives   HELMHOLTZ

$$dA = -S\,dT - p\,dV \tag{2-10}$$

From the definition of Gibbs function we have the relation

$$dG = dH - T\,dS - S\,dT$$

Substituting Eq. (2-9) into this relation gives   Gibbs

$$dG = -S\,dT + V\,dp \tag{2-11}$$

Equations (2-8) through (2-11) are the four basic relations of properties. They are applicable for any process, reversible or irreversible, between equilibrium states of a simple compressible system of fixed mass.

A number of useful partial derivative relations can be readily obtained from the four basic relations. Thus, the total differential of U in terms of S and V is

$$dU = \left(\frac{\partial U}{\partial S}\right)_V dS + \left(\frac{\partial U}{\partial V}\right)_S dV$$

Comparing this equation with Eq. (2-8) gives

$$\left(\frac{\partial U}{\partial S}\right)_V = T \quad \text{and} \quad \left(\frac{\partial U}{\partial V}\right)_S = -p \tag{2-12a}$$

Similar relations are obtained from the other three basic relations. They are as follows:

$$\left(\frac{\partial H}{\partial S}\right)_p = T \quad \text{and} \quad \left(\frac{\partial H}{\partial p}\right)_S = V \tag{2-12b}$$

$$\left(\frac{\partial A}{\partial T}\right)_V = -S \quad \text{and} \quad \left(\frac{\partial A}{\partial V}\right)_T = -p \tag{2-12c}$$

$$\left(\frac{\partial G}{\partial T}\right)_p = -S \quad \text{and} \quad \left(\frac{\partial G}{\partial p}\right)_T = V \tag{2-12d}$$

Since U, H, A, and G are all thermodynamic properties they are point functions. Therefore dU, dH, dA, and dG are exact differentials. Applying the condition of exactness to the four basic relations, Eqs. (2-8) through (2-11), one obtains

$$\left(\frac{\partial T}{\partial V}\right)_S = -\left(\frac{\partial p}{\partial S}\right)_V \qquad \rightarrow U \tag{2-13a}$$

$$\left(\frac{\partial T}{\partial p}\right)_S = \left(\frac{\partial V}{\partial S}\right)_p \qquad \rightarrow H \quad \text{MAXWELL} \atop \text{RELATIONS} \tag{2-13b}$$

$$\left(\frac{\partial S}{\partial V}\right)_T = \left(\frac{\partial p}{\partial T}\right)_V \qquad \rightarrow F \tag{2-13c}$$

$$\left(\frac{\partial S}{\partial p}\right)_T = -\left(\frac{\partial V}{\partial T}\right)_p \qquad \rightarrow G \tag{2-13d}$$

The above equations are known as the four *Maxwell relations*. They are of great usefulness in the transformation of state variables, and particularly in the determination of changes in entropy, which are not experimentally measurable, in terms of the measurable properties P, V, and T.

It is of interest to note that while Eqs. (2-8) and (2-9) apply equally well to a reversible or an irreversible process between equilibrium states, the following three equations apply only to reversible processes:

$$dQ = dU + p\,dV \qquad \text{ONLY FOR} \tag{2-14}$$

$$T\,dS = dU + dW \qquad \text{REVERSIBLE} \tag{2-15}$$

$$dQ = dH - V\,dp \qquad \text{PROCESSES} \tag{2-16}$$

In the derivation of these equations one must use the equations $dW = p\,dV$ and $dQ = T\,dS$ which are for reversible processes only. Unlike Eqs. (2-8) and (2-9), the foregoing three equations are not relations solely among the properties of a system.

If we regard the volume of a system as a function of temperature and pressure, we have

$$dV = \left(\frac{\partial V}{\partial T}\right)_p dT + \left(\frac{\partial V}{\partial p}\right)_T dp \qquad V = V(T, P)$$

The differential coefficients in the preceding equation have important physical significance when defined as follows:

Coefficient of thermal expansion

$$\alpha = \frac{1}{V}\left(\frac{\partial V}{\partial T}\right)_p \tag{2-17}$$

Isothermal compressibility

$$* \quad \kappa = -\frac{1}{V}\left(\frac{\partial V}{\partial p}\right)_T \tag{2-18}$$

We accordingly write

$$dV = \alpha V\, dT - \kappa V\, dp$$

or

$$d\ln V = \alpha\, dT - \kappa\, dp \tag{2-19}$$

Application of the condition of exactness to the foregoing equation results in

$$\left(\frac{\partial \alpha}{\partial p}\right)_T = -\left(\frac{\partial \kappa}{\partial T}\right)_p \tag{2-20}$$

which indicates that the coefficients α and κ are not independent of each other. From the cyclic relation we have

$$\left(\frac{\partial p}{\partial T}\right)_V \left(\frac{\partial T}{\partial V}\right)_p \left(\frac{\partial V}{\partial p}\right)_T = -1$$

Thus

$$\frac{\alpha}{\kappa} = -\frac{(\partial V/\partial T)_p}{(\partial V/\partial p)_T} = \left(\frac{\partial p}{\partial T}\right)_V \tag{2-21}$$

Similar in form to Eq. (2-18) is another coefficient defined as

Adiabatic compressibility

$$* \quad \kappa_s = -\frac{1}{V}\left(\frac{\partial V}{\partial p}\right)_S \tag{2-22}$$

Notice that all equations shown in this chapter involving the partial derivatives $(\partial V/\partial T)_p$, $(\partial V/\partial p)_T$, and/or $(\partial V/\partial p)_s$ may be written in terms of α, κ, and/or κ_s.

EXAMPLE 2-1. Use the first Maxwell relation [Eq. (2-13a)] to demonstrate the correctness of the data given by Keenan, Keyes, Hill, and Moore[1] for superheated steam at 600 psia and 800°F.

Solution.
At 600 psia and 800°F the steam tables give $v = 1.1900$ cu ft/lbm and $s = 1.6343$ Btu/(lbm)(°R). From the steam tables select values of pressure and

[1] J. H. Keenan, F. G. Keyes, P. G. Hill, and J. G. Moore, "Steam Tables (English Units)," John Wiley and Sons, Inc., New York, 1969.

entropy corresponding to a constant specific volume of 1.1900 cu ft/lbm. Plot these data on a p-s diagram and draw the constant volume curve. Then measure the slope of this curve at 600 psia. The value obtained from this measurement is

$$\left(\frac{\partial p}{\partial s}\right)_v = 1{,}670 \; \frac{\text{lbf/sq in.}}{\text{Btu/(lbm)(}^\circ\text{R)}} = 309 \; \frac{^\circ\text{R}}{\text{cu ft/lbm}}$$

Next, from the steam tables select values of temperature and specific volume corresponding to a constant entropy of 1.6343 Btu/(lbm)($^\circ$R). Plot these data on a T-v diagram and draw the constant entropy curve. Then measure the slope of this curve at 800°F. The value obtained from this measurement is

$$\left(\frac{\partial T}{\partial v}\right)_s = -309 \; \frac{^\circ\text{R}}{\text{cu ft/lbm}}$$

These two slopes are equal in magnitude but opposite in sign as predicted by Eq. (2-13a).

EXAMPLE 2-2. The van der Waals equation of state is

$$p = \frac{RT}{v - b} - \frac{a}{v^2}$$

where a and b are two specific constants for a gas, and R is the gas constant. For a gas obeying this equation, find the expressions for the coefficient of thermal expansion α and the isothermal compressibility κ.

Solution.
Differentiating the van der Waals equation gives

$$\left(\frac{\partial p}{\partial T}\right)_v = \frac{R}{v - b} \quad \text{and} \quad \left(\frac{\partial p}{\partial v}\right)_T = \frac{2a}{v^3} - \frac{RT}{(v - b)^2}$$

From the cyclic relation we have

$$\left(\frac{\partial v}{\partial T}\right)_p = -\frac{(\partial p/\partial T)_v}{(\partial p/\partial v)_T} = -\frac{R/(v - b)}{(2a/v^3) - RT/(v - b)^2} = \frac{Rv^3(v - b)}{RTv^3 - 2a(v - b)^2}$$

Therefore

$$\alpha = \frac{1}{v}\left(\frac{\partial v}{\partial T}\right)_p = \frac{Rv^2(v - b)}{RTv^3 - 2a(v - b)^2}$$

$$\kappa = -\frac{1}{v}\left(\frac{\partial v}{\partial p}\right)_T = -\frac{1}{v}\frac{1}{(2a/v^3) - RT/(v - b)^2} = \frac{v^2(v - b)^2}{RTv^3 - 2a(v - b)^2}$$

Also

$$\frac{\alpha}{\kappa} = -\left(\frac{\partial v}{\partial T}\right)_p\left(\frac{\partial p}{\partial v}\right)_T = \left(\frac{\partial p}{\partial T}\right)_v = \frac{R}{v - b}$$

Note that when $a = b = 0$, the van der Waals equation reduces to the ideal gas equation $pv = RT$. Then for an ideal gas we have

$$\alpha = \frac{1}{T} \quad \text{and} \quad \kappa = \frac{v}{RT} = \frac{1}{p}$$

2-4 RELATIONS FOR SPECIFIC HEATS

The *specific heat at constant volume c_v* and the *specific heat at constant pressure c_p* of a homogeneous simple compressible system are defined as

$$c_v = \left(\frac{\partial u}{\partial T}\right)_v \tag{2-23}$$

$$c_p = \left(\frac{\partial h}{\partial T}\right)_p \tag{2-24}$$

Clearly c_v and c_p are functions of state, i.e., they are thermodynamic properties. The specific heats may be expressed either for unit mass or unit mole of a system. They may also be expressed for a system as a whole, in which case the name *heat capacity* is usually used and the symbols C_v and C_p will be employed.

The properties c_v and c_p were originally defined by the expressions

$$c_v = \left(\frac{\mathit{d}q}{\mathit{d}T}\right)_v \quad \text{and} \quad c_p = \left(\frac{\mathit{d}q}{\mathit{d}T}\right)_p \tag{2-25}$$

Inasmuch as $\mathit{d}q$ is not the differential of a property, the use of Eq. (2-25) to define the physical properties c_v and c_p is misleading. This also accounts for the unfortunate choices of the names specific heat and heat capacity for these properties. Of course, for a reversible constant volume process we see from Eq. (2-14) that $\mathit{d}q = du$, and for a reversible constant pressure process we see from Eq. (2-16) that $\mathit{d}q = dh$. Thus, the definitions given by Eqs. (2-23) and (2-24) are equivalent to those given by Eq. (2-25). We nevertheless prefer the definitions given by Eqs. (2-23) and (2-25) for this reason: since specific heats are physical properties they have much more important roles in thermodynamics than simply their roles in the calculation of heat transfer under certain conditions.

The specific heats may be expressed in terms of the entropy through the use of Eqs. (2-12a) and (2-12b). Thus since

$$\left(\frac{\partial u}{\partial s}\right)_v = \left(\frac{\partial u}{\partial T}\right)_v\left(\frac{\partial T}{\partial s}\right)_v = T \quad \text{and} \quad \left(\frac{\partial h}{\partial s}\right)_p = \left(\frac{\partial h}{\partial T}\right)_p\left(\frac{\partial T}{\partial s}\right)_p = T$$

Whereupon we obtain

$$c_v = \left(\frac{\partial u}{\partial T}\right)_v = T\left(\frac{\partial s}{\partial T}\right)_v \tag{2-26}$$

and

$$c_p = \left(\frac{\partial h}{\partial T}\right)_p = T\left(\frac{\partial s}{\partial T}\right)_p \tag{2-27}$$

Since specific heats are related to energy measurements they cannot be calculated from data on p-v-T alone. However, the isothermal changes in c_v with respect to volume and in c_p with respect to pressure can be calculated when adequate p-v-T data are available. We will now derive the required relations that facilitate such calculations. Differentiating Eqs. (2-26) and (2-27) with respect to v and p respectively, while holding T constant in both cases, we have

$$\left(\frac{\partial c_v}{\partial v}\right)_T = T\frac{\partial^2 s}{\partial v\,\partial T} \quad\text{and}\quad \left(\frac{\partial c_p}{\partial p}\right)_T = T\frac{\partial^2 s}{\partial p\,\partial T}$$

Differentiating the Maxwell relations

$$\left(\frac{\partial p}{\partial T}\right)_v = \left(\frac{\partial s}{\partial v}\right)_T \quad\text{and}\quad \left(\frac{\partial v}{\partial T}\right)_p = -\left(\frac{\partial s}{\partial p}\right)_T$$

yields

$$\left(\frac{\partial^2 p}{\partial T^2}\right)_v = \frac{\partial^2 s}{\partial T\,\partial v} \quad\text{and}\quad \left(\frac{\partial^2 v}{\partial T^2}\right)_p = -\frac{\partial^2 s}{\partial T\,\partial p}$$

Therefore we finally have the desired relations

$$\left(\frac{\partial c_v}{\partial v}\right)_T = T\left(\frac{\partial^2 p}{\partial T^2}\right)_v \qquad\qquad\text{DETERMINED USING} \tag{2-28}$$

$$P\text{-}U\text{-}T \text{ data}$$

and

$$\left(\frac{\partial c_p}{\partial p}\right)_T = -T\left(\frac{\partial^2 v}{\partial T^2}\right)_p \tag{2-29}$$

There are two more important equations for specific heats in relation to p-v-T data: one is the ratio of c_p and c_v and the other is their difference. The relation for the ratio of specific heats, denoted by the symbol k, is obtained by dividing Eq. (2-27) by Eq. (2-26). Thus

$$k = \frac{c_p}{c_v} = \frac{(\partial s/\partial T)_p}{(\partial s/\partial T)_v}$$

But by the cyclic relation of Eq. (2-7), we write

$$\left(\frac{\partial s}{\partial T}\right)_p = -\left(\frac{\partial p}{\partial T}\right)_s\left(\frac{\partial s}{\partial p}\right)_T \quad\text{and}\quad \left(\frac{\partial s}{\partial T}\right)_v = -\left(\frac{\partial v}{\partial T}\right)_s\left(\frac{\partial s}{\partial v}\right)_T$$

Consequently

$$k = \frac{c_p}{c_v} = \frac{(\partial p/\partial v)_s}{(\partial p/\partial v)_T} \tag{2-30}$$

The preceding equation may also be written in terms of the isothermal compressibility κ and the adiabatic compressibility κ_s as follows:

$$k = \frac{c_p}{c_v} = \frac{\kappa}{\kappa_s} \tag{2-31}$$

The difference between c_p and c_v is obtained by subtracting Eq. (2-26) from Eq. (2-27). Thus

$$c_p - c_v = T\left(\frac{\partial s}{\partial T}\right)_p - T\left(\frac{\partial s}{\partial T}\right)_v$$

But by Eq. (2-4) we have

$$\left(\frac{\partial s}{\partial T}\right)_p = \left(\frac{\partial s}{\partial T}\right)_v + \left(\frac{\partial s}{\partial v}\right)_T \left(\frac{\partial v}{\partial T}\right)_p$$

Therefore,

$$c_p - c_v = T\left(\frac{\partial s}{\partial v}\right)_T \left(\frac{\partial v}{\partial T}\right)_p$$

Upon using the third Maxwell relation

$$\left(\frac{\partial s}{\partial v}\right)_T = \left(\frac{\partial p}{\partial T}\right)_v$$

we obtain

$$c_p - c_v = T\left(\frac{\partial p}{\partial T}\right)_v \left(\frac{\partial v}{\partial T}\right)_p \tag{2-32}$$

Moreover, from the cyclic relation we have

$$\left(\frac{\partial p}{\partial T}\right)_v = -\left(\frac{\partial v}{\partial T}\right)_p \left(\frac{\partial p}{\partial v}\right)_T$$

Finally,

$$c_p - c_v = -T\left(\frac{\partial v}{\partial T}\right)_p^2 \left(\frac{\partial p}{\partial v}\right)_T = -\frac{T(\partial v/\partial T)_p^2}{(\partial v/\partial p)_T} \tag{2-33}$$

Equations (2-32) or (2-33) can be used for the conversion of c_p into c_v, or vice versa, when adequate p-v-T data are available. In the case of solids and liquids it is very difficult to determine c_v experimentally; it must be converted from measured values of c_p by the use of these equations.

Some observations concerning the relative magnitude of c_p and c_v can be made from Eq. (2-33). Although $(\partial v/\partial T)_p$ is usually positive, it may be zero or negative. For example, at atmospheric pressure $(\partial v/\partial T)_p$ for water is zero at 4°C and is negative between 0°C and 4°C. However, the square of $(\partial v/\partial T)_p$ can never be negative. On the other hand, $(\partial p/\partial v)_T$ is always negative for all known substances. Consequently $(c_p - c_v)$ can never be negative, or c_p can never be smaller than c_v. The two specific heats are equal in the particular case when $(\partial v/\partial T)_p = 0$. Furthermore, $(c_p - c_v)$ approaches zero as T approaches zero, or $c_p = c_v$ at the absolute zero temperature. For solids and liquids both $(\partial v/\partial T)_p$ and $(\partial p/\partial v)_T$ are relatively small and so at low temperatures the difference between c_p and c_v is small although this difference becomes appreciable at high temperatures. Also, because c_p can never be smaller than c_v, we observe from Eq. (2-30) that k can never be smaller than unity.

EXAMPLE 2-3. For a gas obeying the van der Waals equation of state (Example 2-2) find the expression for $(c_p - c_v)$. Show that for this gas c_v is a function of temperature only.

Solution.
Equation (2-33) for $(c_p - c_v)$ can be rewritten in terms of the coefficient of thermal expansion α and isothermal compressibility κ as follows:

$$c_p - c_v = \frac{Tv\left[\frac{1}{v}(\partial v/\partial T)_p\right]^2}{-\frac{1}{v}(\partial v/\partial p)_T} = \frac{Tv\alpha^2}{\kappa}$$

Substituting the expressions for α and (α/κ) as obtained in Example 2-2 leads to the desired expression

$$c_p - c_v = Tv\left[\frac{Rv^2(v - b)}{RTv^3 - 2a(v - b)^2}\right]\left(\frac{R}{v - b}\right)$$

$$= \frac{R}{1 - 2a(v - b)^2/RTv^3}$$

When $a = b = 0$, the van der Waals equation reduces to the ideal gas equation $pv = RT$, for which $c_p - c_v = R$. For a gas obeying the van der Waals equation, $(c_p - c_v)$ is not exactly equal to R, and is not a constant.

Since

$$\left(\frac{\partial p}{\partial T}\right)_v = \frac{R}{v - b}, \quad \left(\frac{\partial^2 p}{\partial T^2}\right)_v = 0$$

it follows that

$$\left(\frac{\partial c_v}{\partial v}\right)_T = T\left(\frac{\partial^2 p}{\partial T^2}\right)_v = 0$$

Hence c_v is a function of temperature only for a gas obeying the van der Waals equation. However, it may be seen from the equation

$$\left(\frac{\partial c_p}{\partial p}\right)_T = -T\left(\frac{\partial^2 v}{\partial T^2}\right)_p$$

$$= R^2 T\left[\frac{2av^{-3} - 6abv^{-4}}{(p - av^{-2} + 2abv^{-3})^3}\right]$$

that c_p for a van der Waals gas is not a function of temperature only.

2-5 RELATIONS FOR ENTROPY, INTERNAL ENERGY, AND ENTHALPY

The entropy change for a simple compressible system may be expressed as a function of any pair of the variables p, v, and T, thus forming three general equations, the derivations of which are as follows. When s is considered as a function of T and v, we have

$$ds = \left(\frac{\partial s}{\partial T}\right)_v dT + \left(\frac{\partial s}{\partial v}\right)_T dv$$

But

$$\left(\frac{\partial s}{\partial T}\right)_v = \frac{c_v}{T} \tag{2-26}$$

and

$$\left(\frac{\partial s}{\partial v}\right)_T = \left(\frac{\partial p}{\partial T}\right)_v \tag{2-13c}$$

Consequently,

$$ds = \frac{c_v}{T}dT + \left(\frac{\partial p}{\partial T}\right)_v dv \qquad\qquad s = s(T, v) \tag{2-34}$$

This is the first ds equation. Now when s is considered as a function of T and p, we have

$$ds = \left(\frac{\partial s}{\partial T}\right)_p dT + \left(\frac{\partial s}{\partial p}\right)_T dp$$

But

$$\left(\frac{\partial s}{\partial T}\right)_p = \frac{c_p}{T} \qquad (2\text{-}27)$$

and

$$\left(\frac{\partial s}{\partial p}\right)_T = -\left(\frac{\partial v}{\partial T}\right)_p \qquad (2\text{-}13d)$$

Consequently,

$$ds = \frac{c_p}{T}\,dT - \left(\frac{\partial v}{\partial T}\right)_p dp \qquad S = S(T,P) \qquad (2\text{-}35)$$

This is the second *ds* equation. Finally when *s* is considered as a function of *p* and *v*, we have

$$ds = \left(\frac{\partial s}{\partial p}\right)_v dp + \left(\frac{\partial s}{\partial v}\right)_p dv$$

But by Eqs. (2-5), (2-26), and (2-27) we have

$$\left(\frac{\partial s}{\partial p}\right)_v = \left(\frac{\partial T}{\partial p}\right)_v \left(\frac{\partial s}{\partial T}\right)_v = \left(\frac{\partial T}{\partial p}\right)_v \frac{c_v}{T}$$

and

$$\left(\frac{\partial s}{\partial v}\right)_p = \left(\frac{\partial T}{\partial v}\right)_p \left(\frac{\partial s}{\partial T}\right)_p = \left(\frac{\partial T}{\partial v}\right)_p \frac{c_p}{T}$$

Consequently,

$$ds = \frac{c_v}{T}\left(\frac{\partial T}{\partial p}\right)_v dp + \frac{c_p}{T}\left(\frac{\partial T}{\partial v}\right)_p dv \qquad S = S(P,v) \qquad (2\text{-}36)$$

This is the third *ds* equation. The three *ds* equations are useful in the evaluation of entropy changes. Since *đq = T ds* in a reversible process, these equations also provide means to calculate heat transfer in a reversible process when specific heats and *p-v-T* data are available.

The three *ds* equations we just derived may be used to yield three equations for *du* as functions of each pair of the variables *p*, *v*, and *T*. Thus the combined first and second laws for a simple compressible system is

$$du = T\,ds - p\,dv \qquad (2\text{-}8)$$

Upon substituting the first *ds* equation into Eq. (2-8), we get the first *du* equation,

$$du = c_v dT + \left[T\left(\frac{\partial p}{\partial T}\right)_v - p \right] dv \tag{2-37}$$

Substituting the second ds equation into Eq. (2-8), and noting that

$$dv = \left(\frac{\partial v}{\partial T}\right)_p dT + \left(\frac{\partial v}{\partial p}\right)_T dp$$

we get the second du equation,

$$du = \left[c_p - p\left(\frac{\partial v}{\partial T}\right)_p \right] dT - \left[p\left(\frac{\partial v}{\partial p}\right)_T + T\left(\frac{\partial v}{\partial T}\right)_p \right] dp \tag{2-38}$$

Substituting the third ds equation into Eq. (2-8), we get the third du equation,

$$du = c_v\left(\frac{\partial T}{\partial p}\right)_v dp + \left[c_p\left(\frac{\partial T}{\partial v}\right)_p - p \right] dv \tag{2-39}$$

It should be noted that out of the three du equations only the one with T and v as independent variables, i.e., Eq. (2-37), is most useful. This equation may be used to form the so-called *energy equation*, which is

$$\left(\frac{\partial u}{\partial v}\right)_T = T\left(\frac{\partial p}{\partial T}\right)_v - p \tag{2-40}$$

As in the case of internal energy, the three ds equations may also be used to yield three equations for dh as functions of each pair of the variables p, v, and T. Thus the basic equation for dh is

$$dh = T\,ds + v\,dp \tag{2-9}$$

Upon substituting the first ds equation into Eq. (2-9) and noting that

$$dp = \left(\frac{\partial p}{\partial T}\right)_v dT + \left(\frac{\partial p}{\partial v}\right)_T dv$$

we get the first dh equation,

$$dh = \left[c_v + v\left(\frac{\partial p}{\partial T}\right)_v \right] dT + \left[T\left(\frac{\partial p}{\partial T}\right)_v + v\left(\frac{\partial p}{\partial v}\right)_T \right] dv \tag{2-41}$$

Substituting the second ds equation into Eq. (2-9), we get the second dh equation,

$$dh = c_p dT + \left[v - T\left(\frac{\partial v}{\partial T}\right)_p \right] dp \tag{2-42}$$

Substituting the third ds equation into Eq. (2-9), we get the third dh equation,

$$dh = \left[v + c_v\left(\frac{\partial T}{\partial p}\right)_v \right]dp + c_p\left(\frac{\partial T}{\partial v}\right)_p dv \tag{2-43}$$

Among the three general equations for enthalpy, the one with T and p as independent variables, i.e., Eq. (2-42), is most useful.

EXAMPLE 2-4. For a gas which obeys the van der Waals equation of state (Example 2-2), determine the relations between T and v and between p and v in a reversible adiabatic process, assuming c_v = constant.

Solution.
For a gas which obeys the van der Waals equation, we have

$$\left(\frac{\partial p}{\partial T}\right)_v = \frac{R}{v - b}$$

Whence from Eq. (2-34),

$$ds = \frac{c_v}{T}dT + \frac{R}{v - b}\,dv$$

For a reversible adiabatic process, this equation becomes

$$\frac{c_v}{T}\,dT + \frac{R}{v - b}\,dv = 0$$

When c_v = constant, integrating gives

$$c_v \ln T + R \ln (v - b) = \text{constant}$$

Therefore

$$T(v - b)^{R/c_v} = \text{constant}$$

Combining this equation with the van der Waals equation leads to

$$\left[\left(p + \frac{a}{v^2}\right)\frac{v - b}{R}\right](v - b)^{R/c_v} = \text{constant}$$

Hence

$$\left(p + \frac{a}{v^2}\right)(v - b)^{1 + R/c_v} = \text{constant}$$

2-6 JOULE-THOMSON COEFFICIENT

The Joule-Thomson expansion consists of allowing a gas to expand through a porous plug from a region of higher pressure to a region of lower pressure as depicted in Fig. 2-1. The process is carried out steadily and adiabatically. When the flow is sufficiently slow, the gas has well-defined pressure and temperature on both sides of the restriction. Such a process, nevertheless, is necessarily irreversible.

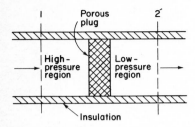

FIG. 2-1. Schematic diagram of the Joule-Thomson experiment.

Since for this process there is no heat transferred, no shaft work done, and no change in potential energy, the steady-flow equation (1-10a) becomes

$$h_1 + \frac{V_1^2}{2} = h_2 + \frac{V_2^2}{2}$$

The kinetic energy terms may be omitted if they are kept equal or approximately equal by making area 2 larger than area 1, or else by simply assuming that the volume rate of flow is small and the difference in kinetic energy is negligible. Accordingly, the steady-flow equation becomes simply

$$h_1 = h_2 \tag{2-44}$$

In other words, in a Joule-Thomson expansion the initial and final enthalpies are equal. Note that since the gas passes through nonequilibrium states on its way from the initial equilibrium state to the final equilibrium state the process is irreversible. The expansion is not at constant enthalpy, but only satisfies the condition that initial enthalpy equals final enthalpy.

A series of Joule-Thomson expansion experiments may be performed. In each experiment the values of p_1 and T_1 for the higher pressure side are kept the same, but different values of p_2, such as p_{2a}, p_{2b}, p_{2c}, etc., are maintained for the lower-pressure side and the corresponding temperature T_2's are measured. The data can then be plotted on a T-p diagram, giving the discrete points 1, 2a, 2b, 2c, etc., as shown in Fig. 2-2a. Since $h_1 = h_{2a} = h_{2b} = h_{2c} = \ldots$, a smooth curve drawn though these points is a curve of constant enthalpy. However, this curve does not represent the process executed by the gas in passing through the restriction. By performing other series of similar experiments, using different

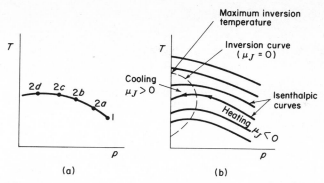

FIG. 2-2. Isenthalpic and inversion curves for Joule-Thomson experiment.

constant values of p_1 and T_1 in each series, a family of curves corresponding to different values of h can be obtained. Such a family of isenthalpic curves is depicted in Fig. 2-2b.

The slope of an isenthalpic curve on a T-p diagram at any point is called the *Joule-Thomson coefficient* μ_J:

$$\mu_J = \left(\frac{\partial T}{\partial p}\right)_h \tag{2-45}$$

The value of μ_J may be positive, negative, or zero. The point at which $\mu_J = 0$ corresponds to the maximum of an isenthalpic curve and is called the *inversion point*. The locus of all the inversion points is called the *inversion curve*. In the region inside the inversion curve where $\mu_J > 0$, the temperature will decrease as the pressure decreases upon throttling—a cooling effect. Whereas in the region outside the inversion curve where $\mu_J < 0$, the temperature will increase as the pressure decreases upon throttling—a heating effect. The temperature at which the inversion curve intersects the temperature axis is called the maximum inversion temperature. When the initial temperature of a Joule-Thomson expansion is higher than the maximum inversion temperature no cooling is possible.

The expression for the Joule-Thomson coefficient μ_J in terms of p, v, T, and c_p is obtained by applying the second dh equation

$$dh = c_p\, dT + \left[v - T\left(\frac{\partial v}{\partial T}\right)_p\right] dp \tag{2-42}$$

Rearranging,

$$dT = \frac{1}{c_p}\, dh + \frac{1}{c_p}\left[T\left(\frac{\partial v}{\partial T}\right)_p - v\right] dp$$

But the total differential of T in terms of h and p is

$$dT = \left(\frac{\partial T}{\partial h}\right)_p dh + \left(\frac{\partial T}{\partial p}\right)_h dp$$

Comparing the coefficients of dp in the foregoing two equations we obtain

$$\mu_J = \left(\frac{\partial T}{\partial p}\right)_h = \frac{1}{c_p}\left[T\left(\frac{\partial v}{\partial T}\right)_p - v\right] \tag{2-46}$$

It is clear from this equation that the Joule-Thomson coefficient and the specific heats are equivalent sources of supplementary data, which when used in connection with p-v-T data facilitate the evaluation of such quantities as internal energy, enthalpy, and entropy.

EXAMPLE 2-5. Derive the expression for the Joule-Thomson coefficient of a gas obeying the van der Waals equation of state (Example 2-2), and determine the equation of the inversion curve.

Solution.
Equation (2-46) for the Joule-Thomson coefficient can be rewritten in terms of the coefficient of thermal expansion α as follows:

$$\mu_J = \left(\frac{\partial T}{\partial p}\right)_h = \frac{v}{c_p}\left[T\frac{1}{v}\left(\frac{\partial v}{\partial T}\right)_p - 1\right] = \frac{v}{c_p}(\alpha T - 1)$$

For a gas obeying the van der Waals equation, the coefficient of thermal expansion α, as obtained in Example 2-2, is

$$\alpha = \frac{Rv^2(v - b)}{RTv^3 - 2a(v - b)^2}$$

Therefore

$$\mu_J = \frac{v}{c_p}\left[\frac{RTv^2(v - b)}{RTv^3 - 2a(v - b)^2} - 1\right]$$

$$= \frac{v}{c_p}\left[\frac{2a(v - b)^2 - RTbv^2}{RTv^3 - 2a(v - b)^2}\right]$$

The equation of the inversion curve is obtained by setting $\mu_J = 0$, which gives

$$T = \frac{2a}{Rb}\left(1 - \frac{b}{v}\right)^2$$

Substituting this equation into the van der Waals equation gives

$$p = \frac{a}{bv}\left(2 - \frac{3b}{v}\right)$$

These two equations are the defining equations for the inversion curve in parametric form. The equation connecting the inversion temperature and the corresponding pressure can be obtained by eliminating v between the preceding two equations.

2-7 CLAPEYRON EQUATION

During a change of phase of a pure substance, such as vaporization, melting, and sublimation, its temperature and pressure remain constant while its entropy and volume undergo changes. The temperature and pressure of a pure substance consisting of two phases in equilibrium are not independent variables. We now proceed to determine how the equilibrium pressure depends on the temperature. The desired relation can be obtained directly from the third Maxwell relation

$$\left(\frac{\partial p}{\partial T}\right)_V = \left(\frac{\partial S}{\partial V}\right)_T \tag{2-13c}$$

since the pressure is a function only of temperature and is independent of volume for coexisting phases, the partial derivative $(\partial p/\partial T)_V$ may be replaced by the total derivative dp/dT. Furthermore, at constant temperature, and incidentally also constant pressure, the entropy and volume are linear functions of each other. We may accordingly write

$$\left(\frac{\partial S}{\partial V}\right)_T = \frac{S^{(\beta)} - S^{(\alpha)}}{V^{(\beta)} - V^{(\alpha)}}$$

where the superscripts (α) and (β) denote the two coexisting phases. Consequently Eq. (2-13c) becomes

$$\frac{dp}{dT} = \frac{S^{(\beta)} - S^{(\alpha)}}{V^{(\beta)} - V^{(\alpha)}}$$

or, in terms of quantities per unit mass or per mole,

$$\frac{dp}{dT} = \frac{s^{(\beta)} - s^{(\alpha)}}{v^{(\beta)} - v^{(\alpha)}}$$

But according to Eq. (2-12b),

$$\left(\frac{\partial h}{\partial s}\right)_p = T$$

We have

$$h^{(\beta)} - h^{(\alpha)} = T(s^{(\beta)} - s^{(\alpha)})$$

Therefore

$$\frac{dp}{dT} = \frac{h^{(\beta)} - h^{(\alpha)}}{T(v^{(\beta)} - v^{(\alpha)})} \qquad (2\text{-}47)$$

This is the famous *Clapeyron equation.*

Another method of arriving at the Clapeyron equation is by utilizing the fact that since

$$dg = -s\,dT + v\,dp \qquad (2\text{-}11)$$

the Gibbs function remains constant during a reversible isothermal-isobaric phase transition. Thus, for a phase transition at T and p we have

$$g^{(\alpha)} = g^{(\beta)}$$

and for a phase transition at $T + dT$ and $p + dp$ we have

$$g^{(\alpha)} + dg^{(\alpha)} = g^{(\beta)} + dg^{(\beta)}$$

whence

$$dg^{(\alpha)} = dg^{(\beta)}$$

Applying Eq. (2-11) we obtain

$$-s^{(\alpha)}\,dT + v^{(\alpha)}\,dp = -s^{(\beta)}\,dT + v^{(\beta)}\,dp$$

or

$$\frac{dp}{dT} = \frac{s^{(\beta)} - s^{(\alpha)}}{v^{(\beta)} - v^{(\alpha)}}$$

which leads to the Clapeyron equation.

The Clapeyron equation is one of the most useful equations in thermodynamics. It can be used in a variety of ways, including the prediction of the effect of pressure on transition temperatures and the computation of latent heats of phase transitions. It is applicable not only to equilibrium transitions between liquids and vapors, but also between solids and vapors, between solids and liquids, and between two different solid phases of a substance.

When the Clapeyron equation is applied to a liquid-vapor (or solid-vapor) equilibrium transition, we can generally neglect the volume of the liquid (or solid) compared with that of the vapor. Furthermore, if we approximate the volume of the vapor by the ideal gas law, the Clapeyron equation simplifies to

$$\frac{dp}{dT} = \frac{h_{fg}}{Tv_g} = \frac{ph_{fg}}{RT^2}$$

or

$$\frac{1}{p}\frac{dp}{dT} = \frac{d(\ln p)}{dT} = \frac{h_{fg}}{RT^2} \qquad (2\text{-}48)$$

This equation is known as the *Clausius-Clapeyron equation.* It can be used to estimate the latent heat of vaporization h_{fg} at low pressures. When this equation is applied to a solid-vapor equilibrium transition (h_{ig} in the equation), it can usually be used to calculate the latent heat of sublimation because in this case the vapor pressure is generally small.

For small temperature changes, the latent heat h_{fg} may be considered constant. Under this circumstance, Eq. (2-48) can be integrated to give

$$\ln p = -\frac{h_{fg}}{RT} + \text{constant} \qquad (2\text{-}49)$$

which means that when $\ln p$ is plotted against $1/T$, one should obtain a straight line with slope equal to $-h_{fg}/R$. In view of the approximations involved, this conclusion cannot be regarded as exact. However, for rough calculations it does provide a means to obtain the vapor pressure of a substance at various temperatures from a small amount of data.

EXAMPLE 2-6. The international edition of Keenan, Keyes, Hill, and Moore[2] gives the following data:

Saturation temperature, °C	99	100	101
Saturation pressure, bars	0.9778	1.0135	1.0502
v_f, cm^3/g	1.0427	1.0435	1.0443
v_g, cm^3/g	1729.9	1672.9	1618.2

Determine the approximate enthalpy of vaporization of water at 100°C.

Solution.
At 100°C the approximate value of dp/dT is

$$\frac{1.0502 - 0.9778}{101 - 99} = 0.0362 \text{ bars/°C} = 3.62 \times 10^3 \text{ newtons/(m}^2\text{)(°C)}$$

By the Clapeyron equation, the enthalpy of vaporization of water at 100°C

[2] J. H. Keenan, F. G. Keyes, P. G. Hill, and J. G. Moore, "Steam Tables (International Edition—Metric Units)," John Wiley and Sons, Inc., New York, 1969.

is then

$$h_{fg} = Tv_{fg}\frac{dp}{dT} = (100 + 273.15)(1672.9 - 1.0435) \times 10^{-6}(3.62 \times 10^{3})$$

$$= 2260 \text{ (newtons)(m)/g} = 2260 \text{ joules/g}$$

as compared with the tabulated value of 2257.0 joules/g.

2-8 BRIDGMAN TABLE FOR THERMODYNAMIC RELATIONSHIPS

In the last few sections we introduced various thermodynamic relations for simple compressible systems involving the partial derivatives of eight thermodynamic properties, namely, p, v, T, s, u, h, a, and g. All the first derivatives are of the type $(\partial x_j/\partial x_k)_{x_i}$, where x_i, x_j, and x_k are three different variables. Among the eight variables there are a total of $8 \times 7 \times 6 = 336$ derivatives of this type. But by the reciprocal theorem,

$$\left(\frac{\partial x_j}{\partial x_k}\right)_{x_i} = \frac{1}{(\partial x_k/\partial x_j)_{x_i}}$$

there are $336/2 = 168$ reciprocal pairs and we need consider only 168 derivatives as actually different. All possible relations between these 168 derivatives would give a complete set of thermodynamic formulas that involve first derivatives. For a simple system with two independent variables, a general thermodynamic relation of practical interest may involve as many as four of the derivatives. Thus the maximum number of possible general relations involving first derivatives would be

$$\frac{168 \times 167 \times 166 \times 165}{4!} = 3.2 \times 10^{7}$$

Fortunately, of this number of possible relations only a few need ever be used in practice.

Bridgman[3] proposed a method for setting up the equations in terms of the three most readily experimentally measurable quantities, namely c_p, $(\partial v/\partial T)_p$, and $(\partial v/\partial p)_T$, by writing each derivative as a ratio of two quantities such as

$$\left(\frac{\partial x_j}{\partial x_k}\right)_{x_i} = \frac{(\partial x_j/\partial \alpha_i)_{x_i}}{(\partial x_k/\partial \alpha_i)_{x_i}} = \frac{(\partial x_j)_{x_i}}{(\partial x_k)_{x_i}}$$

[3] P. W. Bridgman, "A Condensed Collection of Thermodynamic Formulas," Harvard University Press, Cambridge, Mass., 1925; reprinted by Dover Publications, New York, 1961.

where α_i is any new variable, not necessarily one of the eight, and $(\partial x_j)_{x_i}$ and $(\partial x_k)_{x_i}$ are abbreviations for $(\partial x_j/\partial \alpha_i)_{x_i}$ and $(\partial x_k/\partial \alpha_i)_{x_i}$ respectively. Note, however, that the expression $(\partial x_j)_{x_i} = (\partial x_j/\partial \alpha_i)_{x_i}$ is not strictly an equation at all but just a short-hand method of expression. The 168 different partial derivatives are divided into 8 groups, each having the same variable to be held constant. Each group requires 7 expressions of the form $(\partial x_j)_{x_i}$ to form the 21 partial derivatives in that group. Thus a total of $7 \times 8 = 56$ expressions would be needed. This number may be reduced to 28 by choosing the different α's for different groups that satisfy the condition

$$\left(\frac{\partial x_j}{\partial \alpha_k}\right)_{x_k} = -\left(\frac{\partial x_k}{\partial \alpha_j}\right)_{x_j} \quad \text{or} \quad (\partial x_j)_{x_k} = -(\partial x_k)_{x_j}$$

For any arbitrary choice of α_1, the choice of α_2 is fixed by the requirement

$$\left(\frac{\partial x_1}{\partial \alpha_2}\right)_{x_2} = -\left(\frac{\partial x_2}{\partial \alpha_1}\right)_{x_1} \quad \text{or} \quad (\partial x_1)_{x_2} = -(\partial x_2)_{x_1}$$

and the choice of α_3 is fixed by the requirement

$$\left(\frac{\partial x_1}{\partial \alpha_3}\right)_{x_3} = -\left(\frac{\partial x_3}{\partial \alpha_1}\right)_{x_1} \quad \text{or} \quad (\partial x_1)_{x_3} = -(\partial x_3)_{x_1}$$

But by the cyclic relation, we write

$$\left(\frac{\partial x_1}{\partial x_2}\right)_{x_3}\left(\frac{\partial x_2}{\partial x_3}\right)_{x_1}\left(\frac{\partial x_3}{\partial x_1}\right)_{x_2} = -1$$

or

$$\left[\left(\frac{\partial x_1}{\partial \alpha_3}\right)_{x_3}\left(\frac{\partial \alpha_3}{\partial x_2}\right)_{x_3}\right]\left[\left(\frac{\partial x_2}{\partial \alpha_1}\right)_{x_1}\left(\frac{\partial \alpha_1}{\partial x_3}\right)_{x_1}\right]\left[\left(\frac{\partial x_3}{\partial \alpha_2}\right)_{x_2}\left(\frac{\partial \alpha_2}{\partial x_1}\right)_{x_2}\right] = -1$$

Consequently,

$$\left(\frac{\partial x_3}{\partial \alpha_2}\right)_{x_2} = -\left(\frac{\partial x_2}{\partial \alpha_3}\right)_{x_3} \quad \text{or} \quad (\partial x_3)_{x_2} = -(\partial x_2)_{x_3}$$

Thus only the choice of α_1 is left arbitrary. The evaluation of the α's may be quite complicated, but actually none is really needed. They have been used only to demonstrate the feasibility of writing the derivatives as quotients. Bridgman arbitrarily chose $(\partial T)_p = -(\partial p)_T = 1$ for simplicity, and proceeded to prepare the table by writing a sufficient number of derivatives by well-known thermodynamic methods and then splitting these derivatives by inspection into quotients. Table 2-1 is the Bridgman table for the first derivatives as applied to single-phase regions of a simple compressible system.

TABLE 2-1. Bridgman Table for Thermodynamic Relations*

$(\partial T)_p = -(\partial p)_T = 1$

$(\partial v)_p = -(\partial p)_v = \left(\dfrac{\partial v}{\partial T}\right)_p$

$(\partial s)_p = -(\partial p)_s = \dfrac{c_p}{T}$

$(\partial u)_p = -(\partial p)_u = c_p - p\left(\dfrac{\partial v}{\partial T}\right)_p$

$(\partial h)_p = -(\partial p)_h = c_p$

$(\partial a)_p = -(\partial p)_a = -\left[s + p\left(\dfrac{\partial v}{\partial T}\right)_p\right]$

$(\partial g)_p = -(\partial p)_g = -s$

$(\partial v)_T = -(\partial T)_v = -\left(\dfrac{\partial v}{\partial p}\right)_T$

$(\partial s)_T = -(\partial T)_s = \left(\dfrac{\partial v}{\partial T}\right)_p$

$(\partial u)_T = -(\partial T)_u = T\left(\dfrac{\partial v}{\partial T}\right)_p + p\left(\dfrac{\partial v}{\partial p}\right)_T$

$(\partial h)_T = -(\partial T)_h = -v + T\left(\dfrac{\partial v}{\partial T}\right)_p$

$(\partial a)_T = -(\partial T)_a = p\left(\dfrac{\partial v}{\partial p}\right)_T$

$(\partial g)_T = -(\partial T)_g = -v$

$(\partial s)_v = -(\partial v)_s = \dfrac{1}{T}\left[c_p\left(\dfrac{\partial v}{\partial p}\right)_T + T\left(\dfrac{\partial v}{\partial T}\right)_p^2\right]$

$(\partial u)_v = -(\partial v)_u = c_p\left(\dfrac{\partial v}{\partial p}\right)_T + T\left(\dfrac{\partial v}{\partial T}\right)_p^2$

$(\partial h)_v = -(\partial v)_h = c_p\left(\dfrac{\partial v}{\partial p}\right)_T + T\left(\dfrac{\partial v}{\partial T}\right)_p^2 - v\left(\dfrac{\partial v}{\partial T}\right)_p$

$(\partial a)_v = -(\partial v)_a = -s\left(\dfrac{\partial v}{\partial p}\right)_T$

TABLE 2-1. Bridgman Table for Thermodynamic Relations*—Cont'd

$$(\partial g)_v = -(\partial v)_g = -\left[v\left(\frac{\partial v}{\partial T}\right)_p + s\left(\frac{\partial v}{\partial p}\right)_T\right]$$

$$(\partial u)_s = -(\partial s)_u = \frac{p}{T}\left[c_p\left(\frac{\partial v}{\partial p}\right)_T + T\left(\frac{\partial v}{\partial T}\right)_p^2\right]$$

$$(\partial h)_s = -(\partial s)_h = -\frac{vc_p}{T}$$

$$(\partial a)_s = -(\partial s)_a = \frac{1}{T}\left\{p\left[c_p\left(\frac{\partial v}{\partial p}\right)_T + T\left(\frac{\partial v}{\partial T}\right)_p^2\right] + sT\left(\frac{\partial v}{\partial T}\right)_p\right\}$$

$$(\partial g)_s = -(\partial s)_g = -\frac{1}{T}\left[vc_p - sT\left(\frac{\partial v}{\partial T}\right)_p\right]$$

$$(\partial h)_u = -(\partial u)_h = -v\left[c_p - p\left(\frac{\partial v}{\partial T}\right)_p\right] - p\left[c_p\left(\frac{\partial v}{\partial p}\right)_T + T\left(\frac{\partial v}{\partial T}\right)_p^2\right]$$

$$(\partial a)_u = -(\partial u)_a = p\left[c_p\left(\frac{\partial v}{\partial p}\right)_T + T\left(\frac{\partial v}{\partial T}\right)_p^2\right] + s\left[T\left(\frac{\partial v}{\partial T}\right)_p + p\left(\frac{\partial v}{\partial p}\right)_T\right]$$

$$(\partial g)_u = -(\partial u)_g = -v\left[c_p - p\left(\frac{\partial v}{\partial T}\right)_p\right] + s\left[T\left(\frac{\partial v}{\partial T}\right)_p + p\left(\frac{\partial v}{\partial p}\right)_T\right]$$

$$(\partial a)_h = -(\partial h)_a = -\left[s + p\left(\frac{\partial v}{\partial T}\right)_p\right]\left[v - T\left(\frac{\partial v}{\partial T}\right)_p\right] + pc_p\left(\frac{\partial v}{\partial p}\right)_T$$

$$(\partial g)_h = -(\partial h)_g = -v(c_p + s) + Ts\left(\frac{\partial v}{\partial T}\right)_p$$

$$(\partial a)_g = -(\partial g)_a = -s\left[v + p\left(\frac{\partial v}{\partial p}\right)_T\right] - pv\left(\frac{\partial v}{\partial T}\right)_p$$

---

*P. W. Bridgman, "A Condensed Collection of Thermodynamic Formulas," Dover Publications, New York, 1961.

Bridgman has also prepared tables for the first derivatives as applied to two-phase regions and for the second derivatives as applied to single- and two-phase regions. For these tables the readers are referred to Bridgman's original book.

Table 2-1 enables us to set up the equation for any first derivative for a single-phase system by straightforward mathematical manipulations. The following example demonstrates the procedure employed.

EXAMPLE 2-7. Develop the energy equation [Eq. (2-40)] using the Bridgman table.

Solution.

The energy equation is an expression for the derivative $(\partial u/\partial v)_T$. Writing this derivative as a quotient we have

$$\left(\frac{\partial u}{\partial v}\right)_T = \frac{(\partial u)_T}{(\partial v)_T}$$

Expressions for $(\partial u)_T$ and $(\partial v)_T$ are taken from Table 2-1 and substituted into the foregoing expression to give

$$\left(\frac{\partial u}{\partial v}\right)_T = \frac{T(\partial v/\partial T)_p + p(\partial v/\partial p)_T}{-(\partial v/\partial p)_T} = -T\left(\frac{\partial v}{\partial T}\right)_p\left(\frac{\partial p}{\partial v}\right)_T - p$$

But by the cyclic relation we have

$$\left(\frac{\partial p}{\partial v}\right)_T\left(\frac{\partial v}{\partial T}\right)_p\left(\frac{\partial T}{\partial p}\right)_v = -1$$

Consequently we get the energy equation

$$\left(\frac{\partial u}{\partial v}\right)_T = T\left(\frac{\partial p}{\partial T}\right)_v - p$$

2-9 IDEAL GAS RELATIONS

An *ideal gas* is a gas which obeys the relation

$$pv = RT \tag{2-50}$$

for all pressures and temperatures. This equation, known as the ideal gas equation, holds exactly for a real gas only as the pressure approaches zero. It holds approximately for a real gas at higher pressures when the gas is at high temperatures. In this equation, p, v, and T are, respectively, the absolute pressure, molal volume, and absolute temperature, and R is the *universal gas constant* which is the same for all gases. Values of the universal gas constant for different units are given in Appendix Table A-4. When Eq. (2-50) is expressed for unit mass of a gas, R is then the individual gas constant R_i which is different for different gases. The individual gas constant for a gas can be obtained by dividing the universal gas constant by the molecular weight of the gas.

The ideal gas equation is so well known that it hardly needs comment. We devote space here only to use the general equations developed in the previous sections for writing some important ideal gas relations needed for later chapters. First, let us use the energy equation to find out an important property of an ideal gas. Thus from Eq. (2-40) we have for an ideal gas

$$\left(\frac{\partial u}{\partial v}\right)_T = T\left(\frac{\partial p}{\partial T}\right)_v - p = \frac{RT}{v} - p = 0$$

which implies that the internal energy of an ideal gas is a function of temperature only. Furthermore, since $h = u + p\,v = u + R\,T$, it follows that the enthalpy of an ideal gas is also a function of temperature only. Consequently, according to the definitions,

$$c_v = \left(\frac{\partial u}{\partial T}\right)_v \quad \text{and} \quad c_p = \left(\frac{\partial h}{\partial T}\right)_p$$

we see that the specific or molal heats at constant volume and at constant pressure for an ideal gas are also functions of temperature only. We accordingly write for any process of an ideal gas

$$du = c_v dT \tag{2-51}$$

$$dh = c_p dT \tag{2-52}$$

Note that the difference between c_p and c_v for an ideal gas, according to Eq. (2-32), is

$$c_p - c_v = T\left(\frac{\partial p}{\partial T}\right)_v\left(\frac{\partial v}{\partial T}\right)_p = T\left(\frac{R}{v}\right)\left(\frac{R}{p}\right) = R$$

which is invariable.

The entropy, on the other hand, is still a function of two independent variables. Following Eqs. (2-34) to (2-36), the three ds equations for an ideal gas are

$$ds = c_v \frac{dT}{T} + R \frac{dv}{v} \tag{2-53a}$$

$$ds = c_p \frac{dT}{T} - R \frac{dp}{p} \tag{2-53b}$$

$$ds = c_v \frac{dp}{p} + c_p \frac{dv}{v} \tag{2-53c}$$

Integration of the foregoing equations requires a knowledge of the relations between the specific heats and the temperature. Empirical equations for c_p as a function of temperature at zero pressure for various gases are given in Appendix Table A-14.

The Keenan and Kaye gas tables[4] are a valuable aid in making ideal gas computations. They take into account the variations of specific heats with temperature and list thermodynamic variables in terms of temperature alone. The values of internal energy and enthalpy listed are obtained by integrating Eqs. (2-51) and (2-52) from absolute zero to a given temperature, assuming $u = 0$ and $h = 0$ at absolute zero. Since entropy is not a function of temperature alone, it cannot be tabulated directly. It is tabulated indirectly through the introduction of a new variable ϕ as defined by

$$\phi = \int_{T_0}^{T} c_p \frac{dT}{T}$$

where T_0 is some selected reference temperature. ϕ is a function of temperature T only and is tabulated in the gas tables. Then from Eq. (2-53b)

$$\Delta s_{12} = s_2 - s_1 = \int_{T_1}^{T_2} c_p \frac{dT}{T} - R \ln \frac{p_2}{p_1}$$

$$= \int_{T_0}^{T_2} c_p \frac{dT}{T} - \int_{T_0}^{T_1} c_p \frac{dT}{T} - R \ln \frac{p_2}{p_1}$$

$$= \phi_2 - \phi_1 - R \ln \frac{p_2}{p_1} \tag{2-54}$$

For a reversible adiabatic process, from Eq. (2-54) we have

$$\Delta s_{12} = 0 = \phi_2 - \phi_1 - R \ln \frac{p_2}{p_1}$$

Rearranging,

$$\left(\frac{p_2}{p_1}\right)_{s \,=\, \text{constant}} = \exp\left(\frac{\phi_2 - \phi_1}{R}\right) = \frac{\exp(\phi_2/R)}{\exp(\phi_1/R)}$$

This equation suggests the introduction of a new variable p_r, the relative pressure, by the expression

$$p_r = \exp\left(\frac{\phi}{R}\right)$$

which is a function of temperature only and is tabulated in the gas tables. Thus

[4] J. H. Keenan and J. Kaye, "Gas Tables," John Wiley and Sons, Inc., New York, 1948. (An abridged table for air is given in Appendix Table A-11.)

$$\left(\frac{p_2}{p_1}\right)_{s\ =\ constant} = \frac{p_{r_2}}{p_{r_1}} \qquad \text{✳} \qquad (2\text{-}55)$$

for a reversible adiabatic process.

Equation (2-55) can be converted into a volume ratio by substituting $p = RT/v$. Thus

$$\left(\frac{T_2/v_2}{T_1/v_1}\right)_{s\ =\ constant} = \frac{p_{r_2}}{p_{r_1}} \qquad \text{or} \qquad \left(\frac{v_2}{v_1}\right)_{s\ =\ constant} = \frac{T_2/p_{r_2}}{T_1/p_{r_1}}$$

This equation suggests the introduction of a new variable v_r, the relative volume, by the expression

$$v_r = \frac{RT}{p_r}$$

which is a function of temperature only and is tabulated in the gas tables. Thus

$$\left(\frac{v_2}{v_1}\right)_{s\ =\ constant} = \frac{v_{r_2}}{v_{r_1}} \qquad \text{✳} \qquad (2\text{-}56)$$

for a reversible adiabatic process.

Although the specific heats of ideal gases are generally functions of temperature, they are sometimes considered constant in simplified calculations involving small temperature ranges. For such cases, integrating Eqs. (2-51) to (2-53) between two end states 1 and 2 leads to

$$\Delta u_{12} = u_2 - u_1 = c_v(T_2 - T_1) \qquad (2\text{-}57)$$

$$\Delta h_{12} = h_2 - h_1 = c_p(T_2 - T_1) \qquad (2\text{-}58)$$

$$\Delta s_{12} = s_2 - s_1 = c_v \ln \frac{T_2}{T_1} + R \ln \frac{v_2}{v_1} \qquad (2\text{-}59a)$$

$$\Delta s_{12} = s_2 - s_1 = c_p \ln \frac{T_2}{T_1} - R \ln \frac{p_2}{p_1} \qquad (2\text{-}59b)$$

$$\Delta s_{12} = s_2 - s_1 = c_v \ln \frac{p_2}{p_1} + c_p \ln \frac{v_2}{v_1} \qquad (2\text{-}59c)$$

For constant c_p and c_v

Of particular importance are the relations describing a reversible adiabatic process for an ideal gas with constant specific heats. They can be derived directly by assigning $\Delta s_{12} = 0$ in Eq. (2-59c). Thus

$$c_v \ln \frac{p_2}{p_1} + c_p \ln \frac{v_2}{v_1} = 0 \qquad \text{or} \qquad \ln\left[\left(\frac{p_2}{p_1}\right)\left(\frac{v_2}{v_1}\right)^k\right] = 0$$

where

$$k = \frac{c_p}{c_v}$$

Whence

$$\left(\frac{p_2}{p_1}\right)\left(\frac{v_2}{v_1}\right)^k = 1$$

Therefore

$$p_1 v_1^k = p_2 v_2^k$$

or in general terms

$$pv^k = \text{constant} \qquad (2\text{-}60a)$$

Reversible adiabatic process c_v & c_p are constant

Insertion of the ideal gas equation in the preceding equation gives two other equations:

$$Tv^{k-1} = \text{constant} \qquad (2\text{-}60b)$$

and

$$\frac{T}{p^{(k-1)/k}} = \text{constant} \qquad (2\text{-}60c)$$

reversible adiabatic process c_v & c_p are constant

When Eqs. (2-60a) to (2-60c) are written for a process between states 1 and 2, we have

$$\frac{T_2}{T_1} = \left(\frac{p_2}{p_1}\right)^{(k-1)/k} = \left(\frac{v_1}{v_2}\right)^{k-1} \qquad (2\text{-}61)$$

reversible adiabatic process c_v & c_p are constant

Notice that Eqs. (2-60a) to (2-61) hold only for reversible adiabatic processes of ideal gases with constant specific heats.

We should note in passing that the Joule-Thomson coefficient of an ideal gas is zero. It can be seen from Eq. (2-46) that

$$\mu_J = \frac{1}{c_p}\left[T\left(\frac{\partial v}{\partial T}\right)_p - v\right] = \frac{1}{c_p}\left[T\left(\frac{R}{p}\right) - v\right] = 0$$

which means that for an ideal gas there is no temperature change upon throttling between two levels of pressure.

EXAMPLE 2-8.   An ideal gas whose equation of state is $pV = nRT'$ and whose internal energy is a function of T' only is used as the working substance in a Carnot cycle (Fig. 2-3). Show that the ideal gas temperature T' as

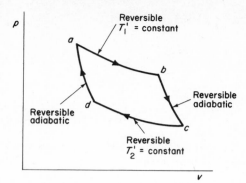

FIG. 2-3. Equivalence of the ideal gas and the thermodynamic temperature scales.

defined in Sec. 1-4 is equal to the thermodynamic temperature T as defined in Sec. 1-9.

Solution.

The first law for a process is

$$Q = \Delta U + W$$

For the isothermal process ab, since $T_1' = $ constant and $\Delta U_{ab} = 0$, we have

$$Q_1 = Q_{ab} = W_{ab} = \int_a^b p\, dV$$

$$= nRT_1' \int_{V_a}^{V_b} \frac{dV}{V} = nRT_1' \ln \frac{V_b}{V_a}$$

Similarly

$$Q_2 = Q_{dc} = nRT_2' \ln \frac{V_c}{V_d}$$

But for a reversible adiabatic process $T'V^{k-1} = $ constant, where $k = c_p/c_v = $ constant. We have for the processes bc and da,

$$\frac{T_1'}{T_2'} = \left(\frac{V_c}{V_b}\right)^{k-1} = \left(\frac{V_d}{V_a}\right)^{k-1}$$

from which

$$\frac{V_b}{V_a} = \frac{V_c}{V_d}$$

Therefore

$$\frac{Q_1}{Q_2} = \frac{nRT_1' \ln (V_b/V_a)}{nRT_2' \ln (V_c/V_d)} = \frac{T_1'}{T_2'}$$

Comparing the above equation with Eq. (1-12) reveals that

$$\frac{T_1}{T_2} = \frac{T_1'}{T_2'}$$

In addition, since the same numerical value was assigned to the fixed point (triple point of water), it follows that

$$T = T'$$

or the thermodynamic temperature is numerically equal to the absolute temperature as measured by an ideal gas thermometer.

EXAMPLE 2-9. An ideal gas with a constant value of c_p = 29.6 joules/ (g mole)($^\circ$K) is made to undergo a cycle consisting of the following reversible processes in a closed system:

> Process 12: The gas expands adiabatically from 5 × 10^6 newtons/m^2 and 550°K to 1 × 10^6 newtons/m^2.
> Process 23: The gas is heated at constant volume until 550°K.
> Process 31: The gas is compressed isothermally back to its initial condition.

Calculate the work, the heat, and the change of entropy per g mole of the gas for each of the three processes.

Solution.
The processes of this example are shown on the pressure-volume and temperature-entropy diagrams of Fig. 2-4. The given data are p_1 = 5 × 10^6 newtons/m^2, p_2 = 1 × 10^6 newtons/m^2, T_1 = T_3 = 550°K.
    For an ideal gas $c_p - c_v = R$ we have

$$c_v = c_p - R = 29.6 - 8.31 = 21.3 \text{ joules/(g mole)}(^\circ\text{K})$$

and

$$k = \frac{c_p}{c_v} = \frac{29.6}{21.3} = 1.39$$

    From Eq. (2-61), we obtain

$$T_2 = T_1\left(\frac{p_2}{p_1}\right)^{(k-1)/k} = 550\left(\frac{1 \times 10^6}{5 \times 10^6}\right)^{(1.39-1)/1.39} = 350^\circ\text{K}$$

    From the ideal gas equation we obtain

$$v_1 = \frac{RT_1}{p_1} = \frac{8.31 \times 550}{5 \times 10^6} = 914 \times 10^{-6} \text{ m}^3/\text{g mole}$$

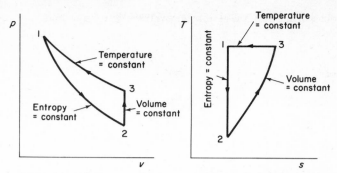

FIG. 2-4. Pressure-volume and temperature-entropy diagrams showing the processes of Example 2-9.

and

$$v_2 = \frac{RT_2}{p_2} = \frac{8.31 \times 350}{1 \times 10^6} = 2910 \times 10^{-6} \text{ m}^3/\text{g mole}$$

Now, because $v_3 = v_2$, we then have

$$p_3 = \frac{RT_3}{v_3} = \frac{8.31 \times 550}{2910 \times 10^{-6}} = 1.57 \times 10^6 \text{ newtons/m}^2$$

The value of p_3 may also be obtained by

$$p_3 = p_2 \frac{T_3}{T_2} = (1 \times 10^6) \times \frac{550}{350} = 1.57 \times 10^6 \text{ newtons/m}^2$$

Since the process from 1 to 2 is reversible and adiabatic, it follows that

$$q_{12} = 0 \quad \text{and} \quad \Delta s_{12} = s_2 - s_1 = 0$$

By the first law, we then have

$$w_{12} = q_{12} - \Delta u_{12} = -\Delta u_{12} = u_1 - u_2 = c_v(T_1 - T_2)$$

$$= 21.3(550 - 350) = 4260 \text{ joules/g mole}$$

Since the process from 1 to 2 follows the relation $pv^k = $ constant, the preceding value of w_{12} can also be obtained by the formula

$$w_{12} = \int_1^2 p \, dv = \frac{p_2 v_2 - p_1 v_1}{1 - k} = \frac{R(T_2 - T_1)}{1 - k}$$

$$= \frac{8.31(350 - 550)}{1 - 1.39} = 4260 \text{ joules/g mole}$$

Since the process from 2 to 3 is at constant volume, it follows that

$$w_{23} = \int_2^3 p \, dv = 0$$

and

$$q_{23} = \Delta u_{23} + w_{23} = \Delta u_{23} = u_3 - u_2 = c_v(T_3 - T_2)$$

$$= 21.3(550 - 350) = 4260 \text{ joules/g mole}$$

The entropy change Δs_{23} is given by Eq. (2-59a),

$$\Delta s_{23} = s_3 - s_2 = c_v \ln \frac{T_3}{T_2} + R \ln \frac{v_3}{v_2} = c_v \ln \frac{T_3}{T_2}$$

$$= 21.3 \ln \frac{550}{350} = 9.62 \text{ joules/(g mole)(}^\circ\text{K)}$$

For the isothermal process from 3 to 1, Eq. (2-59b) gives

$$\Delta s_{31} = s_1 - s_3 = c_p \ln \frac{T_1}{T_3} - R \ln \frac{p_1}{p_3} = -R \ln \frac{p_1}{p_3}$$

$$= -8.31 \ln \frac{5 \times 10^6}{1.57 \times 10^6} = -9.62 \text{ joules/(g mole)(}^\circ\text{K)}$$

which is equal to $- \Delta s_{23}$. The heat rejection is given by

$$q_{31} = T_1(\Delta s_{31}) = 550(-9.62) = -5290 \text{ joules/g mole}$$

Since $\Delta u_{31} = 0$ for the isothermal process, the first law leads to

$$w_{31} = q_{31} = -5290 \text{ joules/g mole}$$

Since the process from 3 to 1 follows the relation $pv = RT = $ a constant, the value of w_{31} can also be obtained by the formula

$$w_{31} = \int_3^1 p \, dv = p_1 v_1 \ln \frac{v_1}{v_3} = RT_1 \ln \frac{p_3}{p_1}$$

$$= 8.31 \times 550 \ln \frac{1.57 \times 10^6}{5 \times 10^6} = -5{,}290 \text{ joules/g mole}$$

The minus sign for q_{31} and w_{31} indicates that heat is removed from the gas and work is done on the gas.

Both the net work and the net heat of the cycle have the same value of

$$4{,}260 - 5{,}290 = -1{,}030 \text{ joules/g mole}$$

PROBLEMS

2-1. Prove the following equations for a simple compressible system:

(a) $\left(\dfrac{\partial c_v}{\partial v}\right)_T = T\left[\dfrac{\partial}{\partial T}\left(\dfrac{\alpha}{\kappa}\right)\right]_v$

(b) $\left(\dfrac{\partial c_p}{\partial p}\right)_T = -T\left[\dfrac{\partial}{\partial T}(\alpha v)\right]_p$

(c) $c_p - c_v = \dfrac{\alpha^2 vT}{\kappa}$

where α and κ are the coefficients of thermal expansion and the isothermal compressibility respectively.

2-2. Prove the following equations for a simple compressible system:

(a) $du = c_v\, dT + \left(\dfrac{\alpha T}{\kappa} - p\right) dv$

(b) $dh = c_p\, dT + (v - \alpha Tv)\, dp$

(c) $ds = \dfrac{\kappa c_v}{\alpha T}\, dp + \dfrac{c_p}{\alpha Tv}\, dv$

(d) $da = -(\alpha pv + s)\, dT + \kappa pv\, dp$

2-3. Find the coefficient of thermal expansion α and the isothermal compressibility κ from each of the following p-v-T relations:

(a) Ideal gas equation, $p = \dfrac{RT}{v}$

(b) Dieterici equation, $p = \dfrac{RT}{v - b}\, e^{-a/RTv}$

(c) Saha-Bose equation, $p = -\dfrac{RT}{2b}\, e^{-a/RTv} \ln\left(\dfrac{v - 2b}{v}\right)$

In the above equations R is the gas constant and a and b are specific constants. Verify that for large values of T and v all the expressions for α and κ go over into the corresponding expressions for an ideal gas.

2-4. Derive the expression for the difference of the isothermal and adiabatic compressibilities $(\kappa_T - \kappa_s)$ in terms of T, v, c_p, and the coefficient of thermal expansion α.

2-5. For a simple compressible system show that

(a) $\left(\dfrac{\partial u}{\partial v}\right)_T = T^2\left[\dfrac{\partial(p/T)}{\partial T}\right]_v$

(b) $\left(\dfrac{\partial h}{\partial p}\right)_T = -T^2\left[\dfrac{\partial(v/T)}{\partial T}\right]_p$

2-6. For a simple compressible system show that

(a) $u = a - T\left(\dfrac{\partial a}{\partial T}\right)_v = -T^2\left[\dfrac{\partial(a/T)}{\partial T}\right]_v$

(b) $h = g - T\left(\dfrac{\partial g}{\partial T}\right)_p = -T^2\left[\dfrac{\partial (g/T)}{\partial T}\right]_p$

(c) $c_v = -T\left(\dfrac{\partial^2 a}{\partial T^2}\right)_v$

(d) $c_p = -T\left(\dfrac{\partial^2 g}{\partial T^2}\right)_p$

2-7. For a simple compressible system show that

(a) $c_p - c_v = \left(\dfrac{\partial v}{\partial T}\right)_p\left[p + \left(\dfrac{\partial u}{\partial v}\right)_T\right] = -\left(\dfrac{\partial p}{\partial T}\right)_v\left[p\left(\dfrac{\partial v}{\partial p}\right)_T + \left(\dfrac{\partial u}{\partial p}\right)_T\right]$

(b) $(c_p - c_v)\dfrac{\partial^2 T}{\partial p\,\partial v} + \left(\dfrac{\partial c_p}{\partial p}\right)_v\left(\dfrac{\partial T}{\partial v}\right)_p - \left(\dfrac{\partial c_v}{\partial v}\right)_p\left(\dfrac{\partial T}{\partial p}\right)_v = 1$

2-8. For a simple compressible system whose volume is independent of temperature and pressure, show that the internal energy and entropy are functions of temperature only.

2-9. The p-v-T relationship for a gas at moderate pressures may be written as

$$\frac{pv}{RT} = 1 + B'p + C'p^2$$

where p is the pressure, v is the molal volume, T is the temperature, R is the universal gas constant, and B' and C' are functions of temperature only.

(a) Show that as the pressure approaches zero,

$$\mu_J c_p \to RT^2\left(\frac{dB'}{dT}\right)$$

(b) Show that the equation of the inversion curve is

$$p = -\frac{dB'/dT}{dC'/dT}$$

2-10. The p-v-T relationship of ammonia at low pressures may be represented by the equation

$$v = \frac{RT}{p} - 0.003 - \frac{0.34}{(0.01\,T)^3} - \frac{60}{(0.01\,T)^{11}}$$

where p is the pressure in atm, v is the specific volume in m^3/kg, T is the temperature in $°K$, and $R = 0.00483$ (atm)$(m^3)/(kg)(°K)$. The Joule-Thomson coefficient for ammonia at $400°K$ and 10 atm is $1.05\ °K/atm$. Estimate the constant-pressure specific heat c_p of ammonia at $400°K$ and 10 atm.

2-11. Determine the effect of pressure on the sublimation temperature of carbon dioxide at 1.328 atm. At this pressure the saturation temperature is

$-75°C$, the enthalpies of saturated solid and vapor are 37.391 kcal/kg and 173.16 kcal/kg respectively, and the volumes of saturated solid and vapor are 0.643×10^{-3} m³/kg and 269.68×10^{-3} m³/kg respectively.

2-12. The liquid-vapor equilibrium curve for nitrogen over the range from the triple point to the normal boiling point may be expressed by the formula

$$\log p = A - BT - \frac{C}{T}$$

where p is the vapor pressure in mm Hg, T is the temperature in °K, and $A = 7.782$, $B = 0.006265$, and $C = 341.6$.

(a) Derive an expression for the enthalpy of vaporization h_{fg} in terms of the constants A, B, C, the temperature T, and the change in specific volume between saturated liquid and saturated vapor v_{fg}.

(b) Calculate h_{fg} for nitrogen at 71.9°K. At this temperature, v_{fg} is approximately 11,530 cm³/g mole.

2-13. An ideal gas with $c_p = 0.254$ cal/(g)(°C) and $c_v = 0.183$ cal/(g)(°C) expands irreversibly from 2 atm, 300°C, to 1 atm, 250°C. Calculate the entropy change Δs_{12} by evaluating $\int (dq/T)$ along three different reversible paths.

2-14. Air in a closed system expands reversibly and adiabatically from 3×10^6 newtons/m² and 200°C to two times its initial volume, and then cools at constant volume until the pressure drops to 8×10^5 newtons/m². Calculate the work done and the heat transferred per kg of air. Use $c_p = 0.243$ kcal/(kg)(°C) and $C_v = 0.174$ kcal/(kg)(°C).

2-15. One pound of air with $c_p = 0.24$ Btu/(lbm)(°R) and $c_v = 0.171$ Btu/(lbm)(°R) is made to undergo the following reversible processes in a closed system:

(1-2) The gas is compressed adiabatically from $p_1 = 14.7$ psia and $T_1 = 500°R$ to $p_2 = 2p_1$.

(2-3) It is then heated at constant pressure p_2 until $v_3 = v_1$.

(3-1) Finally it is cooled at constant volume back to its initial state.

Calculate the work, heat, and change of entropy for each of the three processes and calculate the thermal efficiency for the cycle. Sketch the p-v and T-s diagrams.

2-16. Solve the preceding problem using the gas tables by Keenan and Kaye.

2-17. Nitrogen gas with $c_p = 29.1$ kJ/(kg mole)(°K) is made to undergo the following reversible processes in a closed system:

(1-2) The gas initially at 2 atm and 350°K is heated at constant volume to a temperature of 500°K.

(2-3) It is then expanded adiabatically back to its initial temperature of 350°K.

(3-1) It is finally compressed isothermally back to its initial pressure of 2 atm.

Sketch the p-v and T-s diagrams for the cycle. For one kg mole of nitrogen, calculate the work done, the heat transferred, the internal energy change, and the entropy change for each of the three processes.

Chapter 3
EQUATIONS OF STATE FOR REAL GASES

3-1 VAN DER WAALS EQUATION OF STATE

In the study of a simple compressible system the relation between the pressure p, the specific or molal volume v, and the temperature T is fundamental. An equation expressing this relationship is called an *equation of state*. In the case of an ideal gas the equation of state is expressed by the simple relation

$$pv = RT \qquad\qquad\qquad\qquad\qquad\qquad (2\text{-}50)$$

This equation, essentially an experimental formulation, can nevertheless be demonstrated theoretically on the basis of simple kinetic theory by making certain assumptions including these two major ones: the molecules are point masses and there are no intermolecular forces between molecules. These assumptions are reasonable for gases at vanishing pressures. However, as the pressure increases and the specific volume decreases, the volume occupied by the gas molecules themselves becomes increasingly important and intermolecular forces between molecules become increasingly significant. To take into account these effects, van der Waals in 1873 proposed an equation of state according to the following argument.

Due to the finite size of the molecules the space available for free motion of any molecule is decreased. Consequently the rate of molecular collisions and thereby the pressure are increased over the ideal gas values. Let b denote the volume excluded from molecular motion for each mole of the gas having a molal volume v. The increased impact rate of the molecules in the restricted volume $(v - b)$ will cause the kinetic pressure of the gas due to kinetic energy to increase to

$$\left(\frac{RT}{v}\right)\left(\frac{v}{v-b}\right) = \frac{PT}{v-b}$$

The cohesive force between molecules may be assumed to decrease rapidly with distance between them so that this force is appreciable only at small

85

distances. A molecule far from the containing wall is attracted by its neighboring molecules from all directions with balanced average net effect. However, for a molecule near the wall, while the force components parallel to the wall due to attractions from other molecules are balanced, an average net force normal to the wall tends to pull the molecule from the wall, and the pressure on the wall is thereby reduced. The diminution of pressure should be proportional to the number of molecules being pulled and also proportional to the number of molecules that perform the pulling. Accordingly, the cohesive pressure on account of potential energy should be proportional to the square of the density of the gas and may be written as a/v^2 where a is a constant.

Subtracting the cohesive pressure from the kinetic pressure we have for the net pressure

$$p = \frac{RT}{v-b} - \frac{a}{v^2}$$

Rearranging,

$$\left(p + \frac{a}{v^2}\right)(v - b) = RT \tag{3-1}$$

This is the well known *van der Waals equation of state*, in which p is the pressure, T the absolute temperature, v the molal volume, R the universal gas constant, and a and b are constants for the gas in consistent units.

The van der Waals equation is represented graphically as a family of isotherms on a $p - v$ diagram in Fig. 3-1. At large volumes the isotherms approximate rectangular hyperbolas in agreement with the ideal gas equation. At small volumes the gradients of the isotherms become very steep indicating the behavior of incompressibility of a liquid. Since the equation is a cubic in v it has

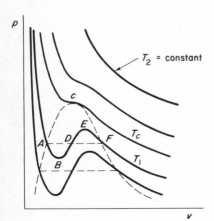

FIG. 3-1. Van der Waals isotherms.

three roots for any given values of p and T. At high temperatures, such as at T_2, only one root is real for all pressures. At lower temperatures, such as at T_1, all three roots are real over a certain pressure range. Thus a maximum and a minimum are formed on the isotherm in the region where experimentally the phenomenon of condensation or vaporization takes place. For a fixed set of values of p and T corresponding to point c in Fig. 3-1, the three real roots become identical, and the isotherm labeled T_c has a point of inflection at c with a horizontal tangent. This point must be identified as the critical point and the isotherm labeled T_c is then the critical isotherm.

The van der Waals isotherms in the normally liquid-vapor two-phase region diverge from the stable paths of isobaric and isothermal phase changes. This divergence is, however, not entirely devoid in meaning. Thus in the isotherm labeled T_1, the portion AB represents a superheated liquid as it is at a temperature higher than the saturation temperature for the existing pressure. The portion FE represents a subcooled vapor as it is at a temperature lower than the saturation temperature for the existing pressure. It is possible to realize experimentally both these portions, but, as we will discuss in Sec. 5-2, they are states in so-called metastable equilibrium and can be easily disturbed by mechanical or other means to change suddenly to stable two-phase states. On the other hand, the portion BDE has a positive slope corresponding to negative compressibility which is not physically possible.

The position of the isobaric-isothermal line AF in Fig. 3-1 that corresponds to states of stable equilibrium in which liquid and vapor coexist may be determined by the fact that the specific Gibbs functions of the two phases in equilibrium have the same value. If A and F are saturated liquid and saturated vapor states we must have

$$g_A = g_F$$

But for an isothermal process,

$$dg = v\,dp$$

Applying these relations to the isotherm $ABDEF$, we have

$$g_F - g_A = \int_A^F v\,dp = 0$$

which is readily seen to mean that the two areas enclosed between the van der Waals isotherm $ABDEF$ and the horizontal line AF are equal. The position of the line AF is thus determined. This procedure for deciding where to draw a horizontal isothermal line to replace the oscillatory section of a van der Waals isotherm is known as the Maxwell construction.

Since the critical isotherm has a point of inflection at the critical point, we have at this point

$$\left(\frac{\partial p}{\partial v}\right)_{T_c} = 0 \quad \text{and} \quad \left(\frac{\partial^2 p}{\partial v^2}\right)_{T_c} = 0$$

Differentiating the van der Waals equation and applying the results to the critical point lead to

$$\left(\frac{\partial p}{\partial v}\right)_{T_c} = -\frac{RT_c}{(v_c - b)^2} + \frac{2a}{v_c^3} = 0 \quad \text{and} \quad \left(\frac{\partial^2 p}{\partial v^2}\right)_{T_c} = \frac{2RT_c}{(v_c - b)^3} - \frac{6a}{v_c^4} = 0$$

Solving these two equations together with the van der Waals equation as applied to the critical point, we obtain the following expressions for the critical properties p_c, T_c, and v_c in terms of the van der Waals constants a and b:

$$p_c = \frac{a}{27b^2} \tag{3-2a}$$

$$T_c = \frac{8a}{27Rb} \tag{3-2b}$$

$$v_c = 3b \tag{3-2c}$$

Conversely, the van der Waals constants a and b can be expressed in terms of the critical properties. Thus the simultaneous solution of the foregoing expressions for p_c and T_c yields

$$a = \frac{27}{64} \frac{(RT_c)^2}{p_c} \tag{3-3a}$$

$$b = \frac{RT_c}{8p_c} \tag{3-3b}$$

It should be noted that the equation $v_c = 3b$ may be used to yield the value of b in terms of v_c. However, since the measurements of p_c and T_c are usually more reliable than that of v_c, the data on p_c and T_c are to be used to determine the values of a and b. Table 3-1 gives the values of a and b for a number of gases as determined from Eqs. (3-3a) and (3-3b).

Despite its capability of describing the general behavior of real gases, the van der Waals equation holds only qualitatively. Any attempt to test the equation quantitatively is doomed to failure. One definite and clear discrepancy is seen in the value of the ratio $p_c v_c / RT_c$ which, according to the van der Waals equation, is

$$\frac{p_c v_c}{RT_c} = \frac{(a/27b^2)(3b)}{R(8a/27Rb)} = \frac{3}{8} = 0.375$$

TABLE 3-1. Constants of the van der Waals Equation of State*

Substance	$a, \dfrac{(\text{liters})^2 (\text{atm})}{(\text{g mole})^2}$	$b, \dfrac{\text{liters}}{\text{g mole}}$
Acetylene	4.417	0.05154
Ammonia	4.197	0.03737
Argon	1.349	0.03227
Benzene	18.50	0.1187
n-Butane	13.70	0.1164
i-Butane	13.14	0.1163
1-Butene	12.60	0.1084
Carbon dioxide	3.606	0.04280
Carbon disulfide	11.10	0.07259
Carbon monoxide	1.456	0.03954
Carbon tetrachloride	19.54	0.1268
Chlorine	6.491	0.05621
n-Decane	52.33	0.3052
Ethane	5.498	0.06500
Ethylene	4.507	0.05749
Freon-12	10.62	0.09964
Helium	0.03414	0.02371
n-Heptane	30.69	0.2051
n-Hexane	24.49	0.1741
Hydrogen	0.2461	0.02668
Hydrogen sulfide	4.460	0.04311
Krypton	2.294	0.03955
Methane	2.256	0.04271
Methyl chloride	7.470	0.06480
Neon	0.2091	0.01697
Nitric oxide	1.438	0.02885
Nitrogen	1.350	0.03864
Nitrous oxide	3.800	0.04430
n-Nonane	44.70	0.2712
n-Octane	37.38	0.2370
Oxygen	1.363	0.03184
n-Pentane	18.82	0.1447
i-Pentane	18.35	0.1437
neo-Pentane	16.93	0.1409
Propane	9.255	0.09033
Propylene	8.340	0.08249
Sulfur dioxide	6.773	0.05678
Sulfur trioxide	8.185	0.06015
Toluene	25.06	0.1523
Water	5.454	0.03042
Xenon	4.112	0.05124

*Calculated by Eqs. (3-3a) and (3-3b) from the critical data given in Appendix Table A-5.

for all gases. But, as given in Appendix Table A-5, the actual values of $p_c v_c / RT_c$ are different for different gases; for most gases they are much smaller than the value predicted by the van der Waals equation. It is noted, however, that van der Waals' simplified reasoning means that his equation is not expected to hold at high densities such as the critical region. Testing of the equation for lower densities should give better correlation.

3-2 BEATTIE-BRIDGEMAN EQUATION OF STATE

Since the days of van der Waals a large number of equations of state have been suggested to account for the behavior of real gases. Some are based on theoretical arguments, others are strictly empirical. In this section we will present the general reasoning of one of the early attempts, the Beattie-Bridgeman equation of state.[1]

The measured pressure of a gas can be written as the sum of the pressure due to kinetic energy and the pressure due to potential energy or, symbolically,

$$p = p_k + p_p$$

where p_k is the kinetic pressure and p_p is the cohesive pressure. For an ideal gas, p_k would be RT/v. Due to the existence of intermolecular forces, some molecules which have just passed through a reference plane drawn anywhere in the midst of the gas will be reflected and pass through again, and thus introduce additional momentum transfer. But since the pressure acting across the reference plane can be defined as the net rate at which momentum normal to it being transmitted across it per unit area in a chosen positive direction, the additional momentum transfer as a result of intermolecular forces would mean additional pressure. If the density of the gas is small so that the molecules act independently as reflectors, one may assume that

$$p_k = \frac{RT}{v}\left(1 + \frac{B}{v}\right) = \frac{RT}{v^2}(v + B)$$

where B is a constant. But if the density is such that the reflecting power per molecule is interfered with by those around it, B will not be constant but will be dependent on the density. The simplest form for this second order correction is a linear function of the density. Accordingly,

$$B = B_0\left(1 - \frac{b}{v}\right)$$

---

[1] J. A. Beattie and O. C. Bridgeman, *Proc. Amer. Acad. Arts & Sci.*, **63**:229 (1928).

Then

$$p_k = \frac{RT}{v^2}\left[v + B_0\left(1 - \frac{b}{v}\right)\right]$$

where B_0 and b are constants.

When slowly moving molecules collide, they tend to move under each other's influence; thus molecular association or aggregation is simulated. This has the same effect as a change in the average molecular weight of the gas and hence the gas constant may be considered to vary. Experimental data indicate that the variation in the number of independent aggregates due to the effect of the average time of encounter is directly proportional to the density and inversely proportional to the cubic power of absolute temperature, so that the gas constant is modified to read

$$R\left(1 - \frac{c}{vT^3}\right)$$

where R is the usual ideal gas constant and c is a constant. Whereupon

$$p_k = \frac{RT(1 - c/vT^3)}{v^2}\left[v + B_0\left(1 - \frac{b}{v}\right)\right]$$

The cohesive pressure is $-A/v^2$ as in the van der Waals equation. However, A should not be a constant but varies with the density. A simple linear variation is assumed for simplicity, thus

$$A = A_0\left(1 - \frac{a}{v}\right)$$

where A_0 and a are constants. Consequently,

$$p_p = -\frac{A_0}{v^2}\left(1 - \frac{a}{v}\right)$$

The final form of the equation is then

$$p = \frac{RT(1 - c/vT^3)}{v^2}\left[v + B_0\left(1 - \frac{b}{v}\right)\right] - \frac{A_0}{v^2}\left(1 - \frac{a}{v}\right) \tag{3-4}$$

The preceding equation is the *Beattie-Bridgeman equation of state*. In addition to the gas constant R this equation has five specific constants (A_0, a, B_0, b, and c) for a gas. These constants have been evaluated for a number of gases by curve-fitting to the experimental data. Table 3-2 gives the values of these constants for a number of gases.

TABLE 3-2. Constants of the Beattie-Bridgeman Equation of State*

Units: atm, liters/g mole, °K

Gas	A_0	a	B_0	b	$c \times 10^{-4}$
He	0.0216	0.05984	0.01400	0.0	0.0040
Ne	0.2125	0.02196	0.02060	0.0	0.101
Ar	1.2907	0.02328	0.03931	0.0	5.99
Kr	2.4230	0.02865	0.05261	0.0	14.89
Xe	4.6715	0.03311	0.07503	0.0	30.02
H_2	0.1975	-0.00506	0.02096	-0.04359	0.0504
N_2	1.3445	0.02617	0.05046	-0.00691	4.20
O_2	1.4911	0.02562	0.04624	0.004208	4.80
Air	1.3012	0.01931	0.04611	-0.01101	4.34
I_2	17.0	0.0	0.325	0.0	4000.
CO_2	5.0065	0.07132	0.10476	0.07235	66.00
NH_3	2.3930	0.17031	0.03415	0.19112	476.87
CH_4	2.2769	0.01855	0.05587	-0.01587	12.83
C_2H_4	6.1520	0.04964	0.12156	0.03597	22.68
C_2H_6	5.8800	0.05861	0.09400	0.01915	90.00
C_3H_8	11.9200	0.07321	0.18100	0.04293	120.00
$1\text{-}C_4H_8$	16.6979	0.11988	0.24046	0.10690	300.00
$iso\text{-}C_4H_8$	16.9600	0.10860	0.24200	0.08750	250.00
$n\text{-}C_4H_{10}$	17.7940	0.12161	0.24620	0.09423	350.00
$iso\text{-}C_4H_{10}$	16.6037	0.11171	0.23540	0.07697	300.00
$n\text{-}C_5H_{12}$	28.2600	0.15099	0.39400	0.13960	400.00
$neo\text{-}C_5H_{12}$	23.3300	0.15174	0.33560	0.13358	400.00
$n\text{-}C_7H_{16}$	54.520	0.20066	0.70816	0.19179	400.00
CH_3OH	33.309	0.09246	0.60362	0.09929	32.03
$(C_2H_5)_2O$	31.278	0.12426	0.45446	0.11954	33.33

*Mostly from H. S. Taylor and S. Glasstone, "A Treatise on Physical Chemistry, vol. II, States of Matter," D. Van Nostrand, New York, 1951.

The Beattie-Bridgeman equation was the best closed equation of state at the time it was proposed. In general, this equation is reasonably accurate when the volumes involved in the calculations are greater than twice the critical volume. In the critical region it is inaccurate.

3-3 OTHER EQUATIONS OF STATE

Among the many equations of state proposed in the past, the *Benedict-Webb-Rubin equation*[2] is certainly one of the best. This equation is entirely empirical

---

[2] M. Benedict, G. B. Webb, and L. C. Rubin, *J. Chem. Phys.*, 8:334 (1940).

and was developed by fitting the experimental data of light hydrocarbons. It is written as follows:

$$p = \frac{RT}{v} + \left(B_0 RT - A_0 - \frac{C_0}{T^2}\right)\frac{1}{v^2} + (bRT - a)\frac{1}{v^3} + \frac{a\alpha}{v^6} + \frac{c(1 + \gamma/v^2)}{T^2}\frac{1}{v^3}e^{-\gamma/v^2}$$

(3-5)

wherein the eight parameters B_0, A_0, C_0, b, a, c, α, and γ are numerical constants which are different for different gases. Their values for a number of gases are given in Table 3-3. These constants were obtained by curve fitting to the experimental p-v-T data. This equation is sufficiently accurate in describing the p-v-T relations of real gases up to densities of 1.8 times the critical density.

Martin and Hou[3] have suggested an equation of state of the form

$$p = \frac{RT}{v - b} + \frac{A_2 + B_2 T + C_2 e^{-5.475\, T/T_c}}{(v - b)^2}$$

$$+ \frac{A_3 + B_3 T + C_3 e^{-5.475\, T/T_c}}{(v - b)^3} + \frac{A_4}{(v - b)^4} + \frac{B_5 T}{(v - b)^5}$$

(3-6)

where b, A_2, B_2, C_2, A_3, B_3, C_3, A_4, and B_5 are nine specific constants. These constants are functions of the critical properties p_c, T_c, $Z_c = p_c v_c/RT_c$, and the slope of the critical isometric $(\partial p/\partial T)_{v_c}$. The values of these constants for several gases are given in Table 3-4. This equation was later revised[4] by adding to it the following term

$$\frac{A_5 + C_5 e^{-5.475\, T/T_c}}{(v - b)^5}$$

making a total of eleven specific constants. The Martin-Hou equation was designed to fit both polar and nonpolar gases up to densities of 1.5 times the critical density. This equation is of great value when one wishes to have an accurate p-v-T representation from a minimum of data because all the constants of this equation can be obtained from data on the critical point and one point on the vapor pressure curve only.

In general, the more specific constants used in an equation of state, the more accurate the equation can be. This explains why the Benedict-Webb-Rubin, Martin-Hou, and other elaborate equations with many specific constants are so often used for calculations of real gas properties which require high precision. However, this is not to infer that all simple equations are inaccurate. The types of gas and the ranges of pressure and temperature are factors to be considered in

[3] J. J. Martin and Y. C. Hou, *A.I.Ch.E. Journal*, 1:142 (1955).
[4] J. J. Martin, R. M. Kapoor, and N. de Nevers, *A.I.Ch.E. Journal*, 5:159 (1959).

TABLE 3-3. Constants of the Benedict-Webb-Rubin Equation of State*

Units: atm, liters/g mole, °K

Substance	a	A_0	b	B_0	$c \times 10^{-6}$	$C_0 \times 10^{-6}$	α	γ
CH_4	0.0494000	1.85500	0.00338004	0.0426000	0.00254500	0.0225700	0.000124359	0.0060000
C_2H_6	0.345160	4.15556	0.0111220	0.0627724	0.0327670	0.179592	0.000243389	0.0118000
C_3H_8	0.947700	6.87225	0.0225000	0.0973130	0.129000	0.508256	0.000607175	0.0220000
$n\text{-}C_4H_{10}$	1.88231	10.0847	0.0399983	0.124361	0.316400	0.992830	0.00110132	0.0340000
$i\text{-}C_4H_{10}$	1.93763	10.23264	0.0424352	0.137544	0.286010	0.849943	0.00107408	0.0340000
$n\text{-}C_5H_{12}$	4.07480	12.1794	0.0668120	0.156751	0.824170	2.12121	0.00181000	0.0475000
$i\text{-}C_5H_{12}$	3.75620	12.7959	0.0668120	0.160053	0.695000	1.74632	0.00170000	0.0463000
$neo\text{-}C_5H_{12}$	3.4905	12.9635	0.0668120	0.170530	0.546	1.273	0.002	0.05
$n\text{-}C_6H_{14}$	7.11671	14.4373	0.109131	0.177813	1.51276	3.31935	0.00281086	0.0666849
$n\text{-}C_7H_{16}$	10.36475	17.5206	0.151954	0.199005	2.47000	4.74574	0.00435611	0.0900000
C_2H_4	0.259000	3.33958	0.0086000	0.0556833	0.021120	0.131140	0.000178000	0.00923000
C_3H_6	0.774056	6.11220	0.0187059	0.0850647	0.102611	0.439182	0.000455696	0.0182900
$i\text{-}C_4H_8$	1.69270	8.95325	0.0348156	0.116025	0.274920	0.927280	0.000910889	0.0295945
C_6H_6	5.570	6.509772	0.07663	0.05030055	1.176418	3.42997	0.0007001	0.02930
NH_3	0.10354029	3.7892819	0.00071958516	0.051646121	0.00015753298	0.17857089	0.00000046521779	0.019805156
Ar	0.0288358	0.823417	0.00215289	0.022282597	0.0007982437	0.01314125	0.00003558895	0.0023382711
CO_2	0.136814	2.73742	0.00721045	0.0499101	0.0149180	0.138567	0.0000847	0.005394
CO	0.03665	1.34122	0.00263158	0.0545425	0.001040	0.00856209	0.000135	0.006
He	-0.00057339	0.040962	-0.000000019727	0.023661	-0.000000005521	-0.00000016227	-0.0000072673	0.00077942
N_2	0.025102	1.053642	0.0023277	0.0407426	0.00072841	0.00805900	0.0001272	0.005300
O_2	0.162689940	0.950851963	0.00358834736	0.0000000035328505	0.0128273741	0.0326435918	-3.927058894	0.0301
SO_2	0.84468	2.12044	0.014653	0.026182	0.11335	0.79384	0.000071955	0.0059236

*H. W. Cooper and J. C. Goldfrank, *Hydrocarbon Processing*, **46:12**:141 (1967).

TABLE 3-4. Constants* of the Martin-Hou Equation of State†

	CO_2	H_2O	C_6H_6	N_2	C_3H_6	H_2S	C_3H_8
T_c	547.5°R	1,165.1°R	562.66°K	126.1°K	364.92°K	672.4°R	666°R
p_c	1069.4 lb/sq in.	32,062 lb/sq in.	48.7 atm	33.5 atm	45.61 atm	1,306.0 lb/sq in.	618 lb/sq in.
v_c	0.03454 cu ft/lb	0.0503 cu ft/lb	3.36 cc/g	90.1 cc/g mole	0.191 liter/g mole	1.565 cu ft/lb mole	3.220 cu ft/lb mole
Z_c	0.27671	0.23246	0.27683	0.29171	0.290932	0.28329	0.278465
b	0.007495	0.0063101	0.730231	22.1466	0.0487575	0.368095	0.714598
R	0.24381	0.59545	1.05052	82.055	0.082055	10.73	10.73
A_2	−8.9273631	−85.7394396	−4,104.138	−1,592,238.2	−10.2233898	−21,206.0776	−44,105.1544
B_2	0.005262476	0.0312961744	2.62510	3,221.616	0.0081804454	9.7631723	21.127023
C_2	−150.97587	−2590.5815	−59,333.234	−22,350,930	−153.061055	−318,608.803	−736,929.321
A_3	0.18907819	3.09248249	8,553.304	89,845,367	1.219453756	21133.013	88,025.6002
B_3	−0.0000704617	−0.00082418321	−3.89165685	−134,024.05	0.00074168283	−7.41557245	−28.1649342
C_3	+0.0831424	113.95968	156,032.7	1,518,821,653	21.7717909	381,344.629	1,846,304.2
A_4	−0.002112459	−0.0567185967	−8,599.2791	−2,467,297,480	−0.068940713	−9897.9772	−89,907.6635
B_5	1.9565593×10^{-8}	4.2388378×10^{-7}	7.53651	244,915,183	5.0385878×10^{-6}	3.3036675	63.416365

*All constants for the equation of state are given in units of T_c, p_c, and v_c for that compound.

†J. J. Martin and Y. C. Hou, *A.I.Ch.E. Journal*, 1:142 (1955).

the selection of equations of state to use for any application. One of the simpler equations, the *Redlich-Kwong equation*[5] with only two specific constants has been shown[6] to have considerable accuracy over a wide range of *p-v-T* conditions. The Redlich-Kwong equation of state is as follows:

$$p = \frac{RT}{v - b} - \frac{a}{T^{1/2}v(v + b)} \tag{3-7}$$

The constants a and b can be evaluated from critical data by the conditions

$$\left(\frac{\partial p}{\partial v}\right)_{T_c} = 0 \quad \text{and} \quad \left(\frac{\partial^2 p}{\partial v^2}\right)_{T_c} = 0$$

In terms of p_c and T_c these constants are

$$a = \frac{0.42748R^2 T_c^{2.5}}{p_c} \tag{3-8a}$$

$$b = \frac{0.08664RT_c}{p_c} \tag{3-8b}$$

The Redlich-Kwong equation is said to furnish satisfactory results above the critical temperature for any pressure. The fair degree of accuracy combined with simplicity in form and in the evaluation of the constants make this equation attractive for industrial use.

3-4 EQUATION OF STATE IN VIRIAL FORM

Kammerlingh Onnes in 1901 introduced the use of power series expansions to express the *p-v-T* relations of gases. The most usual form is an expansion of the product *pv* as a power series in density. Thus

$$pv = A\left(1 + \frac{B}{v} + \frac{C}{v^2} + \frac{D}{v^3} + \cdots\right)$$

We must have $A = RT$ in the foregoing equation to satisfy the ideal gas law at zero density, so that we have

$$\frac{pv}{RT} = 1 + \frac{B}{v} + \frac{C}{v^2} + \frac{D}{v^3} + \cdots \tag{3-9}$$

An equation of this type is known as the *virial equation of state*, and the coefficients *B, C, D*, etc., are called the second, third, fourth, etc., *virial*

[5] O. Redlich and J. N. S. Kwong, *Chem. Rev.*, **44**:233 (1949).
[6] K. K. Shah and G. Thodos, *Ind. Eng. Chem.*, **57**:30 (1965).

coefficients. These coefficients are functions of the temperature, and also, of course, functions of the substance of interest.

At moderately low densities, only two-particle interactions are of significance in explaining departure from ideality of the gas. This effect is expressed by the B/v term in the virial equation. As density is increased, three-particle interactions also become significant; this effect is expressed by the C/v^2 term in the equation. Similarly, higher virial terms are used when there are higher-order molecular interactions. It should be noted that a true virial expansion applies to gases at low and medium densities only. As the series diverges at about liquid density, it is unsuitable for high densities.

The virial equation can also be expressed in terms of a power series in pressure:

$$\frac{pv}{RT} = 1 + B'p + C'p^2 + D'p^3 + \cdots \tag{3-10}$$

where the coefficients B', C', D', etc., are functions of the temperature of the gas. The relationships between these coefficients and those of Eq. (3-9) are

$$B' = \frac{B}{RT} \quad C' = \frac{C - B^2}{(RT)^2} \quad D' = \frac{D - 3BC + 2B^3}{(RT)^3}$$

Equation (3-10) converges less rapidly than does Eq. (3-9) and is useful mainly at low densities.

Since the virial equation is an infinite series, it has to be truncated before determining the virial coefficients. In general, the series is to be truncated at a point where the sum of higher-power terms is estimated to be within the experimental error over the full range studied. The coefficients of the truncated power series are then determined by a least-square or other type of curve-fit of the experimental p-v-T data. In order to obtain an accurate second virial coefficient of a gas, the data used must extend to sufficiently low pressures. Note that the third and higher coefficients are extremely sensitive to the degree of the polynomial chosen, and one must be careful to choose the optimum density range from which to derive the higher virials.

When extremely accurate p-v-T data at very low densities are available, the second and third virials may be obtained by writing the virial equation in the form

$$\left(\frac{pv}{RT} - 1\right) = \frac{B}{v} + \frac{C}{v^2} + \cdots$$

Thus the second virial coefficient is given by

$$B = \lim_{p \to 0} \left(\frac{pv}{RT} - 1 \right) v$$

and the third virial coefficient is given by

$$C = \lim_{p \to 0} \left[\left(\frac{pv}{RT} - 1 \right) v - B \right] v$$

Any of the equations of state discussed in the preceeding sections may be expanded into a virial form. The following is an example.

EXAMPLE 3-1. Expand the van der Waals equation of state into power series in density and in pressure.

Solution.

The van der Waals equation may be transformed to read

$$pv = RT \left(1 - \frac{b}{v} \right)^{-1} - \frac{a}{v}$$

Use of the binomial theorem gives

$$pv = RT \left(1 + \frac{b}{v} + \frac{b^2}{v^2} + \frac{b^3}{v^3} + \cdots \right) - \frac{a}{v}$$

or

$$\frac{pv}{RT} = 1 + \frac{b - (a/RT)}{v} + \frac{b^2}{v^2} + \frac{b^3}{v^3} + \cdots \tag{3-11}$$

This is the van der Waals equation in virial form in terms of density. The virial coefficients are

$$B = b - \frac{a}{RT} \qquad C = b^2 \qquad D = b^3 \qquad \text{etc.}$$

When the equation is written in terms of a power series in pressure we must have

$$B' = \frac{B}{RT} = \frac{b}{RT} - \frac{a}{R^2 T^2}$$

$$C' = \frac{C - B^2}{R^2 T^2} = \frac{2ab}{R^3 T^3} - \frac{a^2}{R^4 T^4}$$

$$D' = \frac{D - 3BC + 2B^3}{R^3 T^3} = -\frac{3ab^2}{R^4 T^4} + \frac{6a^2 b}{R^5 T^5} - \frac{2a^3}{R^6 T^6}$$

Thus

$$\frac{pv}{RT} = 1 + \left(\frac{b}{RT} - \frac{a}{R^2T^2}\right)p + \left(\frac{2ab}{R^3T^3} - \frac{a^2}{R^4T^4}\right)p^2$$

$$+ \left(-\frac{3ab^2}{R^4T^4} + \frac{6a^2b}{R^5T^5} - \frac{2a^3}{R^6T^6}\right)p^3 + \cdots \quad (3\text{-}11a)$$

Since a and b are small corrections in van der Waals equation, the terms involving their product ($a\,b$, a^2, etc.) in Eqs. (3-11) and (3-11a) may be neglected. Thus we may write simply

$$\frac{pv}{RT} = 1 + \left(b - \frac{a}{RT}\right)\frac{1}{v} \quad \text{and} \quad \frac{pv}{RT} = 1 + \left(\frac{b}{RT} - \frac{a}{R^2T^2}\right)p$$

3-5 EVALUATION OF THERMODYNAMIC PROPERTIES FROM AN EQUATION OF STATE

An equation of state may be used not only to calculate pressure, volume, or temperature, but to evaluate other thermodynamic properties such as internal energy, enthalpy, and entropy. The procedure calls for the integration of the differential relations for these properties with the aid of an equation of state and some supplementary data, such as specific heats or Joule-Thomson coefficients. For example, suppose that the change in enthalpy per unit mass of a gas from a reference state at p_0, T_0 to some other state at p, T is to be calculated. We use the second dh equation,

$$dh = c_p\,dT + \left[v - T\left(\frac{\partial v}{\partial T}\right)_p\right]dp \qquad (2\text{-}42)$$

Since h is a property, dh is an exact differential. The line integral of dh is then a function of the end states only and independent of the path. Any process or combination of processes between the two end states can be chosen. Two simple combinations are shown in Fig. 3-2.

For the combination of processes depicted by line oaA in Fig. 3-2, Eq. (2-42) is integrated first at constant pressure p_0 from T_0 to T, then at constant temperature T from p_0 to p. The results are

$$h_a - h_0 = \left[\int_{T_0}^{T} c_p\,dT\right]_{p_0}$$

$$h - h_a = \left\{\int_{p_0}^{p}\left[v - T\left(\frac{\partial v}{\partial T}\right)_p\right]dp\right\}_T$$

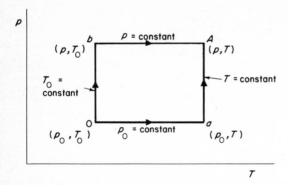

FIG. 3-2. Two combinations of processes connecting states (p_0, T_0) and (p, T).

where h is the enthalpy at the final state (p, T), h_0 is the enthalpy at the initial state (p_0, T_0), and h_a is the enthalpy at the intermediate state (p_0, T). Combining these two equations yields

$$h - h_0 = \left[\int_{T_0}^{T} c_p \, dT\right]_{p_0} + \left\{\int_{p_0}^{p}\left[v - T\left(\frac{\partial v}{\partial T}\right)_p\right] dp\right\}_T \tag{3-12}$$

For the combination of processes obA, Eq. (2-42) is integrated, first at constant temperature T_0 from p_0 to p, then at constant pressure p from T_0 to T. This combination yields

$$h - h_0 = \left\{\int_{p_0}^{p}\left[v - T\left(\frac{\partial v}{\partial T}\right)_p\right] dp\right\}_{T_0} + \left[\int_{T_0}^{T} c_p \, dT\right]_p \tag{3-13}$$

Equation (3-12) requires data on c_p for all temperatures in the specified range at one pressure p_0, whereas Eq. (3-13) requires the c_p data at a higher pressure p. In view of the fact that specific-heat measurements are relatively more convenient to make at low pressures, Eq. (3-12) is preferable.

If the equation of state is in the form $v = f(p, T)$, the second integrand in Eq. (3-12) can be integrated analytically, at least in principle, because the partial derivative $(\partial v/\partial T)_p$ can be obtained by differentiating the equation of state and the variable v can be replaced by substitution. Most equations of state, however, are in the form $p = f(T, v)$ as a consequence of kinetic-theory considerations. With this form of equation of state the second integrand in Eq. (3-12) in the present form is usually difficult to handle. This integrand may be transformed as follows to facilitate easy integration. Since

$$d(pv) = p \, dv + v \, dp$$

we have

$$\left[\int_{p_0}^{p} v\, dp\right]_T = (pv - p_0 v_a) - \left[\int_{v_a}^{v} p\, dv\right]_T$$

where v is the specific volume at p and T, and v_a is the specific volume at p_0 and T. Also, since

$$\left(\frac{\partial v}{\partial T}\right)_p = -\left(\frac{\partial v}{\partial p}\right)_T \left(\frac{\partial p}{\partial T}\right)_v$$

we have

$$\left[\int_{p_0}^{p}\left(\frac{\partial v}{\partial T}\right)_p dp\right]_T = -\left[\int_{v_a}^{v}\left(\frac{\partial p}{\partial T}\right)_v dv\right]_T$$

so that Eq. (3-12) becomes

$$h - h_0 = \left[\int_{T_0}^{T} c_p\, dT\right]_{p_0} + (pv - p_0 v_a) - \left\{\int_{v_a}^{v}\left[p - T\left(\frac{\partial p}{\partial T}\right)_v\right] dv\right\}_T \qquad (3\text{-}14)$$

Let us now consider the integration of the second ds equation,

$$ds = \frac{c_p}{T}\, dT - \left(\frac{\partial v}{\partial T}\right)_p dp \qquad (2\text{-}35)$$

from the reference state (p_0, T_0) to the final state (p, T). Following the combination of processes oaA in Fig. 3-2, we write

$$s - s_0 = \left[\int_{T_0}^{T}\frac{c_p}{T}\, dT\right]_{p_0} - \left[\int_{p_0}^{p}\left(\frac{\partial v}{\partial T}\right)_p dp\right]_T$$

where s is the entropy at the final state (p, T), and s_0 is the entropy at the reference state (p_0, T_0). When the equation of state is in the form $p = f(T, v)$, the foregoing equation is best written as

$$s - s_0 = \left[\int_{T_0}^{T}\frac{c_p}{T}dT\right]_{p_0} + \left[\int_{v_a}^{v}\left(\frac{\partial p}{\partial T}\right)_v dv\right]_T \qquad (3\text{-}15)$$

EXAMPLE 3-2. Using the Benedict-Webb-Rubin equation of state, calculate values of v, h, s, c_v, and c_p for methane at $p = 200$ atm and $T = 400°$K.

Given that $h_0 = 4,600$ cal/g mole, $s_0 = 44.54$ cal/(g mole)($^\circ$K) at $p_0 = 1$ atm, and $T_0 = 300^\circ$K based on the value of zero for the perfect crystal at the absolute zero of temperature. The constant-pressure specific heat for methane at 1 atm may be expressed as

$$c_p = 4.750 + 12.00 \times 10^{-3} T + 3.030 \times 10^{-6} T^2 - 2.630 \times 10^{-9} T^3$$

where c_p is in cal/(g mole)($^\circ$K) and T in $^\circ$K.

Solution.
The Benedict-Webb-Rubin equation is

$$p = \frac{RT}{v} + \left(B_0 RT - A_0 - \frac{C_0}{T^2} \right) \frac{1}{v^2}$$

$$+ (bRT - a)\frac{1}{v^3} + \frac{a\alpha}{v^6} + \frac{c(1 + \gamma/v^2)}{T^2} \frac{1}{v^3} e^{-\gamma/v^2} \quad (3\text{-}5)$$

where p is in atm, T in $^\circ$K, v in liters/g mole, $R = 0.08207$ (liter)(atm)/(g mole)($^\circ$K), and for methane the eight specific constants are

$B_0 = 0.0426000$ $A_0 = 1.85500$ $C_0 = 0.0225700 \times 10^6$

 $b = 0.00338004$ $a = 0.0494000$ $c = 0.00254500 \times 10^6$

 $\alpha = 0.124359 \times 10^{-3}$ $\gamma = 0.60000 \times 10^{-2}$

By differentiating the Benedict-Webb-Rubin equation we obtain

$$\left(\frac{\partial p}{\partial T} \right)_v = \frac{R}{v} + \left(B_0 R + \frac{2C_0}{T^3} \right)\frac{1}{v^2} + \frac{bR}{v^3} - \frac{2c}{T^3}\frac{1}{v^3}\left(1 + \frac{\gamma}{v^2} \right)e^{-\gamma/v^2} \qquad (3\text{-}16a)$$

$$\left(\frac{\partial^2 p}{\partial T^2} \right)_v = -\frac{6C_0}{T^4}\frac{1}{v^2} + \frac{6c}{T^4}\frac{1}{v^3}\left(1 + \frac{\gamma}{v^2} \right)e^{-\gamma/v^2} \qquad (3\text{-}16b)$$

$$\left(\frac{\partial p}{\partial v} \right)_T = -\frac{RT}{v^2} - 2\left(B_0 RT - A_0 - \frac{C_0}{T^2} \right)\frac{1}{v^3} - 3(bRT - a)\frac{1}{v^4} - \frac{6a\alpha}{v^7}$$

$$+ \frac{c}{T^2 v^4} e^{-\gamma/v^2}\left(-3 - \frac{3\gamma}{v^2} + \frac{2\gamma^2}{v^4} \right) \qquad (3\text{-}16c)$$

The state points for this example are shown in Fig. 3-3. The volume of the gas at point A where $p = 200$ atm and $T = 400^\circ$K as determined from the Benedict-Webb-Rubin equation by a trial process is $v = 0.161$ liter/g mole. Since $p_a = p_0 = 1$ atm, the volume of the gas at point a can be obtained with sufficient accuracy by using the ideal gas equation. Thus

$$v_a = \frac{RT}{p_0} = \frac{0.08207 \times 400}{1} = 32.8 \text{ liters/g mole}$$

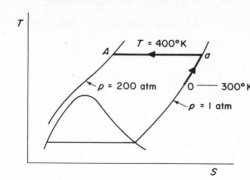

FIG. 3-3. T-S diagram showing state points for Example 3-2.

To evaluate the value of enthalpy, we now calculate each term of Eq. (3-14).

$$\int_{T_0}^{T} c_p \, dT = \int_{T_0}^{T} (4.75 + 12.0 \times 10^{-3} T + 3.03 \times 10^{-6} T^2 - 2.63 \times 10^{-9} T^3) \, dT$$

$$= 4.75 \, (T - T_0) + 6.0 \times 10^{-3}(T^2 - T_0^2) + 1.01 \times 10^{-6}(T^3 - T_0^3)$$
$$- 0.6575 \times 10^{-9}(T^4 - T_0^4)$$

$$\doteq 920.9 \text{ cal/g mole}$$

after substituting numerical values of T and T_0.

$$(pv - p_0 v_a) = 200 \times 0.161 - 1 \times 32.8$$

$$= -0.6 \text{ (liter)(atm)/g mole} = -14.5 \text{ cal/g mole}$$

$$\left[\int_{v_a}^{v} p \, dv\right]_T = \left\{\int_{v_a}^{v} \left[\frac{RT}{v} + \left(B_0 RT - A_0 - \frac{C_0}{T^2}\right)\frac{1}{v^2} + (bRT - a)\frac{1}{v^3} + \frac{a\alpha}{v^6}\right.\right.$$
$$\left.\left. + \frac{c}{T^2}\frac{1}{v^3} e^{-\gamma/v^2} + \frac{c}{T^2}\frac{\gamma}{v^5} e^{-\gamma/v^2}\right] dv\right\}_T$$

wherein

$$\int \frac{1}{v^3} e^{-\gamma/v^2} \, dv = \frac{1}{2\gamma}\int e^{-\gamma/v^2} d\left(-\frac{\gamma}{v^2}\right) = \frac{1}{2\gamma} e^{-\gamma/v^2}$$

$$\int \frac{\gamma}{v^5} e^{-\gamma/v^2} \, dv = \tfrac{1}{2} e^{-\gamma/v^2}\left(\frac{1}{v^2} + \frac{1}{\gamma}\right)$$

The last integration is by the method of integration by parts. Hence we have

$$\left[\int_{v_a}^{v} p \, dv\right]_T = RT \ln \frac{v}{v_a} - \left(B_0 RT - A_0 - \frac{C_0}{T^2}\right)\left(\frac{1}{v} - \frac{1}{v_a}\right)$$

$$- \tfrac{1}{2}(bRT - a)\left(\frac{1}{v^2} - \frac{1}{v_a^2}\right) - \frac{a\alpha}{5}\left(\frac{1}{v^5} - \frac{1}{v_a^5}\right) + \frac{c}{2\gamma T^2}\left(e^{-\gamma/v^2} - e^{-\gamma/v_a^2}\right)$$

$$+ \frac{c}{2T^2}\left[e^{-\gamma/v^2}\left(\frac{1}{v^2} + \frac{1}{\gamma}\right) - e^{-\gamma/v_a^2}\left(\frac{1}{v_a^2} + \frac{1}{\gamma}\right)\right]$$

Also we have

$$\left[\int_{v_a}^{v} T\left(\frac{\partial p}{\partial T}\right)_v dv\right]_T = \left\{\int_{v_a}^{v}\left[\frac{RT}{v} + \left(B_0 RT + \frac{2C_0}{T^2}\right)\frac{1}{v^2} + \frac{bRT}{v^3}\right.\right.$$

$$\left.\left. - \frac{2c}{T^2}\frac{1}{v^3}e^{-\gamma/v^2} - \frac{2c}{T^2}\frac{\gamma}{v^5}e^{-\gamma/v^2}\right]dv\right\}_T$$

$$= RT \ln \frac{v}{v_a} - \left(B_0 RT + \frac{2C_0}{T^2}\right)\left(\frac{1}{v} - \frac{1}{v_a}\right) - \tfrac{1}{2}bRT\left(\frac{1}{v^2} - \frac{1}{v_a^2}\right)$$

$$- \frac{c}{\gamma T^2}\left(e^{-\gamma/v^2} - e^{-\gamma/v_a^2}\right)$$

$$- \frac{c}{T^2}\left[e^{-\gamma/v^2}\left(\frac{1}{v^2} + \frac{1}{\gamma}\right) - e^{-\gamma/v_a^2}\left(\frac{1}{v_a^2} + \frac{1}{\gamma}\right)\right]$$

Combining the last two equations yields

$$\left\{\int_{v_a}^{v}\left[p - T\left(\frac{\partial p}{\partial T}\right)_v\right]dv\right\}_T = \left(A_0 + \frac{3C_0}{T^2}\right)\left(\frac{1}{v} - \frac{1}{v_a}\right) + \frac{a}{2}\left(\frac{1}{v^2} - \frac{1}{v_a^2}\right)$$

$$- \frac{a\alpha}{5}\left(\frac{1}{v^5} - \frac{1}{v_a^5}\right) + \frac{3}{2}\frac{c}{\gamma T^2}\left(e^{-\gamma/v^2} - e^{-\gamma/v_a^2}\right)$$

$$+ \frac{3}{2}\frac{c}{T^2}\left[e^{-\gamma/v^2}\left(\frac{1}{v^2} + \frac{1}{\gamma}\right) - e^{-\gamma/v_a^2}\left(\frac{1}{v_a^2} + \frac{1}{\gamma}\right)\right]$$

where

$$\left(\frac{1}{v} - \frac{1}{v_a}\right) = \frac{1}{0.161} - \frac{1}{32.8} = 6.18$$

$$\left(\frac{1}{v^2} - \frac{1}{v_a^2}\right) = \frac{1}{0.161^2} - \frac{1}{32.8^2} = 38.58$$

$$\left(\frac{1}{v^5} - \frac{1}{v_a^5}\right) = \frac{1}{0.161^5} - \frac{1}{32.8^5} = 9,244$$

$$(e^{-\gamma/v^2} - e^{-\gamma/v_a^2}) = e^{-0.006/0.161^2} - e^{-0.006/32.8^2} = 0.793 - 1 = -0.207$$

and

$$\left[e^{-\gamma/v^2}\left(\frac{1}{v^2} + \frac{1}{\gamma}\right) - e^{-\gamma/v_a^2}\left(\frac{1}{v_a^2} + \frac{1}{\gamma}\right)\right] = 0.793\left(\frac{1}{0.161^2} + \frac{1}{0.006}\right)$$

$$- 1\left(\frac{1}{32.8^2} + \frac{1}{0.006}\right) = -3.91$$

Thus, substituting the numerical values of the specific constants and the temperature gives

$$\left\{\int_{v_a}^{v}\left[p - T\left(\frac{\partial p}{\partial T}\right)_v\right]dv\right\}_T = 14.10 \text{ (liter)(atm)/g mole} = 341.2 \text{ cal/g mole}$$

Then, upon substituting in Eq. (3-14),

$$h - h_0 = 920.9 - 14.5 - 341.2 = 920.9 - 355.7 = 565.2 \text{ cal/g mole}$$

Hence the value of enthalpy at 200 atm and 400°K is

$$h = 4,600 + 565 = 5,165 \text{ cal/g mole}$$

Now we proceed to evaluate the entropy by calculating each term of Eq. (3-15).

$$\int_{T_0}^{T}\frac{c_p}{T}dT = \int_{T_0}^{T}\left(\frac{4.75}{T} + 12.0 \times 10^{-3} + 3.03 \times 10^{-6}T - 2.63 \times 10^{-9}T^2\right)dT$$

$$= 4.75 \ln\frac{T}{T_0} + 12.0 \times 10^{-3}(T - T_0) + 1.515 \times 10^{-6}(T^2 - T_0^2)$$

$$- 0.8767 \times 10^{-9}(T^3 - T_0^3)$$

$$= 2.64 \text{ cal/(g mole)(°K)}$$

after substituting numerical values of T and T_0.

$$\left[\int_{v_a}^{v} \left(\frac{\partial p}{\partial T}\right)_v dv\right]_T = \left\{\int_{v_a}^{v} \left[\frac{R}{v} + \left(B_0 R + \frac{2C_0}{T^3}\right)\frac{1}{v^2} + \frac{bR}{v^3} - \frac{2c}{T^3}\frac{1}{v^3}e^{-\gamma/v^2}\right.\right.$$

$$\left.\left. - \frac{2c}{T^3}\frac{\gamma}{v^5}e^{-\gamma/v^2}\right]dv\right\}_T$$

$$= R \ln\frac{v}{v_a} - \left(B_0 R + \frac{2C_0}{T^3}\right)\left(\frac{1}{v} - \frac{1}{v_a}\right) - \frac{1}{2}bR\left(\frac{1}{v^2} - \frac{1}{v_a^2}\right) - \frac{c}{\gamma T^3}(e^{-\gamma/v^2} - e^{-\gamma/v_a^2})$$

$$- \frac{c}{T^3}\left[e^{-\gamma/v^2}\left(\frac{1}{v^2} + \frac{1}{\gamma}\right) - e^{-\gamma/v_a^2}\left(\frac{1}{v_a^2} + \frac{1}{\gamma}\right)\right]$$

$$= -0.4661 \text{ (liter)(atm)/(g mole)}(^\circ K)$$

$$= -11.28 \text{ cal/(g mole)}(^\circ K)$$

after substituting numerical values. Then upon substituting in Eq. (3-15),

$s - s_0 = 2.64 - 11.28 = -8.64$ cal/(g mole)($^\circ$K)

Hence the value of entropy at 200 atm and 400°K is

$s = 44.54 - 8.64 = 35.90$ cal/(g mole)($^\circ$K)

The value of c_p at 1 atm and 400°K is

$c_{pa} = 4.75 + 12.0 \times 10^{-3} \times 400 + 3.03 \times 10^{-6} \times 400^2 - 2.63 \times 10^{-9} \times 400^3$

$\quad = 9.87$ cal/(g mole)($^\circ$K)

Since at 1 atm the gas may be treated as an ideal gas, it follows that

$c_{va} = c_{pa} - R = 9.87 - 1.99 = 7.88$ cal/(g mole)($^\circ$K)

To calculate the value of c_v at 200 atm and 400°K we use the equation

$$\left(\frac{\partial c_v}{\partial v}\right)_T = T\left(\frac{\partial^2 p}{\partial T^2}\right)_v \qquad\qquad (2\text{-}28)$$

Integration gives

$$c_v - c_{va} = \left[\int_{v_a}^{v} T\left(\frac{\partial^2 p}{\partial T^2}\right)_v dv\right]_T$$

$$= \left[\int_{v_a}^{v}\left(-\frac{6C_0}{T^3}\frac{1}{v^2} + \frac{6c}{T^3}\frac{1}{v^3}e^{-\gamma/v^2} + \frac{6c}{T^3}\frac{\gamma}{v^5}e^{-\gamma/v^2}\right)dv\right]_T$$

$$= \frac{6}{T^3} \left\{ C_0 \left(\frac{1}{v} - \frac{1}{v_a} \right) + \frac{c}{2\gamma} \left(e^{-\gamma/v^2} - e^{-\gamma/v_a^2} \right) \right.$$

$$\left. + \frac{c}{2} \left[e^{-\gamma/v^2} \left(\frac{1}{v^2} + \frac{1}{\gamma} \right) - e^{-\gamma/v_a^2} \left(\frac{1}{v_a^2} + \frac{1}{\gamma} \right) \right] \right\}$$

$$= 0.00849 \text{ (liter)(atm)/(g mole)}(^\circ K) = 0.205 \text{ cal/(g mole)}(^\circ K)$$

after substituting numerical values. Hence the value of c_v at 200 atm and 400°K is

$$c_v = 7.88 + 0.205 = 8.09 \text{ cal/(g mole)}(^\circ K)$$

With the value of c_v at 200 atm and 400°K known, we can now calculate the corresponding value of c_p by using the following equation

$$c_p - c_v = T \left(\frac{\partial p}{\partial T} \right)_v \left(\frac{\partial v}{\partial T} \right)_p \tag{2-32}$$

which, upon using cyclic relation, can be transformed to read

$$c_p - c_v = -T \frac{(\partial p/\partial T)_v^2}{(\partial p/\partial v)_T}$$

At 200 atm, 400°K, the volume of the gas is 0.161 liter/g mole. Substituting these values and the specific constants into Eqs. (3-16a) and (3-16c), we obtain

$$\left(\frac{\partial p}{\partial T} \right)_v = 0.719 \text{ atm/}(^\circ K) \quad \text{and} \quad \left(\frac{\partial P}{\partial v} \right)_T = -1{,}330 \text{ (atm)(g mole)/liter}$$

With these data,

$$c_p - c_v = -400 \times \frac{(0.719)^2}{(-1330)}$$

$$= 0.155 \text{ (liter)(atm)/(g mole)}(^\circ K)$$

$$= 3.75 \text{ cal/(g mole)}(^\circ K)$$

Hence the value of c_p at 200 atm and 400°K is

$$c_p = c_v + 3.75 = 8.09 + 3.75$$

$$= 11.84 \text{ cal/(g mole)}(^\circ K)$$

The results of the preceding calculations are summarized and compared with experimental data as follows:

P = 200 atm $T = 400°K$	v liters/g mole	h cal/g mole	s cal/(g mole)($°$K)	c_v cal/(g mole)($°$K)	c_p cal/(g mole)($°$K)
Calculated	0.161	5,165	35.90	8.09	11.84
Experimental data*	0.160795	5,139	35.80	7.94	11.76

*F. Din, "Thermodynamic Functions of Gases," vol. 3, Butterworth & Co., Ltd., London, 1961.

EXAMPLE 3-3.   Argon at 1 atm and 300°K enters a compressor at a rate of 100 kg/hr. It is compressed reversibly and isothermally to 500 atm. Calculate the power needed to run the compressor and the amount of heat that must be removed per hour from the compressor. The gas is assumed to obey the Redlich-Kwong equation of state.

Solution.
Let us denote the initial and final states of the gas by the subscripts 1 and 2 respectively. Thus p_1 = 1 atm, p_2 = 500 atm, and $T_1 = T_2 = 300°K$.
   The Redlich-Kwong equation is

$$p = \frac{RT}{v - b} - \frac{a}{T^{1/2}v(v + b)} \tag{3-7}$$

where

$$a = 0.42748 \frac{R^2 T_c^{2.5}}{p_c} \tag{3-8a}$$

$$b = 0.08664 \frac{RT_c}{p_c} \tag{3-8b}$$

For argon the critical constants are $T_c = 151°K$ and $p_c = 48$ atm, whence

$$a = 0.42748 \frac{(82.06)^2(151)^{2.5}}{48} = 16.8 \times 10^6 \frac{(\text{atm})(°K^{1/2})(\text{cm}^6)}{(\text{g mole})^2}$$

$$b = 0.08664 \frac{(82.06)(151)}{48} = 22.4 \frac{\text{cm}^3}{\text{g mole}}$$

   Substituting the numerical values of p_2, T_2, a, b, and R into the Redlich-Kwong equation leads to

$$v_2^3 - 49.24v_2^2 + 335.6v_2 - 43,440 = 0$$

from which we obtain

$$v_2 = 56.8 \text{ cm}^3/(\text{g mole})$$

Since $p_1 = 1$ atm, the volume of the gas at the initial state can be obtained with sufficient accuracy by using the ideal gas equation. Thus

$$v_1 = \frac{RT_1}{p_1} = \frac{82.06 \times 300}{1} = 24{,}600 \text{ cm}^3/(\text{g mole})$$

In order to calculate the work done and the heat transfer, we need first to calculate the changes in enthalpy and entropy between the initial and the final states. From the second dh equation, for the isothermal compression process, we have

$$\Delta h_{12} = h_2 - h_1 = \left\{ \int_{p_1}^{p_2} \left[v - T\left(\frac{\partial v}{\partial T}\right)_p \right] dp \right\}_T$$

Now since

$$d(pv) = p\, dv + v\, dp$$

we have

$$\left[\int_{p_1}^{p_2} v\, dp \right]_T = p_2 v_2 - p_1 v_1 - \left[\int_{v_1}^{v_2} p\, dv \right]_T$$

And since

$$\left(\frac{\partial v}{\partial T}\right)_p = -\left(\frac{\partial v}{\partial p}\right)_T \left(\frac{\partial p}{\partial T}\right)_v$$

we have

$$\left[\int_{p_1}^{p_2} \left(\frac{\partial v}{\partial T}\right)_p dp \right]_T = -\left[\int_{v_1}^{v_2} \left(\frac{\partial p}{\partial T}\right)_v dv \right]_T$$

Hence

$$\Delta h_{12} = h_2 - h_1$$

$$= (p_2 v_2 - p_1 v_1) - \left\{ \int_{v_1}^{v_2} \left[p - T\left(\frac{\partial p}{\partial T}\right)_v \right] dv \right\}_T$$

But according to the Redlich-Kwong equation, we have

$$\left(\frac{\partial p}{\partial T}\right)_v = \frac{R}{v - b} + \frac{a}{2T^{3/2} v(v + b)}$$

Thus

$$\Delta h_{12} = h_2 - h_1$$

$$= (p_2 v_2 - p_1 v_1) - \left\{ \int_{v_1}^{v_2} \left[\frac{RT}{v - b} - \frac{a}{T^{1/2} v(v + b)} \right. \right.$$

$$\left. \left. - \frac{RT}{v - b} - \frac{a}{2T^{1/2} v(v + b)} \right] dv \right\}_T$$

$$= (p_2 v_2 - p_1 v_1) - \left\{ \int_{v_1}^{v_2} \left[\frac{-3a}{2T^{1/2} v(v + b)} \right] dv \right\}_T$$

$$= (p_2 v_2 - p_1 v_1) - \frac{1.5a}{T_1^{1/2}} \frac{1}{b} \ln \left[\frac{(v_2 + b)/v_2}{(v_1 + b)/v_1} \right]$$

$$= -17,700 \text{ (atm)(cm}^3)/\text{(g mole)}$$

$$= -1,790 \text{ joules/(g mole)}$$

after substituting numerical values.

We now use the second ds equation to calculate the entropy change. For the isothermal compression process, we have

$$\Delta s_{12} = s_2 - s_1$$

$$= -\left[\int_{p_1}^{p_2} \left(\frac{\partial v}{\partial T} \right)_p dp \right]_T = \left[\int_{v_1}^{v_2} \left(\frac{\partial p}{\partial T} \right)_v dv \right]_T$$

For the Redlich-Kwong equation this becomes

$$\Delta s_{12} = s_2 - s_1$$

$$= \left\{ \int_{v_1}^{v_2} \left[\frac{R}{v - b} + \frac{a}{2T^{3/2} v(v + b)} \right] dv \right\}_T$$

$$= R \ln \left(\frac{v_2 - b}{v_1 - b} \right) - \frac{a}{2bT_1^{3/2}} \ln \left[\frac{(v_2 + b)/v_2}{(v_1 + b)/v_1} \right]$$

$$= -563 \text{ (atm)(cm}^3)/\text{(g mole)(}°\text{K)} = -57.0 \text{ joules/(g mole)(}°\text{K)}$$

after substituting numerical values.

With the changes in entropy and enthalpy known, we are now ready to calculate the rate of heat transfer and the power requirement. The heat transfer for the reversible isothermal compression process is given by

$$\dot{Q}_{12} = \dot{m}T_1 \Delta s_{12}$$

$$= \frac{(10^5 \text{ g/hr})}{39.9 \text{ g/(g mole)}} (300°\text{K}) \left(-57.0 \text{ joules/(g mole)(°K)}\right)$$

$$= -4.29 \times 10^7 \text{ joules/hr}$$

where the minus sign indicates that heat is removed from the gas.

When the changes in kinetic and potential energies are neglected, the steady-flow energy balance gives the power requirement as

$$\dot{W}_{12} = \dot{Q}_{12} + \dot{m}(h_1 - h_2)$$

$$= -4.29 \times 10^7 + \frac{10^5}{39.9} \times 1790$$

$$= -3.84 \times 10^7 \text{ joules/hr} = -10.7 \text{ kW}$$

where the minus sign indicates that work is done on the gas.

EXAMPLE 3-4. Nitrogen at a pressure of 250 atm and a temperature of 400°K expands reversibly and adiabatically in a turbine to an exhaust pressure of 5 atm. The flow rate is 1 kg/sec. Calculate the power output if nitrogen obeys the Redlich-Kwong equation of state. The constant-pressure specific heat of nitrogen at 1 atm is given by

$$c_p = 6.903 - 0.3753 \times 10^{-3} T + 1.930 \times 10^{-6} T^2 - 6.861 \times 10^{-9} T^3$$

where c_p is in cal/(g mole)(°K) and T in °K.

Solution.
The state points of this example are shown in Fig. 3-4.

For nitrogen the critical constants are $T_c = 126.2°$K and $p_c = 33.5$ atm. Substituting these constants into Eqs. (3-8a and b) leads to

$$a = 15.4 \times 10^6 \text{ (atm)(°K}^{1/2}\text{)(cm}^6\text{)/(g mole)}^2$$

$$b = 26.8 \text{ cm}^3/\text{(g mole)}$$

for the Redlich-Kwong equation for nitrogen.

From the Redlich-Kwong equation by a trial-and-error solution, at $p_1 = 250$ atm and $T_1 = 400°$K we obtain

$$v_1 = 143 \text{ cm}^3/\text{(g mole)}$$

Since state 4 is at a low pressure, the ideal gas equation gives

$$v_4 = \frac{RT_4}{p_4} = \frac{82.06 \times 400}{1} = 32,800 \text{ cm}^3/\text{(g mole)}$$

For the reversible and adiabatic process from state 1 to state 2 we must have $\Delta s_{12} = s_2 - s_1 = 0$. We now use the three-step path 1-4-3-2 as depicted in Fig. 3-4 for the evaluation of Δs_{12}. Accordingly we write

$$\Delta s_{12} = \Delta s_{14} + \Delta s_{43} + \Delta s_{32}$$

As in the preceding example, we have

$$\Delta s_{14} = s_4 - s_1$$

$$= \left\{ \int_{v_1}^{v_4} \left[\frac{R}{v - b} + \frac{a}{2T^{3/2} v(v + b)} \right] dv \right\}_T$$

$$= R \ln\left(\frac{v_4 - b}{v_1 - b} \right) - \frac{a}{2bT_1^{3/2}} \ln\left[\frac{(v_4 + b)/v_4}{(v_1 + b)/v_1} \right]$$

$$= 469 \ (\text{atm})(\text{cm}^3)/(\text{g mole})(^\circ K) = 11.35 \ \text{cal}/(\text{g mole})(^\circ K)$$

after substituting numerical values.

If the temperature at state 2 is known, the entropy changes Δs_{43} and Δs_{32} (in cal/g mole($^\circ$K)) can be calculated by the following two equations:

$$\Delta s_{43} = s_3 - s_4 = \int_{T_4 = T_1}^{T_3 = T_2} c_p \frac{dT}{T}$$

$$= \int_{T_1 = 400^\circ K}^{T_2} \left(\frac{6.903}{T} - 0.3753 \times 10^{-3} + 1.93 \times 10^{-6} T \right.$$

$$\left. -6.861 \times 10^{-9} T^2 \right) dT$$

$$= -6.903 \ln \frac{400}{T_2} + 0.3753 \times 10^{-3}(400 - T_2)$$

$$- 0.965 \times 10^{-6}(400^2 - T_2^2) + 2.287 \times 10^{-9}(400^3 - T_2^3)$$

and

$$\Delta s_{32} = s_2 - s_3$$

$$= (0.0242) \left\{ \int_{v_3}^{v_2} \left[\frac{R}{v - b} + \frac{a}{2T^{3/2} v(v + b)} \right] dv \right\}_T$$

$$= (0.0242) \left\{ 82.06 \ln\left(\frac{v_2 - 26.8}{v_3 - 26.8} \right) - \frac{15.4 \times 10^6}{2 \times 26.8 T_2^{3/2}} \ln\left[\frac{(v_2 + 26.8)/v_2}{(v_3 + 26.8)/v_3} \right] \right\}$$

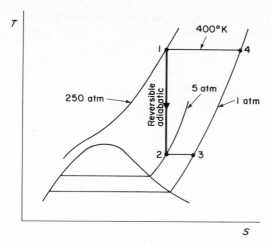

FIG. 3-4. Temperature-entropy diagram showing the state points of Example 3-4.

In order to find T_2 such that $\Delta s_{12} = 0$, a trial-and-error solution will be used. Assume a value of T_2, calculate v_2 by trial-and-error from the Redlich-Kwong equation, calculate v_3 by the ideal gas equation, and then use the preceding two equations to calculate Δs_{43} and Δs_{32}. If $\Delta s_{12} = \Delta s_{14} + \Delta s_{43} + \Delta s_{32} = 0$, the assumed value of T_2 is the correct one. The value of T_2 as determined by the ideal gas relations would be a good initial choice. After a few trials, the correct value of T_2 was determined to be $T_2 = 124°K$.

The following is a final check of the correctness of this assumed value of T_2.

At $p_2 = 5$ atm and $T_2 = 124°K$, the Redlich-Kwong equation gives

$v_2 = 1{,}920$ cm^3/(g mole)

At $p_3 = 1$ atm and $T_3 = T_2 = 124°K$, the ideal gas equation gives

$v_3 = 10{,}200$ cm^3/(g mole)

Whence

$\Delta s_{43} = -7.97$ cal/(g mole)(°K)

$\Delta s_{32} = -3.40$ cal/(g mole)(°K)

Hence

$\Delta s_{12} = 11.35 - 7.97 - 3.40 \approx 0$

within slide-rule accuracy. Therefore $T_2 = 124°K$ is the correct value.

With the value of T_2 known, we are now ready to calculate $\Delta h_{12} = h_2 - h_1$. We again use the three-step path 1-4-3-2 to yield $\Delta h_{12} = \Delta h_{14} + \Delta h_{43} + \Delta h_{32}$. As in Example 3-3, we have

$\Delta h_{14} = h_4 - h_1$

$$= (p_4 v_4 - p_1 v_1) - \frac{1.5a}{T_1^{1/2}} \frac{1}{b} \ln\left[\frac{(v_4 + b)/v_4}{(v_1 + b)/v_1}\right]$$

$$= 4,400 \ (\text{atm})(\text{cm}^3)/(\text{g mole}) = 446 \ \text{joules}/(\text{g mole})$$

and

$\Delta h_{32} = h_2 - h_3$

$$= (p_2 v_2 - p_3 v_3) - \frac{1.5a}{T_2^{1/2}} \frac{1}{b} \ln\left[\frac{(v_2 + b)/v_2}{(v_3 + b)/v_3}\right]$$

$$= -1,450 \ (\text{atm})(\text{cm}^3)/(\text{g mole}) = -147 \ \text{joules}/(\text{g mole})$$

The value of Δh_{43} is given by

$$\Delta h_{43} = h_3 - h_4 = \int_{T_4 = T_1}^{T_3 = T_2} c_p \, dT$$

$$= \int_{T_1 = 400^\circ \text{K}}^{T_2 = 124^\circ \text{K}} (6.903 - 0.3753 \times 10^{-3} T + 1.930 \times 10^{-6} T^2$$

$$- 6.861 \times 10^{-9} T^3) \, dT$$

$$= -1,890 \ \text{cal}/(\text{g mole}) = -7,910 \ \text{joules}/(\text{g mole})$$

Therefore

$$\Delta h_{12} = h_2 - h_1 = \Delta h_{14} + \Delta h_{43} + \Delta h_{32} = 446 - 7,910 - 147$$

$$= -7,610 \ \text{joules}/(\text{g mole})$$

Neglecting the changes in kinetic and potential energies, an energy balance for the adiabatic turbine gives

$$\dot{W}_{12} = \dot{m}(h_1 - h_2) = 1,000 \times \frac{7,610}{28}$$

$$= 2.72 \times 10^5 \ \text{joules}/\text{sec} = 272 \ \text{kW}$$

3-6 PRINCIPLE OF CORRESPONDING STATES

For an ideal gas $pv = RT$ under all conditions. For a real gas a correction factor is introduced so that

$$pv = ZRT \tag{3-17}$$

where Z is called the *compressibility factor*. It expresses the extent of deviation of the gas from an ideal gas. The value of Z is unity for an ideal gas under all conditions. For a real gas Z is a function of its state.

Values of the compressibility factor of any gas may be determined experimentally. There is, however, some generalized information which shall allow us to introduce a very great simplification in the evaluation of approximate values of compressibility factor. To do this let us first define three new variables: the *reduced pressure* p_r, *reduced temperature* T_r, and *reduced volume* v_r as follows:

$$p_r = \frac{p}{p_c} \qquad T_r = \frac{T}{T_c} \qquad v_r = \frac{v}{v_c} \tag{3-18}$$

where p_c, T_c, and v_c are the critical properties.

At the same pressure and temperature the specific or molal volumes of different gases are different. However, experience shows that at the same reduced pressure and reduced temperature, the reduced volumes of different gases are approximately the same. This experimental fact was first used by van der Waals who suggested that

$$v_r = f_1(p_r, T_r) \tag{3-19}$$

for all substances in the gaseous and liquid states. This is known as the van der Waals *principle of corresponding states*.

From the definitions of v_r and Z we have

$$v_r = \frac{v}{v_c} = \frac{ZRTp_c}{Z_c RT_c p} = \frac{Z}{Z_c} \frac{T_r}{p_r} \tag{3-20}$$

where

$$Z_c = \frac{p_c v_c}{RT_c}$$

This is called the critical compressibility factor. From Eqs. (3-19) and (3-20) it follows that

$$Z = f_2(p_r, T_r, Z_c) \tag{3-21}$$

Experimental values of Z_c for most substances fall within a narrow range of from 0.20–0.30. Therefore, as a first approximation, Z_c may be considered as a universal constant so that Eq. (3-21) may be simplified to the form

$$Z = f_3(p_r, T_r) \tag{3-22}$$

This last expression is often called the modified principle of corresponding states.

According to Eq. (3-22), if we plot Z in terms of p_r and T_r, we can use a single graph for all gases. A graph of this kind is called a *generalized compressibility chart*. Figure 3-5 shows a simplified sketch of such a chart. Appendix Figure A-14 shows a general chart for all fluids based on Eq. (3-22) for reduced pressures up to 40. Improved accuracy has been obtained by using Z_c as a third correlating parameter according to Eq. (3-21). Appendix Figures A-15 and A-16 present two different charts for $Z_c = 0.29$ and $Z_c = 0.27$. These two charts are mainly for nonpolar and slightly polar substances. For highly polar substances, such as ammonia and water, it has been suggested that the reduced dipole moment would be a better third parameter.

It is known that the light gases helium and hydrogen do not correlate very well on a generalized compressibility chart. Better correlation may be obtained by redefining the reduced pressure and temperature for these gases as follows:

$$p_r = \frac{p}{p_c + 8} \quad \text{and} \quad T_r = \frac{T}{T_c + 8}$$

where pressure is in atmospheres and temperature is in $^\circ$K.

According to the principle of corresponding states, there is a single relationship between reduced coordinates which holds for all substances. We will now demonstrate the validity of this statement through the van der Waals equation of state. Upon substituting Eq. (3-18) into the van der Waals equation, we get

$$\left(p_r p_c + \frac{a}{v_r^2 v_c^2}\right)(v_r v_c - b) = RT_r T_c$$

But

$$p_c = \frac{a}{27b^2} \qquad T_c = \frac{8a}{27Rb} \qquad v_c = 3b$$

so that

$$\left(p_r + \frac{3}{v_r^2}\right)(3v_r - 1) = 8T_r \tag{3-23}$$

This is the *van der Waals equation in reduced form*. It contains no specific constants for individual gases and is in a sense a universal equation for all gases. Of course, like the van der Waals equation itself, this reduced equation is only an approximate equation for real gases.

EXAMPLE 3-5. Using a generalized compressibility chart, determine the molal volume of methane at 200 atm and 400°K.

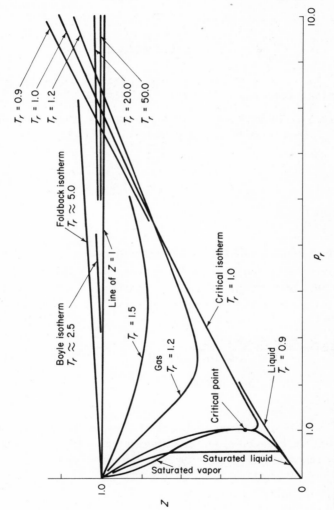

FIG. 3-5. Generalized compressibility chart.

Solution.

For methane, $p_c = 45.8$ atm, $T_c = 190.7°$K, and $Z_c = 0.290$. Hence,

$$p_r = \frac{p}{p_c} = \frac{200}{45.8} = 4.37 \qquad T_r = \frac{T}{T_c} = \frac{400}{190.7} = 2.10$$

From Appendix Fig. A-15 we get $Z = 0.98$, which yields

$$v = \frac{ZRT}{p} = \frac{0.98 \times 0.08206 \times 400}{200} = 0.161 \text{ liters/(g mole)}$$

Note that we obtained the same value of v in Example 3-2 by the use of the Benedict-Webb-Rubin equation of state.

EXAMPLE 3-6. Using a generalized compressibility chart, determine the pressure of carbon dioxide at a temperature of $160°$F and having a specific volume of 0.045 cu ft/lbm.

Solution.

For carbon dioxide, $p_c = 1{,}071$ psia, $T_c = 547.5°$R, and $Z_c = 0.275$. Hence,

$$T_r = \frac{T}{T_c} = \frac{619.7}{547.5} = 1.13$$

Upon substituting numerical values into the equation

$$p = \frac{ZRT}{v} \qquad \text{or} \qquad p_r p_c = \frac{ZRT}{v}$$

we obtain

$$p_r(1{,}071 \times 144) = \frac{Z \times 1{,}545 \times 619.7}{0.045 \times 44.01}$$

or

$$p_r = 3.13Z$$

Since $Z_c = 0.275$, the compressibility chart shown in Appendix Fig. A-16 is to be used. The equation $p_r = 3.13\,Z$ represents a straight line on this chart. The intersection of this line with the reduced isotherm of $T_r = 1.13$ gives the solution as $Z = 0.53$ and $p_r = 1.66$, from which

$$p = p_r p_c = 1.66 \times 1{,}071 = 1{,}780 \text{ psia}$$

Experimental data show that the preceding calculations are reasonably accurate. However, the ideal gas equation would give the following highly inaccurate value

$$p = \frac{RT}{v} = \frac{1{,}545 \times 619.7}{0.045 \times 44.01 \times 144} = 3{,}360 \text{ psia}$$

3-7 GENERALIZED THERMODYNAMIC PROPERTY CHARTS

The principle of corresponding states is useful not only to correlate compressibility data, but also to calculate other thermodynamic properties, such as enthalpy, entropy, and specific heats. We will present in this section the Hougen and Watson's generalized charts for thermodynamic properties of gases and liquids.[7] These charts show the nonideality corrections to various thermodynamic properties. In the following discussions the symbols marked "*" denote properties of ideal gases for which $pv = RT$, and the unmarked symbols denote the properties of real fluids.

The second dh equation is used to obtain the enthalpy correction.

$$dh = c_p \, dT + \left[v - T\left(\frac{\partial v}{\partial T}\right)_p \right] dp \tag{2-42}$$

Integrating this equation isothermally from $p = 0$ to some higher pressure p yields

$$(h^* - h)_T = \left\{ \int_0^p \left[T\left(\frac{\partial v}{\partial T}\right)_p - v \right] dp \right\}_T \tag{3-24}$$

where h^* is the enthalpy of the ideal gas, and h the enthalpy of the real fluid at the given elevated pressure p; both are at the same temperature T. Note that for ideal gases the enthalpy is a function of temperature only. From the equation of state

$$v = \frac{ZRT}{p}$$

and its partial derivative

$$\left(\frac{\partial v}{\partial T}\right)_p = \frac{RZ}{p} + \frac{RT}{p}\left(\frac{\partial Z}{\partial T}\right)_p \tag{3-25}$$

Eq. (3-24) may be rewritten as

$$(h^* - h)_T = \left\{ \int_0^p \left[\frac{RT^2}{p}\left(\frac{\partial Z}{\partial T}\right)_p \right] dp \right\}_T$$

---

[7] O. A. Hougen, K. M. Watson, and R. A. Ragatz, "Chemical Process Principles, part II, Thermodynamics," John Wiley and Sons, Inc., New York, 1959.

In terms of reduced coordinates this equation becomes

$$(h^* - h)_T = \left\{ \int_0^{p_r} \left[\frac{RT_r^2 T_c}{p_r} \left(\frac{\partial Z}{\partial T_r} \right)_{p_r} \right] dp_r \right\}_T$$

or

$$\left(\frac{h^* - h}{T_c} \right)_T = \left[RT_r^2 \int_0^{p_r} \left(\frac{\partial Z}{\partial T_r} \right)_{p_r} d(\ln p_r) \right]_T \qquad (3\text{-}26)$$

Values of the foregoing integral may be obtained by graphical integration, using data from a generalized compressibility chart. Equation (3-26) is the basis of the generalized enthalpy correction chart shown in Appendix Fig. A-17.

A procedure similar to the above is used to construct an entropy correction chart, utilizing the second ds equation.

$$ds = \frac{c_p}{T} dT - \left(\frac{\partial v}{\partial T} \right)_p dp \qquad (2\text{-}35)$$

Integrating this equation isothermally from $p = 0$ to some higher pressure p yields

$$(s_p - s_0^*)_T = - \left[\int_0^p \left(\frac{\partial v}{\partial T} \right)_p dp \right]_T \qquad (3\text{-}27)$$

where s_p is the entropy of the real fluid at p and T, and s_0^* is the entropy of the ideal gas at $p = 0$ and T. Because entropy is a function of both pressure and temperature even for ideal gases, the subscript 0 in the symbol s_0^* is necessary to indicate that p is equal to zero. Since the value of s_0^* is infinite, Eq. (3-27) is not directly useful. To get around this difficulty, Eq. (2-35) is integrated isothermally from $p = 0$ to the given higher pressure p, assuming that the ideal gas equation is satisfied at all times. Thus,

$$(s_p^* - s_0^*)_T = - \left[\int_0^p \left(\frac{\partial v}{\partial T} \right)_p dp \right]_T = - \left(R \int_0^p \frac{dp}{p} \right)_T \qquad (3\text{-}28)$$

where s_p^* is the entropy in the ideal gas state at p and T. Subtracting Eq. (3-27) from Eq. (3-28) yields

$$(s_p^* - s_p)_T = - \left\{ \int_0^p \left[\frac{R}{p} - \left(\frac{\partial v}{\partial T} \right)_p \right] dp \right\}_T$$

Substituting Eq. (3-25) into the preceding equation leads to

$$(s_p^* - s_p)_T = -\left\{ R \int_0^p \left[\frac{1 - Z}{p} - \frac{T}{p} \left(\frac{\partial Z}{\partial T} \right)_p \right] dp \right\}_T$$

In terms of reduced coordinates this equation becomes

$$(s_p^* - s_p)_T = -\left[R \int_0^{p_r} (1 - Z) \frac{dp_r}{p_r} \right]_T + \left[RT_r \int_0^{p_r} \left(\frac{\partial Z}{\partial T_r} \right)_{p_r} \frac{dp_r}{p_r} \right]_T$$

Comparing the last term of this equation with Eq. (3-26) reveals that the term can be written in terms of $(h^* - h)_T$. We accordingly write

$$(s_p^* - s_p)_T = -\left[R \int_0^{p_r} (1 - Z) d(\ln p_r) \right]_T + \left(\frac{h^* - h}{T_r T_c} \right)_T \tag{3-29}$$

Values of the integral in the first term on the right of Eq. (3-29) may be obtained through graphical integration by using data from a generalized compressibility chart. A generalized enthalpy correction chart may be used directly for the last term. Equation (3-29) is the basis of the generalized entropy correction chart shown in Appendix Fig. A-18.

A correction chart for the molal specific heat at constant pressure can be obtained by defining c_p and c_p^* for any substance when it is considered as a real fluid and as an ideal gas respectively, as follows:

$$c_p = \left(\frac{\partial h}{\partial T} \right)_p \quad \text{and} \quad c_p^* = \left(\frac{\partial h^*}{\partial T} \right)_p$$

Whence

$$(c_p - c_p^*)_T = \left[\frac{\partial}{\partial T} (h - h^*) \right]_p$$

In terms of reduced coordinates this equation becomes

$$(c_p - c_p^*)_T = \left[\frac{\partial}{\partial T_r} \left(\frac{h - h^*}{T_c} \right) \right]_p \tag{3-30}$$

where c_p and c_p^* are at the same temperature. Values of $(c_p - c_p^*)$ can be obtained by using data from a generalized enthalpy correction chart. Appendix Fig. A-19 shows a generalized c_p correction chart.

The generalized thermodynamic property charts shown in Figs. A-17 to 19 are for $Z_c = 0.27$. Corrections for Z_c other than 0.27 are available in the reference given in the beginning of this section. In the following examples we will use the charts without correction for Z_c.

EXAMPLE 3-7. Using generalized thermodynamic correction charts, estimate the values of enthalpy and entropy for methane at 200 atm and 400°K. The values of enthalpy and entropy at 1 atm and 400°K are 5512 cal/(g mole) and 47.16 cal/(g mole)(°K) respectively.

Solution.
Referring to Fig. 3-3 we have

$$p_A = 200 \text{ atm} \qquad p_a = 1 \text{ atm} \qquad T_A = T_a = 400°\text{K}$$

$$p_{rA} = \frac{p_A}{p_c} = \frac{200}{45.8} = 4.37$$

$$p_{ra} = \frac{p_a}{p_c} = \frac{1}{45.8} = 0.0218$$

$$T_{rA} = T_{ra} = \frac{T_A}{T_c} = \frac{400}{190.7} = 2.10$$

From the generalized enthalpy correction chart (Fig. A-17),

$$\frac{h_A^* - h_A}{T_c} = 1.78 \text{ cal/(g mole)(°K)} \qquad \text{at} \quad \begin{array}{l} T_r = 2.10 \\ p_r = 4.37 \end{array}$$

$$\frac{h_a^* - h_a}{T_c} = 0 \qquad \text{at} \quad \begin{array}{l} T_r = 2.10 \\ p_r = 0.0218 \end{array}$$

For an ideal gas the enthalpy is a function of temperature only; thus $h_A^* = h_a^*$. Then

$$\frac{h_A - h_a}{T_c} = \frac{h_a^* - h_a}{T_c} - \frac{h_A^* - h_A}{T_c} = -1.78 \text{ cal/(g mole)(°K)}$$

$$h_A - h_a = -1.78 \, T_c = -1.78 \times 190.7 = -340 \text{ cal/(g mole)}$$

Therefore

$$h_A = h_a - 340 = 5,512 - 340 = 5,172 \text{ cal/(g mole)}$$

From the generalized entropy correction chart (Fig. A-18) we obtain

$$(s_A^* - s_A) = 0.72 \text{ cal/(g mole)(°K)} \qquad \text{at} \quad \begin{array}{l} T_r = 2.10 \\ p_r = 4.37 \end{array}$$

$$(s_a^* - s_a) = 0 \qquad \text{at} \quad \begin{array}{l} T_r = 2.10 \\ p_r = 0.0218 \end{array}$$

From Eq. (2-53b)

$$ds = c_p \frac{dT}{T} - R\frac{dp}{p} = -R\frac{dp}{p}$$

for an isothermal process of an ideal gas. Thus

$$s_A^* - s_a^* = -R \ln \frac{p_A}{p_a} = -1.986 \ln 200 = -10.52 \text{ cal/(g mole)}(^\circ\text{K})$$

Then

$$(s_A - s_a) = (s_a^* - s_a) + (s_A^* - s_a^*) - (s_A^* - s_A)$$
$$= 0 - 10.52 - 0.72 = -11.24 \text{ cal/(g mole)}(^\circ\text{K})$$

Therefore

$$s_A = s_a - 11.24 = 47.16 - 11.24 = 35.92 \text{ cal/(g mole)}(^\circ\text{K})$$

Since the given conditions are the same in Examples 3-2 and 3-7, the reader should compare the two results.

EXAMPLE 3-8. A quantity of nitrogen gas contained in a piston-cylinder arrangement is initially at 200 atm and 200°K. Heat is added to the gas reversibly in such a way that the pressure of the gas remains constant until its temperature reaches 300°K. Determine the work and heat transfers per g mole of the gas by the use of generalized charts.

Solution.
For nitrogen, $p_c = 33.5$ atm, $T_c = 126.2°$K, and $Z_c = 0.291$. Hence

$$p_{r1} = p_{r2} = \frac{p_1}{p_c} = \frac{200}{33.5} = 5.97$$

$$T_{r1} = \frac{T_1}{T_c} = \frac{200}{126.2} = 1.58$$

$$T_{r2} = \frac{T_2}{T_c} = \frac{300}{126.2} = 2.38$$

From Fig. A-15 we get

$$Z_1 = 0.90 \text{ at } \begin{array}{c} p_{r1} = 5.97 \\ T_{r1} = 1.58 \end{array} \qquad Z_2 = 1.05 \text{ at } \begin{array}{c} p_{r2} = 5.97 \\ T_{r2} = 2.38 \end{array}$$

which yield

$$v_1 = \frac{Z_1 R T_1}{p_1} = \frac{0.90 \times 0.08206 \times 200}{200} = 0.0739 \text{ liter/(g mole)}$$

$$v_2 = \frac{Z_2 R T_2}{p_2} = \frac{1.05 \times 0.08206 \times 300}{200} = 0.1292 \text{ liter/(g mole)}$$

Therefore the work done by the gas is

$$w_{12} = \int_1^2 p \, dv = p_1(v_2 - v_1) = 200(0.1292 - 0.0739)$$

$$= 11.06 \text{ (liters)(atm)/(g mole)} = 267.7 \text{ cal/(g mole)}$$

From Fig. A-17 we get

$$\frac{h_1^* - h_1}{T_c} = 3.65 \text{ cal/(g mole)(}^\circ\text{K)} \quad \text{at} \quad \begin{aligned} p_{r1} &= 5.97 \\ T_{r1} &= 1.58 \end{aligned}$$

$$\frac{h_2^* - h_2}{T_c} = 1.65 \text{ cal/(g mole)(}^\circ\text{K)} \quad \text{at} \quad \begin{aligned} p_{r2} &= 5.97 \\ T_{r2} &= 2.38 \end{aligned}$$

from which

$$h_1^* - h_1 = 3.65 T_c = 3.65 \times 126.2 = 460.6 \text{ cal/(g mole)}$$

$$h_2^* - h_2 = 1.65 T_c = 1.65 \times 126.2 = 208.2 \text{ cal/(g mole)}$$

The ideal gas state c_p for nitrogen in the temperature range of this problem is fairly constant. Thus the ideal gas state enthalpy change $(h_2^* - h_1^*)$ is simply equal to this constant value of c_p times $(T_2 - T_1)$. Nevertheless for added accuracy let us use Keenan and Kaye's gas tables.[8] From the table for nitrogen, we read

$$h_2^* - h_1^* = 3750.3 - 2498.9 = 1251.4 \text{ Btu/(lb mole)} = 695.2 \text{ cal/(g mole)}$$

Therefore

$$(h_2 - h_1) = (h_1^* - h_1) + (h_2^* - h_1^*) - (h_2^* - h_2)$$

$$= 460.6 + 695.2 - 208.2 = 947.6 \text{ cal/(g mole)}$$

Then, according to the first law, the heat added to the gas is

$$q_{12} = (u_2 - u_1) + w_{12} = (u_2 - u_1) + (p_2 v_2 - p_1 v_1)$$

$$= h_2 - h_1 = 947.6 \text{ cal/(g mole)}$$

Note that if the gas is assumed to behave as an ideal gas the results would be

$$v_1 = \frac{RT_1}{p_1} = \frac{0.08206 \times 200}{200} = 0.08206 \text{ liter/(g mole)}$$

$$v_2 = \frac{RT_2}{p_2} = \frac{0.08206 \times 300}{200} = 0.1231 \text{ liter/(g mole)}$$

---

[8] J. H. Keenan and J. Kaye, "Gas Tables," John Wiley and Sons, Inc., New York, 1948.

$w_{12} = p_1(v_2 - v_1) = 200(0.1231 - 0.08206)$

$\qquad = 8.208 \text{ (liters)(atm)/(g mole)} = 198.6 \text{ cal/(g mole)}$

$q_{12} = h_2 - h_1 = h_2^* - h_1^* = 695.2 \text{ cal/(g mole)}$

which are, of course, all erroneous.

3-8 GENERAL THERMODYNAMIC CONSIDERATIONS ON EQUATIONS OF STATE

There are certain general characteristic behaviors common to all fluids. These common characteristics must be clearly observed in the developing and testing of an equation of state. It is instructive to discuss briefly some of the more important ones at this point.

(1) First of all, any equation of state should reduce to the ideal gas equation as pressure approaches zero at any temperature. This is clearly shown in a generalized compressibility chart in which all isotherms converge to the point where $Z = 1$ at zero pressure. Expressed mathematically,

$$\lim_{p \to 0} \left(\frac{pv}{RT} \right) = 1 \qquad \text{at any temperature}$$

Also, as seen from Fig. 3-5, or from the top part of Appendix Fig. A-15, the reduced isotherms approach the line $Z = 1$ as the temperature approaches infinite, or

$$\lim_{T \to \infty} \left(\frac{pv}{RT} \right) = 1 \qquad \text{at any pressure}$$

(2) The critical isotherm, i.e., the curve relating pressure and volume at the critical temperature, of an equation of state should have a point of inflection at the critical point or, mathematically,

$$\left(\frac{\partial p}{\partial v} \right)_{T_c} = 0 \quad \text{and} \quad \left(\frac{\partial^2 p}{\partial v^2} \right)_{T_c} = 0$$

(3) As demonstrated in Fig. 3-6, the isometrics of an equation of state on a p-T diagram should be essentially straight except at high densities and at low temperatures. In other words all isometrics should approach straight lines with either decreasing density or increasing temperature or, expressed mathematically,

$$\left(\frac{\partial^2 p}{\partial T^2} \right)_v = 0 \quad \text{as } p \to 0 \qquad \text{and} \qquad \left(\frac{\partial^2 p}{\partial T^2} \right)_v = 0 \quad \text{as } T \to \infty$$

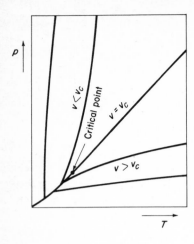

FIG. 3-6. Pressure-temperature diagram with isometric lines.

Also, the critical isometric should be a straight line. That is,

$$\left(\frac{\partial^2 p}{\partial T^2}\right)_v = 0 \quad \text{at} \quad v = v_c$$

An equation of state is expected to predict the slope of the critical isometric of the fluid. It is noted that this slope is identical with the slope of the vapor-pressure curve at the critical point. This equality in slope can be shown analytically from the Clapeyron equation, which gives the slope of the vapor-pressure curve at any temperature as $dp/dT = \Delta s/\Delta v$. At the critical point this equation becomes $dp/dT = (\partial s/\partial v)_{T_c}$. But when the third Maxwell equation is applied at the critical point, we have $(\partial s/\partial v)_{T_c} = (\partial p/\partial T)_{v_c}$. Therefore, the vapor-pressure slope at critical point dp/dT is equal to the slope of critical isometric $(\partial p/\partial T)_{v_c}$.

(4) The slopes of the isotherms of an equation of state on a Z-p compressibility plot as p approaches zero should be negative at lower temperatures and positive at higher temperatures. At a particular temperature called the Boyle temperature (Fig. 3-5), the slope is zero as p approaches zero or, expressed mathematically,

$$\lim_{p \to 0} \left(\frac{\partial Z}{\partial p}\right)_T = 0$$

at the Boyle temperature. An equation of state should predict the Boyle temperature, which is about $2.5\ T_c$ for many fluids.

An isotherm of maximum slope on the Z-p plot as p approaches zero should also be predicted. This isotherm, sometimes called the foldback isotherm

(Fig. 3-5), is about $5\ T_c$ for many fluids. Expressed mathematically,

$$\lim_{p\to 0}\left(\frac{\partial^2 Z}{\partial T\,\partial p}\right) = 0$$

at the foldback temperature. As temperature increases beyond the foldback temperature the slope of the isotherm decreases but always remains positive.

(5) An equation of state should predict the Joule-Thomson coefficient, which is

$$\mu_J = \frac{1}{c_p}\left[T\left(\frac{\partial v}{\partial T}\right)_p - v\right] = \frac{RT^2}{pc_p}\left(\frac{\partial Z}{\partial T}\right)_p \tag{2-46}$$

In particular, it should predict the inversion state, for which $\mu_J = 0$ or

$$T\left(\frac{\partial v}{\partial T}\right)_p - v = 0 \quad\text{or}\quad \left(\frac{\partial Z}{\partial T}\right)_p = 0$$

EXAMPLE 3-9. Based on general thermodynamic considerations test the Beattie-Bridgeman equation of state.

Solution.
The Beattie-Bridgeman equation can be rearranged in virial form to read

$$pv = RT\left(1 + \frac{\beta}{v} + \frac{\gamma}{v^2} + \frac{\delta}{v^3}\right) \tag{3-31}$$

or

$$Z = \frac{pv}{RT} = 1 + \frac{\beta}{v} + \frac{\gamma}{v^2} + \frac{\delta}{v^3} \tag{3-31a}$$

where

$$\beta = -\frac{A_0}{RT} + B_0 - \frac{c}{T^3} \qquad \gamma = \frac{aA_0}{RT} - bB_0 - \frac{cB_0}{T^3} \qquad \delta = \frac{bcB_0}{T^3}$$

The Beattie-Bridgeman equation can also be written in terms of a series in pressure. Thus

$$Z = \frac{pv}{RT} = 1 + \beta'p + \gamma'p^2 + \cdots \tag{3-32}$$

where

$$\beta' = \frac{\beta}{RT} \qquad \gamma' = \frac{\gamma - \beta^2}{(RT)^2}$$

Differentiating Eq. (3-31) we obtain

$$\left(\frac{\partial p}{\partial v}\right)_T = -\frac{RT}{v^2} - \frac{2\beta RT}{v^3} - \frac{3\gamma RT}{v^4} - \frac{4\delta RT}{v^5} \tag{3-33a}$$

$$\left(\frac{\partial^2 p}{\partial v^2}\right)_T = \frac{2RT}{v^3} + \frac{6\beta RT}{v^4} + \frac{12\gamma RT}{v^5} + \frac{20\delta RT}{v^6} \tag{3-33b}$$

$$\left(\frac{\partial p}{\partial T}\right)_v = \frac{R}{v} + \frac{1}{v^2}\left(B_0 R + \frac{2cR}{T^3}\right) - \frac{1}{v^3}\left(bB_0 R - \frac{2cB_0 R}{T^3}\right) - \frac{1}{v^4}\left(\frac{2bcB_0 R}{T^3}\right) \tag{3-33c}$$

$$\left(\frac{\partial^2 p}{\partial T^2}\right)_v = -\frac{1}{v^2}\left(\frac{6cR}{T^4}\right) - \frac{1}{v^3}\left(\frac{6cB_0 R}{T^4}\right) + \frac{1}{v^4}\left(\frac{6bcB_0 R}{T^4}\right) \tag{3-33d}$$

(1) As $p \to 0$, $v \to \infty$, Eqs. (3-31) to (3-32) reduce directly to the ideal gas equation $pv = RT$. Furthermore, it can be shown from Eq. (3-32) that pv/RT approaches unity as T approaches infinity.

(2) From Eqs. (3-33a) and (3-33b), when we set

$$\left(\frac{\partial p}{\partial v}\right)_T = 0 \quad \text{and} \quad \left(\frac{\partial^2 p}{\partial v^2}\right)_T = 0$$

we obtain two simultaneous equations which would give a unique solution for temperature and volume, which, in turn, would give a unique pressure for the critical point.

(3) From Eq. (3-33d) it can be readily shown that the derivative $(\partial^2 p/\partial T^2)_v$ is a small number for most conditions of v and T, approaching zero as $p \to 0$ and $v \to \infty$, or as $T \to \infty$.

(4) Since at low pressures only the first and second virials are of significance, thus from Eq. (3-32) we have

$$\lim_{p \to 0} \left(\frac{\partial Z}{\partial p}\right)_T = \beta' = \frac{\beta}{RT} = \frac{1}{RT}\left(-\frac{A_0}{RT} + B_0 - \frac{c}{T^3}\right)$$

At the Boyle temperature we then have

$$-\frac{A_0}{RT} + B_0 - \frac{c}{T^3} = 0$$

The real root of this equation is the Boyle temperature.

Also at low pressures we have

$$\lim_{p \to 0} \left(\frac{\partial^2 Z}{\partial T \partial p}\right) = \frac{d\beta'}{dT} = \frac{d}{dT}\left(\frac{\beta}{RT}\right) = \frac{1}{(RT)^2}\left(\frac{2A_0}{T} - B_0 R + \frac{4cR}{T^3}\right)$$

The foldback temperature is given by the real root of the equation

$$\frac{2A_0}{T} - B_0 R + \frac{4cR}{T^3} = 0$$

(5) The Joule-Thomson coefficients at low pressures can be found by applying Eq. (3-32) to Eq. (2-46). Thus

$$\mu_J = \frac{RT^2}{pc_p}\left(\frac{\partial Z}{\partial T}\right)_p = \frac{RT^2}{c_p}\left(\frac{d\beta'}{dT} + p\frac{d\gamma'}{dT}\right) = \frac{RT^2}{c_p}\left[\frac{d}{dT}\left(\frac{\beta}{RT}\right) + p\frac{d}{dT}\left(\frac{\gamma - \beta^2}{R^2T^2}\right)\right]$$

The maximum inversion temperature can be found by the expression

$$\lim_{p \to 0} \mu_J = \frac{RT^2}{c_p}\frac{d}{dT}\left(\frac{\beta}{RT}\right) = \frac{1}{Rc_p}\left(\frac{2A_0}{T} - B_0R + \frac{4cR}{T^3}\right) = 0$$

Thus we see that the maximum inversion temperature is the same as the foldback temperature discussed in the preceding paragraph.

PROBLEMS

3-1. Besides the van der Waals equation, there are a few well-known two-constant equations of state as given below. For each of these equations verify the expressions for the specific constants a and b in terms of the critical pressure p_c and the critical temperature T_c.

(a) Dieterici equation (proposed in 1899)

$$p = \frac{RT}{v - b}e^{-a/RTv}$$

for which

$$a = \frac{4R^2T_c^2}{p_ce^2} \quad \text{and} \quad b = \frac{RT_c}{p_ce^2}$$

(b) Berthelot equation (proposed in 1899)

$$\left(p + \frac{a}{Tv^2}\right)(v - b) = RT$$

for which

$$a = \frac{27}{64}\frac{R^2T_c^3}{p_c} \quad \text{and} \quad b = \frac{RT_c}{8p_c}$$

(c) Redlich-Kwong equation (proposed in 1949). (See Eq. (3-7) for the equation of state and Eqs. (3-8a and b) for the constants a and b.)

3-2. The Clausius equation (proposed in 1880) is given as

$$\left[p + \frac{a}{T(v + c)^2}\right](v - b) = RT$$

Verify the following expressions for the specific constants, a, b, and c in terms of the critical properties p_c, T_c, and v_c:

$$a = \frac{27R^2 T_c^3}{64p_c} \qquad b = v_c - \frac{RT_c}{4p_c} \qquad c = \frac{3RT_c}{8p_c} - v_c$$

3-3. Determine the critical compressibility factor $Z_c = p_c v_c / RT_c$ for a gas obeying (a) Dieterici equation, (b) Berthelot equation, and (c) Redlich-Kwong equation. (See Problem 3-1 for the equations.)

3-4. Derive the virial form of the Redlich-Kwong equation of state.

3-5. Show that the Redlich-Kwong equation in reduced form is

$$p_r = \frac{T_r}{Z_c(v_r - 0.08664/Z_c)} - \frac{0.42748}{T_r^{0.5} v_r Z_c^2 (v_r + 0.08664/Z_c)}$$

3-6. Write the Dieterici equation $p = RT(v - b)^{-1} e^{-a/RTv}$ in reduced form. Use this reduced equation to find the pressure dependence of the inversion point of the Joule-Thomson effect.

3-7. To produce liquid oxygen it is desired that the gas be first compressed and cooled to a pressure of 100 atm and a temperature of $-90°$C. The original oxygen gas is at a pressure of 1 atm and a temperature of $22°$C. Calculate the volume (in m^3) of the compressed gas from $100\ m^3$ of the original gas

(a) by using the ideal gas equation

(b) by using a compressibility chart.

3-8. Set up a computer program to calculate the enthalpy h and entropy s of a gas at various pressure p and temperature T, using an equation of state in the virial form

$$\frac{pv}{RT} = 1 + \frac{B(T)}{v} + \frac{C(T)}{v^2} + \frac{D(T)}{v^3}$$

and a constant-pressure specific heat equation

$$c_p = \alpha + \beta T + \gamma T^2 + \delta T^3$$

Use a reference point on the saturated vapor line.

3-9. Using the Martin-Hou equation of state, calculate the values of v (volume), h (enthalpy), s (entropy), u (internal energy), g (Gibbs function), a (Helmholtz function), c_p (constant-pressure specific heat), and c_v (constant-volume specific heat) for one g mole of nitrogen at $300°$K and 300 atm. Given that at $300°$K and 1 atm, $h = 15,590$ joules/(g mole), $s = 191.4$ joules/(g mole) $(°$K), and $c_p = 28.98$ joules/(g mole)$(°$K), where the values of h and s are based on the value of zero for the perfect crystal at $0°$K. Use a computer to carry out the necessary calculations. Give the computer program.

3-10. Calculate the values of v, h, s, c_p, and c_v for carbon monoxide at a pressure of 100 atm and a temperature of $400°$K by the use of the Berthelot equation of state. (See Problem 3-1). Given that at $p_0 = 0.1$ atm and $T_0 = 273°$K, $h_0 = 3834$ cal/(g mole) and $s_0 = 51.23$ cal/(g mole)($°$K) based on the value of zero for the perfect crystal at the absolute zero of temperature. The ideal gas state constant-pressure specific heat for carbon monoxide may be expressed as $c_p = 6.726 + 0.4001 \times 10^{-3}\ T + 1.283 \times 10^{-6}\ T^2 - 0.5307 \times 10^{-9}\ T^3$ where c_p is in cal/(g mole)($°$K) and T in $°$K.

3-11. It is assumed that the following data on steam are available from experimental measurements. From these data, determine for one pound of steam the entropy and enthalpy at a pressure of 550 psia and a temperature of $1,500°$F.

(1) The entropy and enthalpy for saturated liquid at $32.018°$F will be assumed to be zero.

(2) The average c_p for the saturated liquid between $32°$F and $100°$F is approximately 1 Btu/(lbm)($°$F).

(3) The saturation pressure at $100°$F is 0.9503 psia.

(4) The latent heat of vaporization at $100°$F is 1037.0 Btu/lbm.

(5) The c_p values of superheated steam at low pressures are those shown by the curve designated by zero pressure in Fig. 4, p. 122, of Keenan-Keyes-Hill-Moore Steam Tables.[9]

(6) The values of specific volume of superheated steam in the required pressure and temperature ranges are those given in the Steam Tables. (Use pressures, in psia, of 1, 2, 3, 4, 6, 8, 10, 20, 40, 80, 100, 150, 250, 300, 400, 500, 600, 800, and 1,000, and have each at temperatures, in $°$F, of 600, 700, 800, 900, 1,000, 1,100, 1,200, 1,300, 1,400, 1,500, and 1,600.)

3-12. Argon is compressed reversibly and isothermally in a closed system from 1 atm and $300°$K to 120 atm. Determine the work required and the heat transfer per kg of the gas, assuming that argon obeys the Beattie equation of state, which is

$$v = \frac{RT}{p} + \frac{\beta}{RT} + \frac{\gamma}{R^2 T^2}p + \frac{\delta}{R^3 T^3}p^2$$

where

$$\beta = RTB_0 - A_0 - Rc/T^2$$

$$\gamma = -RTB_0 b + A_0 a - RB_0 c/T^2$$

$$\delta = RB_0 bc/T^2$$

in which the constants A_0, B_0, a, b, and c are those of the Beattie-Bridgeman equation as given in Table 3-2.

[9] J. H. Keenan, F. G. Keyes, P. G. Hill, and J. G. Moore, "Steam Tables (English Units)," John Wiley and Sons, Inc., New York, 1969.

3-13. Carbon dioxide expands in a reversible isothermal steady-flow machine from a pressure of 50 atm and a temperature of 100°C to an exhaust pressure of 1 atm. The volume flow rate at the inlet condition is 0.01 m³/sec. Calculate the power output and the rate of heat transfer assuming

    (a) the gas obeys the ideal gas equation

    (b) the gas obeys the Dieterici equation.

3-14. Two kilograms of ammonia vapor at a pressure of 5 atm and a temperature of 300°K are compressed in a closed system reversibly and adiabatically to 50 atm. Calculate the final temperature, the final volume, and the work required if

    (a) the ideal gas equation is obeyed

    (b) the van der Waals equation is obeyed.

Assume $c_v = 8.2$ cal/(g mole)(°K) for both cases. (Hint: The expressions shown in Example 2-4 may be used.)

3-15. Carbon dioxide at a pressure of 800 psia and a temperature of 100°F expands reversibly and adiabatically in a nozzle to a pressure of 500 psia. Calculate the outlet velocity, assuming the inlet velocity to be negligible. The constant-pressure specific heat of carbon dioxide at 14.7 psia may be expressed as

$$c_p = 6.85 + 0.00474T - 7.64 \times 10^{-7} T^2$$

where c_p is in Btu/(lb mole)(°R) and T in °R. Use

    (a) the van der Waals equation

    (b) the Benedict-Webb-Rubin equation.

Use a computer to perform the desired calculations. Give the computer programs.

3-16. Carbon monoxide expands adiabatically in a steady-flow machine from 200 atm, 150°C to 10 atm, 0°C. The constant-pressure specific heat of carbon monoxide at 1 atm may be expressed as

$$c_p = 6.480 + 1.566 \times 10^{-3} T - 0.2387 \times 10^{-6} T^2$$

where c_p is in kcal/(kg mole)(°K) and T in °K. Using the Clausius equation of state (Problem 3-2), determine the work done and the change in entropy per kg of the gas.

3-17. Solve the preceding problem by the use of generalized thermodynamic charts.

3-18. Propane gas at 30 atm and 95°C enters a pipeline at a velocity of 20 m/sec. The pipe is of constant cross section and is perfectly heat insulated. The gas leaves the pipe at a pressure of 8 atm. Use generalized charts to find the temperature and velocity of the gas leaving the pipe.

Chapter 4
MULTICOMPONENT SYSTEMS

4-1 FUNDAMENTAL EQUATIONS

So far our discussions have been limited mainly to single-component simple compressible systems of fixed mass. In this chapter we will extend the treatment to multicomponent simple compressible systems of variable composition. Here we will study only systems where variations in composition are due to the addition or removal of matter, leaving the consideration of chemical reactions to a later chapter.

According to the state principle, two independent properties are required to specify the thermodynamic state of a simple compressible system of fixed mass and composition. However, for a system of variable composition, a specification of the composition is also required. Thus, if entropy and volume are selected as the independent variables, the energy of a closed system of fixed composition is

$$U = U(S,V)$$

In the case of an open system of variable composition we must express U as a function also of the amounts, e.g., the numbers of moles $n_1, n_2, \ldots, n_r$ of the r different components. Thus

$$U = U(S,V,n_1,n_2,\ldots,n_r)$$

The total differential of U is then

$$dU = \left(\frac{\partial U}{\partial S}\right)_{V,n} dS + \left(\frac{\partial U}{\partial V}\right)_{S,n} dV + \sum_{i=1}^{r} \left(\frac{\partial U}{\partial n_i}\right)_{S,V,n_j(j \neq i)} dn_i$$

in which the subscript n implies constancy of all the n's, i.e., constant composition, and the subscript $n_j (j \neq i)$ implies constancy of all other n's except n_i. In the light of Eq. (2-12a), when all the n's are held constant during the differentiation, we have

$$\left(\frac{\partial U}{\partial S}\right)_{V,n} = T \quad \text{and} \quad \left(\frac{\partial U}{\partial V}\right)_{S,n} = -P \tag{4-1}$$

Let us now introduce a new property μ_i to be called the *chemical potential* of component i with the definition

$$\mu_i = \left(\frac{\partial U}{\partial n_i}\right)_{S,V,n_j\ (j \neq i)} \tag{4-2}$$

Accordingly, the total differential of U becomes

$$dU = T\,dS - p\,dV + \sum_{i=1}^{r} \mu_i\,dn_i \tag{4-3}$$

From this equation and the definitions of enthalpy H, Helmholtz function A, and Gibbs function G for simple compressible systems, we obtain

$$dH = T\,dS + V\,dp + \sum_{i=1}^{r} \mu_i\,dn_i \tag{4-4}$$

$$dA = -S\,dT - p\,dV + \sum_{i=1}^{r} \mu_i\,dn_i \tag{4-5}$$

$$dG = -S\,dT + V\,dp + \sum_{i=1}^{r} \mu_i\,dn_i \tag{4-6}$$

Equations (4-3) to (4-6) are the fundamental relations for a homogeneous simple compressible open system for changes between neighboring equilibrium states. They are known as the *Gibbs equations*. From these equations, it follows that μ_i can be related to $U, H, A,$ and G by the expressions

$$
\begin{aligned}
\mu_i &= \left(\frac{\partial U}{\partial n_i}\right)_{S,V,n_j\ (j \neq i)} \\[6pt]
&= \left(\frac{\partial H}{\partial n_i}\right)_{S,p,n_j\ (j \neq i)} \\[6pt]
&= \left(\frac{\partial A}{\partial n_i}\right)_{V,T,n_j\ (j \neq i)} \\[6pt]
&= \left(\frac{\partial G}{\partial n_i}\right)_{T,p,n_j\ (j \neq i)}
\end{aligned}
\tag{4-7}
$$

The last of these expressions is sometimes used alone as the definition of the chemical potential μ_i.

Equations (4-3) to (4-6) can be manipulated to obtain many useful relations. Thus, since dG is an exact differential, it follows from Eq. (4-6) that

$$\left(\frac{\partial \mu_i}{\partial T}\right)_{p,n} = -\left(\frac{\partial S}{\partial n_i}\right)_{T,p,n_j\,(j \neq i)} \tag{4-6a}$$

$$\left(\frac{\partial \mu_i}{\partial p}\right)_{T,n} = \left(\frac{\partial V}{\partial n_i}\right)_{T,p,n_j\,(j \neq i)} \tag{4-6b}$$

$$\left(\frac{\partial \mu_i}{\partial n_k}\right)_{T,p,n_j\,(j \neq k)} = \left(\frac{\partial \mu_k}{\partial n_i}\right)_{T,p,n_j\,(j \neq i)} \tag{4-6c}$$

Similarly, from Eq. (4-5) we obtain

$$\left(\frac{\partial \mu_i}{\partial T}\right)_{V,n} = -\left(\frac{\partial S}{\partial n_i}\right)_{T,V,n_j\,(j \neq i)}$$

$$\left(\frac{\partial \mu_i}{\partial V}\right)_{T,n} = -\left(\frac{\partial p}{\partial n_i}\right)_{T,V,n_j\,(j \neq i)}$$

$$\left(\frac{\partial \mu_i}{\partial n_k}\right)_{T,V,n_j\,(j \neq k)} = \left(\frac{\partial \mu_k}{\partial n_i}\right)_{T,V,n_j\,(j \neq i)}$$

Before returning to the study of fundamental thermodynamic relations we need to state and prove the mathematical theorem known as Euler's theorem on homogeneous functions. This theorem is of great use in thermodynamics. A function $f(z_1, z_2, \ldots, z_r)$ is said to be homogeneous of degree m in the z's if

$$f(\lambda z_1, \lambda z_2, \ldots, \lambda z_r) = \lambda^m f(z_1, z_2, \ldots, z_r)$$

where λ is a constant. *Euler's theorem* states that a homogeneous function of degree m obeys the equation

$$mf(z_1, z_2, \ldots, z_r) = \sum_i z_i \left(\frac{\partial f}{\partial z_i}\right)_{z_j\,(j \neq i)} \tag{4-8}$$

The proof of the theorem is straightforward upon defining the set of variables

$$y_1 = \lambda z_1, y_2 = \lambda z_2, \ldots, y_r = \lambda z_r$$

Whence

$$f(y_1, y_2, \ldots, y_r) = \lambda^m f(z_1, z_2, \ldots, z_r)$$

Differentiating this equation with respect to λ at constant z's results in

$$\sum_i \left(\frac{\partial f}{\partial y_i}\right)_{y_j\,(j \neq i)} (z_i) = m\lambda^{m-1} f(z_1, z_2, \ldots, z_r)$$

This equation holds for all values of λ. If λ is set equal to 1, then $y_i = z_i$ and Eq. (4-8) follows.

Now we observe that U is homogeneous of first degree in S, V, n_i. Applying Euler's theorem results in

$$U = \left(\frac{\partial U}{\partial S}\right)_{V,n} S + \left(\frac{\partial U}{\partial V}\right)_{S,n} V + \sum_{i=1}^{r} \left(\frac{\partial U}{\partial n_i}\right)_{S,V,n_j\,(j \neq i)} n_i$$

Inserting of Eqs. (4-1) and (4-2) in this equation gives

$$U = TS - pV + \sum_{i=1}^{r} \mu_i n_i \tag{4-9}$$

Since by definition $G = U + pV - TS$, the preceding equation is equivalent to

$$G = \sum_{i=1}^{r} \mu_i n_i \tag{4-10}$$

Differentiating this equation gives

$$dG = \sum_{i=1}^{r} \mu_i \, dn_i + \sum_{i=1}^{r} n_i \, d\mu_i$$

Subtracting Eq. (4-6) from the preceding equation yields

$$S\,dT - V\,dp + \sum_{i=1}^{r} n_i \, d\mu_i = 0 \tag{4-11}$$

This is the *Gibbs-Duhem equation*. For changes at constant T and p it simplifies to

$$\sum_{i=1}^{r} n_i \, d\mu_i = 0 \tag{4-12}$$

It is worth noticing that for a single-component system Eq. (4-10) reduces to $G = \mu n$. Hence $\mu = G/n = g$ for a pure component.

It is often convenient to express the equations in terms of one mole of a system and to transform the independent variables from numbers of moles to mole fractions. The mole fraction x_i of component i is defined as the ratio of the number of moles of this component to the total number of moles in the system, or

$$x_i = \frac{n_i}{\displaystyle\sum_{k=1}^{r} n_k}$$

Since the mole fractions must satisfy the identity

$$\sum_{i=1}^{r} x_i = 1$$

the number of mole fractions that are independent is one less than the number of components present. It is usually convenient to choose x_1 as a dependent variable and the other x's as independent variables. As examples, we list the following equations in terms of x's, which have been converted from Eqs. (4-3), (4-11), and (4-12) respectively.

$$du = T\,ds - p\,dv + \sum_{i=2}^{r} \mu_i\,dx_i \tag{4-3a}$$

$$s\,dT - v\,dp + \sum_{i=1}^{r} x_i\,d\mu_i = 0 \tag{4-11a}$$

$$\sum_{i=1}^{r} x_i\,d\mu_i = 0 \qquad \text{at constant } T \text{ and } p \tag{4-12a}$$

where u, s, and v are molal properties.

4-2 PARTIAL MOLAL PROPERTIES

We recall from Eq. (4-7) that the chemical potential is defined by the derivative $(\partial G/\partial n_i)_{T,p,n_j(j \neq i)}$ which, as we will presently see, belongs to a group of properties called partial molal properties. These properties are of great significance in the treatment of multicomponent systems.

Consider any extensive property Y, such as volume, entropy, enthalpy, or Gibbs function which can be expressed as a function of T, p, and the n's. The total differential of Y is then

$$dY = \left(\frac{\partial Y}{\partial T}\right)_{p,n} dT + \left(\frac{\partial Y}{\partial p}\right)_{T,n} dp + \sum_{i=1}^{r} \left(\frac{\partial Y}{\partial n_i}\right)_{T,p,n_j\,(j \neq i)} dn_i$$

This simplifies at constant T and p to

$$dY_{T,p} = \sum_{i=1}^{r} \left(\frac{\partial Y}{\partial n_i}\right)_{T,p,n_j\,(j \neq i)} dn_i$$

We now introduce the *partial molal property* $\overline{Y}_i$ of component i with the definition

$$\overline{Y}_i = \left(\frac{\partial Y}{\partial n_i}\right)_{T,p,n_j \ (j \neq i)} \tag{4-13}$$

Whence we have

$$dY_{T,p} = \sum_{i=1}^{r} \overline{Y}_i \, dn_i \tag{4-14}$$

It is obvious that the partial molal property $\overline{Y}_i$ is an intensive property of the system. It depends not upon the total amount of each component, but only upon the composition, i.e., upon the relative amounts of the components.

Since Y is an extensive property, if all the components of the system are increased in the same proportion at constant T and p, then Y will also increase in the same proportion. Accordingly, if we let the infinitesimal change in n_i be represented by $dn_i = n_i d\lambda$ for all the components, then the corresponding change in Y is $dY_{T,p} = Y \, d\lambda$ where $d\lambda$ is the proportional factor. Inserting these equations in Eq. (4-14) gives

$$Y \, d\lambda = \sum_{i=1}^{r} \overline{Y}_i n_i \, d\lambda$$

Consequently

$$Y = \sum_{i=1}^{r} \overline{Y}_i n_i \tag{4-15}$$

We could, of course, also obtain this equation directly by applying Euler's theorem, because the extensive property Y is homogeneous of first degree in the n's at constant T and p. Equation (4-15) is the general relation between any extensive property and the corresponding partial molal property of a homogeneous system.

Differentiating the foregoing equation results in

$$dY = \sum_{i=1}^{r} \overline{Y}_i \, dn_i + \sum_{i=1}^{r} n_i \, d\overline{Y}_i$$

Comparing this equation with Eq. (4-14) shows that at constant T and p we have

$$\sum_{i=1}^{r} n_i \, d\overline{Y}_i = 0 \tag{4-16}$$

This equation is one of the most useful relations between partial molal properties.

From Eq. (4-16) we may immediately write

$$\sum_{i=1}^{r} n_i \frac{d\bar{Y}_i}{dn_k} = 0$$

Writing the explicit restriction of constant T and p we have

$$\sum_{i=1}^{r} n_i \left(\frac{\partial \bar{Y}_i}{\partial n_k}\right)_{T,p,n_j \, (j \neq k)} = 0 \tag{4-17}$$

It should be emphasized that this equation is valid only at constant T and p. Again we could, of course, obtain this equation by applying Euler's theorem when we observe that partial molal properties are homogeneous of zero degree in the n's at constant T and p.

For one mole of a system we may write the previous equations in terms of mole fractions. Thus

$$y = \frac{Y}{\sum\limits_{k=1}^{r} n_k} = \sum_{i=1}^{r} \bar{Y}_i x_i \tag{4-15a}$$

$$\sum_{i=1}^{r} x_i \, d\bar{Y}_i = 0 \qquad \text{at constant } T \text{ and } p \tag{4-16a}$$

$$\sum_{i=1}^{r} x_i \left(\frac{\partial \bar{Y}_i}{\partial x_k}\right)_{T,p} = 0 \tag{4-17a}$$

In writing a partial derivative one should note that while n_i can vary without any change in $n_j(j \neq i)$, one cannot change x_i without some changes in $x_j(j \neq i)$.

The preceding expressions concerning partial molal property $\bar{Y}_i$ are valid for any extensive property Y. For the Gibbs function, according to Eq. (4-13),

$$\bar{G}_i = \left(\frac{\partial G}{\partial n_i}\right)_{T,p,n_j \, (j \neq i)}$$

Comparing this equation with Eq. (4-7) reveals that

$$\bar{G}_i = \mu_i \tag{4-18}$$

By virtue of this relation, one should notice that Eqs. (4-10) and (4-12) are the particular forms of Eqs. (4-15) and (4-16) as applied to the Gibbs function.

It is worth noticing that the right-hand sides of Eqs. (4-6a) and (4-6b) in Sec. 4-1 are partial molal properties. Accordingly these two equations can be rewritten as

$$\left(\frac{\partial \mu_i}{\partial T}\right)_{p,n} = -\left(\frac{\partial S}{\partial n_i}\right)_{T,p,n_j \, (j \neq i)} = -\bar{S}_i \tag{4-19}$$

$$\left(\frac{\partial \mu_i}{\partial p}\right)_{T,n} = \left(\frac{\partial V}{\partial n_i}\right)_{T,p,n_j \, (j \neq i)} = \bar{V}_i \tag{4-20}$$

Thus the total differential of μ_i can be written as follows:

$$d\bar{G}_i = d\mu_i = -\bar{S}_i \, dT + \bar{V}_i \, dp + \sum_{k=1}^{r} \left(\frac{\partial \mu_i}{\partial n_k}\right)_{T,p,n_j \, (j \neq i)} dn_k \tag{4-21}$$

or

$$d\bar{G}_i = d\mu_i = -\bar{S}_i \, dT + \bar{V}_i \, dp + \sum_{k=2}^{r} \left(\frac{\partial \mu_i}{\partial x_k}\right)_{T,p} dx_k \tag{4-21a}$$

For a system of fixed composition we have

$$d\bar{G}_i = d\mu_i = -\bar{S}_i \, dT + \bar{V}_i \, dp \tag{4-22}$$

which is analogous to Eq. (2-11) for a single-component system.

Any relation among extensive thermodynamic properties that applies to a system as a whole can be converted to one among partial molal properties for a given component by differentiating the given relation with respect to the mole number of this component, and reversing the order of differentiation if the original relation involves derivatives. Thus, consider the relation

$$G = H - Ts$$

Differentiating with respect to n_i at constant T, p, and n_j gives

$$\left(\frac{\partial G}{\partial n_i}\right)_{T,p,n_j \, (j \neq i)} = \left(\frac{\partial H}{\partial n_i}\right)_{T,p,n_j \, (j \neq i)} - T\left(\frac{\partial S}{\partial n_i}\right)_{T,p,n_j \, (j \neq i)}$$

Whence we have the analogous relation

$$\bar{G}_i = \bar{H}_i - T\bar{S}_i \tag{4-23}$$

Similarly, analogous to the relations $H = U + pV$ and $A = U - TS$, we have

$$\bar{H}_i = \bar{U}_i + p\bar{V}_i \tag{4-24}$$

$$\bar{A}_i = \bar{U}_i - T\bar{S}_i \tag{4-25}$$

To derive the analogous expression to the following equation

$$\left(\frac{\partial H}{\partial p}\right)_T = V - T\left(\frac{\partial V}{\partial T}\right)_p$$

we first differentiate this equation with respect to n_i, keeping T, p, and n_j constant.

$$\left[\frac{\partial}{\partial n_i}\left(\frac{\partial H}{\partial p}\right)_{T,n}\right]_{T,p,n_j\,(j\neq i)} = \left(\frac{\partial V}{\partial n_i}\right)_{T,p,n_j\,(j\neq i)} - T\left[\frac{\partial}{\partial n_i}\left(\frac{\partial V}{\partial T}\right)_{p,n}\right]_{T,p,n_j\,(j\neq i)}$$

Next, by reversing the order of differentiation, we have

$$\left[\frac{\partial}{\partial p}\left(\frac{\partial H}{\partial n_i}\right)_{T,p,n_j\,(j\neq i)}\right]_{T,n} = \left(\frac{\partial V}{\partial n_i}\right)_{T,p,n_j\,(j\neq i)} - T\left[\frac{\partial}{\partial T}\left(\frac{\partial V}{\partial n_i}\right)_{T,p,n_j\,(j\neq i)}\right]_{p,n}$$

Whereupon we obtain the desired expression

$$\left(\frac{\partial \bar{H}_i}{\partial p}\right)_{T,n} = \bar{V}_i - T\left(\frac{\partial \bar{V}_i}{\partial T}\right)_{p,n} \tag{4-26}$$

in which the differentiations of $\bar{H}_i$ and $\bar{V}_i$ are carried out at constant composition.

As another example, let us derive the expression analogous to the defining equation of heat capacity at constant pressure.

$$C_p = \left(\frac{\partial H}{\partial T}\right)_p$$

Thus

$$\left(\frac{\partial C_p}{\partial n_i}\right)_{T,p,n_j\,(j\neq i)} = \left[\frac{\partial}{\partial n_i}\left(\frac{\partial H}{\partial T}\right)_{p,n}\right]_{T,p,n_j\,(j\neq i)}$$

or

$$\bar{C}_p = \left[\frac{\partial}{\partial T}\left(\frac{\partial H}{\partial n_i}\right)_{T,p,n_j\,(j\neq i)}\right]_{p,n} = \left(\frac{\partial \bar{H}_i}{\partial T}\right)_{p,n} \tag{4-27}$$

Our initial general treatment on partial molal properties would be incomplete if we neglect to observe that for a single-component system a partial molal property is the same as the corresponding molal property. Thus, for a pure compound Eq. (4-15) reduces to

$$Y = \bar{Y}n \quad \text{or} \quad \bar{Y} = \frac{Y}{n} = y$$

Therefore for a pure compound we must have

$$\bar{G} = \frac{G}{n} = g \qquad \bar{H} = \frac{H}{n} = h \qquad \bar{V} = \frac{V}{n} = v$$

EXAMPLE 4-1. . The partial molal volumes in cu ft/lb mole for a system of binary liquid of *n*-decane ($C_{10}H_{22}$) and carbon dioxide at $160°F$ and 5,000 psia are listed as a function of the mole fraction of carbon dioxide in the following table.

x_{CO_2}	0.2	0.3	0.4	0.5	0.6	0.7	0.8
$\bar{V}_{CO_2}$	0.764	0.773	0.785	0.801	0.837	0.859	0.864
$\bar{V}_{C_{10}H_{22}}$	3.141	3.138	3.133	3.119	3.076	3.028	3.016

Compute the total volume and molal volume of a solution containing 20 lbm of carbon dioxide and 80 lbm of *n*-decane.

Solution.
The molecular weight of CO_2 is 44.01 lbm/lb mole, and that of $C_{10}H_{22}$ is 142.3 lbm/lb mole. Then

$$n_{CO_2} = \frac{20}{44.01} = 0.454 \text{ lb mole}$$

$$n_{C_{10}H_{22}} = \frac{80}{142.3} = 0.562 \text{ lb mole}$$

$$x_{CO_2} = \frac{0.454}{0.454 + 0.562} = \frac{0.454}{1.016} = 0.447$$

$$x_{C_{10}H_{22}} = \frac{0.562}{1.016} = 0.553$$

From the given data a graph of $\bar{V}_{CO_2}$ and $\bar{V}_{C_{10}H_{22}}$ versus x_{CO_2} was prepared, from which the following data were obtained at $x_{CO_2} = 0.447$:

$$\bar{V}_{CO_2} = 0.792 \text{ cu ft/lb mole}$$

$$\bar{V}_{C_{10}H_{22}} = 3.128 \text{ cu ft/lb mole}$$

Substituting these values in Eq. (4-15) yields the total volume

$$V = n_{CO_2}\bar{V}_{CO_2} + n_{C_{10}H_{22}}\bar{V}_{C_{10}H_{22}} = 0.454 \times 0.792 + 0.562 \times 3.128$$

$$= 2.12 \text{ cu ft}$$

The molal volume of the solution is then

$$v = \frac{2.12}{1.016} = 2.09 \text{ cu ft/lb mole}$$

4-3 BINARY SYSTEMS

In order to obtain a somewhat more concrete idea of the partial molal properties and to illustrate the methods of calculating these properties, let us consider the special case of a binary system. We first deal with the volume of a binary system to visualize better the characteristics of partial molal properties.

Let us picture an experiment in which successive amounts of component 2 are added to a fixed amount of component 1 at constant T and p. The resulting volumes of the system are measured. Thus we have the data which shall enable us to write an equation relating the volume V of the system to the number of moles n_2 of component 2. Differentiating of this equation with respect to n_2 gives the value of $\bar{V}_2$ at any n_2. The value of $\bar{V}_2$ can also be obtained graphically by plotting V against n_2 and measuring the slope of the curve at the desired point.

In another graphical method, if we have data on molal volume v of a system tabulated against the mole fraction x_2 of component 2, a plot similar to curve BDB' in Fig. 4-1 can be prepared. We note that

length $(XD) = v$ = molal volume of the system
length $(OB) = v_1^0$ = molal volume of pure component 1
length $(O'B') = v_2^0$ = molal volume of pure component 2

Now, a tangent drawn to the curve at point D, where the mole fraction of component 2 is x_2, cuts the ordinate $x_1 = 1$ at C and the ordinate $x_2 = 1$ at C'. Then for the composition corresponding to point D we must have

length $(OC) = \bar{V}_1$
length $(O'C') = \bar{V}_2$

The truth of the foregoing statement can be seen by applying Eqs. (4-15a) and (4-17a) to a binary solution, noting that $x_1 = 1 - x_2$. Thus

$$v = (1 - x_2)\bar{V}_1 + x_2\bar{V}_2 \tag{4-28}$$

$$(1 - x_2)\left(\frac{\partial \bar{V}_1}{\partial x_2}\right)_{T,p} + x_2\left(\frac{\partial \bar{V}_2}{\partial x_2}\right)_{T,p} = 0 \tag{4-29}$$

Differentiating Eq. (4-28) with respect to x_2 results in

$$\left(\frac{\partial v}{\partial x_2}\right)_{T,p} = -\bar{V}_1 + (1 - x_2)\left(\frac{\partial \bar{V}_1}{\partial x_2}\right)_{T,p} + \bar{V}_2 + x_2\left(\frac{\partial \bar{V}_2}{\partial x_2}\right)_{T,p} \tag{4-30}$$

Subtracting Eq. (4-29) from Eq. (4-30),

$$\left(\frac{\partial v}{\partial x_2}\right)_{T,p} = \bar{V}_2 - \bar{V}_1 \tag{4-31}$$

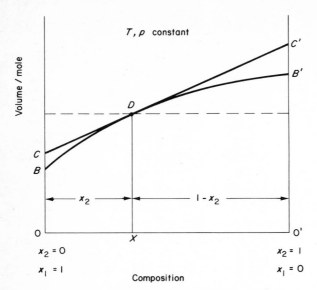

FIG. 4-1. Molal volume-composition diagram for a binary system.

Solving Eqs. (4-28) and (4-31) simultaneously, we get

$$\overline{V}_1 = v - x_2 \left(\frac{\partial v}{\partial x_2} \right)_{T,p} \tag{4-32a}$$

$$\overline{V}_2 = v + (1 - x_2) \left(\frac{\partial v}{\partial x_2} \right)_{T,p} \tag{4-32b}$$

From these equations, by a simple geometrical interpretation in reference to Fig. 4-1, we observe that length $(OC) = \overline{V}_1$ and length $(O'C') = \overline{V}_2$.

Equations (4-32a, b) relate the partial molal volumes to the molal volume of a system as a whole. They can be generalized for any other extensive property Y. Thus

$$\overline{Y}_1 = y - x_2 \left(\frac{\partial y}{\partial x_2} \right)_{T,p} \tag{4-33a}$$

$$\overline{Y}_2 = y + (1 - x_2) \left(\frac{\partial y}{\partial x_2} \right)_{T,p} \tag{4-33b}$$

where y is the property Y per mole of the system. A diagram similar to Fig. 4-1 can be plotted from experimental data of other molal properties, such as molal enthalpy h or molal Gibbs function g, to obtain graphically the corresponding partial molal properties.

For future reference we list below a few useful equations involving chemical potentials for a binary system. These equations can be readily obtained from the corresponding general relations which we already studied.

$$g = (1 - x_2)\mu_1 + x_2\mu_2 \tag{4-34}$$

$$(1 - x_2)\left(\frac{\partial\mu_1}{\partial x_2}\right)_{T,p} + x_2\left(\frac{\partial\mu_2}{\partial x_2}\right)_{T,p} = 0 \tag{4-35}$$

$$\mu_1 = \bar{G}_1 = g - x_2\left(\frac{\partial g}{\partial x_2}\right)_{T,p} \tag{4-36a}$$

$$\mu_2 = \bar{G}_2 = g + (1 - x_2)\left(\frac{\partial g}{\partial x_2}\right)_{T,p} \tag{4-36b}$$

$$d\mu_1 = -\bar{S}_1 dT + \bar{V}_1 dp + \left(\frac{\partial\mu_1}{\partial x_2}\right)_{T,p} dx_2 \tag{4-37a}$$

$$d\mu_2 = -\bar{S}_2 dT + \bar{V}_2 dp + \left(\frac{\partial\mu_2}{\partial x_2}\right)_{T,p} dx_2 \tag{4-37b}$$

4-4 GASEOUS MIXTURES

Before continuing the study of general principles on multicomponent systems, we will explore in this section some common methods of dealing with gas mixtures of constant composition. We begin with a consideration of a rather simple case—mixtures of ideal gases. It is an empirical fact that a mixture of ideal gases behaves also as an ideal gas. Let us consider a homogeneous mixture of inert ideal gases with r components; each component occupies the entire volume V of the container at a common temperature T (Fig. 4-2). Applying the ideal gas equation to the mixture as a whole and to any component i we have

$$pV = nRT \tag{4-38a}$$

$$p_i V = n_i RT \tag{4-38b}$$

where p is the total pressure of the mixture, and p_i is the partial pressure of component i. The partial pressure p_i of component i is defined as the pressure that this component would exert if it occupied the volume alone at the mixture temperature.

Since $n = \displaystyle\sum_{i=1}^{r} n_i$, it follows that

$$p = \sum_{i=1}^{r} p_i \tag{4-39}$$

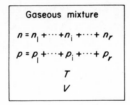

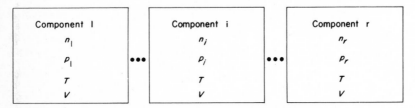

FIG. 4-2. Illustration of Dalton's law of additive pressures.

This is the *Dalton's law of additive pressures*. In words, it states that the total pressure of a mixture of ideal gases is equal to the sum of the partial pressures of the components, if each existed alone at the temperature and volume of the mixture. Taking the ratio between Eqs. (4-38b) and (4-38a) one obtains

$$p_i = \frac{n_i}{n} p = x_i p \tag{4-40}$$

where x_i is the mole fraction of component i.

   An alternate approach in the analysis of ideal gas mixtures is to consider the case where each component exists at the temperature and pressure of the mixture (Fig. 4-3). The partial volume V_i of component i is defined as the volume that would be occupied by this component at the same temperature T and pressure p as that of the mixture. Applying the ideal gas equation then gives

$$pV_i = n_i RT \tag{4-38c}$$

Summing up all V_i's and comparing the result with Eq. (4-38a) leads to

$$V = \sum_{i=1}^{r} V_i \tag{4-41}$$

This is the *Amagat's law of additive volumes*. It states that the volume of a mixture of ideal gases is equal to the sum of the partial volumes of the components when the partial volumes are determined at the pressure and

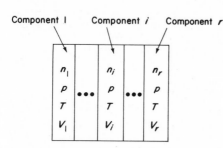

FIG. 4-3. Illustration of Amagat's law of additive volumes.

temperature of the mixture. Taking the ratio between Eqs. (4-38c) and (4-38a) one obtains

$$V_i = \frac{n_i}{n} V = x_i V \tag{4-42}$$

The internal energy, enthalpy, and entropy of a mixture of ideal gases can be evaluated as the sum of the respective properties of the component gases as if each existed alone at the temperature and volume of the mixture. Thus

$$U = nu = \sum_{i=1}^{r} n_i u_i \quad \text{or} \quad u = \sum_{i=1}^{r} x_i u_i \tag{4-43a}$$

$$H = nh = \sum_{i=1}^{r} n_i h_i \quad \text{or} \quad h = \sum_{i=1}^{r} x_i h_i \tag{4-43b}$$

$$S = ns = \sum_{i=1}^{r} n_i s_i \quad \text{or} \quad s = \sum_{i=1}^{r} x_i s_i \tag{4-43c}$$

Since the internal energy and enthalpy for an ideal gas are functions only of temperature, the u_i's and h_i's in the foregoing equations are to be evaluated at the temperature of the mixture. The entropy, however, is a function of two independent properties even for an ideal gas. Therefore the s_i's in Eq. (4-43c) are to be evaluated either at T and V of the mixture, or at T of the mixture and the partial pressure p_i of the component.

The molal specific heats of a mixture are defined as

$$c_v = \left(\frac{\partial u}{\partial T}\right)_{v,x} \quad \text{and} \quad c_p = \left(\frac{\partial h}{\partial T}\right)_{p,x} \tag{4-44}$$

where the subscript x implies constancy of composition. Applying Eqs. (4-43a) and (4-43b) we have

$$c_v = \sum_i x_i \left(\frac{\partial u_i}{\partial T}\right)_v = \sum_i x_i c_{vi} \tag{4-45a}$$

$$c_p = \sum_i x_i \left(\frac{\partial h_i}{\partial T}\right)_p = \sum_i x_i c_{pi} \tag{4-45b}$$

where c_{vi} and c_{pi} are the molal specific heats of component i.

As demonstrated in the previous derivations we note that Dalton's law of additive pressures and Amagat's law of additive volumes hold exactly only for mixtures of ideal gases. However, it was found experimentally that these laws do hold approximately for mixtures of real gases even in some ranges of pressure and temperature where the ideal gas law itself is quite inaccurate. In this case one is not to use the ideal gas equation, but rather some suitable real gas equation of state to find the p_i's or V_i's for the individual components to be used in Eqs. (4-39) or (4-41). Here the p_i's and V_i's are usually renamed as the component pressures and component volumes, respectively. They are no longer defined by Eqs. (4-40) and (4-42).

Often there is a need to have an equation of state for a mixture of real gases. Since there is an infinite variety of compositions for mixtures, it is desirable to devise methods of developing an equation of state for a mixture from the equation of state for the pure components. This may be done empirically by using various combining rules to obtain the constants of an equation of state for a mixture from the constants of the pure components. The most common combining rules are:

Linear combination $\quad k = \sum_i x_i k_i$

Linear square root combination $\quad k = \left(\sum_i x_i k_i^{1/2}\right)^2$

Linear cube root combination $\quad k = \left(\sum_i x_i k_i^{1/3}\right)^3$

Lorentz combination $\quad k = \frac{1}{4}\sum_i x_i k_i + \frac{3}{4}\left(\sum_i x_i k_i^{1/3}\right)\left(\sum_i x_i k_i^{2/3}\right)$

where k represents a constant in an equation of state for a mixture, k_i the corresponding constant in the equation of state for pure component i, and x_i the mole fraction of component i in the mixture.

When accurate p-v-T relations for the components of a mixture are not available, a generalized compressibility chart can be used. One way of using a compressibility chart in this case is to determine the compressibility factors for the pure components from the chart and then to devise a means of combining them to obtain the compressibility factor for the mixture. It is convenient to use a simple linear combination rule of the type

$$Z = \sum_i x_i Z_i \tag{4-46}$$

where Z is the compressibility factor for the mixture, Z_i the compressibility factor for pure component i, and x_i the mole fraction. If the Z_i's are evaluated at p and T of the mixture, Eq. (4-46) reduces to the Amagat's law of additive volumes. Whereas if the Z_i's are evaluated at the V and T of the mixture, this equation then reduces to the Dalton's law of additive pressures.

A general compressibility chart can be used also to evaluate directly the compressibility factor for a mixture. However, as suggested by W. B. Kay,[1] instead of using the true critical point of the mixture, a fictitious point (the *pseudocritical point*) should be employed to calculate the reduced coordinates. Kay first proposed the use of a simple linear combination as follows:

$$p_{c'} = \sum_i x_i p_{ci} \qquad T_{c'} = \sum_i x_i T_{ci} \tag{4-47}$$

where $p_{c'}$ and $T_{c'}$ are the pseudocritical pressure and temperature of the mixture, p_{ci} and T_{ci} are the critical pressure and temperature of component i, and x_i is the mole fraction of component i. A number of more complex combining rules for defining pseudocritical constants have also been proposed.[2]

EXAMPLE 4-2. A number of pure ideal gases at the same temperature T and pressure p fill the different compartments of a well-insulated rigid tank. The thin partitions separating the compartments are broken open and the gases are allowed to mix. Determine the increase in entropy for this mixing process.

Solution.
Since the tank is rigid and insulated, neither work nor heat transfer is involved. From the first law it is apparent that $\Delta U = 0$ for the total system.

[1] W. B. Kay, *Ind. Eng. Chem.*, 28:1014 (1936).
[2] See, for example, J. Joffe, *ibid.*, 39:837 (1947).

Since the initial temperature and pressure of all the gases are the same, the final-mixture temperature and pressure must equal the initial temperature and pressure of the gases.

For each component, from Eq. (2-53b), we have

$$(\Delta S)_i = n_i (\Delta s)_i = n_i \left(-R \ln \frac{p_{i,\text{final}}}{p} \right)$$

where $p_{i,\text{final}}$ is the partial pressure of component i in the final mixture, and p is the initial pressure of each gas which is also the total pressure of the final mixture. From Dalton's law for the final mixture, we have

$$p_{i,\text{final}} = x_i p$$

where x_i is the mole fraction of component i in the mixture. Whence

$$(\Delta S)_i = -Rn_i \ln \frac{x_i p}{p} = -Rn_i \ln x_i$$

For the total system the entropy change for the mixing process is then

$$\Delta S = -R \sum_i n_i \ln x_i \qquad (4\text{-}48)$$

Accordingly, the entropy of the final mixture is given by

$$S_{\text{final mixture}} = \sum_i n_i s_i - R \sum_i n_i \ln x_i \qquad (4\text{-}49)$$

in which the term $\sum_i n_i s_i$ represents the total entropy of the gases before mixing.

Because mixing of gases is an irreversible process, the total entropy change as given by Eq. (4-48) is positive. From this equation it is worth noticing that the increase in entropy due to mixing depends only on the number of moles of the component gases, and is independent of the composition of the gases. Thus, when equal moles of nitrogen and oxygen are mixed, the increase in entropy due to mixing would be exactly the same as when equal moles of carbon dioxide and carbon monoxide are mixed. However, when two portions of the same gas are allowed to mix, there is no increase in entropy. This can be easily seen from Eq. (4-48), because for this case we have $x = 1$ and $\Delta S = 0$.

EXAMPLE 4-3. A gaseous mixture containing 69.5 mole percent carbon dioxide and 30.5 mole percent ethylene (C_2H_4) has a molal volume of 0.111 liter/g mole at a temperature of $100°C$. The experimental value for the pressure of the mixture at the given temperature and molal volume is 175.8 atm. Estimate the pressure of the mixture by the use of

(1) ideal gas law
(2) law of additive pressures and van der Waals equation

(3) law of additive volumes and van der Waals equation
(4) van der Waals equation for the mixture, using linear square root combination for constant a and linear combination for constant b
(5) law of additive pressures and generalized compressibility chart
(6) law of additive volumes and generalized compressibility chart
(7) Kay's rule of pseudocritical point.

Solution.
We denote carbon dioxide by the subscript 1 and ethylene by the subscript 2. The symbols without such a subscript are for the mixture as a whole. Thus the given data are $x_1 = 0.695$, $x_2 = 0.305$, $v = 0.111$ liter/g mole, and $T = 100°C = 373.15°K$. The critical constants are

$$T_{c1} = 304.2°K \qquad p_{c1} = 72.9 \text{ atm} \qquad Z_{c1} = 0.275$$

$$T_{c2} = 283.06°K \qquad p_{c2} = 50.5 \text{ atm} \qquad Z_{c2} = 0.270$$

The van der Waals constants are

$$a_1 = 3.606 \text{ (atm)(liters)}^2/\text{g mole}^2$$

$$b_1 = 0.04280 \text{ liter/g mole}$$

$$a_2 = 4.507 \text{ (atm)(liters)}^2/\text{g mole}^2$$

$$b_2 = 0.05749 \text{ liter/g mole}$$

The universal gas constant is

$$R = 0.08205 \text{ (atm)(liter)/(g mole)(°K)}$$

(1) Applying the ideal gas equation gives

$$p = \frac{RT}{v} = \frac{0.08205 \times 373.15}{0.111} = 276 \text{ atm}$$

Let us define the percentage error in pressure by the equation

$$\text{Percentage error} = \frac{\text{calculated pressure} - \text{experimental pressure}}{\text{experimental pressure}} \times 100$$

Thus, the pressure calculated by the ideal gas equation has an error of $100(276 - 175.8)/175.8 = 57.0\%$.

(2) In order to use the law of additive pressures together with the van der Waals equation, we first calculate the molal volumes of carbon dioxide and ethylene at $T = T_1 = T_2$ and $V = V_1 = V_2$ where V denotes total volume.

$$v_1 = \frac{0.111}{0.695} = 0.160 \text{ liter/g mole}$$

$$v_2 = \frac{0.111}{0.305} = 0.364 \text{ liter/g mole}$$

From the van der Waals equation the component pressures of carbon dioxide and ethylene are then

$$p_1 = \frac{RT}{v_1 - b_1} - \frac{a_1}{v_1^2} = \frac{0.08205 \times 373.15}{0.160 - 0.0428} - \frac{3.606}{0.160^2} = 120.4 \text{ atm}$$

$$p_2 = \frac{RT}{v_2 - b_2} - \frac{a_2}{v_2^2} = \frac{0.08205 \times 373.15}{0.364 - 0.05749} - \frac{4.507}{0.364^2} = 65.9 \text{ atm}$$

Therefore by the law of additive pressures we have

$$p = p_1 + p_2 = 120.4 + 65.9 = 186 \text{ atm}$$

This calculated pressure has an error of $100(186 - 175.8)/175.8 = 5.8\%$.

(3) The law of additive volumes states that the total volume of a mixture is the sum of the component volumes evaluated at the pressure and temperature of the mixture. We accordingly write

$$v = x_1 v_1' + x_2 v_2'$$

or

$$0.111 = 0.695 v_1' + 0.305 v_2' \tag{4-50}$$

wherein v_1' and v_2' are the molal volumes in liters/g mole of carbon dioxide and ethylene respectively, as evaluated at $T = T_1 = T_2$ and $p = p_1 = p_2$. From the van der Waals equation we have

$$p = p_1 = \frac{RT}{v_1' - b_1} - \frac{a_1}{v_1'^2} = \frac{0.08205 \times 373.15}{v_1' - 0.0428} - \frac{3.606}{v_1'^2} \tag{4-51}$$

$$p = p_2 = \frac{RT}{v_2' - b_2} - \frac{a_2}{v_2'^2} = \frac{0.08205 \times 373.15}{v_2' - 0.05749} - \frac{4.507}{v_2'^2} \tag{4-52}$$

The value of p can best be obtained in an iteration process, which consists in assuming values of v_1', obtaining v_2' from Eq. (4-50), and then calculating p_1 and p_2 from Eqs. (4-51) and (4-52). For a correct solution we must have $p_1 = p_2$, which would be the pressure of the mixture. After a few trials the correct value of v_1' was found to be 0.105 liter/g mole. Thus from Eq. (4-50), we have

$$v_2' = \frac{0.111 - 0.695 v_1'}{0.305} = \frac{0.111 - 0.695 \times 0.105}{0.305}$$

$$= 0.125 \text{ liter/g mole}$$

and from Eqs. (4-51) and (4-52) we have

$$p_1 = \frac{0.08205 \times 373.15}{0.105 - 0.0428} - \frac{3.606}{0.105^2} = 165 \text{ atm}$$

$$p_2 = \frac{0.08205 \times 373.15}{0.125 - 0.05749} - \frac{4.507}{0.125^2} = 165 \text{ atm}$$

Hence

$$p = p_1 = p_2 = 165 \text{ atm}$$

This calculated pressure has an error of $100(165 - 175.8)/175.8 = -6.1\%$.

(4) The van der Waals constants for the mixture are

$$a = (x_1 a_1^{1/2} + x_2 a_2^{1/2})^2$$

$$= (0.695 \times 3.606^{1/2} + 0.305 \times 4.507^{1/2})^2 = 3.870 \text{ (atm)(liters)}^2/\text{g mole}^2$$

$$b = x_1 b_1 + x_2 b_2$$

$$= 0.695 \times 0.0428 + 0.305 \times 0.05749 = 0.04728 \text{ liter/g mole}$$

Therefore from the van der Waals equation as applied to the mixture as a whole, we get

$$p = \frac{RT}{v - b} - \frac{a}{v^2} = \frac{0.08205 \times 373.15}{0.111 - 0.04728} - \frac{3.870}{0.111^2} = 166 \text{ atm}$$

This calculated pressure has an error of $100(166 - 175.8)/175.8 = -5.6\%$.

(5) When each gas exists at the temperature and total volume of the mixture, we have as before

$$v_1 = 0.160 \text{ liter/g mole}$$

$$v_2 = 0.364 \text{ liter/g mole}$$

To evaluate the component pressure of carbon dioxide by the use of a generalized compressibility chart, we first write

$$p_1 = p_{r1} p_{c1} = \frac{Z_1 RT}{v_1}$$

or

$$p_{r1} = \frac{Z_1 RT}{p_{c1} v_1} = \frac{0.08205 \times 373.15}{72.9 \times 0.160} Z_1 = 2.625 Z_1$$

Since $Z_{c1} = 0.275$, Appendix Fig. A-16 is to be used. The equation $p_{r1} = 2.625 Z_1$ represents a straight line on this chart. The intersection of this line with the reduced isotherm of

$$T_{r1} = \frac{T}{T_{c1}} = \frac{373.15}{304.2} = 1.23$$

gives the solution as

$$p_{r1} = 1.78 \quad \text{and} \quad Z_1 = 0.68$$

Whence

$$p_1 = p_{r1}p_{c1} = 1.78 \times 72.9 = 130 \text{ atm}$$

Similarly for ethylene, solving the equations

$$p_{r2} = \frac{Z_2 RT}{p_{c2}v_2} = \frac{0.08205 \times 373.15}{50.5 \times 0.364} Z_2 = 1.666 Z_2$$

$$T_{r2} = \frac{T}{T_{c2}} = \frac{373.15}{283.06} = 1.32$$

in connection with Fig. A-16 leads to

$$p_{r2} = 1.36 \quad \text{and} \quad Z_2 = 0.82$$

Whence

$$p_2 = p_{r2}p_{c2} = 1.36 \times 50.5 = 69 \text{ atm}$$

Therefore by the law of additive pressures we have

$$p = p_1 + p_2 = 130 + 69 = 199 \text{ atm}$$

This calculated pressure has an error of $100(199 - 175.8)/175.8 = 13.2\%$.

(6) To conform to the law of additive volumes we again have Eq. (4-50), which is

$$0.111 = 0.695v_1' + 0.305v_2' \tag{4-50}$$

But now this equation shall be used in connection with the following:

$$v_1' = \frac{Z_1 RT}{p} = 0.08205 \times 373.15 \frac{Z_1}{p} \tag{4-53}$$

$$v_2' = \frac{Z_2 RT}{p} = 0.08205 \times 373.15 \frac{Z_2}{p} \tag{4-54}$$

Since p is unknown, an iteration process must be employed. After a few trials it was found that p should be 174 atm. Thus

$$p_{r1} = \frac{p}{p_{c1}} = \frac{174}{72.9} = 2.39$$

$$p_{r2} = \frac{p}{p_{c2}} = \frac{174}{50.5} = 3.45$$

From Fig. A-16 we get

$$Z_1 = 0.61 \text{ at } p_{r1} = 2.39 \text{ and } T_{r1} = 1.23$$

$$Z_2 = 0.69 \text{ at } p_{r2} = 3.45 \text{ and } T_{r2} = 1.32$$

Substituting these values in Eqs. (4-53) and (4-54) leads to

$$v_1' = 0.08205 \times 373.15 \times \frac{0.61}{174} = 0.107 \text{ liter/g mole}$$

$$v_2' = 0.08205 \times 373.15 \times \frac{0.69}{174} = 0.121 \text{ liter/g mole}$$

Thus

$$0.695v_1' + 0.305v_2' = 0.695 \times 0.107 + 0.305 \times 0.121 = 0.111 \text{ liter/g mole}$$

which agrees with Eq. (4-50). Therefore we conclude that

p = 174 atm

This calculated pressure has an error of $100(174 - 175.8)/175.8 = -1.0\%$.

(7) According to Kay's rule the pseudocritical temperature and pressure of the mixture are

$$T_{c'} = x_1 T_{c1} + x_2 T_{c2} = 0.695 \times 304.2 + 0.305 \times 283.06 = 297.8°\text{K}$$

$$p_{c'} = x_1 p_{c1} + x_2 p_{c2} = 0.695 \times 72.9 + 0.305 \times 50.5 = 66.1 \text{ atm}$$

Thus

$$T_r = \frac{T}{T_{c'}} = \frac{373.15}{297.8} = 1.25$$

$$p_r = \frac{p}{p_{c'}} = \frac{ZRT}{p_{c'}v} = \frac{0.08205 \times 373.15}{66.1 \times 0.111} Z = 4.173Z$$

The intersection of the straight line $p_r = 4.173Z$ and the reduced isotherm $T_r = 1.25$ on the compressibility chart (Fig. A-16) gives the solution as

$p_r = 2.59$ and $Z = 0.62$

Therefore

$p = p_r p_{c'} = 2.59 \times 66.1 = 171 \text{ atm}$

This calculated pressure has an error of $100(171 - 175.8)/175.8 = -2.7\%$.

It should be mentioned that one certainly cannot draw any conclusion from a single illustration about the relative accuracy of the various methods employed in this example. However, it is clear that the ideal gas law often gives erroneous solutions when the pressure is not low.

4-5 FUGACITY

For a simple compressible single component system on a molal basis we have the relation

$$dg = -s\,dT + v\,dp \tag{2-11}$$

At constant temperature this becomes

$$dg_T = v\, dp_T \qquad (4\text{-}55)$$

When the ideal gas equation is used we then have

$$dg_T = \frac{RT}{p}\, dp_T = RT d(\ln p)_T \qquad (4\text{-}56)$$

This equation is, of course, not expected to hold for real gases, particularly at relatively high pressures. But it has been found useful to preserve its form and those of other relations derived from it when we deal with real gases. This may be done by introducing a fictitious pressure called the *fugacity* as suggested by G. N. Lewis in 1901. The fugacity f for pure compound is simply that which satisfies the relations

$$(dg)_T = (d\mu)_T = RT\, d(\ln f)_T \qquad (4\text{-}57)$$

and

$$\lim_{p \to 0}\left(\frac{f}{p}\right) = 1 \qquad (4\text{-}58)$$

The implication is that the fugacity f shall play the same role for real gases as does the pressure p for ideal gases.

Integrating Eq. (4-57) for a real gas isothermally from a very low pressure p^*, where ideal gas behavior can be assumed, to a higher pressure p yields

$$g = g^* + RT \ln \frac{f}{f^*} \qquad (4\text{-}59)$$

or

$$\mu = \mu^* + RT \ln \frac{f}{f^*} \qquad (4\text{-}59a)$$

where g, μ, and f are the molal Gibbs function, chemical potential, and fugacity, respectively, at the state (T, p); and g^*, μ^*, and f^* are the corresponding properties at a reference state (T, p^*). The reference state is usually chosen so that $f^* = 1$ at $p^* = 1$, i.e., the reference state for the real gas is chosen to correspond to unit pressure of the ideal gas. The pressure unit is usually the standard atmosphere. However, one must realize that such a reference ideal gas state might be a hypothetical one, because the substance under study might not exist as a gas at the particular temperature and 1 atm.

The fugacity of a real gas at any temperature and pressure can be evaluated with the aid of an equation of state. Thus from Eqs. (4-59) and (4-55),

$$g - g^* = RT \ln \frac{f}{f^*} = \left[\int_{p^*}^{p} v \, dp \right]_T \qquad (4\text{-}60)$$

or

$$RT \ln f = RT \ln f^* + \left[\int_{p^*}^{p} v \, dp \right]_T \qquad (4\text{-}60a)$$

But we have the mathematical relation,

$$RT \ln p = RT \ln p^* + RT \int_{p^*}^{p} \frac{dp}{p}$$

Subtracting this equation from Eq. (4-60a) gives

$$RT \ln \frac{f}{p} = RT \ln \frac{f^*}{p^*} + \left[\int_{p^*}^{p} \left(v - \frac{RT}{p} \right) dp \right]_T$$

When $p^* \to 0$ and $f^*/p^* \to 1$ we have

$$\left(\ln \frac{f}{p} \right)_T = \left[\int_{0}^{p} \left(\frac{v}{RT} - \frac{1}{p} \right) dp \right]_T \qquad (4\text{-}61)$$

where the subscript T indicates that the condition of constant temperature prevails. This equation can be used to evaluate the fugacity of any gas at given T and p when its equation of state is known. Of course, if the gas is ideal, the integral will be zero and f will equal p.

In terms of the compressibility factor $Z = pv/RT$, Eq. (4-61) becomes

$$\left(\ln \frac{f}{p} \right)_T = \left[\int_{0}^{p} (Z - 1) \, d(\ln p) \right]_T$$

Since

$$\frac{dp}{p} = \frac{dp_r}{p_r} \qquad \text{or} \qquad d(\ln p) = d(\ln p_r)$$

we have on reduced coordinates,

$$\left(\ln \frac{f}{p} \right)_T = \left[\int_{0}^{p_r} (Z - 1) \, d(\ln p_r) \right]_T \qquad (4\text{-}61a)$$

The right-hand side of Eq. (4-61a) can be integrated at constant T_r using Z

values found at each p_r from a generalized compressibility chart. The result is shown graphically in Appendix Fig. A-20.

The discussion so far in this section has centered around the fugacity of pure component systems. In practice, however, fugacity of a pure component has little use by itself. The concept of fugacity is chiefly useful in the study of multicomponent systems. We now turn our attention to such cases.

The fugacity f_i of component i in a real gas mixture is defined in terms of partial molal Gibbs function $\bar{G}_i$, or chemical potential μ_i, of this component in the mixture by the equations

$$(d\bar{G}_i)_T = (d\mu_i)_T = RTd(\ln f_i)_T \tag{4-62}$$

$$\lim_{p \to 0} \frac{f_i}{x_i p} = 1 \tag{4-63}$$

in which p is the total pressure of the mixture, and x_i is the mole fraction of component i in the mixture. According to Dalton's law the product $x_i p$ is the partial pressure p_i of component i for an ideal gas mixture. It follows that while the fugacity of a pure component is essentially a fictitious pressure, the fugacity of a component in a mixture can be considered as a fictitious partial pressure. It should be noted that it is not sufficient to write that

$$\lim_{p_i \to 0} \frac{f_i}{p_i} = 1$$

because an infinitely dilute mixture of a gas in other gases does not behave as an ideal gas mixture unless the total pressure approaches zero.

Integrating Eq. (4-62) at constant temperature from a very low pressure p^*, where ideal gas behavior can be assumed for the mixture, to a higher pressure p yields

$$\bar{G}_i = \bar{G}_i^* + RT \ln \frac{f_i}{f_i^*} \tag{4-64a}$$

or

$$\mu_i = \mu_i^* + RT \ln \frac{f_i}{f_i^*} \tag{4-64b}$$

in which $\bar{G}_i$, μ_i, and f_i are the partial molal Gibbs function, chemical potential, and fugacity, respectively, of component i when the mixture is at T, p, and a certain composition; and $\bar{G}_i^*$, μ_i^*, and f_i^* are the corresponding properties when the same mixture is at T and p^*.

Following a procedure corresponding to that used in deriving Eq. (4-61) for a single-component system, we can write a similar equation for a multi-component system. We begin with Eqs. (4-64a) and (4-22),

$$\bar{G}_i - \bar{G}_i^* = RT \ln \frac{f_i}{f_i^*} = \left[\int_{p*}^{p} \bar{V}_i \, dp \right]_T \tag{4-65}$$

and note that when

$$p^* \to 0, \quad \frac{f_i^*}{x_i p^*} = \frac{f_i^*}{p_i^*} \to 1$$

we then obtain

$$\ln f_i = \ln (x_i p) + \left\{ \int_0^p \left[\frac{\bar{V}_i}{RT} - \frac{1}{p} \right] dp \right\}_T \tag{4-66}$$

This equation may be used to calculate the fugacities of the components of a mixture if adequate data are available to yield the $\bar{V}_i$'s.

Since the fugacity of a substance is, in general, a function of both temperature and pressure, we now inquire into these functional relations. For component i in a mixture of fixed composition, we have from Eq. (4-22),

$$\left(\frac{\partial \bar{G}_i}{\partial p} \right)_T = \left(\frac{\partial \mu_i}{\partial p} \right)_T = \bar{V}_i$$

Combining this equation and Eq. (4-62) leads directly to the expression for pressure dependence,

$$\left(\frac{\partial \ln f_i}{\partial p} \right)_T = \frac{\bar{V}_i}{RT} \tag{4-67}$$

Differentiating Eq. (4-64a) at constant pressure gives

$$\left(\frac{\partial \bar{G}_i}{\partial T} \right)_p - \left(\frac{\partial \bar{G}_i^*}{\partial T} \right)_{p*} = R \ln \frac{f_i}{f_i^*} + RT \left[\left(\frac{\partial \ln f_i}{\partial T} \right)_p - \left(\frac{\partial \ln f_i^*}{\partial T} \right)_{p*} \right]$$

But

$$\left(\frac{\partial \bar{G}_i}{\partial T} \right)_p - \left(\frac{\partial \bar{G}_i^*}{\partial T} \right)_{p*} = \bar{S}_i^* - \bar{S}_i$$

$$R \ln \frac{f_i}{f_i^*} = \frac{\bar{G}_i}{T} - \frac{\bar{G}_i^*}{T}$$

$$\left(\frac{\partial \ln f_i^*}{\partial T} \right)_{p*} = 0$$

Therefore

$$\bar{S}_i^* - \bar{S}_i = \frac{\bar{G}_i}{T} - \frac{\bar{G}_i^*}{T} + RT\left(\frac{\partial \ln f_i}{\partial T}\right)_p$$

Rearranging and making use of the relation $\bar{G}_i = \bar{H}_i - T\bar{S}_i$, we have the expression for temperature dependence.

$$\left(\frac{\partial \ln f_i}{\partial T}\right)_p = \frac{\bar{H}_i^* - \bar{H}_i}{RT^2} \tag{4-68}$$

In the case of a pure component i, expressions similar to Eqs. (4-67) and (4-68) can be written in terms of molal properties. Thus

$$\left(\frac{\partial \ln f_i^\circ}{\partial p}\right)_T = \frac{v_i^\circ}{RT} \tag{4-67a}$$

$$\left(\frac{\partial \ln f_i^\circ}{\partial T}\right)_p = \frac{h_i^{\circ*} - h_i^\circ}{RT^2} \tag{4-68a}$$

Notice that $(\bar{H}_i^* - \bar{H}_i)$ in Eq. (4-68) and $(h_i^{\circ*} - h_i^\circ)$ in Eq. (4-68a) are the appropriate enthalpy differences between the initial state of low pressure p^* and the final state of higher pressure p, both at the same temperature. Note also that for an ideal gas $\bar{H}_i^* = h_i^{\circ*}$.

There is an important equation which expresses the relation between fugacity and mole fraction of a multicomponent system. We now derive this equation for the case of a binary system. Applying the Gibbs-Duhem equation for changes between equilibrium states of a binary system at constant temperature T and total pressure p, we have (see Eq. 4-12)

$$n_1 d\mu_1 + n_2 d\mu_2 = 0$$

Substituting Eq. (4-62) into this equation leads to

$$n_1 RTd(\ln f_1) + n_2 RTd(\ln f_2) = 0$$

Dividing through by $nRTdx_2$, where n is the total number of moles in the mixture, and writing explicitly the restriction of constant T and p, we have

$$x_1\left[\frac{\partial(\ln f_1)}{\partial x_2}\right]_{T,p} + x_2\left[\frac{\partial(\ln f_2)}{\partial x_2}\right]_{T,p} = 0$$

But since $x_1 + x_2 = 1$, $dx_1 = -dx_2$, the preceding equation may be written as

$$x_1\left[\frac{\partial(\ln f_1)}{\partial x_1}\right]_{T,p} = x_2\left[\frac{\partial(\ln f_2)}{\partial x_2}\right]_{T,p} \tag{4-69}$$

or

$$\left[\frac{\partial(\ln f_1)}{\partial(\ln x_1)}\right]_{T,p} = \left[\frac{\partial(\ln f_2)}{\partial(\ln x_2)}\right]_{T,p} \tag{4-69a}$$

This is known as the *Gibbs-Duhem relation* for binary systems.

At low total pressures, replacing the fugacities in the foregoing equations by partial vapor pressures gives

$$x_1\left[\frac{\partial(\ln p_1)}{\partial x_1}\right]_{T,p} = x_2\left[\frac{\partial(\ln p_2)}{\partial x_2}\right]_{T,p} \tag{4-70}$$

or

$$\left[\frac{\partial(\ln p_1)}{\partial(\ln x_1)}\right]_{T,p} = \left[\frac{\partial(\ln p_2)}{\partial(\ln x_2)}\right]_{T,p} \tag{4-70a}$$

This is the *Duhem-Margules relation*, where p_1 and p_2 are the partial vapor pressures of components 1 and 2 respectively.

Although we introduced the concept of fugacity to facilitate the treatment of real gases, the definition of fugacity applies to liquids and solids as well. In defining the fugacity of a liquid (or solid) we simply assign to it a fugacity numerically equal to that of the vapor in equilibrium with it. If the ideal gas law is used as an approximation for the vapor, then the fugacity of the liquid (or solid) will be equal to its vapor pressure. Similarly, we define the fugacity of a component in a liquid (or solid) solution as equal to the fugacity of this component in the vapor phase in equilibrium with the liquid (or solid) solution. If the pressure in the vapor phase is low so as to allow the use of ideal gas law for the vapor, then the fugacity of a component in a liquid (or solid) solution will be equal to its partial vapor pressure.

EXAMPLE 4-4. Derive the expression for the fugacity of a gas obeying the van der Waals equation of state.

Solution.
The van der Waals equation is

$$p = \frac{RT}{v - b} - \frac{a}{v^2}$$

Since this equation is explicit in p, we cannot conveniently use Eq. (4-61). Let us go back to Eq. (4-60) and transform the independent variable from p to v. Thus

$$RT \ln \frac{f}{f^*} = \left[\int_{p^*}^{p} v \, dp \right]_T$$

$$= pv - p^*v^* - \left[\int_{v^*}^{v} p \, dv \right]_T$$

where v^* is the molal volume at p^* and T. Substituting the van der Waals equation into the preceding equation and integrating,

$$RT \ln \frac{f}{f^*} = pv - p^*v^* - RT \ln \frac{v - b}{v^* - b} - \frac{a}{v} + \frac{a}{v^*}$$

or

$$\ln f = \ln \frac{f^*}{p^*} + \ln p^*(v^* - b) - \frac{p^*v^*}{RT} + \frac{a}{RTv^*} + \frac{pv}{RT} - \ln (v - b) - \frac{a}{RTv}$$

As $p^* \to 0$, $f^*/p^* \to 1$ and $v^* \to \infty$, whence $p^*v^* \to RT$. Also, from the van der Waals equation we see that

$$\frac{pv}{RT} - 1 = \frac{v}{v - b} - \frac{a}{RTv} - 1 = \frac{b}{v - b} - \frac{a}{RTv}$$

Therefore

$$\ln f = \ln \frac{RT}{v - b} - \frac{2a}{RTv} + \frac{b}{v - b}$$

which is the required expression.

EXAMPLE 4-5. The residual partial molal volume $\overline{\overline{V}}_i$ of component i in a multicomponent system is defined by the equation

$$\overline{\overline{V}}_i = \frac{RT}{p} - \overline{V}_i$$

where $\overline{V}_i$ is the partial molal volume of component i. The residual partial molal volumes for methane in a gaseous mixture of 70 mole percent methane and 30 mole percent ethane at a temperature of $100°F$ are listed in the following table as a function of pressure. Determine the fugacity of methane in the mixture at $100°F$ and 1500 psia.

Pressure, psia	$\overline{\overline{V}}_{CH_4}$, cu ft/lb mole
0	0.800
200	0.752
400	0.700
600	0.651
800	0.612
1,000	0.582
1,250	0.557
1,500	0.535

Solution.

From Eq. (4-66) we have

$$\ln \frac{f_{CH_4}}{x_{CH_4} p} = \left[\int_0^p \left(\frac{\bar{\bar{V}}_{CH_4}}{RT} - \frac{1}{p} \right) dp \right]_T = -\left[\int_0^p \frac{\bar{\bar{V}}_{CH_4}}{RT} dp \right]_T$$

The values of $\bar{\bar{V}}_{CH_4}/RT$ in the unit of $psia^{-1}$ were first calculated from the given data on $\bar{\bar{V}}_{CH_4}$ and $R = 10.73$ (psia)(cu ft)/(lb mole)($^\circ$R), $T = 559.67^\circ$R. These values were then plotted against the pressure in psia. From this graph by graphical integration we get

$$\left[\int_0^{1,500} \frac{\bar{\bar{V}}_{CH_4}}{RT} dp \right]_T = 0.159$$

Thus

$$\frac{f_{CH_4}}{x_{CH_4} p} = e^{-0.159} = 0.853$$

Whence

$$f_{CH_4} = 0.853 x_{CH_4} p = 0.853 \times 0.70 \times 1,500$$

$$= 896 \text{ psia}$$

EXAMPLE 4-6. Show that the fugacity of a pure compressed liquid at pressures moderately greater than its vapor pressure is approximately equal to its fugacity at its vapor pressure at the same temperature, or expressed symbolically,

$$f^{(L)}(p,T) \approx f^{(\text{sat liq})}(T)$$

where $f^{(L)}(p, T)$ denotes the fugacity of the pure compressed liquid at the pressure p and temperature T, and $f^{(\text{sat liq})}(T)$ denotes the fugacity of its saturated liquid at the same temperature T.

Solution.

For a pure substance, from Eqs. (4-55) and (4-57), we have

$$dg_T = RTd(\ln f)_T = v \, dp_T$$

Integrating at constant temperature between the saturated liquid state and the compressed liquid state, we have

$$RT \ln \frac{f^{(L)}(p,T)}{f^{(\text{sat liq})}(T)} = \int_{p^{(\text{sat})}}^p v \, dp$$

where $p^{(\text{sat})}$ denotes the saturation pressure at the temperature T. Since for a liquid the volume v is approximately a constant, we can then write

$$RT \ln \frac{f^{(L)}(p,T)}{f^{(\text{sat liq})}(T)} \approx v(p - p^{(\text{sat})})$$

Furthermore, since v is small for a liquid, the value of $v(p - p^{(\text{sat})})$ will be small for a moderate pressure difference $(p - p^{(\text{sat})})$. Therefore, for a pure compressed liquid at pressures moderately greater than its vapor pressure, we conclude that

$$f^{(L)}(p,T) \approx f^{(\text{sat liq})}(T)$$

A similar expression can be obtained for a pure compressed solid.

4-6 IDEAL SOLUTIONS

In the preceding section we presented Eq. (4-66) as a rigorous thermodynamic expression for the evaluation of the fugacity of a component in a mixture. In using this equation it is necessary to have data on $\overline{V}_i$ for the composition of interest as a function of pressure. Such data, however, are not readily available for most cases. A simplified method that can be easily applied is often desirable. One way is to assume a simple relation between the fugacity of a component in a mixture to its fugacity in a pure state. A widely used simple relation is

$$f_i = f_i^\circ x_i \tag{4-71}$$

where f_i is the fugacity of component i in a mixture, f_i° is the fugacity of pure component i at the same temperature and pressure as the mixture, and x_i is the mole fraction of this component in the mixture.

A system defined by Eq. (4-71) is usually referred to as an ideal solution. It applies to liquid and solid solutions as well as mixtures of real gases. Accordingly, we will generalize its definition with the statement: *An ideal solution* is a solution in which the fugacity of each component is equal to the product of its mole fraction and the fugacity of the pure component at the same temperature, pressure, and state of aggregation as of the solution. It has been found experimentally that gaseous mixtures exhibit ideal solution behavior at low pressures, and nonelectrolyte liquid solutions become ideal at high dilutions. A nonelectrolyte liquid solution is a solution where all the components exist as uncharged species rather than as charged ions. The concept of ideal solution is extensively used in treating liquid solutions of nonelectrolytes.

As consequences of the definition of an ideal solution certain characteristics are inherent in the nature of such a solution. First let us see if there is any volume change upon mixing pure components to form an ideal solution. From

Eqs. (4-67) and (4-67a) we write immediately

$$\left[\frac{\partial \ln (f_i/f_i^\circ)}{\partial p}\right]_T = \frac{\bar{V}_i - v_i^\circ}{RT} \tag{4-72}$$

in which f_i and $\bar{V}_i$ are the fugacity and partial molal volume of component i in the solution; and f_i° and v_i° are the fugacity and molal volume of pure component i at the same temperature, pressure, and state of aggregation as of the solution. For an ideal solution, since $f_i/f_i^\circ = x_i$ which is a constant for constant composition, it follows that

$$\bar{V}_i - v_i^\circ = 0 \quad \text{or} \quad \bar{V}_i = v_i^\circ \tag{4-73}$$

Thus the partial molal volume of each component in an ideal solution is equal to the molal volume of the corresponding pure component at T and p of the solution. In accordance with Eq. (4-15) the total volume of the solution is

$$V_{\text{after mixing}} = \sum_i n_i \bar{V}_i$$

The total volume of all the components before mixing is, of course,

$$V_{\text{before mixing}} = \sum_i n_i v_i^\circ$$

The change in total volume due to mixing is then

$$\Delta V_{\text{mixing}} = V_{\text{after mixing}} - V_{\text{before mixing}} = \sum_i n_i (\bar{V}_i - v_i^\circ) = 0$$

Thus, there is no volume change in forming an ideal solution from the pure components originally at the temperature, pressure, and state of aggregation as of the solution.

Similarly we shall find that there is no enthalpy change or heat evolved in forming an ideal solution. This can be seen readily from Eqs. (4-68) and (4-68a) that

$$\left[\frac{\partial \ln (f_i/f_i^\circ)}{\partial T}\right]_p = \frac{h_i^\circ - \bar{H}_i}{RT^2} \tag{4-74}$$

For an ideal solution, applying the condition $f_i/f_i^\circ = x_i$ we conclude that

$$\bar{H}_i - h_i^\circ = 0 \quad \text{or} \quad \bar{H}_i = h_i^\circ \tag{4-75}$$

The change in enthalpy or heat evolved due to mixing at constant temperature and pressure is then

$$\Delta H_{\text{mixing}} = \sum_i n_i (\bar{H}_i - h_i^\circ) = 0$$

For a binary ideal solution, since $\overline{V}_1 = v_1^\circ$ and $\overline{V}_2 = v_2^\circ$, then from Eq. (4-15a) we have

$$
\begin{aligned}
v &= (1 - x_2)\overline{V}_1 + x_2 \overline{V}_2 \\
&= (1 - x_2)v_1^\circ + x_2 v_2^\circ
\end{aligned} \tag{4-76}
$$

Thus a molal volume versus composition plot is a straight line for an ideal solution. The same is true for a molal enthalpy versus composition plot. These are depicted in Fig. 4-4.

We have seen from the foregoing discussions that there are no changes in volume and enthalpy, and consequently no change in internal energy, in forming an ideal solution from the pure components which are originally at the temperature, pressure, and state of aggregation of the solution. That is, volume, enthalpy, and internal energy are additive properties for ideal solutions. However, the same is not true in the case of entropy and any other property based on entropy.

Applying the definition of fugacity to the case of an ideal solution we write

$$
\begin{aligned}
(d\overline{G}_i)_T &= RT\, d\,[\ln (f_i^\circ x_i)]_T \\
&= RT\, d\,[\ln f_i^\circ + \ln x_i]_T
\end{aligned}
$$

But at a given T and p, f_i° is a constant so that $d(\ln f_i^\circ)_T = 0$. Whence

$$
(d\overline{G}_i)_T = RT\, d\,(\ln x_i)_T
$$

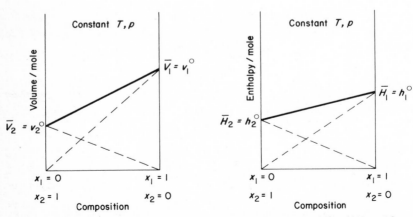

FIG. 4-4. Molal volume-composition and molal enthalpy-composition diagrams for a binary ideal solution.

Integrating this equation,

$$\int_{g_i^o}^{\bar{G}_i} (d\bar{G}_i)_T = RT \int_{x_i=1}^{x_i} d(\ln x_i)_T$$

results in

$$\bar{G}_i - g_i^o = RT \ln x_i \tag{4-77}$$

where $\bar{G}_i$ is the partial molal Gibbs function of component i with a mole fraction of x_i, and g_i^o is the molal Gibbs function of pure component i at T and p of the solution.

Since $\bar{G}_i = \mu_i$ (Eq. 4-18) and $g_i^o = \mu_i^o$, Eq. (4-77) can be rewritten as

$$\mu_i = \mu_i^o + RT \ln x_i \tag{4-78}$$

where μ_i is the chemical potential of component i with a mole fraction of x_i, and μ_i^o is the chemical potential of pure component i at T and p of the solution.

In accordance with the relations

$$\bar{G}_i = \bar{H}_i - T\bar{S}_i \quad \text{for component } i$$
$$g_i^o = h_i^o - Ts_i^o \quad \text{for pure component } i$$

we write

$$\bar{S}_i - s_i^o = \frac{\bar{H}_i - h_i^o}{T} - \frac{\bar{G}_i - g_i^o}{T}$$

But for an ideal solution $\bar{H}_i - h_i^o = 0$ whence

$$\bar{S}_i - s_i^o = -\frac{\bar{G}_i - g_i^o}{T} = -R \ln x_i \tag{4-79}$$

It should be noted that since x_i is less than 1, we must have $\bar{S}_i > s_i^o$. The increase in entropy due to mixing at constant temperature and pressure is given by

$$\Delta S_{\text{mixing}} = \sum_i n_i (\bar{S}_i - s_i^o) = -\sum_i n_i R \ln x_i \tag{4-79a}$$

Accordingly, the entropy of an ideal solution is then

$$S = \sum_i n_i \bar{S}_i = \sum_i n_i s_i^o - \sum_i n_i R \ln x_i \tag{4-80}$$

Note that in Example 4-2 we obtained these same equations for an ideal gas mixture.

EXAMPLE 4-7. A tank having a total volume of 1 m^3 contains a gaseous mixture of 70 mole percent of ethane and 30 mole percent of nitrogen at 400°K and 200 atm. Determine the mass, enthalpy, and entropy of the mixture in the tank, assuming it to be an ideal solution. The following table gives the data for pure ethane and pure nitrogen at 400°K and 200 atm:

	volume, $v°$ cm^3/g mole	enthalpy, $h°$ joules/g mole	entropy, $s°$ joules/(g mole)(°K)
ethane	113.4	31,390	190.2
nitrogen	179.6	18,090	154.0

In this table the values of enthalpy and entropy are based on the value of zero for the perfect crystal at the absolute zero of temperature.

Solution.
According to Eq. (4-76) the molal volume of the mixture is

$$v = (x\bar{V})_{C_2H_6} + (x\bar{V})_{N_2} = (xv°)_{C_2H_6} + (xv°)_{N_2}$$

$$= 0.70 \times 113.4 + 0.30 \times 179.6 = 133 \text{ cm}^3/\text{g mole}$$

Similarly, the molal enthalpy of the mixture is

$$h = (x\bar{H})_{C_2H_6} + (x\bar{H})_{N_2} = (xh°)_{C_2H_6} + (xh°)_{N_2}$$

$$= 0.70 \times 31,390 + 0.30 \times 18,090 = 27,400 \text{ joules/g mole}$$

From Eq. (4-80) the molal entropy of the mixture is

$$s = \left[(xs°)_{C_2H_6} + (xs°)_{N_2}\right] - R\left[(x \ln x)_{C_2H_6} + (x \ln x)_{N_2}\right]$$

$$= (0.70 \times 190.2 + 0.30 \times 154.0) - 8.314(0.70 \ln 0.70 + 0.30 \ln 0.30)$$

$$= 184 \text{ joules/(g mole)(°K)}$$

The apparent or average molecular weight of the mixture is

$$M = (xM)_{C_2H_6} + (xM)_{N_2}$$

$$= 0.70 \times 30 + 0.30 \times 28 = 29.4 \text{ g/g mole}$$

Therefore the specific volume of the mixture is

$$v' = \frac{v}{M} = \frac{133}{29.4} = 4.52 \text{ cm}^3/\text{g}$$

and the mass of the mixture is

$$m = \frac{V}{v'} = \frac{10^6}{4.52} = 0.221 \times 10^6 \text{g} = 221 \text{ kg}$$

The number of moles of the mixture is

$$n = \frac{m}{M} = \frac{0.221 \times 10^6}{29.4} = 7{,}520 \text{ g moles}$$

Therefore the enthalpy of the mixture is

$$H = nh = 7{,}520 \times 27{,}400 = 2.06 \times 10^8 \text{ joules}$$

and the entropy of the mixture is

$$S = ns = 7{,}520 \times 184 = 1.38 \times 10^6 \text{ joules/}^\circ\text{K}$$

EXAMPLE 4-8. A low-grade natural gas mixture consisting of 70 mole percent of methane and 30 mole percent of nitrogen is compressed reversibly and isothermally in a steady-flow compressor from 10 atm and 220°K to 100 atm. The flow rate is 5 kg/min. Calculate the power requirement and the heat transfer rate, assuming the mixture to behave as (a) an ideal gas and (b) an ideal solution. The following table gives the enthalpy and entropy data for pure methane and pure nitrogen, based on the value of zero for the perfect crystal at 0°K.

Pure gas	State 1 10 atm and 220°K		State 2 100 atm and 220°K	
	enthalpy, h° joules/g mole	entropy, s° joules/(g mole)($^\circ$K)	enthalpy, h° joules/g mole	entropy, s° joules/(g mole)($^\circ$K)
methane	16,250	155.8	12,930	125.4
nitrogen	13,180	162.9	12,170	140.2

Solution.

The apparent or average molecular weight of the mixture is

$$M = (xM)_{\text{CH}_4} + (xM)_{\text{N}_2}$$

$$= 0.70 \times 16 + 0.30 \times 28 = 19.6 \text{ g/g mole}$$

(1) Consider the mixture as an ideal gas.

Since from Eq. (2-52), $dh = c_p \, dT$ for an ideal gas, there will be no change in enthalpy for the isothermal process, i.e.,

$$\Delta h_{12} = h_2 - h_1 = 0$$

Meanwhile, from Eq. (2-53b), for an isothermal process we have

$$ds = c_p \frac{dT}{T} - R \frac{dp}{p} = -R \frac{dp}{p}$$

Therefore the change in entropy of the mixture is

$$\Delta s_{12} = s_2 - s_1 = -R \ln \frac{p_2}{p_1} = -8.314 \ln \frac{100}{10}$$

$$= -19.1 \text{ joules/(g mole)(}^\circ\text{K)}$$

$$= -\frac{19.1}{19.6} = -0.974 \text{ joules/(g)(}^\circ\text{K)}$$

The heat transfer rate for the reversible isothermal compression process is given by

$$\dot{Q}_{12} = \dot{m} T_1 \Delta s_{12} = (5,000)(220)(-0.974)$$

$$= -1.07 \times 10^6 \text{ joules/min}$$

where the minus sign indicates that heat is removed from the mixture.

When the changes in kinetic and potential energies are neglected, the steady-flow energy balance yields

$$\dot{W}_{12} = \dot{Q}_{12} + \dot{m}(h_1 - h_2)$$

$$= \dot{Q}_{12} = -1.07 \times 10^6 \text{ joules/min} = -17.8 \text{ kw}$$

where the minus sign indicates that work is done on the mixture.

(2) Consider the mixture as an ideal solution.

From Eq. (4-15a) and (4-75), we have for the ideal solution

$$h = (x\bar{H})_{CH_4} + (x\bar{H})_{N_2} = (xh^\circ)_{CH_4} + (xh^\circ)_{N_2}$$

Since the mole fractions do not change during the change of state, it follows that

$$\Delta h_{12} = h_2 - h_1 = [x(h_2^\circ - h_1^\circ)]_{CH_4} + [x(h_2^\circ - h_1^\circ)]_{N_2}$$

$$= 0.70(12,930 - 16,250) + 0.30(12,170 - 13,180)$$

$$= -2,630 \text{ joules/g mole} = -\frac{2,630}{19.6} = -134 \text{ joules/g}$$

From Eq. (4-80), the molal entropy of the mixture is given by

$$s = \left[(xs^\circ)_{CH_4} + (xs^\circ)_{N_2} \right] - R \left[(x \ln x)_{CH_4} + (x \ln x)_{N_2} \right]$$

Whence the change in entropy of the mixture is given by

$$\Delta s_{12} = s_2 - s_1 = [x(s_2^\circ - s_1^\circ)]_{CH_4} + [x(s_2^\circ - s_1^\circ)]_{N_2}$$

$$= 0.70(125.4 - 155.8) + 0.30(140.2 - 162.9)$$

$$= -28.1 \text{ joules/(g mole)(}^\circ\text{K)} = -\frac{28.1}{19.6} = -1.43 \text{ joules/(g)(}^\circ\text{K)}$$

The heat transfer rate for the reversible isothermal compression process is then

$$\dot{Q}_{12} = \dot{m}T_1 \Delta s_{12} = (5,000)(220)(-1.43) = -1.57 \times 10^6 \text{ joules/min}$$

With the changes in kinetic and potential energies neglected, the power requirement is given by

$$\dot{W}_{12} = \dot{Q}_{12} + \dot{m}(h_1 - h_2) = -1.57 \times 10^6 + 5,000(134)$$

$$= -0.90 \times 10^6 \text{ joules/min} = -15.0 \text{ kw}$$

EXAMPLE 4-9. Solve part (2) of Example 4-8 by using generalized thermodynamic charts instead of the given enthalpy and entropy data for the pure components.

Solution.

The critical temperature and pressure of pure methane and pure nitrogen are as follows:

$$T_{c(CH_4)} = 190.7°K \qquad p_{c(CH_4)} = 45.8 \text{ atm}$$

$$T_{c(N_2)} = 126.2°K \qquad p_{c(N_2)} = 33.5 \text{ atm}$$

Since the mixture is assumed to behave as an ideal solution, as in Example 4-8, we have

$$\Delta h_{12} = h_2 - h_1 = [x(h_2° - h_1°)]_{CH_4} + [x(h_2° - h_1°)]_{N_2}$$

$$\Delta s_{12} = s_2 - s_1 = [x(s_2° - s_1°)]_{CH_4} + [x(s_2° - s_1°)]_{N_2}$$

We now use the generalized thermodynamic correction charts (Appendix Figs. A-17 and A-18) to evaluate the enthalpy and entropy changes for the pure components. For pure methane and pure nitrogen at the mixture temperature and pressure, the reduced coordinates are

$$T_{r1(CH_4)} = T_{r2(CH_4)} = \frac{T_1}{T_{c(CH_4)}} = \frac{220}{190.7} = 1.15$$

$$p_{r1(CH_4)} = \frac{p_1}{p_{c(CH_4)}} = \frac{10}{45.8} = 0.218$$

$$p_{r2(CH_4)} = \frac{p_2}{p_{c(CH_4)}} = \frac{100}{45.8} = 2.18$$

$$T_{r1(N_2)} = T_{r2(N_2)} = \frac{T_1}{T_{c(N_2)}} = \frac{220}{126.2} = 1.74$$

$$p_{r1(N_2)} = \frac{p_1}{p_{c(N_2)}} = \frac{10}{33.5} = 0.299$$

$$p_{r2(N_2)} = \frac{p_2}{p_{c(N_2)}} = \frac{100}{33.5} = 2.99$$

From the generalized enthalpy correction chart (Fig. A-17) we have

$$\left[\frac{h_1^{*o} - h_1^o}{T_c}\right]_{CH_4} = 0.34 \text{ cal/(g mole)(}^\circ K) \quad \text{at} \quad \begin{matrix} T_{r1(CH_4)} = 1.15 \\[2mm] p_{r1(CH_4)} = 0.218 \end{matrix}$$

where the superscript "*o" denotes a pure component in the ideal gas state. Also we have

$$\left[\frac{h_2^{*o} - h_2^o}{T_c}\right]_{CH_4} = 5.05 \text{ cal/(g mole)(}^\circ K) \quad \text{at} \quad \begin{matrix} T_{r2(CH_4)} = 1.15 \\[2mm] p_{r2(CH_4)} = 2.18 \end{matrix}$$

$$\left[\frac{h_1^{*o} - h_1^o}{T_c}\right]_{N_2} = 0.16 \text{ cal/(g mole)(}^\circ K) \quad \text{at} \quad \begin{matrix} T_{r1(N_2)} = 1.74 \\[2mm] p_{r1(N_2)} = 0.299 \end{matrix}$$

$$\left[\frac{h_2^{*o} - h_2^o}{T_c}\right]_{N_2} = 1.86 \text{ cal/(g mole)(}^\circ K) \quad \text{at} \quad \begin{matrix} T_{r2(N_2)} = 1.74 \\[2mm] p_{r2(N_2)} = 2.99 \end{matrix}$$

Since in ideal gas states the enthalpy is a function of temperature only, it follows that

$$h_{1(CH_4)}^{*o} = h_{2(CH_4)}^{*o}$$

$$h_{1(N_2)}^{*o} = h_{2(N_2)}^{*o}$$

Therefore

$$\left[\frac{h_2^o - h_1^o}{T_c}\right]_{CH_4} = \left[\frac{h_1^{*o} - h_1^o}{T_c}\right]_{CH_4} - \left[\frac{h_2^{*o} - h_2^o}{T_c}\right]_{CH_4}$$

$$= 0.34 - 5.05 = -4.71 \text{ cal/(g mole)(}^\circ K)$$

Whence

$$[h_2^o - h_1^o]_{CH_4} = -4.71 T_{c(CH_4)} = -4.71 \times 190.7 = -898 \text{ cal/g mole}$$

Similarly

$$\left[\frac{h_2^o - h_1^o}{T_c}\right]_{N_2} = 0.16 - 1.86 = -1.70 \text{ cal/(g mole)(}^\circ K)$$

Whence

$$[h_2^o - h_1^o]_{N_2} = -1.70 T_{c(N_2)} = -1.70 \times 126.2 = -215 \text{ cal/g mole}$$

Hence, for the mixture we have

$$\Delta h_{12} = h_2 - h_1 = [x(h_2^\circ - h_1^\circ)]_{CH_4} + [x(h_2^\circ - h_1^\circ)]_{N_2}$$

$$= 0.70(-898) + 0.30(-215) = -693 \text{ cal/g mole}$$

$$= -\frac{693}{19.6} = -35.4 \text{ cal/g} = -148 \text{ joules/g}$$

From the generalized entropy correction chart (Fig. A-18) we have

$$[s_1^{*\circ} - s_1^\circ]_{CH_4} = 0.22 \text{ cal/(g mole)(}^\circ\text{K)} \quad \text{at} \quad \begin{aligned} T_{r1(CH_4)} &= 1.15 \\ p_{r1(CH_4)} &= 0.218 \end{aligned}$$

$$[s_2^{*\circ} - s_2^\circ]_{CH_4} = 3.34 \text{ cal/(g mole)(}^\circ\text{K)} \quad \text{at} \quad \begin{aligned} T_{r2(CH_4)} &= 1.15 \\ p_{r2(CH_4)} &= 2.18 \end{aligned}$$

$$[s_1^{*\circ} - s_1^\circ]_{N_2} = 0.06 \text{ cal/(g mole)(}^\circ\text{K)} \quad \text{at} \quad \begin{aligned} T_{r1(N_2)} &= 1.74 \\ p_{r1(N_2)} &= 0.299 \end{aligned}$$

$$[s_2^{*\circ} - s_2^\circ]_{N_2} = 0.82 \text{ cal/(g mole)(}^\circ\text{K)} \quad \text{at} \quad \begin{aligned} T_{r2(N_2)} &= 1.74 \\ p_{r2(N_2)} &= 2.99 \end{aligned}$$

From Eq. (2-53b)

$$ds = c_p \frac{dT}{T} - R \frac{dp}{p} = -R \frac{dp}{p}$$

for an isothermal process of an ideal gas. Thus

$$[s_2^{*\circ} - s_1^{*\circ}]_{CH_4} = [s_2^{*\circ} - s_1^{*\circ}]_{N_2} = -R \ln \frac{p_2}{p_1} = -1.986 \ln \frac{100}{10}$$

$$= -4.57 \text{ cal/(g mole)(}^\circ\text{K)}$$

Therefore

$$[s_2^\circ - s_1^\circ]_{CH_4} = [s_1^{*\circ} - s_1^\circ]_{CH_4} + [s_2^{*\circ} - s_1^{*\circ}]_{CH_4} - [s_2^{*\circ} - s_2^\circ]_{CH_4}$$

$$= 0.22 - 4.57 - 3.34 = -7.69 \text{ cal/(g mole)(}^\circ\text{K)}$$

Similarly

$$[s_2^\circ - s_1^\circ]_{N_2} = 0.06 - 4.57 - 0.82 = -5.33 \text{ cal/(g mole)(}^\circ\text{K)}$$

Hence, for the mixture we have

$$\Delta s_{12} = s_2 - s_1 = [x(s_2^\circ - s_1^\circ)]_{CH_4} + [x(s_2^\circ - s_1^\circ)]_{N_2}$$

$$= 0.70(-7.69) + 0.30(-5.33)$$

$$= -6.98 \text{ cal/(g mole)}(^\circ K) = -\frac{6.98}{19.6}$$

$$= -0.356 \text{ cal/(g)}(^\circ K) = -1.49 \text{ joules/(g)}(^\circ K)$$

With the changes in entropy and enthalpy known, we are now ready to calculate the rate of heat transfer and the power requirement. The heat transfer rate for the reversible isothermal compression process is given by

$$\dot{Q}_{12} = \dot{m}T_1 \Delta s_{12} = (5,000)(220)(-1.49) = -1.64 \times 10^6 \text{ joules/min}$$

With the changes in kinetic and potential energies neglected, the steady-flow energy balance yields

$$\dot{W}_{12} = \dot{Q}_{12} + \dot{m}(h_1 - h_2) = -1.64 \times 10^6 + 5,000(148)$$
$$= -0.90 \times 10^6 \text{ joules/min} = -15.0 \text{ kw}$$

4-7 DILUTE LIQUID SOLUTIONS

Having examined the characteristics of ideal solutions in general, we now seek to study the particular case of dilute nonelectrolyte liquid solutions. One component predominates in a dilute solution. This component is called the solvent and the other components are called solutes. We will denote the solvent by the subscript 1 and the solutes by the subscripts 2, 3, etc.

There are several laws which describe the behavior of infinitely dilute solutions. Although these laws are essentially empirical in nature and may be stated independently, they are, nevertheless, thermodynamically interrelated. Let us begin with the empirical statement called *Henry's law*: the fugacity of a solute of a dilute liquid solution is proportional to its mole fraction. That is to say

$$f_2 = k_2 x_2 \tag{4-81}$$

where k_2 is the proportional constant. Originally Henry's law was stated in terms of partial-vapor pressure instead of fugacity. Thus

$$p_2 = k_2 x_2 \tag{4-82}$$

With Henry's law as a starting point, we can now deduce another law for an infinitely dilute solution. For a binary solution we have the equation

$$\left(\frac{\partial \ln f_1}{\partial \ln x_1}\right)_{T,p} = \left(\frac{\partial \ln f_2}{\partial \ln x_2}\right)_{T,p} \tag{4-69a}$$

Now, if Henry's law holds for the solute, that is, $f_2 = k_2 x_2$, then

$$\ln f_2 = \ln k_2 + \ln x_2$$

or

$$\left(\frac{\partial \ln f_2}{\partial \ln x_2}\right)_{T,p} = 1$$

substituting this equation in Eq. (4-69a) results in

$$\left(\frac{\partial \ln f_1}{\partial \ln x_1}\right)_{T,p} = 1$$

Upon integration at constant T and p,

$$\int_{f_1^\circ}^{f_1} d \ln f_1 = \int_{x_1 = 1}^{x_1} d \ln x_1$$

we have

$$\ln \frac{f_1}{f_1^\circ} = \ln x_1$$

or

$$f_1 = f_1^\circ x_1 \tag{4-83}$$

where f_1° is the fugacity of the pure solvent at the same temperature and pressure as of the solution. If the vapor in equilibrium with the liquid solution may be considered as an ideal gas, we may substitute vapor pressure for fugacity, thus

$$p_1 = p_1^\circ x_1 \tag{4-84}$$

This is known as *Raoult's law*, where p_1 is the partial vapor pressure of the solvent, and p_1° is the vapor pressure of pure solvent at the temperature and pressure of the solution. Since p_1° is a strong function of the temperature and a weak function of the pressure, it may be considered as a function of temperature alone for the conditions we will be concerned with.

We see from the foregoing derivation that in an infinitely dilute solution when the solute obeys Henry's law, the solvent obeys Raoult's law, and vice versa. As a matter of fact, this statement can be extended to include any number of solutes in the same dilute solution.

It is clear from Eqs. (4-81) and (4-83) that in an infinitely dilute solution the fugacities of the solutes and the solvent are proportional to their respective mole fraction. Hence in this respect every infinitely dilute solution is an ideal solution. In the case of a solute, however, the proportional constant can not be directly determined from a knowledge of its pure state alone.

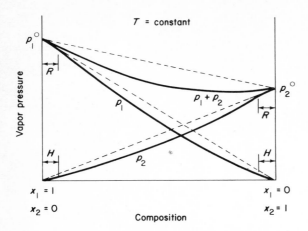

FIG. 4-5. Vapor pressure-composition diagram for a binary solution showing deviations from Raoult's law.

Figure 4-5 is a vapor-pressure versus composition diagram of a binary solution showing deviations from Raoult's law. The solid lines depict experimental curves, and the dotted lines depict the behavior of a solution which obeys Raoult's law throughout all compositions. The letters R and H indicate the respective regions for the actual solution in which Raoult's law and Henry's law hold. The curves in Fig. 4-5 represent a system that exhibits negative deviations from Raoult's law, meaning that actual values of vapor pressure are below those predicted by the law. Other systems may exhibit positive deviations.

4-8 ACTIVITY AND NONIDEAL LIQUID SOLUTIONS

In order to describe the behavior of real gas mixtures a new property called the fugacity was introduced in Sec. 4-5. This concept has also been extended to liquid (or solid) solutions by defining the fugacity in connection with the vapor in equilibrium with the liquid (or solid) solutions. However, the fugacities of condensed phases are often difficult to determine, particularly when dealing with almost involatile substances because of the smallness of their vapor pressures under ordinary conditions. Fortunately we are interested only in the ratio of fugacities of a substance in two different states, not the absolute value of either fugacity. We shall now define another new property in terms of the ratio of fugacities, which can readily be applied to liquid (or solid) solutions.

The ratio of the fugacity f_i of component i in any given state to its fugacity $f_i^{\oplus}$ in some standard state, taken at the same temperature, is defined as its *activity* a_i. Thus

$$a_i = \left(\frac{f_i}{f_i^{\oplus}}\right)_T \tag{4-85}$$

Integrating Eq. (4-62) at constant temperature between the given state and the standard state, and substituting Eq. (4-85) into it results in

$$\bar{G}_i - \bar{G}_i^{\oplus} = \mu_i - \mu_i^{\oplus} = RT \ln\left(\frac{f_i}{f_i^{\oplus}}\right) = RT \ln a_i \tag{4-86}$$

Another parameter commonly used in the description of nonideal solutions is the *activity coefficient* γ, which for any component is defined as the ratio of its activity to mole fraction in the given state, or

$$\gamma_i = \frac{a_i}{x_i} \tag{4-87}$$

The choice of a standard state is entirely arbitrary. One standard state is chosen for each component at each state of aggregation. The temperature of the system is always chosen as the temperature of the standard state. Thus temperature is not to be a fixed reference value although the pressure or concentration of the standard state for each state of aggregation shall be arbitrarily fixed. We now give the conventional standard states which are generally adopted for the sake of convenience.

For a single-component system it is convenient to choose the pure component in its appropriate state of aggregation (gas, liquid, or solid) at the system temperature and 1 atm pressure as the standard state. In the case of liquids (or solids) this standard state becomes hypothetical when the vapor pressure of the pure liquid (or pure solid) exceeds 1 atm.

For gaseous components in a mixture there are two common choices for a standard state. The pure component gas either at 1 atm pressure or at the system pressure can be chosen. When the pure component gas at the system temperature and 1 atm pressure is chosen, the standard state shall have a fugacity of 1 atm. Accordingly, the activity and fugacity of the gas component in the mixture are numerically equal. For ideal gas mixtures, since the fugacity and partial pressure of a component are identical, its activity and partial pressure are therefore numerically equal. On the other hand, when the pure-component gas at the system temperature and pressure is chosen as the standard state, then according to Eqs. (4-71) and (4-85) $f_i^{\circ} = f_i^{\oplus}$ and the activity of each component gas becomes equal to its mole fraction in ideal solutions.

For the solvent of a liquid solution, the standard state is usually taken as the pure solvent (or, in other words, in the limit of the infinitely dilute solution) at the temperature and pressure of the solution. Thus by definition,

$$f_1^\oplus = f_1^\circ \tag{4-88}$$

where f_1° is the fugacity of pure solvent at the temperature and pressure of the solution. Figure 4-6 is a fugacity-composition diagram, indicating the standard state as the limiting case of ideality. If the solution itself is ideal, then from Eqs. (4-71), (4-85) and (4-88), we have

$$a_1 = \frac{f_1}{f_1^\oplus} = \frac{f_1^\circ x_1}{f_1^\circ} = x_1 \tag{4-89}$$

That is, in an ideal solution the activity of the solvent equals its mole fraction. Substituting Eq. (4-89) into Eq. (4-87) we have

$$\gamma_1 = \frac{a_1}{x_1} = 1$$

That is, the activity coefficient of the solvent is always unity in an ideal solution. Consequently, departures of the activity coefficient from unity then indicate nonideal behavior of the solution.

For a solute in a liquid solution, the standard state is usually chosen in such a way that the activity shall be equal to mole fraction in very dilute solutions where Henry's law is applicable. Thus,

$$\lim_{x_2 \to 0} \left(\frac{a_2}{x_2} \right) = \lim_{x_2 \to 0} \left(\frac{f_2}{f_2^\oplus x_2} \right) = 1$$

$$\lim_{x_2 \to 0} \left(\frac{f_2}{x_2} \right) = \lim_{x_2 \to 0} \left(\frac{k_2 x_2}{x_2} \right) = k_2 \quad \text{or} \quad \lim_{x_2 \to 0} \left(\frac{f_2}{k_2 x_2} \right) = 1$$

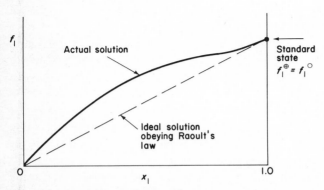

FIG. 4-6. Standard state for the solvent.

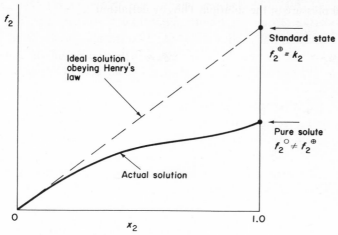

FIG. 4-7. Standard state for the solute.

where k_2 is the constant in Henry's law. In order to satisfy the preceding two equations simultaneously, the standard state should be that state in which $f_2^\oplus$ and k_2 are numerically equal. This standard state can be found graphically (Fig. 4-7) by extrapolation of the line which represents Henry's law for the dilute solution to a concentration of $x_2 = 1$. It should be understood that the standard state so defined is hypothetical and is not an actual state of the pure solute.

In Sec. 4-5 we derived the Gibb-Duhem relation between fugacity and mole fraction for a binary system. This relationship can also be expressed in terms of activity and activity coefficient. Since $a_i = f_i/f_i^\oplus$ and $f_i^\oplus$ is a constant at a given temperature and pressure, Eq. (4-69) can be written as

$$x_1 \left[\frac{\partial(\ln a_1)}{\partial x_1} \right]_{T,p} = x_2 \left[\frac{\partial(\ln a_2)}{\partial x_2} \right]_{T,p} \tag{4-90}$$

In terms of activity coefficient, since $a_i = \gamma_i x_i$, the preceding equation becomes

$$x_1 \left[\frac{\partial(\ln \gamma_1)}{\partial x_1} \right]_{T,p} + x_1 \left[\frac{d(\ln x_1)}{dx_1} \right] = x_2 \left[\frac{\partial(\ln \gamma_2)}{\partial x_2} \right]_{T,p} + x_2 \left[\frac{d(\ln x_2)}{dx_2} \right]$$

or

$$x_1 \left[\frac{\partial(\ln \gamma_1)}{\partial x_1} \right]_{T,p} = x_2 \left[\frac{\partial(\ln \gamma_2)}{\partial x_2} \right]_{T,p} \tag{4-91}$$

The various forms of the Gibbs-Duhem relation are useful in correlating thermodynamic properties of binary solutions. Since the activity coefficient is a direct measure of nonideality of solutions, Eq. (4-91) is the most useful form. This equation is particularly valuable when the activity coefficient of one component of a binary solution is to be calculated from that of the other component.

Since the activity of a substance is, in general, a function of both temperature and pressure, we now proceed to develop the equations expressing the temperature and pressure effects on activity for a solution of fixed composition. These equations will enable us to calculate the activity at one temperature or pressure from data at another temperature or pressure. The effect of temperature on activity is readily obtained by applying Eq. (4-68) to both f_i and $f_i^\oplus$. Thus

$$\left(\frac{\partial \ln a_i}{\partial T}\right)_p = \left[\frac{\partial \ln (f_i/f_i^\oplus)}{\partial T}\right]_p = \frac{\bar{H}_i^* - \bar{H}_i}{RT^2} - \frac{\bar{H}_i^* - \bar{H}_i^\oplus}{RT^2} = -\frac{\bar{H}_i - \bar{H}_i^\oplus}{RT^2} \qquad (4\text{-}92)$$

Likewise the effect of pressure on activity is readily obtained by applying Eq. (4-67) to both f_i and $f_i^\oplus$. Thus

$$\left(\frac{\partial \ln a_i}{\partial p}\right)_T = \left[\frac{\partial \ln (f_i/f_i^\oplus)}{\partial p}\right]_T = \frac{\bar{V}_i - \bar{V}_i^\oplus}{RT} \qquad (4\text{-}93)$$

Since $\gamma_i = a_i/x_i$ and x_i is a constant for a solution of fixed composition, the foregoing two equations give also the effects of temperature and pressure on activity coefficient.

With the activity and activity coefficient defined we turn now to express other thermodynamic properties of nonideal solutions in terms of these newly defined properties. Differentiating Eq. (4-86) with respect to temperature, keeping pressure and composition constant, results in

$$\left(\frac{\partial \mu_i}{\partial T}\right)_{p,x} = \left(\frac{\partial \mu_i^\oplus}{\partial T}\right)_{p,x} + R \ln a_i + RT\left[\frac{\partial(\ln a_i)}{\partial T}\right]_{p,x}$$

where the subscript x implies constancy of composition. But from Eq. (4-19) we have

$$\left(\frac{\partial \mu_i}{\partial T}\right)_{p,x} = -\bar{S}_i \quad \text{and} \quad \left(\frac{\partial \mu_i^\oplus}{\partial T}\right)_{p,x} = -\bar{S}_i^\oplus$$

Hence

$$\bar{S}_i = \bar{S}_i^\oplus - R \ln a_i - RT \left[\frac{\partial(\ln a_i)}{\partial T}\right]_{p,x} \tag{4-94}$$

In terms of activity coefficient, this equation becomes

$$\bar{S}_i = \bar{S}_i^\oplus - R \ln (\gamma_i x_i) - RT \left[\frac{\partial(\ln \gamma_i)}{\partial T}\right]_{p,x} \tag{4-94a}$$

Multiplying Eq. (4-94) by T and adding the resulting expression to Eq. (4-86) lead to

$$\mu_i + T\bar{S}_i = \mu_i^\oplus + T\bar{S}_i^\oplus - RT^2 \left[\frac{\partial(\ln a_i)}{\partial T}\right]_{p,x}$$

But from Eqs. (4-18) and (4-23) we have

$$\bar{H}_i = \bar{G}_i + T\bar{S}_i = \mu_i + T\bar{S}_i$$

$$\bar{H}_i^\oplus = \bar{G}_i^\oplus + T\bar{S}_i^\oplus = \mu_i^\oplus + T\bar{S}_i^\oplus$$

Hence

$$\bar{H}_i = \bar{H}_i^\oplus - RT^2 \left[\frac{\partial(\ln a_i)}{\partial T}\right]_{p,x} \tag{4-95}$$

In terms of activity coefficient this equation becomes

$$\bar{H}_i = \bar{H}_i^\oplus - RT^2 \left[\frac{\partial(\ln \gamma_i)}{\partial T}\right]_{p,x} \tag{4-95a}$$

Differentiating Eq. (4-86) with respect to pressure, keeping temperature and composition constant, results in

$$\left(\frac{\partial \mu_i}{\partial p}\right)_{T,x} = \left(\frac{\partial \mu_i^\oplus}{\partial p}\right)_{T,x} + RT \left[\frac{\partial(\ln a_i)}{\partial p}\right]_{T,x}$$

But from Eq. (4-20) we have

$$\left(\frac{\partial \mu_i}{\partial p}\right)_{T,x} = \bar{V}_i \quad \text{and} \quad \left(\frac{\partial \mu_i^\oplus}{\partial p}\right)_{T,x} = \bar{V}_i^\oplus$$

Hence

$$\bar{V}_i = \bar{V}_i^\oplus + RT \left[\frac{\partial(\ln a_i)}{\partial p}\right]_{T,x} \tag{4-96}$$

In terms of activity coefficient this equation becomes

$$\bar{V}_i = \bar{V}_i^\oplus + RT \left[\frac{\partial(\ln \gamma_i)}{\partial p}\right]_{T,x} \tag{4-96a}$$

In Eqs. (4-92) to (4-96) we introduced the new properties $\bar{H}_i^\oplus$, $\bar{V}_i^\oplus$, and $\bar{S}_i^\oplus$. Some comments are in order concerning these new properties. When the equations are applied to the solvent ($i = 1$) of a solution, the values of $\bar{H}_1^\oplus$, $\bar{V}_1^\oplus$, and $\bar{S}_1^\oplus$ are simply the values for the pure solvent at the temperature and pressure of the solution. On the other hand, when the equations are applied to a solute ($i = 2$), since the solute standard state is defined in such a way that $\gamma_2 = 1$ in infinite dilute solutions, the values of $\bar{H}_2^\oplus$ and $\bar{V}_2^\oplus$ are the values at infinite dilution and at the temperature and pressure of the solution. This can be seen from Eqs. (4-95a) and (4-96a) by imposing $\gamma_2 = 1$ at $x_2 \to 0$. Thus

$$\bar{H}_2^\oplus = \lim_{x_2 \to 0} \bar{H}_2 \quad \text{and} \quad \bar{V}_2^\oplus = \lim_{x_2 \to 0} \bar{V}_2$$

However, from Eq. (4-94a), letting $\gamma_2 = 1$ at $x_2 \to 0$, we have

$$\bar{S}_2^\oplus = \lim_{x_2 \to 0} (\bar{S}_2 + R \ln x_2)$$

It follows that $\bar{S}_2^\oplus$ is not equal to $\bar{S}_2$ at infinite dilution.

It is worthwhile drawing attention here to notice that when the definition of an ideal solution, namely $f_i = f_i^\circ x_i$, is imposed, and when the pure component at the temperature and pressure of the solution is chosen as the standard state, we have

$$a_i = \frac{f_i}{f_i^\oplus} = \frac{f_i^\circ x_i}{f_i^\circ} = x_i$$

so that Eqs. (4-94), (4-95), and (4-96) reduce to Eqs. (4-79), (4-75), and (4-73) respectively for ideal solutions.

PROBLEMS

4-1. By the use of Euler's theorem derive Eq. (4-17).

4-2. Derive the following relations for several partial molal properties from their equivalent relations that apply to the multicomponent system as a whole:

(a) $\bar{S}_i = -\left(\dfrac{\partial \mu_i}{\partial T}\right)_{p,n}$

(b) $\bar{V}_i = \left(\dfrac{\partial \mu_i}{\partial p}\right)_{T,n}$

(c) $\bar{H}_i = \mu_i - T\left(\dfrac{\partial \mu_i}{\partial T}\right)_{p,n}$

(d) $\bar{A}_i = \mu_i - p\left(\dfrac{\partial \mu_i}{\partial p}\right)_{T,n}$

(e) $\bar{U}_i = \mu_i - T\left(\dfrac{\partial \mu_i}{\partial T}\right)_{p,n} - p\left(\dfrac{\partial \mu_i}{\partial p}\right)_{T,n}$

4-3. Verify that
   (a) For a single-component system

$$\left[\frac{\partial(\mu/T)}{\partial T}\right]_p = -\frac{h}{T^2}$$

   (b) For the ith component in a multicomponent system

$$\left[\frac{\partial(\mu_i/T)}{\partial T}\right]_{p,n} = -\frac{\bar{H}_i}{T^2}$$

4-4. The following data are for a system of binary liquid solution of ethane (C_2H_6) and n-pentane (C_5H_{12}) at $220°F$ and 2,000 psia.

mole fraction of ethane	0.2	0.3	0.4	0.5	0.6	0.7	0.8
molal volume, cu ft/lb mole	1.913	1.851	1.798	1.754	1.717	1.708	1.757

Determine the partial molal volumes of ethane and n-pentane for a solution of 50 mole percent ethane and 50 mole percent of n-pentane at $220°F$ and 2,000 psia.

4-5. The compressibility factor Z for a system of binary liquid solution of methane (CH_4) and n-pentane (C_5H_{12}) at $100°F$ and 4,000 psia are listed as a function of the mole fraction of methane in the following table.

x_{CH_4}	0.2	0.3	0.4	0.5	0.6	0.7	0.8	0.9
Z	1.0775	1.0182	0.9623	0.9097	0.8651	0.8358	0.8291	0.8491

Compute the partial molal volumes of the components for a solution of 60 mole percent of methane and 40 mole percent of n-pentane at $100°F$ and 4,000 psia.

4-6. The partial molal enthalpies for a system of binary liquid solution consisting of 40 mole percent of methane (CH_4) and 60 mole percent of n-butane (C_4H_{10}) at $100°F$ are as follows:

At 1,500 lbf/in.^{2} abs $\overline{H}_{CH_4} = 3,790$ Btu/lb mole

$\overline{H}_{C_4H_{10}} = 1,290$ Btu/lb mole

At 3,000 lbf/in.^{2} abs $\overline{H}_{CH_4} = 3,380$ Btu/lb mole

$\overline{H}_{C_4H_{10}} = 1,700$ Btu/lb mole

Compute the change in enthalpy of the solution when it is compressed isothermally at $100°F$ from 1,500 to 3,000 psia.

4-7. The equations of state of a gaseous mixture and of each of its component gases are all assumed to be in the following virial form

$$pv = RT + \frac{B}{v} + \frac{C}{v^2} + \frac{D}{v^3} + \cdots,$$

where v denotes the molal volume. Derive the expressions for the virial coefficients in the equation of state of the mixture in terms of the virial coefficients in the equation of state of the component gases, assuming that the Dalton's law holds for the mixture.

4-8. A gaseous mixture contains 70 mole percent of methane and 30 mole percent of nitrogen. Calculate the specific volume of the mixture at $0°F$ and 1,500 psia by using (a) the ideal gas equation, and (b) the pseudocritical concept. The experimental value of the compressibility factor of this mixture at $0°F$ and 1,500 psia is 0.82. What is the true value of the specific volume?

4-9. Using the seven methods as stated in Example 4-3, estimate the pressure of a natural gas mixture containing 70 mole percent of methane (CH_4), 20 mole percent of ethane (C_2H_6), and 10 mole percent of nitrogen, and having a temperature of $70°C$ and a molal volume of 0.2 liter/g mole.

4-10. A rigid insulated tank of 0.2 cu m capacity contains a natural gas mixture of 80 mole percent methane and 20 mole percent ethane at a pressure of 30 bars and a temperature of $40°C$. Due to leakage through a faulty valve, the pressure drops to 20 bars before the valve is repaired. By the use of the Kay's rule and generalized thermodynamic charts, calculate the mass of mixture that leaks out from the tank, assuming that the mass that remains in the tank expands reversibly and adiabatically.

4-11. A tank having a total volume of 1 m^3 contains a gaseous mixture of 70 mole percent of ethane and 30 mole percent of nitrogen at 311°K and 170 atm. Determine the mass of the mixture in the tank assuming it to be (a) an ideal gas and (b) an ideal solution. The compressibility factors of pure ethane and pure nitrogen at 311°K and 170 atm are 0.510 and 1.045, respectively. The experimental value of the compressibility factor of this mixture at 311°K and 170 atm is 0.629. What is the real mass in the tank?

4-12. A mixture of 70 mole percent methane and 30 mole percent nitrogen is compressed irreversibly in an adiabatic steady-flow compressor from 5 atm, −100°C to 15 atm, 10°C. The flow rate is 1 kg/sec. Calculate the rate of increase in entropy and the power requirement, assuming the mixture to behave as an ideal solution. (a) Use generalized thermodynamic charts. (b) Use the following enthalpy and entropy data for the pure components.

Pure gas	State 1 5 atm and − 100°C		State 2 15 atm and 10°C	
	enthalpy, $h°$ joules/g mole	entropy, $s°$ joules/(g mole)(°K)	enthalpy, $h°$ joules/g mole	entropy, $s°$ joules/(g mole)(°K)
methane	14,730	153.6	18,420	161.3
nitrogen	11,840	161.8	15,000	166.9

The constant-pressure specific heats of pure methane and pure nitrogen at 1 atm may be assumed to be 34.4 joules/(g mole)(°K) and 29.0 joules/(g mole)(°K) respectively.

4-13. Solve the preceding problem, assuming the gas mixture to obey (a) the ideal gas equation, and (b) the Redlich-Kwong equation. Use the linear square root combination for the constant a and the linear combination for the constant b.

4-14. A gas mixture of 70 mole percent methane and 30 mole percent nitrogen at 100 atm and 300°K expands with a negligible initial velocity through an insulated nozzle to a final pressure of 35 atm. The final temperature is reported to be 255°K. Determine the possibility of this process and calculate the exit velocity by the use of (a) ideal gas relations and (b) Kay's pseudocritical rule and generalized thermodynamic charts.

4-15. For dilute liquid solutions, considering Raoult's law as an empirical rule, derive Henry's law.

Chapter 5

MULTICOMPONENT PHASE EQUILIBRIUM

5-1 CRITERIA OF EQUILIBRIUM

We recall that one of the corollaries of the second law is the principle of increase of entropy, expressed symbolically

$$(dS)_{\text{isolated system}} \geqslant 0 \qquad (1\text{-}23)$$

where the inequality holds for all natural processes where irreversible effects are inevitable. This equation provides a criterion for thermodynamic equilibrium. When an isolated system undergoes a natural process, only those states having higher entropies than the initial state are possible to attain. As a matter of fact, an isolated system always proceeds continuously from states of lower entropy to states of higher entropy until it attains a state of maximum entropy. At this state no further natural process is possible and the system is in a state of equilibrium.

Let us consider a closed simple compressible system, either homogeneous or heterogeneous, containing various substances that can transport between phases and interact chemically. Such a system is not in chemical equilibrium. However, let us suppose that the system has already attained equality of temperature and pressure throughout; thus it is in thermal and mechanical equilibrium. This system and its environment constitute an isolated system. Suppose that the environment is at a uniform temperature T and undergoes nothing but reversible processes. Let a quantity of heat dQ be transferred from the environment into the system, causing a change in entropy of the environment

$$dS_0 = \frac{-dQ}{T} \qquad (5\text{-}1)$$

Meanwhile, this heat flowing into the system causes the system to undergo some process, including phase or chemical changes, with a resulting increase in entropy

dS. Applying Eq. (1-23) to the system-environment combination gives

$$dS + dS_0 \geqslant 0$$

When incorporated with Eq. (5-1), this equation becomes

$$dS - \frac{dQ}{T} \geqslant 0 \quad \text{or} \quad dQ - T\,dS \leqslant 0$$

During the infinitesimal process, the system has a change in internal energy dU and performs an amount of work $p\,dV$ on the environment, where p is the pressure exerted by the environment on the system. Hence by the first law,

$$dQ = dU + p\,dV$$

Therefore we finally have

$$dU + p\,dV - T\,dS \leqslant 0 \qquad (5\text{-}2)$$

where the equality holds if the system is already in complete equilibrium, and the inequality holds if it is not. Equation (5-2) is the general condition of equilibrium for closed simple compressible systems.

Upon combining Eq. (5-2) and the defining equations of enthalpy, Helmholtz function, and Gibbs function, we have

$$dH - V\,dp - T\,dS \leqslant 0 \qquad (5\text{-}3)$$

$$dA + p\,dV + S\,dT \leqslant 0 \qquad (5\text{-}4)$$

$$dG - V\,dp + S\,dT \leqslant 0 \qquad (5\text{-}5)$$

The condition of equilibrium as expressed in Eqs. (5-2) to (5-5) is too general to be of direct utility. More useful special forms, applicable to a system subject to additional restrictions, are given below.

For a closed system at constant energy U and constant volume V, Eq. (5-2) reduces to

$$(dS)_{U,V} \geqslant 0 \qquad (5\text{-}6)$$

meaning that for all processes at constant U and V the entropy increases or, in the limit, remains constant. Therefore when a closed system is in equilibrium at constant U and V, its entropy must be at a maximum. Note that Eq. (5-6) is equivalent to Eq. (1-23).

On the other hand, for a closed system at constant entropy S and constant volume V, Eq. (5-2) reduces to

$$(dU)_{S,V} \leqslant 0 \qquad (5\text{-}7)$$

Thus the energy of a closed system at constant S and V decreases during a natural process, becoming a minimum at the final equilibrium state for which the equality sign in Eq. (5-7) holds.

Similarly, from Eqs. (5-3) to (5-5) we have the following conditions:

$$(dS)_{H,p} \geqslant 0 \tag{5-8}$$

$$(dH)_{S,p} \leqslant 0 \tag{5-9}$$

$$(dA)_{T,V} \leqslant 0 \tag{5-10}$$

$$(dG)_{T,p} \leqslant 0 \tag{5-11}$$

It is imperative to note that the inequality expressions in Eqs. (5-2) to (5-11) specify the conditions for natural processes to occur, thus carrying the system toward equilibrium. After the state of equilibrium has been reached, the equality sign in these equations holds, and no more natural process is possible. Thus, for example, the expression $(dS)_{U,V} > 0$ is the condition for natural processes to occur at constant U and V for a system not in equilibrium. On the other hand, if we find $(\delta S)_{U,V} < 0$ for all possible variations away from a state, these variations are not really possible and no change in the system can really take place; this state must be an equilibrium state of the system. Therefore the criterion for equilibrium under constant U and V may be rewritten in the following alternative formulation:

$$(\delta S)_{U,V} \leqslant 0 \tag{5-6a}$$

Here we use the symbol δ to denote small but finite hypothetical or virtual variations away from equilibrium to distinguish from the symbol d for differential changes toward equilibrium.

Similarly we rewrite the other criteria for equilibrium in the following alternative formulations in terms of virtual variations:

$$\delta U + p\,\delta V - T\,\delta S \geqslant 0 \tag{5-2a}$$

$$(\delta U)_{S,V} \geqslant 0 \tag{5-7a}$$

$$(\delta S)_{H,p} \leqslant 0 \tag{5-8a}$$

$$(\delta H)_{S,p} \geqslant 0 \tag{5-9a}$$

$$(\delta A)_{T,V} \geqslant 0 \tag{5-10a}$$

$$(\delta G)_{T,p} \geqslant 0 \tag{5-11a}$$

Since virtual variations are not necessarily infinitesimal, we need to retain the second-order terms in writing the differential relations for virtual variations based on the definitions of H, A, and G. For instance, from the definition that $A = U - TS$, we write

$$\delta A = \delta U - T \delta S - S \delta T - \delta S \delta T \quad \text{or} \quad \delta U - T \delta S = \delta A + S \delta T + \delta S \delta T$$

Substituting this relation into Eq. (5-2a) leads to

$$\delta A + p \delta V + S \delta T + \delta S \delta T \geqslant 0 \qquad (5\text{-}4a)$$

This is the alternative expression for Eq. (5-4). Notice that the second-order term introduced in this expression does not affect the condition expressed in Eq. (5-10a).

According to Eq. (5-6a), when we are to testify whether a system is in a state of equilibrium or not, the procedure is to choose a neighboring new state which has the same U and V as the state under test but differs in some other variables. Then the virtual variation $(\delta S)_{U,V}$ for the transition from the state under test to the proposed neighboring new state is calculated. This calculation is repeated for all possible neighboring states. If $(\delta S)_{U,V} < 0$ for all such neighboring states, then the state under test is truly an equilibrium state. In such calculations one should note that unless the state under test and the neighboring state used in the testification are both equilibrium states, their thermodynamic properties are not defined. Hence one must imagine that the needed equilibrium conditions in the states can be established by some form of additional constraints, such as the insertion of semipermeable membranes or impervious partitions. The state under test is a true equilibrium state if $(\delta S)_{U,V} < 0$ for all possible variations after the additional contraints are removed.

5-2 STABILITY

The criteria as expressed in Eqs. (5-6a) to (5-11a) require that as a condition for equilibrium the first-order variation of a given parameter under certain constraints be equal to zero. However, in order to insure the *stability* of equilibrium it is necessary that the condition of minimum or maximum be satisfied for second- (or higher-) order variations. Let us consider a homogeneous single-component simple compressible closed system and start from the general condition that

$$\delta U + p \delta V - T \delta S > 0 \qquad (5\text{-}2a)$$

A virtual change of entropy and volume in the system would lead to a virtual change in its energy. To insure stability we need to include higher-order differentials in writing the virtual variation δU. Now a Taylor expansion of δU in powers of δS and δV gives

$$\delta U = \delta^1 U + \delta^2 U + \delta^3 U + \cdots,$$

$$= (T \, \delta S - p \, \delta V) + \frac{1}{2} \left[\left(\frac{\partial^2 U}{\partial S^2} \right)_V (\delta S)^2 + 2 \frac{\partial^2 U}{\partial S \, \partial V} \delta S \, \delta V + \left(\frac{\partial^2 U}{\partial V^2} \right)_S (\delta V)^2 \right]$$

$$+ \frac{1}{3!} \left[\left(\frac{\partial^3 U}{\partial S^3} \right)_V (\delta S)^3 + 3 \frac{\partial^3 U}{\partial S^2 \, \partial V} (\delta S)^2 \, \delta V \right.$$

$$\left. + 3 \frac{\partial^3 U}{\partial S \, \partial V^2} \delta S (\delta V)^2 + \left(\frac{\partial^3 U}{\partial V^3} \right)_S (\delta V)^3 \right] + \cdots,$$

where $\delta^1 U$, $\delta^2 U$, and $\delta^3 U$ represent the first-, second-, and third-order terms in the virtual variation δU and

$$\left(\frac{\partial U}{\partial S} \right)_V = T \quad \text{and} \quad \left(\frac{\partial U}{\partial V} \right)_S = -p$$

Then from Eq. (5-2a) it follows that

$$\delta^2 U + \delta^3 U + \cdots > 0$$

When we choose the virtual changes δS and δV in a way such that the second-order terms outweigh all the higher terms in the Taylor expansion, we have

$$\delta^2 U > 0 \tag{5-12}$$

or

$$\left(\frac{\partial^2 U}{\partial S^2} \right)_V (\delta S)^2 + 2 \frac{\partial^2 U}{\partial S \, \partial V} \delta S \, \delta V + \left(\frac{\partial^2 U}{\partial V^2} \right)_S (\delta V)^2 > 0 \tag{5-12a}$$

The preceding expression is the criterion of stability. It requires that the second-order terms in a homogeneous quadratic differential form be positive for any conceivable virtual process with all possible pairs of δS and δV (except the trivial pair $\delta S = \delta V = 0$). This condition is referred to in mathematical terms that the quadratic form be positive definite, for which we must have

$$\left(\frac{\partial^2 U}{\partial S^2} \right)_V > 0 \quad \left(\frac{\partial^2 U}{\partial V^2} \right)_S > 0 \quad \text{and} \quad \left(\frac{\partial^2 U}{\partial S^2} \right)_V \left(\frac{\partial^2 U}{\partial V^2} \right)_S - \left(\frac{\partial^2 U}{\partial S \, \partial V} \right)^2 > 0 \tag{5-13}$$

The condition for stable equilibrium as expressed by the first inequality in Eq. (5-13) can be rewritten as

$$\left(\frac{\partial^2 U}{\partial S^2} \right)_V = \left(\frac{\partial T}{\partial S} \right)_V > 0 \quad \text{or} \quad \left(\frac{\partial S}{\partial T} \right)_V = \frac{C_v}{T} > 0 \tag{5-14}$$

The physical significance of this stability criterion is that heat added at constant volume must increase the temperature of a stable system. Since T is positive, it follows that the heat capacity at constant volume is positive, or $C_v > 0$.

Similarly, the condition as expressed by the second inequality in Eq. (5-13) can be rewritten as

$$\left(\frac{\partial^2 U}{\partial V^2}\right)_S = -\left(\frac{\partial p}{\partial V}\right)_S > 0 \quad \text{or} \quad \left(\frac{\partial p}{\partial V}\right)_S < 0 \tag{5-15}$$

Thus, the adiabatic compressibility as defined by the expression $\kappa_s = -1/V$ $(\partial V/\partial p)_S$ must be positive.

It is worth noting that there is no need to examine the third inequality in Eq. (5-13), because the physical result arising from it will be derived later in the consideration of other stability criteria. (See Problem 5-2.)

Conditions for stable equilibrium imposed on other thermodynamic functions can be obtained by the use of alternative stability criteria. Thus, the condition that involves the variation of the Helmholtz function can be obtained by the inequality of Eq. (5-4a),

$$\delta A + p\, \delta V + S\, \delta T + \delta S\, \delta T > 0$$

Now a Taylor expansion of δA in powers of δV and δT gives

$$\delta A = \text{(first-order terms } \delta^1 A) + \text{(second-order terms } \delta^2 A) + \cdots$$

$$= (-p\, \delta V - S\, \delta T) + \frac{1}{2}\left[\left(\frac{\partial^2 A}{\partial V^2}\right)_T (\delta V)^2 + 2\frac{\partial^2 A}{\partial V\, \partial T}\, \delta V\, \delta T\right.$$

$$\left. + \left(\frac{\partial^2 A}{\partial T^2}\right)_V (\delta T)^2\right] + \cdots$$

where

$$\left(\frac{\partial A}{\partial V}\right)_T = -p \quad \text{and} \quad \left(\frac{\partial A}{\partial T}\right)_V = -S$$

Then from Eq. (5-4a) it follows that

$$\delta S\, \delta T + \frac{1}{2}\left(\frac{\partial^2 A}{\partial V^2}\right)_T (\delta V)^2 + \frac{\partial^2 A}{\partial V\, \partial T}\, \delta V\, \delta T + \frac{1}{2}\left(\frac{\partial^2 A}{\partial T^2}\right)_V (\delta T)^2 > 0$$

in which the third- and higher-order terms have been neglected.

Since

$$S = -\left(\frac{\partial A}{\partial T}\right)_V \qquad \left(\frac{\partial S}{\partial V}\right)_T = -\frac{\partial^2 A}{\partial V\, \partial T} \qquad \left(\frac{\partial S}{\partial T}\right)_V = -\left(\frac{\partial^2 A}{\partial T^2}\right)_V$$

we have therefore

$$\delta S \,\delta T = \left(\frac{\partial S}{\partial V}\right)_T \delta V \,\delta T + \left(\frac{\partial S}{\partial T}\right)_V (\delta T)^2 = -\frac{\partial^2 A}{\partial V \,\partial T}\,\delta V \,\delta T - \left(\frac{\partial^2 A}{\partial T^2}\right)_V (\delta T)^2$$

Consequently, the criterion of stability is then

$$\left(\frac{\partial^2 A}{\partial V^2}\right)_T (\delta V)^2 - \left(\frac{\partial^2 A}{\partial T^2}\right)_V (\delta T)^2 > 0 \tag{5-16}$$

for which we must have

$$\left(\frac{\partial^2 A}{\partial V^2}\right)_T > 0 \quad \text{and} \quad \left(\frac{\partial^2 A}{\partial T^2}\right)_V < 0 \tag{5-17}$$

The condition for stable equilibrium as expressed by the second inequality in Eq. (5-17) leads to

$$-\left(\frac{\partial^2 A}{\partial T^2}\right)_V = \left(\frac{\partial S}{\partial T}\right)_V = \frac{C_v}{T} > 0$$

which is Eq. (5-14) as obtained previously. However, the first inequality in Eq. (5-17) leads to the following new condition:

$$\left(\frac{\partial^2 A}{\partial V^2}\right)_T = -\left(\frac{\partial p}{\partial V}\right)_T > 0 \quad \text{or} \quad \left(\frac{\partial p}{\partial V}\right)_T < 0 \tag{5-18}$$

Thus, the isothermal compressibility as defined by the expression $\kappa = -1/V\,(\partial V/\partial p)_T$ must be positive.

The conditions as expressed by Eqs. (5-14) and (5-18) are the two basic conditions for stability for a homogeneous single-component simple compressible closed system. They are sometimes referred to as the conditions for thermal and mechanical stabilities respectively. Other conditions, such as Eq. (5-15), are not independent, and can be derived from the two basic ones. We will illustrate this derivation in an example.

So far in this section we are chiefly concerned with the conditions for stable equilibrium. There are, however, three other kinds of equilibrium which satisfy the equilibrium conditions, such as $(\delta^1 U)_{S,\,V} = 0$, for continuous changes of state. In a continuous change of state all the intensive properties of a system are continuous functions of the independent variables. On the other hand, a discontinuous change of state involves a change of phase, where discontinuities appear in some of the intensive properties of the system. We will now specify the conditions for all the four kinds of equilibrium for continuous changes of state, using internal energy as the parameter.

(a) Stable equilibrium: $(\delta^1 U)_{S,V} = 0$ and $\delta^2 U > 0$ for any conceivable virtual change with all possible pairs of δS and δV (except the trivial pair $\delta S = \delta V = 0$). These conditions are necessary but not sufficient for absolute stability.

(b) Neutral equilibrium: $(\delta^1 U)_{S,V} = 0$ and $\delta^2 U = 0$

(c) Unstable equilibrium: $(\delta^1 U)_{S,V} = 0$ and $\delta^2 U < 0$

(d) Metastable equilibrium: $(\delta^1 U)_{S,V} = 0$ and $\delta^2 U > 0$. These conditions are the same as for stable equilibrium. The system is stable with respect to continuous changes, but not with respect to discontinuous changes. Superheated liquids and subcooled vapors (for example, the portions AB and EF in Fig. 3-1 in a van der Waals isotherm) are in metastable states.

It is instructive to note the similarity between equilibrium states in thermodynamics and in mechanics. A marble in four types of mechanical equilibrium is depicted in Fig. 5-1. In Fig. 5-1(a) the marble is in mechanical stable equilibrium, because it will always revert to it after having been disturbed. In Fig. 5-1(b) the marble is in mechanical neutral equilibrium because it will move to a new equilibrium position after disturbance. In Fig. 5-1(c) the marble is in mechanical unstable equilibrium, because it will leave its position under the influence of even a small disturbance. In Fig. 5-1(d) the marble is in mechanical metastable equilibrium, because it is stable with respect to small disturbances, but it will move to a new state of equilibrium when the disturbance exceeds a certain magnitude.

EXAMPLE 5-1. From the two basic conditions expressed by Eqs. (5-14) and (5-18), show that for stability of a homogeneous single-component simple compressible closed system we must have

$$\left(\frac{\partial S}{\partial T}\right)_p > 0 \quad \text{and} \quad \left(\frac{\partial p}{\partial V}\right)_S < 0$$

Solution.

From Eq. (2-33) we have

$$C_p - C_v = -\frac{T(\partial V/\partial T)_p^2}{(\partial V/\partial p)_T}$$

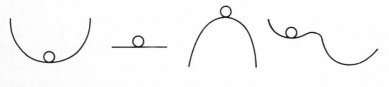

| (a) Stable | (b) Neutral | (c) Unstable | (d) Metastable |

FIG. 5-1. Four types of mechanical equilibrium.

Since $T \geqslant 0$, $(\partial V/\partial T)_p^2 \geqslant 0$, and $(\partial V/\partial p)_T < 0$ (see Eq. (5-18)), we must have

$$C_p - C_v \geqslant 0 \quad \text{or} \quad \frac{C_p}{C_v} \geqslant 1$$

Consequently, from Eqs. (2-26) and (2-27),

$$\frac{C_p}{C_v} = \frac{(\partial S/\partial T)_p}{(\partial S/\partial T)_V} \geqslant 1$$

But according to Eq. (5-14), $(\partial S/\partial T)_V = C_v/T > 0$. Therefore we conclude that

$$\left(\frac{\partial S}{\partial T}\right)_p = \frac{C_p}{T} > 0 \tag{5-19}$$

meaning that heat added at constant pressure increases the temperature of a stable system; in other words, the heat capacity at constant pressure is positive for positive T.

From the cyclic relation we have

$$\left(\frac{\partial p}{\partial V}\right)_S = -\frac{(\partial S/\partial V)_p}{(\partial S/\partial p)_V} \tag{5-20a}$$

$$\left(\frac{\partial p}{\partial V}\right)_T = -\frac{(\partial T/\partial V)_p}{(\partial T/\partial p)_V} \tag{5-20b}$$

Dividing Eq. (5-20a) by Eq. (5-20b),

$$\frac{(\partial p/\partial V)_S}{(\partial p/\partial V)_T} = \frac{(\partial S/\partial V)_p}{(\partial T/\partial V)_p}\frac{(\partial T/\partial p)_V}{(\partial S/\partial p)_V} = \frac{(\partial S/\partial T)_p}{(\partial S/\partial T)_V}$$

Substituting Eqs. (5-14), (5-18), and (5-19) into the preceding equation we conclude that

$$\left(\frac{\partial p}{\partial V}\right)_S < 0$$

This is Eq. (5-15).

5-3 HETEROGENEOUS EQUILIBRIUM

A heterogeneous system consists of two or more phases which are separated from each other by surfaces of phase transition. We now examine the conditions of equilibrium for such a system. In our present discussions we will neglect any surface effect as compared with the effects of bulk phases. Furthermore, let the system as a whole be a closed simple compressible system, allowing no chemical reactions. Assume that the system has already attained equality in temperature and pressure throughout and thus it is in thermal and mechanical equilibrium. Our purpose here is to find the conditions that will insure chemical equilibrium, thereby achieving complete equilibrium.

Let the heterogeneous system under consideration have φ homogeneous phases and r components. In the absence of chemical reactions the number of components of a system is identical with the number of molecular species present in the system. Let us assume, at first, that every component is present in every phase. Although the system as a whole is closed, the phases are open because transportation of components between phases is allowed. For any one of the homogeneous simple compressible open phase the differential of the Gibbs function for changes between neighboring equilibrium states is expressed by the equation

$$dG^{(\alpha)} = -S^{(\alpha)}dT + V^{(\alpha)}dp + \sum_{i=1}^{r} \mu_i^{(\alpha)} \, dn_i^{(\alpha)} \tag{4-6}$$

Applying this equation to each phase of the system and summing up we get the change in the Gibbs function of the system,

$$dG = \sum_{\alpha=1}^{\varphi} \left(-S^{(\alpha)}dT + V^{(\alpha)}dp + \sum_{i=1}^{r} \mu_i^{(\alpha)} \, dn_i^{(\alpha)} \right)$$

or

$$dG = -S \, dT + V \, dp + \sum_{\alpha=1}^{\varphi} \left(\sum_{i=1}^{r} \mu_i^{(\alpha)} \, dn_i^{(\alpha)} \right) \tag{5-21}$$

where the subscripts denote components and the superscripts denote phases; S is the entropy and V the volume of the whole system. Since the system is assumed to be in thermal and mechanical equilibrium, it is obvious that temperature and pressure must be the same throughout the system.

At constant temperature and pressure the condition for equilibrium, according to Eq. (5-11), is

$$(dG)_{T,p} = 0$$

Applying this condition to Eq. (5-21) results in

$$(dG)_{T,p} = \sum_{\alpha=1}^{\varphi} \left(\sum_{i=1}^{r} \mu_i^{(\alpha)} \, dn_i^{(\alpha)} \right) = 0 \tag{5-22}$$

Since the system as a whole is closed and no chemical reactions occur, the changes in the n_i's are due solely to the transfer of components between phases.

Hence the n_i's are subjected to the following equations of constraint,

$$\sum_{\alpha=1}^{\varphi} n_i^{(\alpha)} = \text{constant} \qquad i = 1, 2, \ldots, r \tag{5-23}$$

Our task is to find the equations of phase equilibrium which satisfy the condition as expressed by Eq. (5-22), subject to the constraints as expressed by Eq. (5-23). To do this it is convenient to use *Lagrange's method of undetermined multipliers*.* Thus, differentiating each equation of constraint and multiplying each by a different Lagrangian multiplier lead to the relations

$$\sum_{\alpha=1}^{\varphi} \lambda_i \, dn_i^{(\alpha)} = 0 \qquad i = 1, 2, \ldots, r$$

Adding this equation and Eq. (5-22) results in

$$\sum_{i=1}^{r} \sum_{\alpha=1}^{\varphi} (\mu_i^{(\alpha)} + \lambda_i) \, dn_i^{(\alpha)} = 0$$

Equating each coefficient to zero results in a set of equations,

$$\mu_i^{(\alpha)} = -\lambda_i \qquad i = 1, 2, \ldots, r \qquad \alpha = 1, 2, \ldots, \varphi$$

meaning that when a heterogeneous system is at equilibrium at constant temperature and pressure, the chemical potential of any given component must have a common value in all phases. Therefore we conclude that

$$\mu_i^{(1)} = \mu_i^{(2)} = \cdots = \mu_i^{(\varphi)} \qquad i = 1, 2, \ldots, r \tag{5-24}$$

a total of $(\varphi - 1)r$ equations. These are the *equations of phase equilibrium.*

In view of the preceding discussion, it is evident that the chemical potential is the driving force for mass transfer between phases or between portions of a given phase, just as the temperature is the driving force for heat transfer and the pressure is the driving force for "work transfer." In other words, the chemical potential plays a role in chemical equilibrium similar to those played by the temperature in thermal equilibrium and the pressure in mechanical equilibrium.

Since the fugacity of component i is directly related to its chemical potential by the equation

$$(d\mu_i)_T = RTd(\ln f_i)_T \tag{4-62}$$

fugacity can also be used as criterion of phase equilibrium. Thus from Eq. (5-24) we can write

$$f_i^{(1)} = f_i^{(2)} = \cdots = f_i^{(\varphi)} \qquad i = 1, 2, \ldots, r \tag{5-25}$$

*See Appendix 1.

In arriving at the series of Eq. (5-24) we have assumed that every component is present in every phase. However, if a component is not actually present in a phase but is present in an adjoining phase, the equilibrium condition would be seen from the virtual variation that $(\delta G)_{T,p} \geqslant 0$. For illustrative purposes, let us consider a component i in equilibrium between two phases. For this case, since $\delta n_i^{(1)} = -\delta n_i^{(2)}$, we have

$$(\mu_i^{(1)} - \mu_i^{(2)})\delta n_i^{(1)} \geqslant 0$$

If component i is present in phase 2 but not in phase 1, it is possible for $\delta n_i^{(1)}$ to be positive but not negative. Therefore we must have for equilibrium,

$$\mu_i^{(1)} \geqslant \mu_i^{(2)} \tag{5-26}$$

EXAMPLE 5-2. Liquid and vapor phases of a substance may coexist in equilibrium at different pressures. For example, if a liquid-vapor system is confined in a vessel that also contains some inert gas which is insoluble in the liquid, the pressure applied to the liquid can be appreciably greater than its partial vapor pressure. In order to study the effect of applied pressure on the vapor pressure of a liquid, consider the idealized setup shown in Fig. 5-2, where the single-component liquid and vapor phases under different pressures are separated by a nondeformable, heat-conducting membrane permeable to vapor only. (a) Show that $(\partial p^{(\beta)}/\partial p^{(\alpha)})_T = v^{(\alpha)}/v^{(\beta)}$ where the superscripts (α) and (β) denote the liquid and vapor phases respectively. (b) Liquid water at 300 bars and 200°C is in equilibrium with water vapor according to the setup shown in Fig. 5-2. Find the pressure of the water vapor.

Solution.
(a) Because of the presence of the membrane, mechanical equilibrium is maintained in the system. To insure phase equilibrium, according to Eq. (5-24), we must have $\mu^{(\alpha)} = \mu^{(\beta)}$ whence $d\mu^{(\alpha)} = d\mu^{(\beta)}$. But for a pure component, $\mu = g$, and

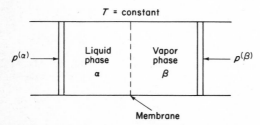

FIG. 5-2. Effect of applied pressure on vapor pressure.

$$d\mu^{(\alpha)} = dg^{(\alpha)} = -s^{(\alpha)}dT + v^{(\alpha)}dp^{(\alpha)}$$
$$d\mu^{(\beta)} = dg^{(\beta)} = -s^{(\beta)}dT + v^{(\beta)}dp^{(\beta)}$$

Thus

$$-s^{(\alpha)}dT + v^{(\alpha)}dp^{(\alpha)} = -s^{(\beta)}dT + v^{(\beta)}dp^{(\beta)}$$

At constant temperature this equation becomes

$$v^{(\alpha)}dp^{(\alpha)} = v^{(\beta)}dp^{(\beta)}$$

Therefore

$$\left(\frac{\partial p^{(\beta)}}{\partial p^{(\alpha)}}\right)_T = \frac{v^{(\alpha)}}{v^{(\beta)}}$$

This equation, known as the *Poynting relation*, describes how the vapor pressure changes with applied pressure at constant temperature. Since at the same temperature, the specific volume of a liquid is much smaller than that of a vapor, we see from this equation that the effect of an applied pressure on the vapor pressure is very weak at ordinary conditions.

(b) Since the liquid and vapor are in thermal equilibrium, the vapor is also at 200°C. According to the Poynting relation we have

$$\left[\int_{p_{\text{sat}}}^{p^{(\alpha)}} v^{(\alpha)}dp^{(\alpha)}\right]_T = \left[\int_{p_{\text{sat}}}^{p^{(\beta)}} v^{(\beta)}dp^{(\beta)}\right]_T$$

where $p_{\text{sat}} = 15.538$ bars[1] which is the saturation pressure corresponding to 200°C. Integration of the foregoing equation can be performed by assuming the liquid to be incompressible and the vapor to behave as an ideal gas. Thus $v^{(\alpha)} = v_f = 1.1565$ cm^3/g which is the specific volume of saturated liquid at 200°C; and $v^{(\beta)} = p_{\text{sat}} v_g/p^{(\beta)}$, where $v_g = 127.36$ cm^3/g which is the specific volume of saturated vapor at 200°C. We accordingly write

$$v_f(p^{(\alpha)} - p_{\text{sat}}) = p_{\text{sat}}v_g \ln \frac{p^{(\beta)}}{p_{\text{sat}}}$$

Hence

$$p^{(\beta)} = p_{\text{sat}} \exp\left[\frac{v_f}{v_g}\left(\frac{p^{(\alpha)}}{p_{\text{sat}}} - 1\right)\right] = 15.538 \exp\left[\frac{1.1565}{127.36}\left(\frac{300}{15.538} - 1\right)\right]$$

$$= 18.3 \text{ bars}$$

Note that at 18.3 bars the saturation temperature is 207.6°C which is greater

---

[1] Data from J. H. Keenan, F. G. Keyes, P. G. Hill, and J. G. Moore, "Steam Tables (International Edition)," John Wiley and Sons, Inc., New York, 1969.

than the existing temperature of $200°C$. The vapor is therefore in a supersaturated metastable state.

5-4 BINARY VAPOR-LIQUID SYSTEMS

It is clear that there are basic differences in phase behavior between a single-component system and a multicomponent system. First of all, in a single-component system the compositions of different phases are always the same. In a multicomponent system, however, the compositions of different phases in equilibrium, in general, are different. Secondly, during a change of phase, e.g., in vaporization, in a single-component system the temperature and the pressure always remain constant. However, for a mixture of two or more substances of different volatilities the pressure does not remain constant but falls continuously during isothermal vaporization. These and other features in the phase behavior of multicomponent systems can be illustrated by studying a binary vapor-liquid system with a miscible liquid.

The properties of a binary vapor-liquid system can be most readily seen by means of a three-dimensional surface with pressure, temperature, and composition as the coordinates. A plane of constant composition cutting through a p-T-x surface would give us a p-T phase diagram as depicted in Fig. 5-3. The vapor-liquid two-phase region in the figure is bounded by a *border curve* (*ABCDE*). This curve is made up of two portions. The one on the left is the bubble-point line (*ABC*), and the one on the right is the dew-point line (*EDC*). A *bubble-point line* is the locus of states of initial vaporization, whereas a *dew-point line* is the locus of states of initial condensation. The point where the bubble-point and dew-point lines meet, or where the vapor and liquid phases become indistinguishable, is the *critical point* (*C*) of the mixture. It does not necessarily correspond to the maximum pressure and temperature at which the mixture can exist in two phases. Point *B* on the border curve represents the

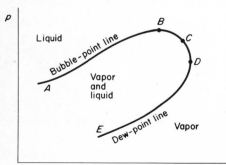

FIG. 5-3. Pressure-temperature phase diagram at constant composition for a binary mixture.

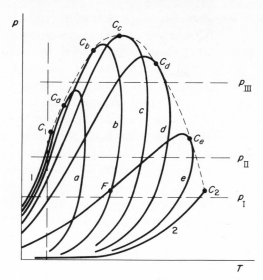

FIG. 5-4. Pressure-temperature border curves for a binary system at various compositions. [After W. B. Kay, *Ind. Eng. Chem.*, **30**:459 (1938); and *J. Chem. Eng. Data*, **15**:46 (1970).]

maximum pressure at which a system of the given composition can exist in two phases. Point D represents the maximum temperature at which the system can exist in two phases. Note that in the case of a single-component system, lines ABC and EDC coincide to form a single vapor-pressure line, and the points $B,C,$ and D coincide to form a single critical point.

The border curve shown in Fig. 5-3 is for a single composition. To show the border curves of various compositions in a two-dimensional plane, we cut the p-T-x surface at a number of different constant compositions and project all the border curves onto one single p-T plane, as illustrated in Fig. 5-4. In this figure each of the five loops represents the p-T border curve of a mixture of definite composition. These five mixtures are labeled $a, b, \ldots, e$. The border curves converge at each end to the single curves 1 and 2, representing the vapor-pressure curves of the two pure components. Any point, such as point F, where the vapor branch of one curve intersects the liquid branch of another, represents coexisting phases in equilibrium. The critical points of the five mixtures are labeled $C_a,$ $C_b, \ldots, C_e$ in the figure, and the critical points of the two pure components are labeled C_1 and C_2. Line $(C_1 C_a \ldots C_e C_2)$ represents the critical envelope, which is the locus of all critical points for all possible mixtures of the two substances.

An inspection of Fig. 5-4 discloses that depending on the composition of a mixture, its critical point may lie on either side of the maximum pressure

point on the border curve. Figure 5-5(a) shows an enlargement of the critical
region in Fig. 5-3 where the critical point C lies between the maximum pressure
point B and the maximum temperature point D. Figure 5-5(b) shows the critical
region of a mixture for which the critical point C lies on the left side of the
maximum pressure point B. Figure 5-5(c) shows the critical region of a mixture
for which the critical point C lies below the maximum temperature point D. In
addition to the border curves in these figures, new lines are added to indicate the
percentage, e.g., by volume of the substance which is in the liquid phase at any
state in the two-phase region. In this representation, the dew-point line is then a
line of zero percent liquid and the bubble-point line is a line of 100 percent
liquid.

A glance at Fig. 5-5 reveals a very interesting phenomenon, known as
retrograde condensation. In Fig. 5-5(a), if a vapor is compressed isothermally at

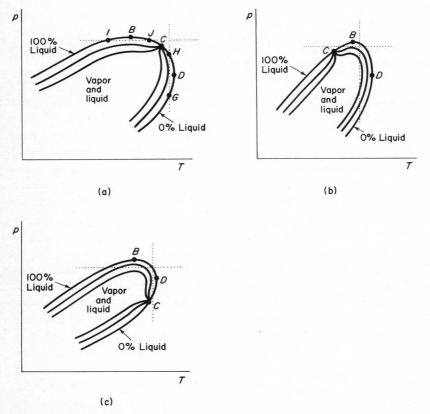

FIG. 5-5. Retrograde condensation.

constant total composition, as depicted by the vertical dotted line, condensation begins at point G on the dew-point line. As compression is continued, the amount of liquid at first gradually increases to a maximum, then gradually decreases, disappearing entirely as the dew-point line is again reached at point H. A similar phenomenon happens when the pressure is lowered from H to G. This is called the isothermal retrograde condensation. On the other hand, if a liquid is heated isobarically at constant total composition, as depicted by the horizontal dotted line in Fig. 5-5(a), vaporization begins at point I on the bubble-point line. As heating is continued, the amount of vapor at first gradually increases to a maximum, then gradually decreases, disappearing entirely as the bubble-point line is again reached at point J. A similar phenomenon occurs when the mixture is cooled from J to I. This is called the isobaric retrograde vaporization.

In Fig. 5-5(b), the vertical dotted line intersects the dew-point line twice, depicting a process of isothermal retrograde condensation. The horizontal dotted line also intersects the dew-point line twice, depicting a process of isobaric retrograde condensation.

If the critical point C of a mixture is located below the maximum temperature point D as shown in Fig. 5-5(c), the vertical- and horizontal-dotted lines respectively would depict processes of isothermal and isobaric retrograde vaporization. Such situations are possible but unusual.

Since many industrial processes are conducted under constant pressures, it is of great utility to have diagrams relating temperature and composition at constant pressures. Typical forms of T-x relations for a binary system at various pressures are shown in Fig. 5-6. These curves can be obtained by cutting a p-T-x surface by planes of constant pressures. The curves marked p_{I}, p_{II}, and p_{III} in Fig. 5-6 correspond to the indicated constant pressures in Fig. 5-4. The p_{I} curve is typical for a pressure which is below the critical pressures of both pure components; the p_{II} curve is for a pressure which is above the critical pressure of the less volatile component (component 2), and the p_{III} curve is for a pressure which is above the critical pressures of both pure components.

To make the discussion easier, the p_{I} curve in Fig. 5-6 is duplicated in Fig. 5-7. The lens-shaped vapor-liquid two-phase area is bounded by the dew-point line on the top and the bubble-point line on the bottom. The area above the dew-point line represents states of superheated vapor mixtures, whereas the area below the bubble-point line represents states of subcooled liquid solutions. If heat is added at constant pressure to a liquid solution at state a, the temperature of the solution will increase while its composition remains unchanged until the bubble-point line at state b, is reached. Additional heat addition will cause vaporization of both components in such a proportion that the vapor appears in a composition corresponding to point b'. During vaporization the composition of

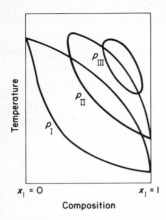

FIG. 5-6. Temperature-composition diagram for a binary system at various pressures.

the vapor moves from b' towards d, while the composition of the liquid moves from b towards d''. At any state c in the two-phase region, the vapor has a composition corresponding to point c' and the liquid has a composition corresponding to point c''. If vaporization is carried to completion, the final vapor mixture at state d will have the same composition as the originial liquid solution. Further heat addition beyond the dew-point line will, of course, superheat the vapor, say to state e.

When processes are conducted under constant temperature condition, it is desirable to have a pressure vs. composition diagram at constant temperature as shown in Fig. 5-8. A p-x diagram may be obtained by cutting a p-T-x surface by a plane of constant temperature. The one shown in Fig. 5-8 is for a constant temperature which is below the critical temperatures of both pure components, corresponding to the constant temperature line shown by the vertical-dotted line in Fig. 5-4. In Fig. 5-8 the vapor-liquid two-phase area is bounded by the bubble-point line on the top and the dew-point line on the bottom. The area below the dew-point line represents states of superheated

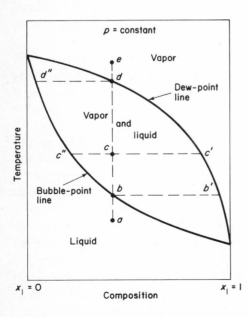

FIG. 5-7. Temperature-composition diagram for a binary system at a constant pressure lower than the critical pressures of both pure components.

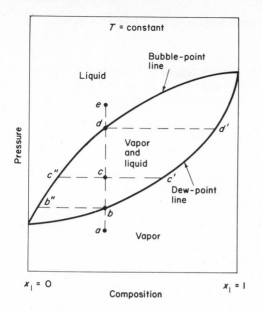

FIG. 5-8. Pressure-composition diagram for a binary system at a constant temperature lower than the critical temperatures of both pure components.

vapor mixtures, while the area above the bubble-point line represents states of compressed liquid solutions. Line (*abcde*) represents a constant temperature compression process. During condensation between states *b* and *d* the vapor composition moves from *b* towards *d'*, while the liquid composition moves from *b"* towards *d*.

EXAMPLE 5-3. The vapor pressures of pure propane and pure *n*-pentane can be calculated from the Antoine equation:

$$\log p^0 = A - \frac{B}{C + t}$$

where p^0 is the vapor pressure in mm Hg, t is the temperature in $^\circ$C, and A, B, and C are constants. For propane[2] $A = 6.82973$, $B = 813.20$, and $C = 248.00$. For *n*-pentane $A = 6.85221$, $B = 1064.63$, and $C = 232.00$. Calculate the compositions of the liquid and vapor phases of the propane–*n*-pentane system vs. temperature for a total pressure of 760 mm Hg, assuming that Raoult's and Dalton's laws are obeyed.

Solution.

We denote propane by the subscript 1 and *n*-pentane by the subscript 2.

[2] Data from F. D. Rossini et al., "Selected Values of Physical and Thermodynamic Properties of Hydrocarbons and Related Compounds," Carnegie Press, Carnegie Institute of Technology, Pittsburgh, 1953.

From Raoult's law the partial vapor pressures of the two components are

$$p_1 = p_1^0 x_1^{(\alpha)} \qquad p_2 = p_2^0 x_2^{(\alpha)} \tag{5-27}$$

where p_1^0 and p_2^0 are the vapor pressures of pure components 1 and 2 respectively, and $x_1^{(\alpha)}$ and $x_2^{(\alpha)}$ are the mole fractions of components 1 and 2 respectively in the liquid phase of the system. The total pressure of the system is then

$$p = p_1 + p_2 = p_1^0 x_1^{(\alpha)} + p_2^0 x_2^{(\alpha)} = x_1^{(\alpha)}(p_1^0 - p_2^0) + p_2^0$$

Therefore

$$x_1^{(\alpha)} = \frac{p - p_2^0}{p_1^0 - p_2^0} \tag{5-28}$$

From Dalton's law the partial vapor pressure of component 1 can also be written as

$$p_1 = p x_1^{(\beta)} \tag{5-29}$$

wherein p is the total pressure of the system, and $x_1^{(\beta)}$ the mole fraction of component 1 in the vapor phase. Combining Eqs. (5-27) and (5-29) we obtain

$$p_1^0 x_1^{(\alpha)} = p x_1^{(\beta)}$$

Therefore

$$x_1^{(\beta)} = \frac{p_1^0 x_1^{(\alpha)}}{p} \tag{5-30}$$

With $p = 760$ mm Hg and p_1^0 and p_2^0 calculated from the Antoine equation, Eqs. (5-28) and (5-30) are now used to calculate the compositions of the liquid and vapor phases of the system. The results are shown in the following table, from which a T-x diagram similar to Fig. 5-7 can be drawn.

Temp. °C	p_1^0 mm Hg	p_2^0 mm Hg	$x_1^{(\alpha)}$	$x_2^{(\alpha)}$	$x_1^{(\beta)}$	$x_2^{(\beta)}$
36.1	–	760	0	1.000	0	1.000
30	8,027	615	0.020	0.980	0.211	0.789
20	6,243	424	0.058	0.942	0.476	0.524
10	4,762	284	0.106	0.894	0.664	0.336
0	3,554	183	0.171	0.829	0.800	0.200
– 10	2,588	114	0.261	0.739	0.889	0.111
– 20	1,833	68	0.392	0.608	0.945	0.055
– 30	1,257	38	0.592	0.408	0.979	0.021
– 40	832	20	0.911	0.089	0.997	0.003
– 42.1	760	–	1.000	0	1.000	0

EXAMPLE 5-4.   A vapor mixture of 20 mole percent ethane (C_2H_6) and 80 mole percent propylene (C_3H_6), initially at 14.7 psia and 40°F, is compressed isothermally to higher pressures. (a) What is the pressure when the first drop of liquid appears, and what is the composition of that drop of liquid? (b) What is the pressure when the last portion of vapor condenses, and what is the composition of that portion of vapor? Assume that Raoult's and Dalton's laws are obeyed. The vapor pressures of ethane and propylene at 40°F are 385 psia and 96.4 psia respectively.

Solution.

We denote ethane by the subscript 1 and propylene by the subscript 2. Figure 5-8 shows schematically the pressure-composition diagram, except that in this example, owing to the simplified assumption of Raoult's law, the bubble-point line would be a straight line.

(a) When the first drop of liquid appears at the pressure p_b corresponding to point b, the vapor phase will be at point b with the given composition of $x_1^{(\beta)} = 0.20$ and $x_2^{(\beta)} = 0.80$. The liquid phase will be at point b'', the composition of which is to be determined. From Raoult's and Dalton's laws, the partial vapor pressures of the two components are

$$p_1 = x_1^{(\alpha)} \, p_1^0 = x_1^{(\beta)} \, p_b$$

$$p_2 = x_2^{(\alpha)} \, p_2^0 = x_2^{(\beta)} \, p_b = (1 - x_1^{(\alpha)}) \, p_2^0 = (1 - x_1^{(\beta)}) \, p_b$$

Dividing, we have

$$\frac{x_1^{(\alpha)}}{1 - x_1^{(\alpha)}} = \frac{p_2^0}{p_1^0} \frac{x_1^{(\beta)}}{1 - x_1^{(\beta)}} = \frac{96.4}{385} \frac{0.20}{0.80} = 0.0626$$

from which we obtain

$$x_1^{(\alpha)} = 0.059 \quad \text{and} \quad x_2^{(\alpha)} = 1 - x_1^{(\alpha)} = 0.941$$

Whence

$$p_b = \frac{x_1^{(\alpha)}}{x_1^{(\beta)}} p_1^0 = \frac{0.059}{0.20} \times 385 = 113.6 \text{ psia}$$

or

$$p_b = p_1 + p_2 = x_1^{(\alpha)} p_1^0 + x_2^{(\alpha)} p_2^0$$

$$= 0.059 \times 385 + 0.941 \times 96.4 = 113.4 \text{ psia}$$

(b) When the last portion of vapor condenses at the pressure p_d corresponding to point d, the liquid phase will be at point d with the given composition of $x_1^{(\alpha)} = 0.20$ and $x_2^{(\alpha)} = 0.80$ while the vapor phase will be at point d', the composition of which is to be determined. Analogous to Part (a), we have

$$\frac{x_1^{(\beta)}}{1 - x_1^{(\beta)}} = \frac{p_1^0}{p_2^0} \frac{x_1^{(\alpha)}}{1 - x_1^{(\alpha)}} = \frac{385}{96.4} \frac{0.20}{0.80} = 0.998$$

from which we obtain

$$x_1^{(\beta)} = 0.50 \quad \text{and} \quad x_2^{(\beta)} = 1 - x_1^{(\beta)} = 0.50$$

Whence

$$p_d = \frac{x_1^{(\alpha)}}{x_1^{(\beta)}} p_1^0 = \frac{0.20}{0.50} \times 385 = 154 \text{ psia}$$

or

$$p_d = p_1 + p_2 = x_1^{(\alpha)} p_1^0 + x_2^{(\alpha)} p_2^0 = 0.20 \times 385 + 0.80 \times 96.4 = 154 \text{ psia}$$

5-5 AZEOTROPE

It is significant to note that although in a multicomponent system the compositions of liquid and vapor phases in equilibrium are generally different, there is an important exception where these compositions are the same. A solution whose equilibrium vapor phase has the same composition as that of the liquid phase is known as an *azeotrope*. This identical composition is termed the azeotropic composition. At azeotropic composition the solution behaves like a pure component and boils at constant temperature and pressure.

Azeotropic solutions are nonideal solutions which deviate from Raoult's law. There are two common types of azeotropes. A positive azeotrope, shown in Fig. 5-9 for a two-component system, results from large positive deviations from

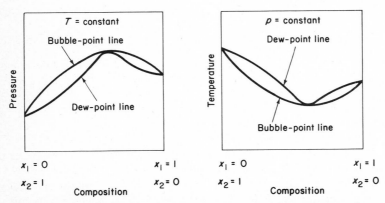

FIG. 5-9. Pressure-composition and temperature-composition diagrams showing positive azeotrope.

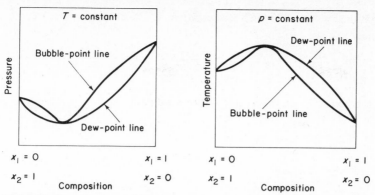

FIG. 5-10. Pressure-composition and temperature-composition diagrams showing negative azeotrope.

Raoult's law. It is characterized by a maximum vapor pressure and a minimum boiling point. On the other hand, a negative azeotrope, shown in Fig. 5-10 for a two-component system, results from large negative deviations from Raoult's law. It is characterized by a minimum vapor pressure and a maximum boiling point.

The special characteristics of azeotropic solutions can be seen from the Gibbs-Duhem relation (Eq. 4-11a) as applied to a binary liquid-vapor system. Thus for the respective liquid and vapor phases we have

$$s^{(\alpha)}dT - v^{(\alpha)}dp + (1 - x_2^{(\alpha)})d\mu_1 + x_2^{(\alpha)}d\mu_2 = 0 \qquad (5\text{-}31)$$

$$s^{(\beta)}dT - v^{(\beta)}dp + (1 - x_2^{(\beta)})d\mu_1 + x_2^{(\beta)}d\mu_2 = 0 \qquad (5\text{-}32)$$

where the subscripts denote components and the superscripts (α) and (β) denote respectively the liquid and vapor phases. Subtracting Eq. (5-31) from Eq. (5-32) yields

$$(s^{(\beta)} - s^{(\alpha)})dT - (v^{(\beta)} - v^{(\alpha)})dp + (x_2^{(\alpha)} - x_2^{(\beta)})(d\mu_1 - d\mu_2) = 0 \qquad (5\text{-}33)$$

At the azeotropic composition $x_2^{(\alpha)} = x_2^{(\beta)}$ and the preceding equation becomes

$$(s^{(\beta)} - s^{(\alpha)})dT - (v^{(\beta)} - v^{(\alpha)})dp = 0$$

or

$$\frac{dp}{dT} = \frac{s^{(\beta)} - s^{(\alpha)}}{v^{(\beta)} - v^{(\alpha)}}$$

Thus if a solution remains azeotropic the simultaneous variation of temperature and pressure is governed by the Clapeyron equation.

For variations of pressure and composition at constant temperature, Eq. (5-33) becomes

$$(v^{(\beta)} - v^{(\alpha)}) \frac{\partial p}{\partial x_2^{(\alpha)}} = (x_2^{(\alpha)} - x_2^{(\beta)}) \left(\frac{\partial \mu_1}{\partial x_2^{(\alpha)}} - \frac{\partial \mu_2}{\partial x_2^{(\alpha)}} \right)$$

Applying this equation to an azeotropic solution we have

$$(v^{(\beta)} - v^{(\alpha)}) \frac{\partial p}{\partial x_2^{(\alpha)}} = 0$$

Since $v^{(\beta)} \neq v^{(\alpha)}$, we must have

$$\left(\frac{\partial p}{\partial x_2^{(\alpha)}} \right)_T = 0$$

meaning that at a given temperature the total vapor pressure of a binary solution is a maximum or a minimum at the azeotropic composition.

Similarly, for variations of temperature and composition at constant pressure, Eq. (5-33) becomes

$$(s^{(\beta)} - s^{(\alpha)}) \frac{\partial T}{\partial x_2^{(\alpha)}} = -(x_2^{(\alpha)} - x_2^{(\beta)}) \left(\frac{\partial \mu_1}{\partial x_2^{(\alpha)}} - \frac{\partial \mu_2}{\partial x_2^{(\alpha)}} \right)$$

Applying this equation to an azeotropic solution we have

$$(s^{(\beta)} - s^{(\alpha)}) \frac{\partial T}{\partial x_2^{(\alpha)}} = 0$$

Since $s^{(\beta)} \neq s^{(\alpha)}$, we must have

$$\left(\frac{\partial T}{\partial x_2^{(\alpha)}} \right)_p = 0$$

meaning that at a given pressure the boiling temperature of a binary solution is a maximum or a minimum at the azeotropic composition.

5-6 GIBBS PHASE RULE

The thermodynamic state at equilibrium for a heterogeneous system of φ phases, each containing r components that do not react chemically, is a function of the temperature, the pressure, and the $(r - 1)\varphi$ values of mole fractions expressing the compositions of all the phases. Therefore, the total number of variables defining the intensive state of the system as a whole is $(r - 1)\varphi + 2$. However, these variables are not all independent. The equations of phase equilibrium as written in Eq. (5-24) impose a total of $(\varphi - 1)r$ conditions that must be satisfied.

Therefore, the number of independent intensive variables required to fix the equilibrium state of a multicomponent, multiphase system is then

$$F = [(r - 1)\varphi + 2] - (\varphi - 1)r$$

or

$$F = r - \varphi + 2 \tag{5-34}$$

where F is called the variance or the degree of freedom of the system. Equation (5-34) is the *phase rule* first enunciated in 1875 by J. Willard Gibbs.

In arriving at the phase rule we have assumed that each of the φ phases of the system contains every one of the r components. Nevertheless, if any component is missing from any phase, there will be one less composition variable for that phase, and also there will be one less condition equation to be satisfied, so that the difference is the same and the phase rule remains unchanged.

A system for which there is no independent intensive property is said to be invariant, one with one independent property is said to be univariant, one with two independent properties is said to be divariant, and so on. We will now apply the phase rule to determine the variance of a few systems.

For a single-component system the phase rule reduces to $F = 3 - \varphi$. Thus, in a single-phase region, $F = 2$ and the system is divariant. On the other hand, in a two-phase region, $F = 1$ and the system is univariant. If either pressure or temperature is fixed, all the other intensive properties of each of the two phases in equilibrium are also fixed. At the triple point, $F = 0$ and the system is invariant, i.e., none of its intensive properties can be varied as long as the three phases exist together in equilibrium.

For a two-component system, such as the vapor-liquid system shown in Figs. 5-7 and 5-8, the phase rule reduces to $F = 4 - \varphi$. In a single-phase region, $F = 3$ and the temperature, pressure, and composition may vary independently. In a two-phase region, $F = 2$, and either the temperature and pressure, or the temperature (or pressure) and composition of one of the phases can be chosen as the independent variables.

So far in our discussions on vapor-liquid equilibrium of binary systems we treat only miscible liquids. Two liquids are said to be miscible if when mixed in any concentration they form a single liquid phase. Many organic liquid pairs are miscible. On the other hand, if when mixed two liquids remain in two phases, they are said to be immiscible. Liquid mercury and water are practically immiscible. Intermediate between these two extremes are liquid pairs which are miscible for some concentrations and immiscible for the others, these liquids are said to be partially miscible. Most partially miscible mixtures are those with water or some other inorganic liquid as one of the pair. Some organic liquid pairs are also partially miscible over certain ranges. The degree of miscibility of two liquids may change greatly with temperature.

A phase diagram for two immiscible liquids is shown in Fig. 5-11, in which line AB is the bubble-point line for the mixture of two liquid phases, and lines DE and GE are the respective dew-point lines in relation to pure liquids 2 and 1. In the liquid-liquid two-phase region, $F = 2$ and the system is divariant. A liquid mixture at state a is actually a mixture of pure liquid 1 at state a' and pure liquid 2 at state a''. When heat is added at constant pressure to the system initially at state a, the system will go from state a toward state b, always consisting of two liquid phases, one of pure 1 and the other of pure 2. At state b there are three phases coexisting (pure 1 liquid phase at state B, pure 2 liquid phase at state A, and vapor phase at state E), and the system is univariant. Since the vapor formed at state E is richer in component 1 than the system as a whole, more liquid 1 will be evaporated as more heat is added. As the last trace of liquid 1 is evaporated, the system is comprised of two phases (pure 2 liquid phase at state A and vapor phase at state E) and the system is again divariant. When the system changes from state b to state c, the pure 2 liquid phase changes from state A to state c'' and the vapor phase changes from state E to state c'. When the system reaches state d, the last trace of pure liquid 2 is in state d''. Further heating will super-heat the vapor mixture toward state e.

A typical form of phase diagram for systems involving two partially miscible liquids is shown in Fig. 5-12, in which liquids I and II denote liquids which are

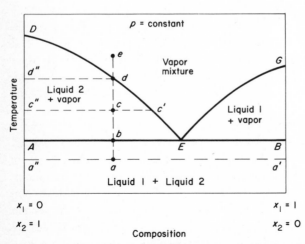

FIG. 5-11. Temperature-composition diagram for two immiscible liquids.

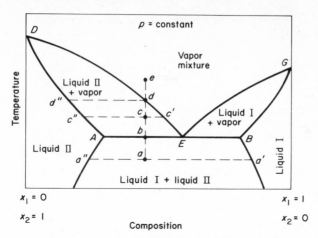

FIG. 5-12. Temperature-composition diagram for two partially miscible liquids.

rich in component 1 and 2, respectively. The lines DA and DE represent equilibrium between liquid II and vapor, with horizontal tie lines such as $c''c'$ connecting the two phases that coexist in equilibrium. Similarly, the lines GB and GE represent equilibrium between liquid I and vapor. The equilibrium between liquids I and II in the absence of a vapor phase is represented by the lines Aa'' and Ba', with a tie line $a''a'$ as shown. Along line AB there are three phases coexisting: liquid II at state A, liquid I at state B, and vapor mixture at state E. In the single-phase regions (liquid I, liquid II, and vapor mixture regions) the system is trivariant, in the two-phase regions (liquid I + liquid II, liquid I + vapor, and liquid II + vapor regions) the system is divariant, and in the three-phase region (along line AB) the system is univariant. The line $(abcde)$ represents a constant-pressure heating process.

All the examples given in the preceding discussions are for vapor-liquid equilbrium. An example for solid-liquid equilibrium is shown in Fig. 5-13, a temperature-composition diagram for a binary system with miscible liquids and immiscible solids. The eutectic point E is the only state in which a liquid can coexist in equilibrium with both solid phases at a given pressure. If a liquid having the composition of state d is cooled at constant pressure, when state d is reached the component 2 will start to crystallize to form a solid phase at state d''. Component 1 will not crystallize until state b at the eutectic temperature is reached. It is clear that at a given pressure only when the composition is at the eutectic point will the two components freeze uniformly. This is why a eutectic solution makes good alloy castings.

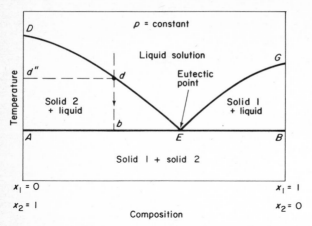

FIG. 5-13. Temperature-composition diagram for binary solid-liquid equilibrium of miscible liquids and immiscible solids.

EXAMPLE 5-5. A binary solid-liquid system consists of two metals which are miscible in the liquid phase but immiscible in the solid phase. Derive the equations for the temperature-composition phase diagram at constant pressure (similar to Fig. 5-13) and the equation which determines the eutectic point. Assume that the liquid solution is ideal and the latent heats of fusion of the pure metals are not sensitive to temperature changes.

Solution.
The subscripts 1 and 2 denote the two components and the superscripts (α) and (β) denote the respective liquid and solid phases. In addition the superscript 0 denotes pure component state. Since the liquid solution is ideal, in accordance with Eq. (4-78) we have

$$\mu_1^{(\alpha)} = \mu_1^{(\alpha),0} + RT \ln x_1^{(\alpha)} \tag{5-35}$$

$$\mu_2^{(\alpha)} = \mu_2^{(\alpha),0} + RT \ln x_2^{(\alpha)} \tag{5-35a}$$

Furthermore, for equilibrium between pure solid phase 1 and the liquid phase we must have

$$\mu_1^{(\alpha)} = \mu_1^{(\beta),0} \tag{5-36}$$

Similarly, for equilibrium between pure solid phase 2 and the liquid phase we must have

$$\mu_2^{(\alpha)} = \mu_2^{(\beta),0} \tag{5-37}$$

Combining Eqs. (5-35) and (5-36) leads to

$$\ln x_1^{(\alpha)} = \frac{\mu_1^{(\beta),0} - \mu_1^{(\alpha),0}}{RT} \tag{5-38}$$

Making use of the relation (see Problem 4-3),

$$\left[\frac{\partial(\mu/RT)}{\partial T}\right]_p = -\frac{h}{RT^2}$$

and noting that at the melting point of pure solid 1, $\mu_1^{(\beta),0} = \mu_1^{(\alpha),0}$ we then have

$$\ln x_1^{(\alpha)} = -\int_{T_1}^{T} \frac{h_1^{(\beta),0} - h_1^{(\alpha),0}}{RT^2}\, dT = \int_{T_1}^{T} \frac{L_1}{RT^2}\, dT$$

$$= \frac{L_1}{R}\left(\frac{1}{T_1} - \frac{1}{T}\right) = \frac{L_1}{RT_1}\left(1 - \frac{T_1}{T}\right) \tag{5-39}$$

where T_1 is the melting point of pure solid 1 at pressure p (point G in Fig. 5-13); T is the freezing point of the liquid solution with mole fraction $x_1^{(\alpha)}$; and $L_1 = h_1^{(\alpha),0} - h_1^{(\beta),0}$ is the latent heat of fusion of pure solid 1. In the preceding integration L_1 has been assumed to be a constant. Rearrangement of Eq. (5-39) gives

$$x_1^{(\alpha)} = \exp\left[\frac{L_1}{RT_1}\left(1 - \frac{T_1}{T}\right)\right] \tag{5-40}$$

Similarly, starting from Eqs. (5-35a) and (5-37) we have

$$x_2^{(\alpha)} = \exp\left[\frac{L_2}{RT_2}\left(1 - \frac{T_2}{T}\right)\right] \tag{5-41}$$

where T_2 is the melting point of pure solid 2 at pressure p (point D in Fig. 5-13); T is the freezing point of the liquid solution with mole fraction $x_2^{(\alpha)}$; and $L_2 = h_2^{(\alpha),0} - h_2^{(\beta),0}$ is the latent heat of fusion of pure solid 2 (assumed to be a constant).

In Fig. 5-13, Eq. (5-40) is the equation for curve GE, and Eq. (5-41) is the equation for curve DE. At the eutectic point E, where three phases (solid 1, solid 2, and liquid solution) are in equilibrium, Eqs. (5-40) and (5-41) must be satisfied simultaneously. Thus at the eutectic point we must have

$$\exp\left[\frac{L_1}{RT_1}\left(1 - \frac{T_1}{T_E}\right)\right] + \exp\left[\frac{L_2}{RT_2}\left(1 - \frac{T_2}{T_E}\right)\right] = 1$$

where T_E is the eutectic temperature. Putting T_E into Eq. (5-40) or (5-41) we have the composition at the eutectic point.

5-7 BOILING-POINT ELEVATION AND FREEZING-POINT DEPRESSION

A direct result of the lowering of vapor pressure according to Raoult's law is the phenomenon of boiling-point elevation for dilute solutions containing non-volatile solutes. The boiling point of a liquid is defined as the temperature at which the vapor pressure is equal to the pressure exerted by its environment, which is usually one atmosphere. Since the solute is nonvolatile, it does not contribute any vapor pressure. Thus the vapor pressure is due entirely to the solvent. As the solute is dissolved, the vapor pressure of the solvent decreases. Consequently, when a solute such as sugar is dissolved in a solvent like water, the solution boils at a higher temperature than the pure solvent.

To derive the equation for boiling-point elevation, we first realize that since the vapor pressure p_1° of the pure solvent may be considered as a function of temperature alone, Raoult's law (Eq. 4-84) can be stated formally as

$$p_1 = f(T, x_1)$$

The total differential of p_1 is then

$$dp_1 = \left(\frac{\partial p_1}{\partial T}\right)_{x_1} dT + \left(\frac{\partial p_1}{\partial x_1}\right)_T dx_1$$

From Raoult's law we have

$$\left(\frac{\partial p_1}{\partial x_1}\right)_T = p_1^\circ$$

and from the Clapeyron relation, by neglecting the molal volume of liquid and approximating the molal volume of vapor by the ideal gas law, we have

$$\left(\frac{\partial p_1}{\partial T}\right)_{x_1} = \left(\frac{\partial p_1^\circ x_1}{\partial T}\right)_{x_1} = x_1 \frac{dp_1^\circ}{dT} = x_1 \frac{h_{1fg}^0}{T(v_{1g}^0 - v_{1f}^0)} = x_1 p_1^\circ \frac{h_{1fg}^0}{RT^2}$$

where h_{1fg}^0 is the latent heat of vaporization per mole of the pure solvent. Whence we have

$$dp_1 = \frac{h_{1fg}^0}{RT^2} x_1 p_1^\circ \, dT + p_1^\circ \, dx_1$$

Since we are to vary T and x_1 so that the vapor pressure p_1 is unchanged at the environmental value, we accordingly write

$$p_1^\circ \, dx_1 = -\frac{h_{1fg}^0}{RT^2} x_1 p_1^\circ \, dT$$

Separating the variables and integrating, assuming h°_{1fg} constant,

$$\int_{x_1 = 1}^{x_1} \frac{dx_1}{x_1} = -\frac{h^{\circ}_{1fg}}{R} \int_{T_0}^{T} \frac{dT}{T^2}$$

$$\ln x_1 = \frac{h^{\circ}_{1fg}}{R}\left(\frac{1}{T} - \frac{1}{T_0}\right)$$

where T_0 is the boiling point of the pure solvent and T the boiling point of the solution. Since this equation is for dilute solutions with $x_1 \to 1$ and $x_2 \to 0$, hence $T_0 \approx T$ and $\ln x_1 = \ln(1 - x_2) \approx -x_2$. Accordingly we may write

$$x_2 = \frac{h^{\circ}_{1fg}}{R}\frac{\Delta T}{T_0^2}$$

or $\qquad \Delta T = \frac{RT_0^2}{h^{\circ}_{1fg}} x_2$ $\qquad\qquad\qquad\qquad\qquad$ (5-42)

where $\Delta T = T - T_0$ is the elevation in boiling point corresponding to the mole fraction x_2 of the solute. This is known as the *van't Hoff law of boiling-point elevation*. This equation may be used to determine the latent heat of vaporization of a solvent or the molecular weight of a nonvolatile solute.

In the preceding discussions we learned that a solution containing nonvolatile solutes boils at a higher temperature than the pure solvent. In the same manner a solution will freeze at a lower temperature as compared to the pure solvent, provided that when the solution starts to freeze it is the pure solvent which forms the solid phase and the solute is confined to the liquid phase. The treatment of freezing-point depression is so closely analogous to that of boiling-point elevation that it will not be necessary to reproduce it in detail. The equation analogous to Eq. (5-42) is

$$\Delta T = \frac{RT_0^2}{h^{\circ}_{1if}} x_2 \qquad\qquad\qquad\qquad\qquad (5\text{-}43)$$

where $\Delta T = T_0 - T$ is the depression in freezing point corresponding to the mole fraction x_2 of the solute, T_0 is the freezing point of the pure solvent, T is the freezing point of the solution, and h°_{1if} is the latent heat of fusion per mole of the pure solvent. Equation (5-43) is known as the *van't Hoff law of freezing-point depression*. Freezing-point depressions may also be used in molecular weight determinations.

EXAMPLE 5-6. When 7.6 g of a substance is dissolved in 1000 g of benzene (C_6H_6), the boiling point at 1 atm is raised from 80.10°C to 80.24°C. Estimate the molecular weight of the solute. The enthalpy of vaporization of pure benzene at 1 atm is 7353 cal/g mole.

Solution.
Inserting the given data in Eq. (5-42) yields

$$x_2 = \frac{h^0_{1fg}}{R} \frac{\Delta T}{T^2_0} = \frac{7,353}{1.986} \frac{(80.24 - 80.10)}{(80.10 + 273.15)^2} = 0.00415$$

On the other hand, by definition,

$$x_2 = \frac{n_2}{n_1 + n_2}$$

where

n_1 = number of moles of benzene = $\dfrac{1,000}{78.1}$ = 12.8 g moles

n_2 = number of moles of the solute

Whence

$$n_2 = \frac{x_2 n_1}{1 - x_2} = \frac{0.00415 \times 12.8}{1 - 0.00415} = 0.0533 \text{ g moles}$$

Therefore the molecular weight of the solute is

$$\frac{7.6}{0.0533} = 143 \text{ g/g mole}$$

EXAMPLE 5-7. Ethylene glycol ($C_2 H_4 (OH)_2$) is sold as Prestone for anti-freeze. It has a density of 1.12 g/cm^3. How many liters of pure ethylene glycol must be added to a 16-liter automobile radiator to prevent freezing at $-15°C$?

Solution.
From the Keenan-Keyes-Hill-Moore's steam tables (international edition) for water at $0°C$ we obtain

$$h^0_{1if} = h^0_{1ig} - h^0_{1fg} = 2834.8 - 2501.4 = 333.4 \text{ joules/g}$$

which will be assumed to be constant in the temperature range of this problem. Then by the van't Hoff law of freezing-point depression, we have

$$x_2 = \frac{(\Delta T)h^0_{1if}}{RT^2_0} = \frac{[0 - (-15)](333.4 \times 18)}{8.314(273.15)^2}$$

$$= 0.145 \text{ g mole of Prestone/g mole of solution}$$

Whence

$$x_1 = 1 - 0.145 = 0.855 \text{ g mole of water/g mole of solution}$$

Then the mass ratio of Prestone and water is

$$\frac{0.145 \times 62}{0.855 \times 18} = 0.584 \text{ g of Prestone/g of water}$$

and the volume ratio of Prestone and water is

$$\left(0.584 \ \frac{\text{g of Prestone}}{\text{g of water}}\right) \frac{1.00 \text{ g of water/cm}^3}{1.12 \text{ g of Prestone/cm}^3}$$

$$= 0.521 \text{ liter of Prestone/liter of water}$$

Let y = liters of Prestone added to the 16-liter radiator. We have then

$$y + \frac{y}{0.521} = 16$$

Hence

$$y = 5.5 \text{ liters}$$

5-8 OSMOTIC EQUILIBRIUM

In addition to boiling-point elevation and freezing-point depression there is another phenomenon, known as osmotic equilibrium, which is related to the lowering of vapor pressure that results from the dissolution of a solute in a solvent. Consider the system depicted in Fig. 5-14, where the single-component liquid (α phase) and the two-component liquid solution (β phase) are separated by a nondeformable, heat-conducting membrane permeable to component 1 only. Here component 1 is the solvent and component 2 the solute. The chemical potential of the solvent in the solution is lower than that of the pure solvent at the same temperature and pressure. If both sides of the membrane are kept at the same pressure, a flow of solvent from the α phase to the β phase will be observed. The flow of solvent can be prevented and phase equilibrium

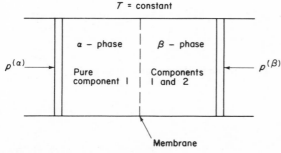

FIG. 5-14. Osmotic equilibrium.

restored if the pressure $p^{(\beta)}$ is isothermally increased to a value so that

$$\mu_1^{(\alpha)} = \mu_1^{(\beta)}$$

The excess pressure that must be applied to a solution to bring the chemical potential of its solvent up to that of pure solvent at the same temperature is called the *osmotic pressure* π of the solution. Expressed symbolically,

$$\pi = p^{(\beta)} - p^{(\alpha)} \tag{5-44}$$

Let us proceed now to see how the osmotic pressure π is related to other thermodynamic properties. As usual, the subscripts denote components and the superscripts denote phases. In addition, the superscript 0 denotes pure component state. Applying Eq. (4-21) to component 1 in each phase, we have

$$d\mu_1^{(\alpha)} = v_1^{(\alpha),0} dp^{(\alpha)} - s_1^{(\alpha),0} dT \tag{5-45}$$

and

$$d\mu_1^{(\beta)} = \bar{V}_1^{(\beta)} dp^{(\beta)} - \bar{S}_1^{(\beta)} dT + \left(\frac{\partial \mu_1^{(\beta)}}{\partial n_1^{(\beta)}}\right)_{T,p^{(\beta)},n_2^{(\beta)}} dn_1^{(\beta)} + \left(\frac{\partial \mu_1^{(\beta)}}{\partial n_2^{(\beta)}}\right)_{T,p^{(\beta)},n_1^{(\beta)}} dn_2^{(\beta)} \tag{5-46}$$

Since at equilibrium $\mu_1^{(\alpha)} = \mu_1^{(\beta)}$, we must have

$$d\mu_1^{(\alpha)} = d\mu_1^{(\beta)} \tag{5-47}$$

Now let us assume that both the α and β phases consist initially of pure solvent 1 at the same pressure $p^{(\alpha)}$, so that π is initially zero. Then $n_2^{(\beta)}$ moles of solute 2 are added to the β phase, and the pressure on the β phase is increased isothermally to $p^{(\beta)}$ so that the negative effect on chemical potential of the solvent in the solution due to the presence of solute is just balanced by the positive effect due to the increase in pressure on the solution. Now since neither temperature nor pressure is changed in the α phase, $d\mu_1^{(\alpha)} = 0$. Therefore from Eq. (5-47) we must have $d\mu_1^{(\beta)} = 0$. Furthermore, since $dT = 0$, $dn_1^{(\beta)} = 0$, Eq. (5-46) becomes

$$\bar{V}_1^{(\beta)} dp^{(\beta)} = -\left(\frac{\partial \mu_1^{(\beta)}}{\partial n_2^{(\beta)}}\right)_{T,p^{(\beta)},n_1^{(\beta)}} dn_2^{(\beta)}$$

This equation may be integrated under the initial condition that

$$p^{(\beta)} = p^{(\alpha)} \qquad n_2^{(\beta)} = 0$$

The result is

$$\pi = p^{(\beta)} - p^{(\alpha)} = -\int_0^{n_2^{(\beta)}} \frac{1}{\bar{V}_1^{(\beta)}} \left(\frac{\partial \mu_1^{(\beta)}}{\partial n_2^{(\beta)}}\right)_{T,p^{(\beta)},n_1^{(\beta)}} dn_2^{(\beta)} \tag{5-48}$$

For simplicity we will now consider the β phase as an ideal solution. From Eq. (4-78) it follows that

$$\mu_1^{(\beta)} = \mu_1^{(\beta),0} + RT \ln x_1^{(\beta)} = \mu_1^{(\beta),0} + RT \ln n_1^{(\beta)} - RT \ln (n_1^{(\beta)} + n_2^{(\beta)})$$

where $\mu_1^{(\beta),0}$ is the chemical potential of pure solvent 1 at T and $p^{(\beta)}$. Consequently,

$$\left(\frac{\partial \mu_1^{(\beta)}}{\partial n_2^{(\beta)}}\right)_{T,p^{(\beta)},n_1^{(\beta)}} = -\frac{RT}{n_1^{(\beta)} + n_2^{(\beta)}} \tag{5-49}$$

Also from Eq. (4-73) for an ideal solution we have

$$\bar{V}_1^{(\beta)} = v_1^{(\beta),0} \tag{5-50}$$

where $v_1^{(\beta),0}$ is the molal volume of pure solvent 1 at T and $p^{(\beta)}$. Substituting Eqs. (5-49) and (5-50) into Eq. (5-48) leads to

$$\pi = \frac{RT}{v_1^{(\beta),0}} \int_0^{n_2^{(\beta)}} \frac{dn_2^{(\beta)}}{n_1^{(\beta)} + n_2^{(\beta)}} = \frac{RT}{v_1^{(\beta),0}} \ln \left(\frac{n_1^{(\beta)} + n_2^{(\beta)}}{n_1^{(\beta)}}\right)$$

or

$$\pi = -\frac{RT}{v_1^{(\beta),0}} \ln x_1^{(\beta)} \tag{5-51}$$

where $v_1^{(\beta),0}$ has been assumed to be independent of pressure.

If the solution is infinitely dilute, $x_2^{(\beta)} = 1 - x_1^{(\beta)} \ll 1$ and $\ln x_1^{(\beta)} = \ln (1 - x_2^{(\beta)}) \approx -x_2^{(\beta)}$. Equation (5-51) can then be rewritten as

$$\pi = \frac{RT}{v_1^{(\beta),0}} x_2^{(\beta)} = \frac{RT}{v_1^{(\beta),0}} \frac{n_2^{(\beta)}}{n_1^{(\beta)} + n_2^{(\beta)}} \tag{5-52}$$

Furthermore, since the solution is dilute, $n_1^{(\beta)} + n_2^{(\beta)} \approx n_1^{(\beta)}$, and $v_1^{(\beta),0}(n_1^{(\beta)} + n_2^{(\beta)}) \approx v_1^{(\beta),0} n_1^{(\beta)} \approx V^{(\beta)} =$ total volume of β phase. Equation (5-52) can be simplified to read

$$\pi = \frac{n_2^{(\beta)} RT}{V^{(\beta)}} \tag{5-53}$$

This equation is known as the *van't Hoff law of osmotic pressure*. It is identical in form to the ideal gas equation, with the osmotic pressure π playing the role of the pressure p. Notice that Eq. (5-53) applies to infinitely dilute solutions only. It is often employed for molecular weight determination for high molecular weight molecules. For such calculations Eq. (5-53) is rewritten as

$$\pi = \frac{cRT}{M_2} \quad \text{or} \quad M_2 = RT\left(\frac{c}{\pi}\right)$$

where c is the concentration of solute in grams per unit volume, and M_2 is the molecular weight of solute in grams per g mole. Since the van't Hoff law is for infinitely dilute solutions, it is important in molecular weight calculations to use data for limiting values of c/π as $c \to 0$.

EXAMPLE 5-8. A liquid solution contains 5 grams of ordinary sugar (sucrose, $C_{12}H_{22}O_{11}$) in 100 grams of water. (a) Estimate the vapor pressure of the solution at $100°C$. (b) Estimate the boiling point of the solution at 1 atm. (c) Estimate the osmotic pressure of the solution at $30°C$.

Solution.

The molecular weight of water is 18 g/g mole, and that of sugar is 342 g/g mole. The mole fractions of water and sugar are

$$x_1 = x_{water} = \frac{100/18}{100/18 + 5/342} = 0.9974$$

$$x_2 = x_{sugar} = 1 - 0.9974 = 0.0026$$

(a) The vapor pressure of pure water at $100°C$ is $p_1^o = 1$ atm = 14.696 psia. According to Raoult's law, the partial pressure of water in the solution is

$$p_1 = p_1^o x_1 = 1 \times 0.9974 = 0.9974 \text{ atm} = 14.658 \text{ psia}$$

Since sugar is nonvolatile, the vapor pressure of the solution at $100°C$ is also 0.9974 atm = 14.658 psia. Thus the vapor pressure of water is lowered by 0.038 psi due to the presence of sugar.

(b) The boiling point of pure water at 1 atm is $T_0 = 100°C$. The latent heat of vaporization of pure water at 1 atm is $h_{1fg}^o = 970.4$ Btu/1 bm = 539.1 cal/g. From Eq. (5-42), the elevation in boiling point is

$$\Delta T = \frac{RT_0^2}{h_{1fg}^o} x_2 = \frac{1.986(100 + 273.15)^2}{539.1 \times 18} \times 0.0026 = 0.074°C$$

Thus, the boiling point of the solution at 1 atm is

$$T = T_0 + \Delta T = 100 + 0.074 = 100.074°C$$

(c) The specific volume of pure water at $30°C$ is

$$v_1^o = 1.0043 \text{ cm}^3/\text{g} = 18.08 \text{ cm}^3/\text{g mole}$$

From Eq. (5-51), the osmotic pressure is then

$$\pi = -\frac{RT}{v_1^o} \ln x_1 = -\frac{82.06(30 + 273.15)}{18.08} \ln 0.9974 = 3.58 \text{ atm}$$

Since the solution is very dilute, Eq. (5-53) can also be used to obtain approximately the same answer.

5-9 HIGHER-ORDER PHASE TRANSITIONS

In a change of phase, such as vaporization, fusion, or sublimation of a pure substance, the temperature and pressure remain constant while entropy and volume undergo finite changes. Since from Eq. (2-11) for a unit mass we have

$$dg = -s\,dT + v\,dp$$

it is apparent that there will be no change in Gibbs function during such a phase transition. However, since

$$\left(\frac{\partial g}{\partial T}\right)_p = -s \quad \text{and} \quad \left(\frac{\partial g}{\partial p}\right)_T = v \tag{2-12d}$$

it follows that the first-order derivatives of the Gibbs function must undergo finite changes. Such a transition is therefore called a *first-order phase transition*. The left part of Fig. 5-15 illustrates the main characteristics of first-order transitions.

In contrast to first-order transitions there are phase changes taking place at constant temperature and pressure with no entropy and volume changes. Thus for such a phase transition the first-order derivatives of the Gibbs function exhibit no discontinuity as illustrated in the right part of Fig. 5-15. However, if the second-order derivatives of the Gibbs function undergo finite changes during the transition, such a transition is defined as a *second-order phase transition*. Since from Eqs. (2-12d), (2-14), (2-15), and (2-27) we have

$$\left(\frac{\partial^2 g}{\partial T^2}\right)_p = -\left(\frac{\partial s}{\partial T}\right)_p = -\frac{c_p}{T}, \quad \left(\frac{\partial^2 g}{\partial p^2}\right)_T = \left(\frac{\partial v}{\partial p}\right)_T = -\kappa v, \quad \left(\frac{\partial^2 g}{\partial T\partial p}\right) = \left(\frac{\partial v}{\partial T}\right)_p = \alpha v$$

a true second-order phase transition will show finite changes in the specific heat, the isothermal compressibility, and the coefficient of thermal expansion.

In a first-order phase transition, the slope of the $p\text{-}T$ equilibrium curve is given by the Clapeyron equation

$$\frac{dp}{dT} = \frac{s^{(f)} - s^{(i)}}{v^{(f)} - v^{(i)}}$$

where the superscripts (i) and (f) denote the initial and final phases of a transition. In a second-order transition the right side of this equation becomes indeterminate. To determine the slope dp/dT for a second-order phase transition we now use the conditions that there are no changes in entropy and volume. Starting from the condition that $s^{(i)} = s^{(f)}$ or $ds^{(i)} = ds^{(f)}$ and through the use of the second ds equation (Eq. 2-35), we obtain

$$c_p^{(i)}dT - Tv\alpha^{(i)}dp = c_p^{(f)}dT - Tv\alpha^{(f)}dp$$

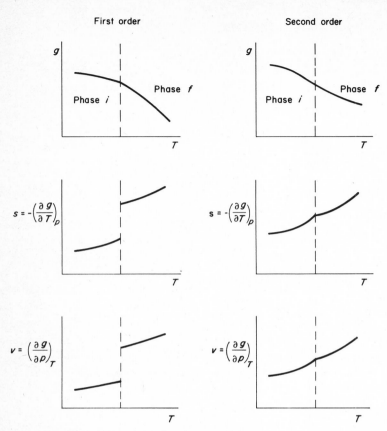

FIG. 5-15. First- and second-order phase transitions.

where we have used $v = v^{(i)} = v^{(f)}$. Therefore

$$\frac{dp}{dT} = \frac{c_p^{(f)} - c_p^{(i)}}{Tv(\alpha^{(f)} - \alpha^{(i)})} \tag{5-54}$$

Alternately, starting from the condition that $v^{(i)} = v^{(f)}$ or $dv^{(i)} = dv^{(f)}$ and noting that

$$dv = \left(\frac{\partial v}{\partial T}\right)_p dT + \left(\frac{\partial v}{\partial p}\right)_T dp = v\alpha dT - v\kappa\, dp$$

we obtain

$$v\alpha^{(i)}dT - v\kappa^{(i)}dp = v\alpha^{(f)}dT - v\kappa^{(f)}dp$$

Therefore

$$\frac{dp}{dT} = \frac{\alpha^{(f)} - \alpha^{(i)}}{\kappa^{(f)} - \kappa^{(i)}} \tag{5-55}$$

Equations (5-54) and (5-55) are known as *Ehrenfest's equations* for second-order phase transitions.

The transition from normal to superconductivity of a type I superconductor in the
absence of a magnetic field is an example of
second-order phase transition. Other transitions,
such as from ferromagnetism to paramagnetism
at the Curie point and from ordinary liquid
helium to superfluid liquid helium at λ point,
were originally thought to be second order.
These transitions satisfy the conditions that T, p,
g, s, and v (and therefore u, h, and a) remain
constant. But more precise experimental data indicate that the conditions of finite changes in c_p,
κ, and α are not satisfied. Instead, these prop-

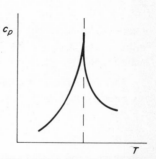

FIG. 5-16. Specific heat vs. temperature diagram for a λ transition.

erties would become infinite at the transition temperature. A diagram of c_p vs. T
for such a transition is shown in Fig. 5-16. It was the resemblance of this curve
to the Greek letter λ that led to the name λ *transition* for such a transition. We
will come back to the subject of λ transition when we study the phase behavior
of helium.

PROBLEMS

5-1. Prove the equivalence of the following two criteria for equilibrium:

$$(\delta S)_{U,V} < 0 \quad \text{and} \quad (\delta U)_{S,V} > 0$$

5-2. Show that

$$\left(\frac{\partial^2 U}{\partial S^2}\right)_V \left(\frac{\partial^2 U}{\partial V^2}\right)_S - \left(\frac{\partial^2 U}{\partial S \partial V}\right)^2 = -\frac{T}{C_v}\left(\frac{\partial p}{\partial V}\right)_T$$

and then verify that the condition of stability as expressed by the third inequality in Eq. (5-13) is equivalent to that of Eqs. (5-14) and (5-18).

5-3. Show that the van der Waals equation of state does not satisfy the
stability criterion for $(\partial p/\partial v)_T$ for all values of the parameters. Plot the van der
Waals isotherms on a pressure-volume plot, and show the region of instability.

5-4. It is known that $(\partial p/\partial v)_T = 0$ for a pure substance at the critical point.
From the condition of stability imposed on the Helmholtz function, show that
$(\partial^2 p/\partial v^2)_T = 0$ and $(\partial^3 p/\partial v^3)_T < 0$.

5-5. For a single-component, two-phase system, show that the latent heat of transition L from phase i to phase f varies with temperature T according to the relation

$$\frac{dL}{dT} = c_p^{(f)} - c_p^{(i)} + \frac{L}{T} - L\,\frac{v^{(f)}\alpha^{(f)} - v^{(i)}\alpha^{(i)}}{v^{(f)} - v^{(i)}}$$

where c_p denotes the constant-pressure specific heat, v denotes the specific volume, and α denotes the coefficient of thermal expansion. Simplify the preceding relation if phase i is either a solid or a liquid and phase f is an ideal gas.

5-6. In a binary liquid-vapor mixture, the relative volatility ω is defined as the ratio of the vapor pressures of the two pure components, or

$$\omega = \frac{p_1^0}{p_2^0}$$

Assuming Raoult's and Dalton's laws hold, show that

$$x_1^{(\beta)} = \frac{\omega x_1^{(\alpha)}}{1 + (\omega - 1)x_1^{(\alpha)}}$$

where $x_1^{(\beta)}$ and $x_1^{(\alpha)}$ are the mole fractions of component 1 in the vapor and liquid phases respectively.

5-7. At extremely high pressures graphite and diamond exist in equilibrium according to the requirement that their specific Gibbs functions are equal. Estimate the pressure required to make diamonds from graphite at a temperature of $25°C$. At $25°C$ and 1 atm the following data are available:

	Graphite	Diamond
Specific Gibbs function, joules/kg	0	2.37×10^5
Specific volume, m^3/kg	4.45×10^{-4}	2.85×10^{-4}
Isothermal compressibility κ, atm^{-1}	3.0×10^{-6}	0.16×10^{-6}

5-8. For a binary vapor-liquid system at 1 atm, the dew-point line and bubble-point line on the temperature-composition diagram (Fig. 5-7) are assumed to obey the following equations

$$T = T_2 - (T_2 - T_1)[x_1^{(\beta)}]^2 \quad \text{and} \quad T = T_2 - (T_2 - T_1)x_1^{(\alpha)}(2 - x_1^{(\alpha)})$$

respectively. In these equations, T_1 and T_2 are the saturation temperatures of pure components 1 and 2 at 1 atm, and $x_1^{(\beta)}$ and $x_1^{(\alpha)}$ are the mole fractions of component 1 in the vapor and liquid phases respectively. What will be the composition of the vapor phase when a liquid solution of 40 mole percent of component 1 is heated to its boiling temperature?

5-9. Calculate and draw the temperature-composition diagram for the system carbon tetrachloride (CCl_4) and tin tetrachloride ($SnCl_4$) at a total pressure of 760 mm Hg, assuming that Raoult's and Dalton's laws are obeyed. The following table gives the vapor-pressure data of the pure components.

Temperature, °C	77	80	90	100	110	114
$p°$ of CCl_4, mm Hg	760	836	1,112	1,450	1,880	—
$p°$ of $SnCl_4$, mm Hg	—	258	362	490	673	760

5-10. Using the vapor pressure data for carbon tetrachloride and tin tetrachloride given in Prob. 5-9, calculate and draw the pressure-composition diagram at $100°C$. Assume that Raoult's and Dalton's laws are obeyed.

5-11. Assuming both liquid and vapor phases to behave as ideal solutions in connection with the general liquid-vapor equilibrium conditions that $f_1^{(\alpha)} = f_1^{(\beta)}$ and $f_2^{(\alpha)} = f_2^{(\beta)}$, and evaluating the fugacities by the use of the generalized fugacity chart (Appendix Fig. A-20), solve the problem as given in Example 5-4. Describe clearly how you obtain the fugacities of the hypothetical pure liquid and pure vapor states of this problem.

5-12. In a process of evaporative sea-water desalination, the sea water with 1.4 mole percent of salt (NaCl) is heated to its boiling point at 1 atm. The evolved vapor, which is essentially pure water, is subsequently condensed to give the required fresh water supply. Estimate the boiling temperature of the salt solution.

5-13. When five liters of ethyl alcohol (C_2H_5OH) are poured into a 20-liter automobile radiator in winter, how low will the temperature be before the solution freezes? The density of ethyl alcohol is 0.789 g/milliliter.

5-14. It is experimentally determined that when 52.5 g of benzil ($C_{14}H_{10}O_2$) are dissolved in 1000 g of toluene (C_7H_8), the boiling point is raised from 110.62 to $111.40°C$. Estimate the enthalpy of vaporization of pure toluene.

5-15. For a dilute solution containing a volatile solute, derive an expression of boiling-point elevation, assuming that the solvent obeys Raoult's law and the solute obeys Henry's law.

Chapter 6
ELASTIC AND INTERFACIAL-
TENSION SYSTEMS

6-1 WORK DONE IN STRETCHING AN ELASTIC WIRE

We derived in Sec. 1-6 the work done by a compressible system on the moving boundary due to volume change ($p\,dV$ work). There are a number of other work modes that occur frequently in thermodynamic analyses. The objective of this chapter is to introduce two specialized mechanical systems with work modes other than $p\,dV$.

First consider the work done in stretching an elastic thin solid rod or wire of cross-sectional area A and natural length L, as illustrated in Fig. 6-1. If the stretching force F acts through an elongation dL, the work done is

$$\bar{d}W = -F\,dL \tag{6-1}$$

where the negative sign indicates that work input is required to increase the elongation of a rod, in accordance with our sign convention.

It is appropriate in the study of elastic solids to express work in terms of the *stress σ* and the *strain ϵ*, which are defined as

$$\sigma = \frac{F}{A} \tag{6-2}$$

$$d\epsilon = \frac{dL}{L} \tag{6-3}$$

In Eq. (6-2) tensile stress will be taken as positive and compressive stress as negative. We are to neglect the lateral strain when the rod is under axial stress. Within the elastic limit, the process of loading and unloading can be performed reversibly. Upon substituting Eqs. (6-2) and (6-3) into Eq. (6-1), we get the work of elastic stretching

$$\bar{d}W = -A\,L\,\sigma\,d\epsilon$$

or

$$dW = -V_0\, \sigma\, d\epsilon \tag{6-4}$$

where V_0 = A L = volume of the rod at the unstrained state. It is convenient to introduce specific quantities referred to a unit of volume of the system. Thus, the work per unit unstrained volume (for which the symbol w is used) is

$$dw = -\sigma\, d\epsilon \tag{6-4a}$$

where the minus sign indicates that work input is required to increase the strain of an elastic solid.

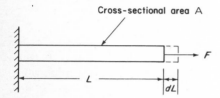

Cross-sectional area A

FIG. 6-1. Elastic thin rod or wire under tension.

6-2 SIMPLE ELASTIC SYSTEMS

The length and lateral dimensions of an elastic rod under tension change. The volume change resulting from the stretching has associated with it some $p\, dV$ work owing to the presence of an ambient pressure p. However, since the change in V is generally very small, this $p\, dV$ work can be neglected with respect to strain work so long as the stress σ is very much greater than the ambient pressure p. A system for which the only reversible work mode is one-dimensional elastic stretching is called a *simple elastic system.*

For a simple elastic system the reversible work per unit unstrained volume of the system is given by

$$dw = -\sigma\, d\epsilon \tag{6-4a}$$

The first law for a reversible process per unit untrained volume of the system is then

$$dq = du - \sigma\, d\epsilon \tag{6-5}$$

The combined first and second laws becomes

$$T\, ds = du - \sigma\, d\epsilon$$

or

$$du = T\,ds + \sigma\,d\epsilon \tag{6-6}$$

which is the very basic property relationship for a simple elastic system.

Since the quantity $-\sigma\,d\epsilon$ for a simple elastic system is analogous to $p\,dv$ for a simple compressible system, it is a simple matter to derive a series of relevant thermodynamic relations for a simple elastic system which are analogous to those derived for a simple compressible system by simply replacing p by $-\sigma$ and v by ϵ. For example, the four *Maxwell relations* for a simple elastic system are

$$\left(\frac{\partial T}{\partial \epsilon}\right)_S = \left(\frac{\partial \sigma}{\partial S}\right)_\epsilon \qquad \text{U} \tag{6-7a}$$

$$\left(\frac{\partial T}{\partial \sigma}\right)_S = -\left(\frac{\partial \epsilon}{\partial S}\right)_\sigma \qquad \text{H} \tag{6-7b}$$

$$\left(\frac{\partial \sigma}{\partial T}\right)_\epsilon = -\left(\frac{\partial S}{\partial \epsilon}\right)_T \qquad \text{F} \qquad \text{MAXWELL RELATIONS} \tag{6-7c}$$

$$\left(\frac{\partial \epsilon}{\partial T}\right)_\sigma = \left(\frac{\partial S}{\partial \sigma}\right)_T \qquad \text{G} \tag{6-7d}$$

The first two ds equations are

$$ds = \frac{c_\epsilon}{T}\,dT - \left(\frac{\partial \sigma}{\partial T}\right)_\epsilon d\epsilon \qquad \text{FIRST 2} \tag{6-8}$$

$$ds = \frac{c_\sigma}{T}\,dT + \left(\frac{\partial \epsilon}{\partial T}\right)_\sigma d\sigma \qquad \text{ds EQUATIONS} \tag{6-9}$$

and the first du equation is

$$du = c_\epsilon\,dT + \left[\sigma - T\left(\frac{\partial \sigma}{\partial T}\right)_\epsilon\right]d\epsilon \qquad \text{du EQ'N} \tag{6-10}$$

in which c_ϵ and c_σ are the specific heats per unit volume of the system at constant strain and constant stress respectively.

An equilibrium state of a simple elastic system within the elastic limit can be described by the three variables: stress σ, strain ϵ, and temperature T. Thus its equation of state will take the form

$$f(\sigma, \epsilon, T) = 0$$

It is convenient to choose the strain ϵ as the dependent variable, so that

$$\epsilon = \epsilon(\sigma, T)$$

The differential form of this equation of state is

$$d\epsilon = \left(\frac{\partial \epsilon}{\partial \sigma}\right)_T d\sigma + \left(\frac{\partial \epsilon}{\partial T}\right)_\sigma dT$$

where the differential coefficients have been given these names:

Young's isothermal modulus of elasticity $Y = \left(\dfrac{\partial \sigma}{\partial \epsilon}\right)_T$

Coefficient of thermal strain $\alpha = \left(\dfrac{\partial \epsilon}{\partial T}\right)_\sigma$

The equation of state in differential form is then

$$d\epsilon = \frac{1}{Y} d\sigma + \alpha \, dT \qquad\qquad\qquad EQ'N \ OF \ STATE \qquad\qquad (6\text{-}11)$$

This equation indicates that the increase in strain of the system is the sum of the elastic strain $d\sigma/Y$ and the thermal strain $\alpha \, dT$.

Since ϵ is a state property of the system in the elastic range, $d\epsilon$ is an exact differential. Applying the condition of exactness to Eq. (6-11) results in

$$\left[\frac{\partial(1/Y)}{\partial T}\right]_\sigma = \left(\frac{\partial \alpha}{\partial \sigma}\right)_T$$

which indicates that the Young's modulus Y and the coefficient of thermal strain α are not independent of each other.

From the cyclic relation of Eq. (2-7), we have

$$\left(\frac{\partial \sigma}{\partial \epsilon}\right)_T \left(\frac{\partial \epsilon}{\partial T}\right)_\sigma \left(\frac{\partial T}{\partial \sigma}\right)_\epsilon = -1$$

or

$$\beta = \left(\frac{\partial \sigma}{\partial T}\right)_\epsilon = -\left(\frac{\partial \sigma}{\partial \epsilon}\right)_T \left(\frac{\partial \epsilon}{\partial T}\right)_\sigma = -Y\alpha$$

where β is called the *coefficient of thermal stress*. With β so defined we can rewrite Eq. (6-11) to read

$$d\sigma = Y \, d\epsilon + \beta \, dT \qquad\qquad\qquad\qquad\qquad\qquad (6\text{-}12)$$

Note that when the material passes the elastic conditions into plastic states, loading and unloading are irreversible, and the thermal equation of state $\epsilon = \epsilon(\sigma, T)$ ceases to be single-valued.

Within the elastic limit there is a range in which the stress is directly proportional to the strain at constant temperature, or

$$\sigma = Y\epsilon \quad \text{at constant } T \tag{6-13}$$

This is known as *Hooke's law*, or Hooke's equation of state. It holds only for sufficiently small strains up to a point on the stress-strain diagram known as the limit of proportionality. In many materials, particularly in metals, the limit of proportionality and the elastic limit are so close to each other that the distinction between them can be ignored.

A stress-strain diagram showing isotherms for a one-dimensional elastic system, such as a metal rod, with positive coefficient of thermal strain is depicted in Fig. 6-2(a). The isotherms are linear in the range shown where Hooke's law is obeyed. The slope of such an isothermal line is equal to Young's modulus which decreases with increasing temperature. From this figure it is seen that the coefficient of thermal stress is negative, meaning that for a metal increasing temperature at constant strain causes a tensile stress to decrease or a compressive stress to increase.

For comparison, a stress-strain diagram for a rubber band is shown in Fig. 6-2(b). It is seen that the coefficient of thermal stress of rubber is positive, meaning that increasing temperature at constant elongation causes its tensile stress to increase.

EXAMPLE 6-1. The tensile stress in a steel rod 5 m long and 5 cm^2 cross-sectional area is increased reversibly from zero to 3×10^5 kN/m^2. (a) If the process occurs at a constant 300°K temperature, calculate the work done, the heat transferred, and the change in internal energy. (b) If the process

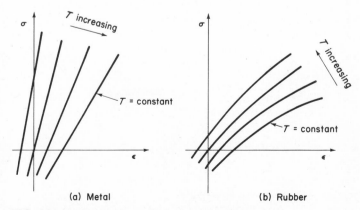

FIG. 6-2.   Isotherms of one-dimensional elastic systems.

occurs adiabatically from an initial $300°K$ temperature, calculate the change in temperature. Assume constant values of $\rho = 7.86 \ g/cm^3$, $c_\sigma = 0.482$ joules/(g)($°K$), $Y = 2.07 \times 10^8 \ kN/m^2$, and $\alpha = 1.20 \times 10^{-5}°K^{-1}$.

Solution.
From Eq. (6-9) for a unit volume of the rod we have

$$dq = T \, ds = c_\sigma \, dT + T\left(\frac{\partial \epsilon}{\partial T}\right)_\sigma d\sigma = c_\sigma \, dT + T \, \alpha \, d\sigma$$

(a) For a reversible isothermal process we have

$$dq = T \, ds = T \, \alpha \, d\sigma$$

Integrating and substituting numerical values lead to

$$q_{12} = T \, \alpha \, (\sigma_2 - \sigma_1)$$
$$= (300°K)(1.20 \times 10^{-5}°K^{-1})(3 \times 10^5 \, kilonewtons/m^2 - 0)$$
$$= 1{,}080 \ kilojoules/m^3$$

Now, the volume of the rod is

$$V = (5m)(5 \times 10^{-4} m^2) = 25 \times 10^{-4} m^3$$

which may be considered constant. Thus

$$Q_{12} = (1.08 \times 10^3 \, kilojoule/m^3)(25 \times 10^{-4} m^3) = 2.70 \ kilojoules$$

Since $Y = (\partial\sigma/\partial\epsilon)_T$, it follows that at constant temperature

$$d\epsilon = \frac{d\sigma}{Y}$$

Substituting this equation in Eq. (6-4) yields

$$dW = - \, V_0 \, \sigma \, d\epsilon = - \frac{V_0}{Y} \, \sigma \, d\sigma$$

Integrating and substituting numerical values lead to

$$W_{12} = - \frac{V_0}{2Y} \, (\sigma_2^2 - \sigma_1^2)$$
$$= - \frac{(25 \times 10^{-4} m^3)}{2(2.07 \times 10^8 \, kilonewtons/m^2)} \, [(3 \times 10^5 \, kilonewtons/m^2)^2 - 0]$$
$$= - \, 0.543 \ kilojoule$$

From the first law we then obtain

$$\Delta U_{12} = Q_{12} - W_{12} = 2.70 - (-0.543) = 3.24 \ kilojoules$$

(b) For a reversible adiabatic process we have

$$dq = T \, ds = 0 = c_\sigma \, dT + T \, \alpha \, d\sigma$$

Hence

$$\left(\frac{\partial T}{\partial \sigma}\right)_S = -\frac{T\alpha}{c_\sigma}$$

If T, α, and c_σ are insensitive to stress we may write

$$\Delta T = -\frac{T\alpha}{c_\sigma}(\Delta\sigma)$$

Substituting numerical values gives

$$\Delta T = -\frac{(300°K)(1.20 \times 10^{-5}°K^{-1})}{(0.482 \text{ joule/g}°K)(7.86 \times 10^6 \text{g/m}^3)}(3 \times 10^8 \text{newton/m}^2 - 0)$$

$$= -0.285°C$$

Thus, the temperature of the steel rod decreases by 0.285°C due to the reversible adiabatic increase of stress.

EXAMPLE 6-2. A steel rod having a cross-sectional area of 4 cm² and a free length of 2 m is used to hang a portion of a steam pipe line. The load on the rod is 5 kilonewtons when the pipe line is full of steam. (a) As steam begins to flow through the pipe the temperature of the rod increases slowly from 15 to 55°C. What is the work done during this heating process? (b) When the steam is shut off the rod cools back from 55 to 15°C. If the pipe line is somehow jammed and cannot move as the rod cools, what will be the final stress in the rod? Assume that $Y = 2.07 \times 10^8$ kilonewtons/m² and $\alpha = 1.26 \times 10^{-5}$ °K⁻¹.

Solution.

(a) When the pipe is full of steam, the stress in the rod is

$$\sigma = \frac{F}{A} = \frac{5 \text{ kN}}{4 \times 10^{-4} \text{m}^2} = 1.25 \times 10^4 \text{kilonewtons/m}^2$$

which does not vary during heating of the rod. Thus from Eq. (6-11), for the heating process since $d\sigma = 0$ it follows that

$$d\epsilon = \frac{1}{Y}d\sigma + \alpha\, dT = \alpha\, dT$$

Substituting this relation into Eq. (6-4) gives

$$dW = -V_0\, \sigma\, d\epsilon = -V_0\, \alpha\, \sigma\, dT$$

Integrating at constant σ yields

$W = - V_0 \, \alpha \, \sigma \, (\Delta T)$

$= - (4 \times 10^{-4} \, m^2)(2 \, m)\left(1.26 \times 10^{-5} \, \dfrac{1}{^\circ K}\right)$

$\left(1.25 \times 10^4 \, \dfrac{kilonewtons}{m^2}\right) [(55 - 15)^\circ K]$

$= - 5.04 \times 10^{-3} \, kilonewton\text{-}m = - 5.04 \, joules$

where the minus sign indicates that work is done on the rod.

(b) Since the pipe line is jammed when steam is shut off, the stress in the rod at the beginning of the cooling process is $\sigma_{initial} = 1.25 \times 10^4$ kilonewtons/m^2 as obtained in Part (a). Now inasmuch as the pipe does not move during cooling of the rod, $d\epsilon = 0$. From Eq. (6-11) we then have

$$d\epsilon = \frac{1}{Y} \, d\sigma + \alpha \, dT = 0 \qquad d\sigma = - Y \alpha \, dT$$

Integrating this equation for the cooling process, we obtain the final stress in the rod as follows:

$\sigma_{final} = \sigma_{initial} - Y \alpha \, (\Delta T)$

$= \left(1.25 \times 10^4 \, \dfrac{kilonewtons}{m^2}\right)$

$- \left(2.07 \times 10^8 \, \dfrac{kilonewtons}{m^2}\right) \left(1.26 \times 10^{-5} \, \dfrac{1}{^\circ K}\right) [(15 - 55)^\circ K]$

$= 1.168 \times 10^5 \, kilonewtons/m^2$

6-3 THREE-DIMENSIONAL ELASTIC SYSTEMS

So far the discussion on elastic systems has been restricted to the simple case of one-dimensional stretching. We now study the thermodynamics of elasticity for three-dimensional general case.

Let ϵ_i $(i = 1, 2, \ldots, 6)$ be the six strain components in a strained elastic solid. The strain components ϵ_1, ϵ_2, and ϵ_3 are longitudinal strains and are the respective fractional increases in the length of elements located initially in the x-, y-, and z-directions. The strain components ϵ_4, ϵ_5, and ϵ_6 are shearing strains and are the respective decreases in the angles between two elements initially in the x- and y-, x- and z-, and y- and z-directions.

The six stress components σ_i $(i = 1, 2, \ldots, 6)$ are defined mathematically by the relation

$$\sigma_i = \frac{1}{V_0} \left(\frac{\partial U}{\partial \epsilon_i}\right)_{S, \, \epsilon_j (j \neq i)} \tag{6-14}$$

where U is the internal energy of the system, and V_0 is the volume of the system in the unstrained state. The volume V_0 is, of course, constant for the system at a given temperature. The derivative in the defining equation for σ_i is to be evaluated at constant entropy S and constant strain components other than ϵ_i. Physically, the stress component σ_1 is defined as the force in the x-direction per unit area normal to the x-direction; the stress components σ_2 and σ_3 are similarly defined for y- and z-directions respectively. The stress component σ_4 is defined as the force in the y-direction per unit area normal to the x-direction or the force in the x-direction per unit area normal to the y-direction. The stress components σ_5 and σ_6 are similarly defined in relation to the x- and z-directions and the y- and z-directions respectively. Thus the stress components σ_1, σ_2, and σ_3 are normal stresses, and σ_4, σ_5, and σ_6 are shearing stresses. These six stress components are depicted pictorially in Fig. 6-3, wherein the following more illustrative double-index notations are used:

$$\sigma_{xx} = \sigma_1 \qquad \sigma_{yy} = \sigma_2 \qquad \sigma_{zz} = \sigma_3$$

$$\sigma_{xy} = \sigma_{yx} = \sigma_4 \qquad \sigma_{xz} = \sigma_{zx} = \sigma_5 \qquad \sigma_{yz} = \sigma_{zy} = \sigma_6$$

The strain components for an elastic solid may vary from point to point within the solid. For the purpose of thermodynamic analyses we are to assume uniformity of the strain components. For such a uniformly strained solid the six quantities $V_0 \epsilon_i (i = 1, 2, \ldots, 6)$ will play the same role in the description of the thermodynamic states of the solid system as the volume V does for a fluid system. Thus the internal energy of a uniformly strained solid system of a single

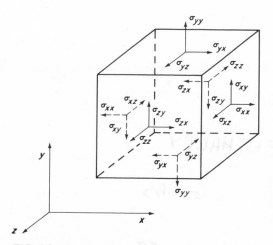

FIG. 6-3. Stress components.

chemical constituent may be written as

$$U = U[S, V_0 \epsilon_i (i = 1, 2, \ldots, 6)]$$

based on the unstrained state as reference. The differential of the energy is then

$$dU = \left(\frac{\partial U}{\partial S}\right)_\epsilon dS + \sum_{i=1}^{6} \left(\frac{\partial U}{\partial \epsilon_i}\right)_{S, \epsilon_j (j \neq i)} d\epsilon_i$$

in which the subscript ϵ implies constancy of all six strain components, and the subscript ϵ_j $(j \neq i)$ implies constancy of all other strain components except ϵ_i. Inasmuch as the constancy of all the strain components in the evaluation of the derivative $(\partial U/\partial S)_\epsilon$ implies the constancy of the total volume, from Eq. (2-12a) we may write

$$\left(\frac{\partial U}{\partial S}\right)_\epsilon = T$$

And from Eq. (6-14) we have

$$\left(\frac{\partial U}{\partial \epsilon_i}\right)_{S, \epsilon_j (j \neq i)} = V_0 \sigma_i$$

Consequently the expression for dU then becomes

$$dU = T \, dS + V_0 \sum_{i=1}^{6} \sigma_i \, d\epsilon_i \tag{6-15}$$

This is the expression for the combined first and second laws for a general elastic system. The last term in the preceding equation is the work done on the system, viz.,

$$dW = - V_0 \sum_{i=1}^{6} \sigma_i \, d\epsilon_i \tag{6-16}$$

where the minus sign indicates that work input is required to increase the strains of an elastic solid.

At this point it is helpful to define two properties, the elastic enthalpy H and the elastic Gibbs function G:

$$H = U - V_0 \sum_{i=1}^{6} \sigma_i \epsilon_i \qquad \text{ENTHALPY} \tag{6-17}$$

$$G = H - TS = U - TS - V_0 \sum_{i=1}^{6} \sigma_i \epsilon_i \qquad \text{GIBBS} \tag{6-18}$$

Note that the Helmholtz function A takes the usual definition

$$A = U - TS \qquad \text{HELMHOLTZ}$$

From these definitions it follows that

$$dH = T\,dS - \sum_{i=1}^{6} \epsilon_i\,d\sigma_i \tag{6-19}$$

$$dA = -S\,dT + V_0 \sum_{i=1}^{6} \sigma_i\,d\epsilon_i \tag{6-20}$$

$$dG = -S\,dT - V_0 \sum_{i=1}^{6} \epsilon_i\,d\sigma_i \tag{6-21}$$

From Eqs. (6-15) and (6-19) to (6-21) we have the following set of *Maxwell relations* as applied to a general elastic system of a single chemical constituent:

$$\frac{1}{V_0}\left(\frac{\partial T}{\partial \epsilon_i}\right)_{S,\,\epsilon_j(j\neq i)} = \left(\frac{\partial \sigma_i}{\partial S}\right)_{\epsilon} \qquad \text{U} \tag{6-22a}$$

$$\left(\frac{\partial T}{\partial \sigma_i}\right)_{S,\,\sigma_j(j\neq i)} = -V_0\left(\frac{\partial \epsilon_i}{\partial S}\right)_{\sigma} \qquad \text{H} \qquad \left.\begin{array}{c}\\ \\ \end{array}\right\} \text{MAXWELL} \tag{6-22b}$$

$$\frac{1}{V_0}\left(\frac{\partial S}{\partial \epsilon_i}\right)_{T,\,\epsilon_j(j\neq i)} = -\left(\frac{\partial \sigma_i}{\partial T}\right)_{\epsilon} \qquad \text{·F} \qquad \left.\begin{array}{c}\\ \end{array}\right\} \text{RELATIONS} \tag{6-22c}$$

$$\left(\frac{\partial S}{\partial \sigma_i}\right)_{T,\,\sigma_j(j\neq i)} = V_0\left(\frac{\partial \epsilon_i}{\partial T}\right)_{\sigma} \qquad \text{G} \tag{6-22d}$$

In the preceding Maxwell relations, the subscripts ϵ and ϵ_j $(j \neq i)$ are defined as before; whereas the subscript σ implies constancy of all six stress components, and the subscript σ_j $(j \neq i)$ implies constancy of all other stress components except σ_i.

Analogous to C_v and C_p, the heat capacity at constant strain C_ϵ and the heat capacity at constant stress C_σ are defined as follows:

$$C_\epsilon = \left(\frac{\partial U}{\partial T}\right)_\epsilon = T\left(\frac{\partial S}{\partial T}\right)_\epsilon \qquad \text{Heat capacity at constant strain} \tag{6-23}$$

$$C_\sigma = \left(\frac{\partial H}{\partial T}\right)_\sigma = T\left(\frac{\partial S}{\partial T}\right)_\sigma \qquad \text{Heat capacity at constant stress} \tag{6-24}$$

where the subscripts ϵ and σ imply that all six strain components and all six stress components are to be held constant.

In a general elastic system, there are six *coefficients of thermal strain* defined as

$$\alpha_i = \left(\frac{\partial \epsilon_i}{\partial T}\right)_\sigma \tag{6-25}$$

and six *coefficients of thermal stress* defined as

$$\beta_i = \left(\frac{\partial \sigma_i}{\partial T}\right)_\epsilon \tag{6-26}$$

There are 36 *isothermal elastic stiffness coefficients* $Y_{T,ij}$ and 36 *adiabatic elastic stiffness coefficients* $Y_{S,ij}$ defined as

$$Y_{T,ij} = \left(\frac{\partial \sigma_i}{\partial \epsilon_j}\right)_{T,\,\epsilon_i(i \neq j)} \tag{6-27}$$

$$Y_{S,ij} = \left(\frac{\partial \sigma_i}{\partial \epsilon_j}\right)_{S,\,\epsilon_i(i \neq j)} \tag{6-28}$$

Furthermore, there are 36 *isothermal elastic compliance coefficients* $\kappa_{T,ij}$ and 36 *adiabatic elastic compliance coefficients* $\kappa_{S,ij}$ defined as

$$\kappa_{T,ij} = \left(\frac{\partial \epsilon_i}{\partial \sigma_j}\right)_{T,\,\sigma_i(i \neq j)} \tag{6-29}$$

$$\kappa_{S,ij} = \left(\frac{\partial \epsilon_i}{\partial \sigma_j}\right)_{S,\,\sigma_i(i \neq j)} \tag{6-30}$$

The coefficients defined here for an elastic system are analogous to those defined for a fluid system. Thus, the coefficient of thermal strain α_i is analogous to the coefficient of thermal expansion $\alpha = \frac{1}{V}(\partial V/\partial T)_p$; the elastic compliance coefficient $\kappa_{T\,\text{or}\,S,\,ij}$ is analogous to the compressibility $\kappa_{T\,\text{or}\,S} = -\frac{1}{V}(\partial V/\partial P)_{T\,\text{or}\,S}$; the elastic stiffness coefficient $Y_{T\,\text{or}\,S}$ is analogous to the bulk modulus $-V(\partial P/\partial V)_{T\,\text{or}\,S}$, and the coefficient of thermal stress β_i is analogous to the coefficient $\alpha/\kappa = (\partial P/\partial T)_V$.

From Eqs. (6-15) and (6-19) to (6-21) the following *Maxwell-like equations* can be obtained:

$$\left(\frac{\partial \sigma_i}{\partial \epsilon_j}\right)_{T,\,\epsilon_i(i \neq j)} = \left(\frac{\partial \sigma_j}{\partial \epsilon_i}\right)_{T,\,\epsilon_j(j \neq i)}$$

$$\left(\frac{\partial \sigma_i}{\partial \epsilon_j}\right)_{S,\,\epsilon_i(i \neq j)} = \left(\frac{\partial \sigma_j}{\partial \epsilon_i}\right)_{S,\,\epsilon_j(j \neq i)}$$

$$\left(\frac{\partial \epsilon_i}{\partial \sigma_j}\right)_{T,\,\sigma_i(i \neq j)} = \left(\frac{\partial \epsilon_j}{\partial \sigma_i}\right)_{T,\,\sigma_j(j \neq i)}$$

$$\left(\frac{\partial \epsilon_i}{\partial \sigma_j}\right)_{S,\,\sigma_i(i \neq j)} = \left(\frac{\partial \epsilon_j}{\partial \sigma_i}\right)_{S,\,\sigma_j(j \neq i)}$$

Whence it follows that $Y_{T,ij} = Y_{T,ji}$, $Y_{S,ij} = Y_{s,ji}$, $\kappa_{T,ij} = \kappa_{T,ji}$, and $\kappa_{S,ij} = \kappa_{s,ji}$.

As a result of these symmetry relations, the number of independent coefficients for Y_T, Y_S, κ_T, and κ_S are reduced from 36 to 21 each.

Since the Helmholtz function may be written as

$$A = A \ [T, V_0 \ \epsilon_i \ (i = 1, 2, \ldots, 6)]$$

it follows from Eq. (6-20) that

$$\sigma_i = \frac{1}{V_0} \left(\frac{\partial A}{\partial \epsilon_i} \right)_{T, \ \epsilon_j(j \neq i)}$$

whereupon we may write

$$\sigma_i = \sigma_i \ [T, V_0 \ \epsilon_j \ (j = 1, 2, \ldots, 6)]$$

A power series expansion of σ_i in terms of ϵ_j at constant temperature, neglecting for small strains all terms of higher order than linear in the expansion, leads to the following general form of isothermal *Hooke's law*,

$$\sigma_i = \sum_{j=1}^{6} Y_{T,ij} \epsilon_j \tag{6-31}$$

In this equation the constant term does not appear because the stresses are so defined as to vanish in the unstrained state, and the coefficients $Y_{T,ij}$ are considered as functions of the temperature only and are independent of the strain.

Hooke's law as described above applies only to a slightly strained elastic system in an isothermal environment. This formulation may be generalized for use in a general thermal environment of variable temperature by including the effect due to thermal expansion. With the simple assumption that the stress varies linearly with the temperature at constant strain, Duhamel and von Neumann proposed the following *generalized form of Hooke's law*.

$$\sigma_i = \sum_{j=1}^{6} Y_{T_0,ij} \epsilon_j + \beta_i (T - T_0) \tag{6-32}$$

where T_0 is the reference temperature and the coefficients $Y_{T_0,ij}$ are to be evaluated at this temperature. This equation is useful for small strains and small temperature changes. Note that Eq. (6-32) is nothing more than an integrated form of Eq. (6-12) when generalized to the three-dimensional case.

The various coefficients defined in this section are mostly interrelated. Two of the relationships are shown in the following example.

EXAMPLE 6-3. Show that

$$\alpha_i = - \sum_{j=1}^{6} \kappa_{T,ij} \beta_j \tag{6-33}$$

$$\beta_i = - \sum_{j=1}^{6} Y_{T,ij} \alpha_j \tag{6-34}$$

Solution.

From the functional relation

$$\epsilon_i = \epsilon_i [T, \sigma_j (j = 1, 2, \ldots, 6)]$$

it follows that

$$d\epsilon_i = \left(\frac{\partial \epsilon_i}{\partial T}\right)_\sigma dT + \sum_{j=1}^{6} \left(\frac{\partial \epsilon_i}{\partial \sigma_j}\right)_{T, \sigma_i (i \neq j)} d\sigma_j$$

Substituting Eqs. (6-25) and (6-29) into the preceding equation leads to

$$d\epsilon_i = \alpha_i \, dT + \sum_{j=1}^{6} \kappa_{T,ij} \, d\sigma_j$$

Dividing this equation by dT and applying it to the constant-strain condition result in

$$\alpha_i + \sum_{j=1}^{6} \kappa_{T, ij} \left(\frac{\partial \sigma_j}{\partial T}\right)_\epsilon = 0$$

Then by using Eq. (6-26), Eq. (6-33) is obtained.

Similarly, starting from the functional relation

$$\sigma_i = \sigma_i [T, V_0 \, \epsilon_j (j = 1, 2, \ldots, 6)]$$

and utilizing Eqs. (6-25), (6-26), and (6-27), one obtains Eq. (6-34).

EXAMPLE 6-4. Derive the equation for the difference in the heat capacities for an elastic system.

Solution.

Consider the entropy of an elastic system as a function of the temperature and the strain components

$$S = S[T, V_0 \, \epsilon_i (i = 1, 2, \ldots, 6)]$$

The differential of S is then

$$dS = \left(\frac{\partial S}{\partial T}\right)_\epsilon dT + \sum_{i=1}^{6} \left(\frac{\partial S}{\partial \epsilon_i}\right)_{T, \epsilon_j (j \neq i)} d\epsilon_i = \frac{C_\epsilon}{T} dT + \sum_{i=1}^{6} \left(\frac{\partial S}{\partial \epsilon_i}\right)_{T, \epsilon_j (j \neq i)} d\epsilon_i$$

Dividing this equation by dT and applying it to the constant-stress condition,

$$\left(\frac{\partial S}{\partial T}\right)_\sigma = \frac{C_\epsilon}{T} + \sum_{i=1}^{6} \left(\frac{\partial S}{\partial \epsilon_i}\right)_{T, \epsilon_j (j \neq i)} \left(\frac{\partial \epsilon_i}{\partial T}\right)_\sigma$$

or

$$C_\sigma = C_\epsilon + T \sum_{i=1}^{6} \left(\frac{\partial S}{\partial \epsilon_i}\right)_{T,\,\epsilon_j(j \neq i)} \alpha_i$$

But by Eq. (6-22c) we have

$$\left(\frac{\partial S}{\partial \epsilon_i}\right)_{T,\,\epsilon_j(j \neq i)} = -V_0 \left(\frac{\partial \sigma_i}{\partial T}\right)_\epsilon = -V_0\,\beta_i$$

The desired equation is then

$$C_\sigma - C_\epsilon = -TV_0 \sum_{i=1}^{6} \beta_i\,\alpha_i$$

This equation is the analog of Eq. (2-32) for a fluid system. Notice that $C_\sigma \to 0$, $C_\epsilon \to 0$, and $C_\sigma - C_\epsilon \to 0$ as $T \to 0$.

6-4 RUBBER ELASTICITY

The thermodynamics of rubber elasticity furnishes an example for the study of elastic systems. While synthetic rubbers have a wide variety of chemical constitutions, natural rubber is essentially hydrocarbon polyisoprene $(C_5 H_8)_n$, built up in the form of a network of long continuous chains. With links capable of free rotation these long-chain molecules are subject to micro-Brownian motion of the chain elements so as to take up a random configuration. Cross links between molecules are provided by a chemical process known as the vulcanization. As a result of its molecular flexibility, rubber possesses the physical property of a high degree of elastic extensibility under relatively small stresses. Moderate long extensions, say up to 300 percent, can occur in most rubbers more or less reversibly at ordinary temperatures without causing permanent structural changes. We are to confine our discussions here to this reversible elastic range.

Consider a bar of rubber of length L. When a tensile force F is applied at each end of the sample, its length will increase and its cross-sectional area will decrease. It is an experimental fact that the changes in length and area are such that the total volume of the sample remains essentially constant. This means that rubber is approximately incompressible just like a liquid. We will accordingly neglect the $p\,dV$ work and consider $-F\,dL$ as the only external work term.

The combined first and second laws for the sample of rubber is

$$dU = T\,dS + F\,dL \tag{6-35}$$

With the usual definition of the Helmhotz function

$$A = U - TS$$

we have

$$dA = F\,dL - S\,dT \qquad (6\text{-}36)$$

Whence it follows that

$$F = \left(\frac{\partial A}{\partial L}\right)_T = \left(\frac{\partial U}{\partial L}\right)_T - T\left(\frac{\partial S}{\partial L}\right)_T \qquad (6\text{-}37)$$

Using the condition of exactness to Eq. (6-36) we have the Maxwell equation

$$\left(\frac{\partial S}{\partial L}\right)_T = -\left(\frac{\partial F}{\partial T}\right)_L \qquad (6\text{-}38)$$

Inserting this equation in Eq. (6-37) gives

$$\left(\frac{\partial U}{\partial L}\right)_T = F - T\left(\frac{\partial F}{\partial T}\right)_L \qquad (6\text{-}39)$$

N.B.
IMPORTANT
EQ'NS

Equations (6-38) and (6-39) provide the fundamental background for determining experimentally both the entropy and internal energy changes associated with the elastic deformation of rubbers.

It is known that the partial derivative $(\partial U/\partial L)_T$ is approximately zero in the extension ratio range we are interested in. For the sake of convenience let us define a rubber for which $(\partial U/\partial L)_T$ is exactly equal to zero as an *ideal rubber*, but bear in mind that all rubbers may be classified as ideal at low and moderate extensions. According to this definition, the internal energy of an ideal rubber is a function of temperature only and is independent of extension. This behavior is in direct analogy to that of an ideal gas for which internal energy is a function of temperature only and is independent of volume. In ideal gases there is no intermolecular potential; this explains the independence of U on V. In the case of an ideal rubber, there is surely intermolecular potential energy. But this potential energy remains unchanged as the polymer chains uncoil and straighten under the action of external forces while there is no alteration in the system total volume.

According to Eq. (6-37) we see that the restoring force that tends to bring a sample of stretched rubber back to its original length is in general associated with changes in entropy and internal energy. However, for an ideal rubber Eq. (6-37) reduces to

$$F = -T\left(\frac{\partial S}{\partial L}\right)_T \qquad (6\text{-}40)$$

which implies that ideal rubber elasticity is entirely an entropy effect. This is in direct contrast to the case of an ordinary hard solid where the elasticity is primarily associated with changes in internal energy.

Since there is no internal energy change in the isothermal deformation of an ideal rubber then, according to the first law, heat must be evolved from the rubber when work is done in stretching it isothermally. From the second law it is apparent that the entropy of the rubber decreases on isothermal stretching.

It is a well-known fact that when an ideal rubber is stretched adiabatically its temperature rises. This is seen from the following analysis. Applying the first law to the system, we write for a reversible process,

$$dQ = dU - F \, dL$$

When the process is adiabatic we have

$$dU = F \, dL \tag{6-41}$$

But since U is a function of T only for an ideal rubber, it follows that

$$dU = \left(\frac{\partial U}{\partial T}\right)_L dT = C_L \, dT \tag{6-42}$$

where C_L is the heat capacity at constant length. Combining Eqs. (6-41) and (6-42) gives for the reversible adiabatic stretching process

$$dT = \frac{1}{C_L} F \, dL \tag{6-43}$$

Since C_L is positive, when F is also positive (tensile force), dL is positive (extension), we see that dT must be positive and that the temperature rises.

EXAMPLE 6-5. The thermal equation of state of a strip of an ideal rubber of a certain initial length and cross-sectional area is given by the expression

$$F = K \, T \left(\frac{L}{L_0} - \frac{L_0^2}{L^2}\right)$$

where F is tensile force in newtons, T is temperature in $^\circ$K, L is length in meters, L_0 is the value of L at zero tension, and $K = 1.33 \times 10^{-2}$ newtons/$^\circ$K. (a) Calculate the work done and the heat transfer when the rubber strip is stretched slowly from $L_0 = 1$ m to $L = 2 \, L_0 = 2$ m at a constant temperature of $T = 300^\circ$K. (b) Calculate the final temperature when the rubber strip is stretched reversibly and adiabatically from $T_0 = 300^\circ$K, $L_0 = 1$ m to $L = 2 \, L_0 = 2$ m, assuming $C_L = 1.2$ joules/$^\circ$K for the rubber strip.

Solution.

(a) Reversible isothermal stretching: From Eq. (6-1) we have

$$dW = - F \, dL = - K \, T \left(\frac{L}{L_0} - \frac{L_0^2}{L^2}\right) dL$$

Integrating at constant temperature gives

$$W = -KT \left[\frac{1}{2 L_0} (L^2 - L_0^2) - L_0^2 \left(\frac{1}{L_0} - \frac{1}{L} \right) \right]$$

$$= -(1.33 \times 10^{-2})(300) \left[\frac{1}{2} (4 - 1) - \left(1 - \frac{1}{2} \right) \right] = -3.99 \text{ joules}$$

where the minus sign indicates that work is done on the rubber strip.

Since for an ideal rubber the internal energy is a function of temperature only, $\Delta U = 0$ for the isothermal stretching process. By the first law we have then

$$Q = W = -3.99 \text{ joules}$$

where the minus sign indicates that heat is rejected from the rubber strip.

(b) Reversible adiabatic stretching: From Eq. (6-43) we have

$$dT = \frac{1}{C_L} F\, dL = \frac{1}{C_L} KT \left(\frac{L}{L_0} - \frac{L_0^2}{L^2} \right) dL$$

or

$$\frac{dT}{T} = \frac{K}{C_L} \left(\frac{L}{L_0} - \frac{L_0^2}{L^2} \right) dL$$

Integrating the preceding equation yields

$$\ln \frac{T}{T_0} = \frac{K}{C_L} \left[\frac{1}{2 L_0} (L^2 - L_0^2) - L_0^2 \left(\frac{1}{L_0} - \frac{1}{L} \right) \right]$$

$$= \frac{1.33 \times 10^{-2}}{1.2} \left[\frac{1}{2} (4 - 1) - \left(1 - \frac{1}{2} \right) \right] = 0.01108$$

or

$$\frac{T}{T_0} = 1.01115$$

Therefore the final temperature is

$$T = 1.01115\, T_0 = 1.01115 \times 300 = 303.3^\circ \text{K}$$

The temperature of the rubber strip is increased by 3.3°C in the reversible adiabatic stretching.

6-5 WORK MODE INVOLVING INTERFACIAL TENSION

In the remainder of this chapter we seek to study another mechanical system— a surface film. A liquid and a gas that coexist in equilibrium are separated by a finite but extremely thin layer of phase boundary of a magnitude comparable with molecular dimensions across which the transition between the two phases takes place. A molecule in the interior of the liquid phase experiences, in general, no resultant intermolecular attractions from its neighbors. However, a molecule in the phase boundary is attracted more strongly by the closely oriented liquid molecules on one side than by the very distant gas molecules on the other side. Work must be done against the resulting unbalanced attraction force to bring a molecule from the interior of the liquid to the surface. A molecule in the phase boundary can thus be considered as having a greater potential energy than a molecule in the interior of the liquid. The total additional potential energy is obviously proportional to the surface area between the phases. The surface tends to assume a shape of minimum area to conform with the conditions of stable equilibrium at minimum potential energy. In other words, the net inward attraction from the liquid causes the surface to diminish in area until the surface is the smallest possible for a given volume, subject to the external conditions or forces acting on the system. For a given volume a sphere has the minimum surface. This is the reason why liquid droplets in a gas and gas bubbles in a liquid are spherical.

In ordinary thermodynamics, surface work effects are usually small compared to such bulk contributions as volume changes. A system chosen to demonstrate surface work must have a geometry which will emphasize surface effects. In addition, the disturbing effect of gravity should be eliminated. Let us consider as a system a double soap film with a small amount of liquid in between. This is stretched across a rectangular wire frame, one side of which is movable, as illustrated in Fig. 6-4. During a change in the film area there are mass transfers between the interior of the liquid and the surface film. Since the surface film tends to assume a minimum area, it tends to pull the movable wire to the left, and a force F as indicated is required to keep the wire in position. When the film

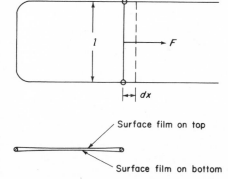

Surface film on top

Surface film on bottom

FIG. 6-4. Surface films stretched on a wire frame.

is stretched quasistatically to the right a distance dx, the film area A will increase by an amount $dA = 2\,l\,dx$. Thus the work required is

$$dW = -F\,dx = -\frac{F}{2 \times l}\,dA$$

Here we neglect the small work done due to volume change. The work required per unit increase in surface area is defined as the *interfacial tension* γ, which is also the force acting perpendicularly to a line of unit length in the surface. That is,

$$\gamma = \frac{F\,dx}{2 \times l \times dx} = \frac{F}{2 \times l}$$

Hence the work required becomes

$$dW = -\gamma\,dA \tag{6-44}$$

where, in accordance with our sign convention, the negative sign indicates that work input is required to increase the area of an interface.

EXAMPLE 6-6. An atomizer slowly shoots out minute water droplets of an average radius of 10^{-4} cm into the air. Estimate the work required when 1 kg of water is atomized isothermally at 25°C. The interfacial tension of water in contact with air at 25°C is 7.2×10^{-4} newtons/cm.

Solution.
From the Keenan-Keyes-Hill-Moore steam tables, for water at 25°C the specific volume is 1.0029 cm^3/g. Thus the number of droplets formed by 1 kg of water is

$$\frac{(1.0029 \text{ cm}^3/\text{g})(10^3\,\text{g/kg})}{\frac{4}{3}\,\pi(10^{-4})^3\,\text{cm}^3} = 2.394 \times 10^{14}$$

The total surface area of all the droplets is

$$A = 4\pi(10^{-4})^2(2.394 \times 10^{14}) = 3.009 \times 10^7\,\text{cm}^2/\text{kg}$$

From Eq. (6-44), $dW = -\gamma\,dA$. We then have

$$W = -\gamma(A - A_0)$$

where A_0 is the initial surface area of 1 kg of water before being atomized. Since $A \gg A_0$, we may write

$$W = -\gamma A$$

Therefore

$$W = -\left(7.2 \times 10^4 \, \frac{\text{newtons}}{\text{cm}}\right)\left(3.009 \times 10^7 \, \frac{\text{cm}^2}{\text{kg}}\right)$$

$$= -2.166 \times 10^4 \, \frac{\text{newton-cm}}{\text{kg}} = -2.166 \times 10^2 \, \text{joules/kg}$$

where the negative sign indicates that work input is required to form the droplets.

6-6 PLANE INTERFACE

The Gibbs' classical treatment of surface thermodynamics centers on the introduction of a mathematical surface of zero thickness, which can be somehow arbitrarily located in the interfacial region. Each of the two bulk phases on both sides of the mathematical surface is assumed to remain homogeneous in composition and physical properties right up to the mathematical surface. Each extensive property of the system is imagined to be the total contributions of these two hypothetical homogeneous bulk phases and the mathematical surface. The contribution from the mathematical surface to any extensive property is termed surface excess of that property and is defined by the expression

$$X^{(\Delta)} = X - X^{(\alpha)} - X^{(\beta)}$$

where $X^{(\Delta)}$ is the surface excess of any bulk extensive property, X is the actual value of the property of the entire system including the interface, while $X^{(\alpha)}$ and $X^{(\beta)}$ are the respective values of the property of the two hypothetical homogeneous bulk phases α and β without an interface.

Although Gibbs' treatment of surfaces is mathematically elegant and easy to use, it is difficult to visualize physically, because the imaginary mathematical surface of zero thickness does not have the physical correspondence of the real interfacial layer of finite thickness. We will follow an alternative formulation developed by Guggenheim,[1] treating the surface as a layer of small but finite uniform thickness.

As depicted in Fig. 6-5, two homogeneous phases α and β coexist in equilibrium, with a plane interface Δ lying between the two bulk phases. The parallel planes AA' and BB' form the boundaries between the surface layer Δ and the bulk phases α and β respectively. It is desirable, though not essential, to place one plane just inside each of the homogeneous bulk phases, so that the space between the planes AA' and BB' corresponds as nearly as possible to the actual physically inhomogeneous surface region. Since an actual surface layer is usually

[1] E. A. Guggenheim, *Trans. Faraday Soc.*, **36**:397 (1940).

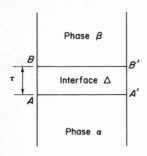

FIG. 6-5. Two-phase system
with plane interface.

only a few molecular dimensions (about 10^{-7} cm), the distance τ between the planes AA' and BB' is submicroscopic. The volume of a surface layer of area A is $V^{(\Delta)} = \tau$ A. The work done by the surface layer is increasing its volume by $dV^{(\Delta)}$ is $p\, dV^{(\Delta)}$, and in increasing its area by dA is $-\gamma\, d$A. The total work is then

$$d W^{(\Delta)} = p\, dV^{(\Delta)} - \gamma\, d\text{A} \tag{6-45}$$

where the superscript (Δ) denotes the surface layer, and p is the pressure and γ is the interfacial tension.

For the homogeneous bulk phases α and β we have the usual fundamental equation for energy (see Eq. 4-3).

$$dU^{(\alpha)} = T\, dS^{(\alpha)} - p\, dV^{(\alpha)} + \sum_i \mu_i dn_i^{(\alpha)} \tag{6-46a}$$

$$dU^{(\beta)} = T\, dS^{(\beta)} - p\, dV^{(\beta)} + \sum_i \mu_i dn_i^{(\beta)} \tag{6-46b}$$

For the surface layer Δ, however, the term $-p\, dV$ must be replaced by the expression $-p\, dV^{(\Delta)} + \gamma\, d$A, and the equation for energy becomes

$$dU^{(\Delta)} = T\, dS^{(\Delta)} - p\, dV^{(\Delta)} + \gamma\, d\text{A} + \sum_i \mu_i dn_i^{(\Delta)} \tag{6-47}$$

Because of the thermal, mechanical, and chemical equilibrium between the α, β, and surface phases, we must have

$$T^{(\alpha)} = T^{(\beta)} = T^{(\Delta)} = T, \qquad p^{(\alpha)} = p^{(\beta)} = p^{(\Delta)} = p, \qquad \mu_i^{(\alpha)} = \mu_i^{(\beta)} = \mu_i^{(\Delta)} = \mu_i$$

so that there is no need to add superscripts to T, p, and μ_i in Eqs. (6-46a), (6-46b), and (6-47). Notice that the term $\Sigma \mu_i dn_i$ appears in each of these equations, because even for a closed system with constant n_i's in the entire system, there may be variations in the n_i's in the individual subsystems—phases α, β, and surface layer. Adding Eqs. (6-46a), (6-46b), and (6-47) together, one obtains the equation for energy of the entire system, including the two bulk phases and the surface layer.

$$dU = T\, dS - p\, dV + \gamma\, d\text{A} + \sum_i \mu_i dn_i \tag{6-48}$$

where U, S, V, and n_i are the energy, entropy, volume, and number of moles of component i respectively for the entire system.

Analogous to the procedure used in the case of a bulk phase (see Sec. 4-1), through the use of Euler's theorem for the surface layer, we obtain

$$U^{(\Delta)} = T S^{(\Delta)} - p V^{(\Delta)} + \gamma A + \sum_i \mu_i n_i^{(\Delta)} \tag{6-49}$$

The Helmholtz function $A^{(\Delta)}$ and the Gibbs function $G^{(\Delta)}$ of the surface layer are now defined as

$$A^{(\Delta)} = U^{(\Delta)} - T S^{(\Delta)} = - p V^{(\Delta)} + \gamma A + \sum_i \mu_i n_i^{(\Delta)} \tag{6-50}$$

$$G^{(\Delta)} = A^{(\Delta)} + p V^{(\Delta)} - \gamma A = \sum_i \mu_i n_i^{(\Delta)} \tag{6-51}$$

Differentiating the first parts of the foregoing two equations and making use of Eq. (6-47), one obtains

$$dA^{(\Delta)} = dU^{(\Delta)} - T \, dS^{(\Delta)} - S^{(\Delta)} \, dT$$
$$= - S^{(\Delta)} \, dT - p \, dV^{(\Delta)} + \gamma \, dA + \sum_i \mu_i \, dn_i^{(\Delta)} \tag{6-52}$$

$$dG^{(\Delta)} = dA^{(\Delta)} + p \, dV^{(\Delta)} + V^{(\Delta)} \, dp - \gamma \, dA - A \, d\gamma$$
$$= - S^{(\Delta)} \, dT + V^{(\Delta)} \, dp - A \, d\gamma + \sum_i \mu_i \, dn_i^{(\Delta)} \tag{6-53}$$

Differentiating the second part of Eq. (6-50),

$$dA^{(\Delta)} = - p \, dV^{(\Delta)} - V^{(\Delta)} \, dp + \gamma \, dA + A \, d\gamma$$
$$+ \sum_i \mu_i \, dn_i^{(\Delta)} + \sum_i n_i^{(\Delta)} \, d\mu_i$$

Subtracting Eq. (6-52) from the preceding equation yields

$$S^{(\Delta)} \, dT - V^{(\Delta)} \, dp + A \, d\gamma + \sum_i n_i^{(\Delta)} \, d\mu_i = 0 \tag{6-54}$$

This is the *Gibbs-Duhem equation* applied to a surface layer. Dividing this equation by the surface area results in

$$d\gamma = - s^{(\Delta)} \, dT + \tau \, dp - \sum_i \Gamma_i \, d\mu_i \tag{6-55}$$

where

$s^{(\Delta)} = S^{(\Delta)}/A$ = entropy per unit area of surface layer

$\tau = V^{(\Delta)}/A$ = thickness of the surface layer

$\Gamma_i = n_i^{(\Delta)}/A$

which is the number of moles of component i per unit area of the surface layer and is named the surface concentration of component i. Equation (6-55) is the basic equation for the interfacial tension.

We now turn our attention to the special case of a single-component system. In a single-component system the only possible interface between two fluid phases is that between a liquid and its vapor. The name *surface tension of the liquid* is usually used for γ in this system. A single-component, two-phase system has one degree of freedom, i.e., there is only one independent intensive variable. The temperature is usually the most convenient choice for this variable. Thus the surface tension γ may be considered as a function of temperature alone. We will explore how the surface tension varies with the temperature.

For a single-component system, Eq. (6-55) simplifies to

$$d\gamma = -s^{(\Delta)}\, dT + \tau\, dp - \Gamma\, d\mu \tag{6-56}$$

Meanwhile from the condition of phase equilibrium between the liquid phase α and the vapor phase β of the single-component system, we must have

$$d\mu = d\mu^{(\alpha)} = d\mu^{(\beta)}$$

or

$$d\mu = -s^{(\alpha)}\, dT + v^{(\alpha)}\, dp = -s^{(\beta)}\, dT + v^{(\beta)}\, dp \tag{6-57}$$

where $s^{(\alpha)}$ and $v^{(\alpha)}$ are the molal entropy and molal volume in the liquid phase, and $s^{(\beta)}$ and $v^{(\beta)}$ are those in the vapor phase. Elimination of $d\mu$ and dp from Eqs. (6-56) and (6-57) leads to

$$-\frac{d\gamma}{dT} = (s^{(\Delta)} - \Gamma s^{(\alpha)}) - (\tau - \Gamma v^{(\alpha)})\, \frac{s^{(\beta)} - s^{(\alpha)}}{v^{(\beta)} - v^{(\alpha)}} \tag{6-58}$$

The preceding equation can be transformed to another form which involves energy changes instead of entropy changes. By applying Eq. (6-49) to the unit area of the surface layer we have

$$\gamma = u^{(\Delta)} - Ts^{(\Delta)} + p\tau - \Gamma\mu \tag{6-59}$$

where $u^{(\Delta)} = U^{(\Delta)}/A =$ energy per unit area of surface layer and, as defined before, $s^{(\Delta)} = S^{(\Delta)}/A$, $\Gamma = n^{(\Delta)}/A$, and $\tau = V^{(\Delta)}/A$. Meanwhile, for the two bulk phases α and β we have on a molal basis,

$$\mu = g^{(\alpha)} = u^{(\alpha)} - T s^{(\alpha)} + p\, v^{(\alpha)} \tag{6-60}$$

$$\mu = g^{(\beta)} = u^{(\beta)} - T s^{(\beta)} + p\, v^{(\beta)} \tag{6-61}$$

By the use of Eqs. (6-59) to (6-61) the entropy terms in Eq. (6-58) can be eliminated. Equation (6-58) then becomes

$$\gamma - T \frac{d\gamma}{dT} = (\mu^{(\Delta)} - \Gamma u^{(\alpha)}) - (\tau - \Gamma v^{(\alpha)}) \frac{u^{(\beta)} - u^{(\alpha)}}{v^{(\beta)} - v^{(\alpha)}} \tag{6-62}$$

Note that the numerical values of τ, Γ, $s^{(\Delta)}$, and $u^{(\Delta)}$ naturally depend on the exact positions assigned to the planes AA' and BB' which form the boundaries of the surface layer. Nevertheless it can be shown that Eqs. (6-58) and (6-62) are invariant with respect to these positions. This must be true, of course, since γ is by definition invariant with respect to the choice of these planes.

When the temperature is well below the critical temperature, $v^{(\alpha)} \ll v^{(\beta)}$, and τ/Γ is roughly of the same order as $v^{(\alpha)}$. Thus, the term containing the factor $(\tau - \Gamma v^{(\alpha)})/(v^{(\beta)} - v^{(\alpha)})$ in Eqs. (6-58) and (6-62) may be ignored for practical purposes. We may therefore write the following simplified equations:

$$-\frac{d\gamma}{dT} = s^{(\Delta)} - \Gamma s^{(\alpha)} \tag{6-63}$$

$$\gamma - T \frac{d\gamma}{dT} = u^{(\Delta)} - \Gamma u^{(\alpha)} \tag{6-64}$$

The right-hand side of Eq. (6-63) is the entropy of a unit area of surface minus the entropy of the same material content of liquid. Thus $- T(d\gamma/dT) = T(s^{(\Delta)} - \Gamma s^{(\alpha)})$ is the heat absorbed by the system when a unit area of surface is created isothermally and reversibly. Furthermore, since γ is the work done on the system, and $(u^{(\Delta)} - \Gamma u^{(\alpha)})$ is the energy gain by the system, Eq. (6-64) is a form of the first law for the process in which a unit area of surface is created isothermally and reversibly by the transfer of material from the liquid phase to the surface.

Since the distinction between the liquid and vapor phases disappears at the critical point, the surface tension must vanish at the critical temperature. However, although γ must vanish at the critical temperature, it is found empirically that at the neighborhood of the critical temperature, γ of all liquids also vanishes. An empirical equation for the temperature dependence of γ of a pure liquid in equilibrium with its vapor has the form

$$\gamma = \gamma_0 \left(1 - \frac{t}{t'_c}\right)^n \tag{6-65}$$

wherein γ is the surface tension at $t°C$, γ_0 is the surface tension at $0°C$, t'_c is a temperature in $°C$ ranging from 6 to $8°C$ below the critical temperature, and n is a constant between 1 and 2. For water, $\gamma_0 = 7.55 \times 10^{-4}$ newtons/cm, $t'_c = 368°C$, and $n = 1.2$.

6-7 CURVED INTERFACE

We turn now to the study of curved interfaces and consider the effects particularly caused by curvature. Guggenheim[2] shows that the equations derived for a plane interface may be applied to a curved interface with an accuracy adequate for practical purposes, provided that the thickness of the surface layer is small compared with its radius of curvature. As mentioned in Sec. 6-6, the actual thickness of a surface layer under most conditions is about 10^{-7} cm. Accordingly, we will restrict our considerations to curved interfaces with radii of curvature much greater than 10^{-7} cm.

A typical curved interface is seen in a liquid droplet suspended in a gas. The droplet is spherical in shape due to the effect of surface tension. Let us consider such a system for illustrative purposes. Figure 6-6 shows an element of a spherical interface of mean radius R with liquid phase α on the concave side and vapor phase β on the convex side. For simplicity we will ignore the thickness of the surface layer and ignore the variation of γ with curvature. For mechanical equilibrium of the segment of surface film, a force balance along the center line CC results in

$$(p^{(\alpha)} - p^{(\beta)})\pi R^2 (\sin d\theta)^2 = 2 \pi R \gamma (\sin d\theta)^2$$

from which

$$p^{(\alpha)} - p^{(\beta)} = \frac{2}{R} \gamma \qquad\qquad\qquad (6\text{-}66)$$

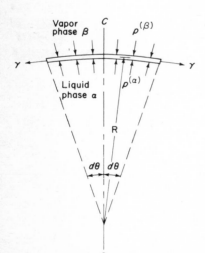

FIG. 6-6. Two-phase system with spherical interface.

[2] *Ibid.*

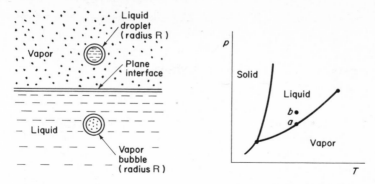

FIG. 6-7. Thermodynamic states of liquid droplet and vapor bubble.

This is the *Laplace-Kelvin relation*, which shows that the pressure is greater on the concave side of the surface than on the convex side. If the surface is not spherical, it can be characterized by two principal radii of curvature R_1 and R_2, and the general formula for the pressure difference becomes

$$p^{(\alpha)} - p^{(\beta)} = \gamma \left(\frac{1}{R_1} + \frac{1}{R_2} \right) \tag{6-67}$$

where $p^{(\alpha)}$ is always the pressure on the concave side and $p^{(\beta)}$ that on the convex side. This is the fundamental equation of capillarity and is the basis for the experimental determination of surface tension γ.

Equation (6-66) applies equally well for vapor bubbles immersed in a large expanse of liquid and for liquid droplets immersed in a large expanse of vapor. However, for a vapor bubble contained inside a thin spherical film of liquid and immersed in a large expanse of vapor, the excess pressure is twice as much as indicated by Eq. (6-66), because both the inner and outer surfaces of the thin liquid film contribute separate surface tensions.

The thermodynamic states of liquid droplets and vapor bubbles may be clearly understood by referring to Fig. 6-7, which shows a large mass of liquid and vapor in equilibrium across a plane interface, together with a liquid droplet immersed in the vapor phase and a vapor bubble immersed in the liquid phase. The bulk liquid and vapor phases are in equilibrium at temperature T and pressure p. Their thermodynamic states are represented by point a on the vapor-pressure curve in the adjacent phase diagram. According to Eq. (6-66), the pressure in the liquid droplet exceeds that in the bulk vapor phase by the amount $2\gamma/R$. Thus the liquid in the droplet is compressed, its thermodynamic state is represented by point b on the phase diagram (the scale between points a and b being grossly exaggerated). Similarly, according to the same equation, the pressure in the vapor bubble exceeds that in the bulk liquid phase also by the amount $2\gamma/R$. If the radius of the vapor bubble is the same as that of the liquid droplet, the thermodynamic state of the vapor in the bubble is also represented by point b. Thus the vapor in the bubble is in a metastable equilibrium state of the so-called supersaturated vapor.

EXAMPLE 6-7. A common method of determining the surface tension of a liquid is to observe the position of the meniscus in a capillary. Consider the case shown in Fig. 6-8 where water in a glass tube of radius r rises to a height z with a contact angle θ between water and glass. (a) Show that the surface tension γ is given by the equation

$$\gamma = \frac{r}{2 \cos \theta} (\rho^{(\beta)} - \rho^{(\alpha)}) g z = \frac{R}{2} (\rho^{(\beta)} - \rho^{(\alpha)}) g z$$

where $\rho^{(\alpha)}$ and $\rho^{(\beta)}$ are the densities of the α phase (air) and the β phase (water), g is the gravitational acceleration, and R is the radius of curvature of the surface of the meniscus (assumed spherical in shape). (b) In a capillary experiment as depicted in Fig. 6-8, the following data are recorded: temperature of water and air = $20°C$, radius of glass tube $r = 0.3$ cm, capillary rise $z = 0.427$cm, and contact angle $\theta = 30°$. Determine the surface tension of water in contact with air at $20°C$.

Solution.
(a) The vertical component of the surface tension force acting on the rim of the meniscus is $2\pi r\gamma \cos \theta$. This force component is balanced by the relative weight of the column of water in air. Thus we have

$$2 \pi r \gamma \cos \theta = \pi r^2 z (\rho^{(\beta)} - \rho^{(\alpha)}) g$$

Whence

$$\gamma = \frac{r}{2 \cos \theta} (\rho^{(\beta)} - \rho^{(\alpha)}) g z \tag{6-68}$$

As the radius of curvature of the surface of the meniscus (assumed spherical in shape) is $R = r/\cos \theta$, Eq. (6-68) may be rewritten as

$$\gamma = \frac{R}{2} (\rho^{(\beta)} - \rho^{(\alpha)}) g z \tag{6-69}$$

Note that Eq. (6-69) can be obtained directly from the Laplace-Kelvin relation. Let p_0 denote the pressure at the plane interface AA, and let $p^{(\alpha)}$ and $p^{(\beta)}$ denote the pressures at the curved interface BB in the α phase (air) and β phase (water). From fluid statics we have

$$p^{(\alpha)} = p_0 - \rho^{(\alpha)} g z \quad \text{and} \quad p^{(\beta)} = p_0 - \rho^{(\beta)} g z$$

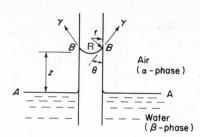

FIG. 6-8. Capillary action.

Substituting these equations into Eq. (6-66) leads to

$$p^{(\alpha)} - p^{(\beta)} = (p_0 - \rho^{(\alpha)} g\, z) - (p_0 - \rho^{(\beta)} g\, z) = \frac{2}{R}\, \gamma$$

or

$$\gamma = \frac{R}{2}\, (\rho^{(\beta)} - \rho^{(\alpha)}) g\, z$$

(b) When the density of air is neglected as compared with that of water, Eq. (6-68) reduces to

$$\gamma = \frac{r}{2 \cos \theta}\, \rho^{(\beta)} g\, z$$

At $20°C$ for water, $\rho^{(\beta)} = (1.0018)^{-1}$ g/cm^3. Thus

$$\gamma = \frac{(0.3 \text{ cm})}{2 \cos 30°} \left(\frac{10^{-3}}{1.0018} \text{ kg/cm}^3 \right) (9.80 \text{ m/sec}^2)(0.427 \text{ cm})$$

$$= 7.24 \times 10^4 \text{ newtons/cm}$$

PROBLEMS

6-1. A simple elastic system undergoes a reversible adiabatic process from an initial state (T_0, ϵ_0) to a final state (T, ϵ). Derive the relationship among T_0, T, ϵ_0, and ϵ.

6-2. A steel wire of cross-sectional area 0.01 cm^2 at a temperature of $25°C$ and under a tension of 20 newtons is stretched between two rigid supports 1.5 m apart. When the temperature is reduced to $15°C$, what is the tension in the wire? Assume that $Y = 2.05 \times 10^8$ kilonewtons/m^2 and $\alpha = 1.5 \times 10^{-5}\ °K^{-1}$.

6-3. A 1.5 mm diameter steel wire is clamped at $40°C$ between two supports 2 m apart. Calculate the force at its ends when the temperature is reduced to $0°C$

(a) if the supports are very rigid

(b) if the supports may yield a maximum distance of 0.5 mm.

Assume that $Y = 2.07 \times 10^8$ kilonewtons/m^2 and $\alpha = 1.2 \times 10^{-5}$ $^\circ$K^{-1}.

6-4. The tensile force in a steel wire of 1 m long and 1 mm diameter is increased reversibly at constant temperature of 300°K from zero to 100 newtons. Calculate the work done, the heat transferred, and the change in internal energy of the wire. Assume that $Y = 2.00 \times 10^8$ kilonewtons/m^2 and $\alpha = 1.20 \times 10^{-5}$ $^\circ$K^{-1}.

6-5. The tensile force in a steel wire 1 m long and 1 mm diameter is increased reversibly and adiabatically from zero to 100 newtons. The initial temperature of the wire is 300°K. What is the change in temperature? Assume constant values of $\alpha = 1.20 \times 10^{-5}$ $^\circ$K^{-1}, $\rho = 7.86$ g/cm^3, and $c_\sigma = 0.482$ joules/(g)($^\circ$K).

6-6. The tensile force in a steel wire 2 ft long and 0.05 in. diameter is increased reversibly at constant temperature of 80°F from zero to 25 lbf. Calculate the work done, the heat transferred, and the change in internal energy of the wire. Assume that $Y = 30 \times 10^6$ lbf/in.^{2} and $\alpha = 7 \times 10^{-6}$ $^\circ$R^{-1}.

6-7. The tensile force in a steel wire 2 ft long and 0.05 in. diameter is increased reversibly and adiabatically from zero to 25 lbf. The initial temperature of the wire is 80°F. What is the change in temperature? Assume constant values of $\alpha = 7 \times 10^{-6}$ $^\circ$R^{-1}, $\rho = 0.284$ lbm/in.^{3}, and $c_\sigma = 0.115$ Btu/(lbm)($^\circ$R).

6-8. For a strip of an ideal rubber prove that

(a) $\left(\dfrac{\partial F}{\partial T}\right)_L = \dfrac{F}{T}$

(b) $F = T \, \Psi \, (L)$, where $\Psi \, (L)$ is a function of L only.

(c) L at zero tension is independent of T.

6-9. For a strip of rubber show that the difference in heat capacities at constant force and at constant length is given by the expression

$$C_F - C_L = -T\left(\frac{\partial F}{\partial T}\right)_L \left(\frac{\partial L}{\partial T}\right)_F$$

If the rubber strip is an ideal rubber, show that

$$C_F - C_L = \frac{F^2}{T\left(\dfrac{\partial F}{\partial L}\right)_T}$$

6-10. For a bar of nonideal rubber the thermal equation of state is given by the expression

$$F = K\,T\left(\frac{L}{L_0} - \frac{L_0^2}{L^2}\right)$$

and the linear expansivity at zero tension α_0 is given by the expression

$$\alpha_0 = \frac{1}{L_0} \frac{dL_0}{dT}$$

where F is tensile force, T is temperature, L is length, L_0 is length at zero tension, and K is a constant.

(a) Show that for a reversible adiabatic stretching process,

$$\left(\frac{\partial T}{\partial L}\right)_S = \frac{T}{C_L}\left(\frac{\partial F}{\partial T}\right)_L = \frac{KT}{C_L}\left[\left(\frac{L}{L_0} - \frac{L_0^2}{L^2}\right) - \alpha_0 T\left(\frac{L}{L_0} + 2\frac{L_0^2}{L^2}\right)\right]$$

where C_L is the heat capacity at constant length for the rubber bar.

(b) If $T = 300°K$, $K = 1.33 \times 10^{-2}$ newtons/°K, $C_L = 2.5$ joules/°K, and $\alpha_0 = 5 \times 10^{-4} \, °K^{-1}$, calculate $(\partial T/\partial L)_S$ for values of L/L_0 equal to 1 and 2.

6-11. An atomizer slowly shoots out minute water droplets of an average radius of 10^{-4} cm into air. Estimate the surface work required and the heat transferred when 1 kg of water is atomized isothermally at 20°C. The temperature variation of the surface tension of water against air is:

Temperature, °C	10	15	20	25	30
Surface tension, dyne/cm	74.22	73.49	72.75	71.97	71.18

6-12. Liquid water is atomized reversibly and isothermally at 100°C into droplets of an average radius of 10^{-4} cm in equilibrium with its vapor. Estimate the work required and the heat transferred per kg of water atomized. The temperature-surface tension relationship is given by Eq. (6-65).

6-13. For a reversible adiabatic expansion of a single-component liquid-vapor interface, show that

$$A\left(\frac{d\gamma}{dT}\right) = \text{constant}$$

6-14. The surface tension of a water-air interface is 7.2×10^{-4} newtons/cm at 25°C. Calculate the pressure of water inside a spherical droplet of 10^{-5} cm radius in contact with air which is at 1 atm and 25°C.

6-15. Derive the expression for the work required to blow up slowly a soap bubble of radius R into air which is at atmospheric pressure p_{atm}, if γ is the surface tension of the soap solution in contact with air.

Chapter 7

SYSTEMS INVOLVING
EXTERNAL FORCE FIELDS

7-1 WORK OF MAGNETIZATION

Just as an electron current in a small loop produces a magnetic field, an electron revolving in its orbit around the nucleus and rotating about its own axis has associated a magnetic dipole with its motion. In the absence of an external magnetic field all such dipoles cancel each other. In the presence of an external field, the frequencies and senses of orbiting and spinning of the electrons will be changed in such a manner as to oppose the external field. This is the *diamagnetic* nature of all material substances. In some materials, however, there are permanent magnetic dipoles owing to unbalanced electron orbits or spins. These atoms behave like elementary dipoles which tend to align with an external field and strengthen it. When this effect in a material is greater than the diamagnetic tendency common to all atoms, this material is termed *paramagnetic*. Paramagnetism is temperature dependent. When the temperature is lowered sufficiently, the atomic elementary dipoles are magnetically aligned within microscopic domains which can be readily aligned by a relatively small external field to form a large induction. This is referred to as *ferromagnetic*. A ferromagnetic material becomes paramagnetic above a temperature known as the *Curie temperature*. The Curie temperatures of iron and nickel are far above room temperature; thus they are usually referred to as ferromagnetic. However, the Curie temperatures of some metallic salts are below one degree Kelvin; they are usually referred to as paramagnetic. A paramagnetic material is not a magnet if there is no external magnetic field applied to it. In the presence of an external field it becomes slightly magnetized in contradistinction to a ferromagnetic material which would show very strong magnetic effects.

The magnetic effects in a ferromagnetic material are not reversible because the reverse process of demagnetization forms a hysteresis loop with the forward process of magnetization. Thus the state of a ferromagnetic system depends not only on its present condition but also on its past history. A ferromagnetic system is therefore not amenable to thermodynamic analyses. On the other hand, for a paramagnetic system (such as a paramagnetic salt) or a diamagnetic system (such as a superconducting material), the process of magnetization is reversible and the state of the system can be described in terms of a few thermodynamic variables.

According to the principles of magnetism, the magnetic induction inside the vacuum space of an infinite solenoid or a solenoid wound in the form of a toroid is given by

$$\mathbf{B} = \mu_0 \frac{N}{L} i = \mu_0 \mathbf{H}$$

wherein $\mathbf{B}$ is the *magnetic induction* or *flux density*, N/L is the number of turns of winding per unit axial length of the solenoid, $\mathbf{H}$ is the *magnetic field* or *intensity* set up by the flow of current i in the winding, and μ_0 is the *permeability of free space*. When the space inside the toroid is filled with a magnetic material, the magnetic induction in the core then becomes

$$\mathbf{B} = \mu_0(\mathbf{H} + \mathbf{M}) \tag{7-1}$$

where $\mathbf{M}$ is the *magnetization* or *magnetic moment per unit volume* of the material. This magnetic moment arises from the reorientation of originally random distribution of orbital and spin motions of electrons under the influence of an external magnetic field.

Equation (7-1) is written for the rationalized MKSA units which now form part of the System International (SI) units. In this system, $\mathbf{B}$ is in weber/m^2, $\mathbf{H}$ is in amp/m, $\mathbf{M}$ is in amp/m, and μ_0 has a value of $4\pi \times 10^{-7}$ weber/(amp)(m) or newton/amp^2.

Consider now the reversible magnetization of a magnetic solid in the form of a thin toroidal ring of uniform cross-sectional area A and mean circumference L, with a winding of N closely spaced turns applied completely around its outside, as shown in Fig. 7-1. When the small change in volume caused by the field (the so-called magnetostriction) is neglected, the volume of the magnetic material $V =$ AL shall be a constant. The magnetic field set up by the current i is

$$\mathbf{H} = \frac{N}{L} i = \frac{N \text{A}}{V} i \tag{7-2}$$

which in turn produces an induction $\mathbf{B}$. If the current i is changed, and in time $d\tau$ the magnetic induction is changed by $d\mathbf{B}$, then according to Faraday's law of electromagnetic induction, there is induced in the winding a back emf E according to the relation

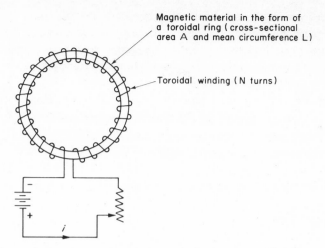

FIG. 7-1. Reversible magnetization of a magnetic material.

$$E = -NA\frac{d\mathbf{B}}{d\tau}$$

A quantity of electricity $d\mathbf{Z}$ is transferred in the circuit during the time interval $d\tau$, so that the work done by the system is

$$dW = E\,d\mathbf{Z} \tag{7-3}$$

or

$$dW = -NA\frac{d\mathbf{B}}{d\tau}\,d\mathbf{Z} = -NA\frac{d\mathbf{Z}}{d\tau}\,d\mathbf{B}$$

But

$$\frac{d\mathbf{Z}}{d\tau} = i$$

then

$$dW = -NAi\,d\mathbf{B}$$

Combining the preceding equation with Eq. (7-2) gives

$$dW = -V\mathbf{H}\,d\mathbf{B} \tag{7-4}$$

Substituting Eq. (7-1) into Eq. (7-4) results in

$$dW = -\mu_0\,V\mathbf{H}\,d\mathbf{H} - \mu_0\,V\mathbf{H}\,d\mathbf{M} \tag{7-5}$$

When there is no material within the toroidal winding, M would be zero, and the right-hand side of the preceding equation reduces to the first term only. This means that the quantity $(-\mu_0 \, V H \, dH)$ is the work required to increase the magnetic field of the empty space of volume V by an amount dH. However, it is noted that this quantity is equal to the differential of $(-\frac{1}{2} \mu_0 \, V H^2)$ and is therefore an exact differential. But the differential of work in thermodynamics is necessarily inexact. Accordingly the term $-d(\frac{1}{2} \mu_0 \, V H^2)$ should be rightfully classified as the magnetic energy of the empty space of volume V and be included in the total internal energy term when the first law of thermodynamics is written. It is reasonable, therefore, to regard the second term of Eq. (7-5) alone as the expression for reversible work in the magnetization of a magnetic material. Thus we have finally

$$dW = -\mu_0 \, V H \, dM \tag{7-6}$$

where the minus sign indicates that work input is required to increase the magnetization of a substance. The preceding equation can be written in terms of the *total magnetic moment* $I = VM$. Thus

$$dW = -\mu_0 \, H \, dI \tag{7-6a}$$

EXAMPLE 7-1. The $M - H - T$ relationship for a paramagnetic solid at small values of the ratio H/T can be expressed by the Curie equation which is:

$$I = V M = C_c \frac{H}{T}$$

where C_c is a characteristic constant known as the Curie constant. Iron ammonium alum is a paramagnetic salt that obeys the Curie equation. Determine the work done per kg of the salt when its temperature changes from $2°K$ to $1°K$ under a constant magnetic field of 0.8×10^6 amp/m. For iron ammonium alum the Curie constant in rationalized MKSA (or SI) units is 1.14×10^{-4} $(m^3)(°K)/kg$.

Solution.
For one kg of the salt, the initial and final total magnetic moments are

$$I_1 = C_c \frac{H_1}{T_1} = \left(1.14 \times 10^{-4} \; \frac{(m^3)(°K)}{kg}\right)(1 \text{ kg}) \frac{(0.8 \times 10^6 \text{ amp/m})}{(2°K)}$$

$$= 45.6 \text{ amp. m}^2$$

$$I_2 = C_c \frac{H_2}{T_2} = \left(1.14 \times 10^{-4} \; \frac{(m^3)(°K)}{kg}\right)(1 \text{ kg}) \frac{(0.8 \times 10^6 \text{ amp/m})}{(1°K)}$$

$$= 91.2 \text{ amp. m}^2$$

From Eq. (7-6) the work done is

$$W_{12} = -\mu_0 \, H_1(I_2 - I_1)$$

$$= -\left(4\,\pi \times 10^{-7}\,\frac{\text{newton}}{\text{amp}^2}\right)\left(0.8 \times 10^6\,\frac{\text{amp}}{\text{m}}\right)[(91.2 - 45.6)(\text{amp})(\text{m}^2)]$$

$$= -45.8(\text{newton})(\text{m}) = -45.8 \text{ joules}$$

where the minus sign indicates that work is done on the salt.

7-2 SIMPLE MAGNETIC SYSTEMS

In thermodynamic analyses of magnetic solids, all changes in volume are small and may be ignored. Furthermore, the effects of pressure on the solids are also small and may also be ignored. The only work interaction is that due to the magnetization of the material, i.e.,

$$dW = -\mu_0 \, V \, H \, dM \tag{7-6}$$

A system for which the only reversible work mode is the magnetization of the magnetic material is called a *simple magnetic system.*

The first law for a reversible process in a simple magnetic system is

$$dQ = dU - \mu_0 \, V \, H \, dM \tag{7-7}$$

Combination of this equation and the second law leads to

$$T \, dS = dU - \mu_0 \, V \, H \, dM \quad \text{or} \quad dU = T \, dS + \mu_0 \, V \, H \, dM \tag{7-8}$$

This is the very basic equation which combines the first and second laws as applied to a simple magnetic system. Note that this equation is a relation solely among the properties of the system.

It is helpful to define two new properties, the magnetic enthalpy H and magnetic Gibbs function G:

$$H = U - \mu_0 \, V \, H \, M \qquad \text{MAGNETIC ENTHALPY} \tag{7-9}$$

$$G = H - TS = U - TS - \mu_0 \, V \, H \, M \qquad \text{MAGNETIC GIBBS FUNCTION} \tag{7-10}$$

The Helmholtz function A takes the usual definition:

$$A = U - TS \qquad \text{HELMHOLTZ}$$

From these definitions and Eq. (7-8) it follows that

$$dH = T \, dS - \mu_0 \, V \, M \, dH \qquad \text{ENTHALPY} \tag{7-11}$$

$$dA = -S \, dT + \mu_0 \, V \, H \, dM \qquad \text{HELMHOLTZ} \tag{7-12}$$

$$dG = -S \, dT - \mu_0 \, V \, M \, dH \qquad \text{GIBBS} \tag{7-13}$$

Applying the condition of exactness to the four basic relations, Eqs. (7-8) and (7-11) to (7-13), leads to the following four Maxwell relations for a simple magnetic system:

$$\left(\frac{\partial T}{\partial M}\right)_S = \mu_0 \; V \left(\frac{\partial H}{\partial S}\right)_M \qquad U \tag{7-14a}$$

$$\left(\frac{\partial T}{\partial H}\right)_S = - \mu_0 \; V \left(\frac{\partial M}{\partial S}\right)_H \qquad H \tag{7-14b}$$

$$\left(\frac{\partial S}{\partial M}\right)_T = - \mu_0 \; V \left(\frac{\partial H}{\partial T}\right)_M \qquad F \tag{7-14c}$$

$$\left(\frac{\partial S}{\partial H}\right)_T = \mu_0 \; V \left(\frac{\partial M}{\partial T}\right)_H \qquad G \tag{7-14d}$$

MAXWELL RELATIONS

In an analogy to C_v and C_p, let us define the heat capacity at constant magnetic moment C_M and the heat capacity at constant magnetic field C_H for a simple magnetic system by the following equations:

$$C_M = \left(\frac{\partial U}{\partial T}\right)_M = T \left(\frac{\partial S}{\partial T}\right)_M \tag{7-15}$$

$$C_H = \left(\frac{\partial H}{\partial T}\right)_H = T \left(\frac{\partial S}{\partial T}\right)_H \tag{7-16}$$

Taking the difference between the preceding two equations one obtains

$$C_H - C_M = T \left(\frac{\partial S}{\partial T}\right)_H - T \left(\frac{\partial S}{\partial T}\right)_M$$

But by Eq. (2-4), we have

$$\left(\frac{\partial S}{\partial T}\right)_H = \left(\frac{\partial S}{\partial T}\right)_M + \left(\frac{\partial S}{\partial M}\right)_T \left(\frac{\partial M}{\partial T}\right)_H$$

Therefore,

$$C_H - C_M = T \left(\frac{\partial S}{\partial M}\right)_T \left(\frac{\partial M}{\partial T}\right)_H$$

Substituting Eq. (7-14c) into this equation leads to

$$C_H - C_M = - T \mu_0 \; V \left(\frac{\partial H}{\partial T}\right)_M \left(\frac{\partial M}{\partial T}\right)_H$$

But by the cyclic relation we have

$$\left(\frac{\partial H}{\partial T}\right)_M = -\left(\frac{\partial M}{\partial T}\right)_H \left(\frac{\partial H}{\partial M}\right)_T$$

Consequently,

$$C_H - C_M = \mu_0\, TV\left(\frac{\partial M}{\partial T}\right)_H^2 \left(\frac{\partial H}{\partial M}\right)_T$$

$$= \frac{\mu_0\, TV\left(\dfrac{\partial M}{\partial T}\right)_H^2}{\left(\dfrac{\partial M}{\partial H}\right)_T} \tag{7-17}$$

The equations of entropy and internal energy for a simple magnetic system can be easily derived. Thus when we consider $S = S(T, \mathbf{M})$ we have

$$dS = \left(\frac{\partial S}{\partial T}\right)_M dT + \left(\frac{\partial S}{\partial M}\right)_T d\mathbf{M}$$

Substituting Eqs. (7-15) and (7-14c) into this equation leads to the first dS equation:

$$dS = \frac{C_M}{T}\, dT - \mu_0\, V\left(\frac{\partial H}{\partial T}\right)_M d\mathbf{M} \qquad\qquad S = S(T, m) \tag{7-18}$$

When we consider $S = S(T, \mathbf{H})$, we have

$$dS = \left(\frac{\partial S}{\partial T}\right)_H dT + \left(\frac{\partial S}{\partial H}\right)_T d\mathbf{H}$$

Substituting Eqs. (7-16) and (7-14d) into this equation leads to the second dS equation:

$$dS = \frac{C_H}{T}\, dT + \mu_0\, V\left(\frac{\partial M}{\partial T}\right)_H d\mathbf{H} \qquad\qquad S = S(T, H) \tag{7-19}$$

Upon substituting the first dS equation into Eq. (7-8), we obtain the first dU equation:

$$dU = C_M dT + \mu_0\, V\left[\mathbf{H} - T\left(\frac{\partial H}{\partial T}\right)_M\right] d\mathbf{M} \qquad u = u(T, m) \tag{7-20}$$

Note that all the preceding equations for a simple magnetic system can be readily obtained from the corresponding equations for a simple compressible system by observing that $\mathbf{H}$ corresponds to p and $(-\mu_0\, V\, d\mathbf{M})$ corresponds to dV. The reader is urged to be familiar with the direct changeover of these equations.

EXAMPLE 7-2. The $M - H - T$ relationship for a paramagnetic solid can be expressed by the Curie-Weiss equation:

$$I = V M = C_{cw} \frac{H}{T - \Theta}$$

where I is the total magnetic moment, and C_{cw} and Θ are characteristic constants of the material. This equation is intended to apply for temperatures in excess of Θ. When the temperature is reduced, the material will acquire more complicated magnetic conditions that often lead to ferromagnetism for which the Curie-Weiss equation no longer applies. For a paramagnetic solid that obeys the Curie-Weiss equation, write the expressions for the changes in entropy S and internal energy U between two states.

Solution.
From the Curie-Weiss equation we have

$$\left(\frac{\partial H}{\partial T}\right)_I = \frac{I}{C_{cw}}$$

Then from Eq. (7-18) we have

$$dS = C_M \frac{dT}{T} - \mu_0 \left(\frac{\partial H}{\partial T}\right)_I dI = C_M \frac{dT}{T} - \frac{\mu_0}{C_{cw}} I\, dI$$

Whence

$$\Delta S_{12} = \int_{T_1}^{T_2} C_M \frac{dT}{T} - \frac{\mu_0}{2 C_{cw}} (I_2^2 - I_1^2)$$

And from Eq. (7-20) we have

$$dU = C_M dT + \mu_0 \left[H - T \left(\frac{\partial H}{\partial T}\right)_I \right] dI = C_M dT + \mu_0 \left[\frac{I}{C_{cw}} (T - \Theta) - T \frac{I}{C_{cw}} \right] dI$$

$$= C_M dT - \frac{\mu_0 \Theta}{C_{cw}} I\, dI$$

Whence

$$\Delta U_{12} = \int_{T_1}^{T_2} C_M\, dT - \frac{\mu_0 \Theta}{2 C_{cw}} (I_2^2 - I_1^2)$$

EXAMPLE 7-3. Nickel at $700°K$ obeys the Curie-Weiss equation (Example 7-2), and the characteristic constants are $C_{cw} = 5.05 \times 10^{-3}$ $(m^3)(°K)/kg$ mole and $\Theta = 538°K$. (a) Calculate the entropy change when one kg of nickel is placed in a magnetic field of 8×10^5 amp/m at a constant temperature of 700°K. (b) Calculate the change in temperature when the nickel is then adiabatically demagnetized, using $C_M = 30.8$ kJ/(kg mole)(°K).

Solution.

(a) Reversible isothermal magnetization (process 12):

$I_1 = 0$

$I_2 = C_{cw} \dfrac{H_2}{T_2 - \Theta} = \left(5.05 \times 10^{-3} \dfrac{(m^3)(°K)}{kg\ mole}\right) \dfrac{\left(8 \times 10^5 \dfrac{amp}{m}\right)}{(700 - 538)°K}$

$= 24.9(amp)(m^2)/kg\ mole$

From Example 7-2, for the isothermal process we have

$\Delta S_{12} = -\dfrac{\mu_0}{2C_{cw}} (I_2^2 - I_1^2)$

$= -\dfrac{\left(4\pi \times 10^{-7} \dfrac{N}{amp^2}\right)}{2\left(5.05 \times 10^{-3} \dfrac{(m^3)(°K)}{kg\ mole}\right)}\left[\left(24.9 \dfrac{(amp)(m^2)}{kg\ mole}\right)^2 - 0\right]$

$= -0.0771\ joule/(kg\ mole)(°K) = -\dfrac{0.0771\ joule/(kg\ mole)(°K)}{58.71\ kg/kg\ mole}$

$= -0.00131\ joule/(kg)(°K)$

(b) Reversible adiabatic demagnetization (process 23):

From Example 7-2, for the reversible adiabatic process we have

$\Delta S_{23} = 0 = C_M \ln \dfrac{T_3}{T_2} - \dfrac{\mu_0}{2C_{cw}} (I_3^2 - I_2^2)$

Rearranging and substituting numerical values lead to

$\ln \dfrac{T_3}{T_2} = \dfrac{\mu_0}{2C_{cw}C_M} (I_3^2 - I_2^2)$

$= \dfrac{(4\pi \times 10^{-7}\ N/amp^2)}{2\left(5.05 \times 10^{-3} \dfrac{(m^3)(°K)}{kg\ mole}\right)\left(30.8 \times 10^3 \dfrac{(newtons)(m)}{(kg\ mole)(°K)}\right)}$

$\cdot \left[0 - \left(24.9 \dfrac{(amp)(m^2)}{kg\ mole}\right)^2\right] = -2.505 \times 10^{-6}$ or $\ln \dfrac{T_2}{T_3} = 2.505 \times 10^{-6}$

Therefore

$\dfrac{T_2}{T_3} = 1.000,002,505$ or $\dfrac{T_3 - T_2}{T_2} = \dfrac{1}{1.000,002,505} - 1 \approx -0.000,002,505$

Hence the change in temperature is

$$\Delta T = T_3 - T_2 = -0.000,002,505 \times 700 = -0.0018°C$$

7-3 WORK OF POLARIZATION

In contrast with an electric conductor which has a sufficiently large number of free electrons, a dielectric or electric insulator has none or only a relatively small number of free electrons. The major effect of an electric field on a dielectric is the polarization of the electric dipoles. Work is done by the electric field on the dielectric material during the polarization process.

For the purpose of deriving the equation of work in polarizing a dielectric, let us consider a parallel-plate *capacitor* or *condenser* as illustrated in Fig. 7-2. The two plates, each of area A with a distance of separation L, are charged with equal and opposite charges ± **Z**. For plates of a size large compared with the separation, the edge effects are negligible and electric lines are supposed to go perpendicularly from one plate to the other. According to electrostatics, when the space between the plates is a vacuum, the *electric field* or *intensity* **E** created by the charges is given by

$$\mathbf{E} = \frac{\mathbf{Z}}{A\epsilon_0}$$

where ϵ_0 is the *permittivity of free space* and has a value of 8.854×10^{-12} coul²/(newton)(m²) in rationalized MKSA units (or SI units). The *potential difference* E between the plates is

$$\mathsf{E} = \mathbf{E}L = \frac{\mathbf{Z}}{A\epsilon_0}\, L$$

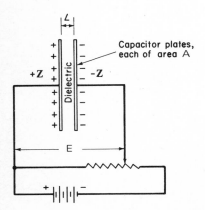

FIG. 7-2. Parallel-plate capacitor.

and the *capacitance* **C** of the capacitor is

$$C = \frac{\mathbf{Z}}{\mathsf{E}} = \frac{A\epsilon_0}{L}$$

Let a dielectric material be inserted between the plates. In the absence of an electric field, in spite of the atomic irregularities we can imagine the dielectric to be composed of generally uniform distributions of positive and negative charges. Under the influence of an external electric field a rearrangement of the charges in the dielectric takes place and it thus becomes polarized. The positive charges are displaced slightly in the direction of the field, while the negative charges are displaced in the opposite direction. Thus, because of the presence of the dielectric between the capacitor plates, the effective charge on each plate is reduced by a factor κ, known as the *dielectric constant*. Hence the electric field and the potential difference between the plates are reduced by the factor κ, while the capacitance is increased by the same factor. Thus

$$\mathbf{E} = \frac{\mathbf{Z}}{A\epsilon_0\kappa} \qquad \mathsf{E} = \mathbf{E}L = \frac{\mathbf{Z}}{A\epsilon_0\kappa}L \qquad \mathbf{C} = \frac{\mathbf{Z}}{\mathsf{E}} = \frac{A\epsilon_0\kappa}{L}$$

The *electric polarization* **P** of a dielectric is defined as the electric dipole moment per unit volume and is related to the applied electric field **E** by the equation

$$\mathbf{P} = \epsilon_0(\kappa - 1)\mathbf{E} \tag{7-21}$$

While **P** and **E** in turn are related to the *electric displacement* **D** by the equation

$$\mathbf{D} = \epsilon_0\mathbf{E} + \mathbf{P} \tag{7-22}$$

Whence

$$\mathbf{D} = \kappa\epsilon_0\mathbf{E} \quad \text{and} \quad \frac{\mathbf{Z}}{A} = \mathbf{D}$$

The reversible work done in charging a capacitor is

$$dW = -\mathsf{E}\,d\mathbf{Z}$$

which may be transformed to read

$$dW = -(\mathbf{E}L)d(A\mathbf{D}) = -(AL)\mathbf{E}\,d\mathbf{D}$$

or

$$dW = -V\mathbf{E}\,d\mathbf{D} \tag{7-23}$$

where $V = AL$ = volume of the dielectric. Substituting Eq. (7-22) into Eq. (7-23) results in

$$dW = -V\epsilon_0 \mathbf{E}\, d\mathbf{E} - V\mathbf{E}\, d\mathbf{P}$$

When there is no material between the capacitor plates, $\mathbf{P}$ would be zero, and the right-hand side of the preceding equation reduces to the first term only. Thus, the quantity $(-V\epsilon_0 \mathbf{E}\, d\mathbf{E}) = -d(\frac{1}{2} V\epsilon_0 \mathbf{E}^2)$ is the work required to increase the electric field of the free space between the capacitor plates by an amount $d\mathbf{E}$. This quantity is additive to the internal energy when the first law is written. Therefore the reversible work in the polarization of a dielectric material is

$$dW = -V\mathbf{E}\, d\mathbf{P} \tag{7-24}$$

where the minus sign indicates that work input is required to increase the polarization of a dielectric. The preceding equation can be written in terms of the *total electric moment* $\mathbf{P}' = V\mathbf{P}$. Thus

$$dW = -\mathbf{E}\, d\mathbf{P}' \tag{7-24a}$$

If the polarization of a dielectric material is the only reversible work mode, such a system is called a *simple dielectric system*. The thermodynamic relations for a simple dielectric system can be readily obtained from the corresponding relations for a simple compressible system by observing that $\mathbf{E}$ corresponds to p and $(-V d\mathbf{P})$ corresponds to dV. For example, in an analogy to Eq. (2-8) we have

$$dU = T\, dS + V\mathbf{E}\, d\mathbf{P} \tag{7-25}$$

or

$$dU = T\, dS + \mathbf{E}\, d\mathbf{P}' \tag{7-25a}$$

For a compressible dielectric material, there are two reversible work modes, namely, one for expansion and the other for polarization. The basic relation corresponding to Eq. (7-25) then becomes

$$dU = T\, dS - p\, dV + V\mathbf{E}\, d\mathbf{P} \tag{7-26}$$

or

$$dU = T\, dS - p\, dV + \mathbf{E}\, d\mathbf{P}' \tag{7-26a}$$

EXAMPLE 7-4. The static values of the dielectric constant of liquid nitrobenzene ($C_6H_5NO_2$) as a function of temperature are as follows:

Temperature t, °C	90	110	130
Dielectric constant κ	24.9	22.7	20.8

Calculate the work done per cm^3 of nitrobenzene when the electric field is increased from zero to 10^4 volts/m while the temperature is increased from 90 to $130°C$, assuming that the temperature varies linearly with the electric field during the process.

Solution.

The given data of dielectric constant can be fitted by the following linear equation:

$$\kappa - 1 = 33.2 - 0.104\ t \tag{7-27}$$

where t is in $°C$. Since the temperature t is assumed to vary linearly with the electric field during the process, we write

$$t = a + b\ \mathbf{E}$$

Upon substituting the given data that $t_1 = 90°C$, $t_2 = 130°C$, $\mathbf{E}_1 = 0$, and $\mathbf{E}_2 = 10^4$ volts/m, we have

$$t = 90 + 4 \times 10^{-3}\ \mathbf{E} \tag{7-28}$$

where t is in $°C$ and $\mathbf{E}$ is in volt/m. Inserting Eq. (7-28) into Eq. (7-27) gives

$$\kappa - 1 = 23.8 - 4.16 \times 10^{-4}\ \mathbf{E} \tag{7-29}$$

where $\mathbf{E}$ is in volt/m.

Substituting Eq. (7-29) into Eq. (7-21), we obtain the expression for the electric polarization

$$\mathbf{P} = \epsilon_0(23.8 - 4.16 \times 10^{-4}\ \mathbf{E})\ \mathbf{E} \tag{7-30}$$

where $\mathbf{P}$ is in $coul/m^2$, $\mathbf{E}$ is in volt/m (which is newton/coul), and $\epsilon_0 = 8.854 \times 10^{-12}\ coul^2/(newton)(m^2)$ in rationalized MKSA units (or SI units).

Differentiating Eq. (7-30),

$$d\mathbf{P} = \epsilon_0(23.8 - 4.16 \times 10^{-4} \times 2\ \mathbf{E})d\mathbf{E}$$

Inserting this equation in Eq. (7-24), we obtain

$$W = -V \int \mathbf{E}\ d\mathbf{P} = -V\ \epsilon_0 \int_{\mathbf{E}_1}^{\mathbf{E}_2} (23.8\mathbf{E} - 8.32 \times 10^{-4}\ \mathbf{E}^2)d\mathbf{E}$$

$$= -V\ \epsilon_0 \left[\frac{23.8}{2}\ (\mathbf{E}_2^2 - \mathbf{E}_1^2) - \frac{8.32 \times 10^{-4}}{3}\ (\mathbf{E}_2^3 - \mathbf{E}_1^3) \right]$$

$$= -(10^{-6})(8.854 \times 10^{-12}) \cdot \left[\frac{23.8}{2}\ (10^4)^2 - \frac{8.32 \times 10^{-4}}{3}\ (10^4)^3 \right]$$

$$= -8.08 \times 10^{-9}\ (newton)(m) = -8.08 \times 10^{-9}\ \text{joules}$$

where the minus sign indicates that work is done on the system.

7-4 PIEZOELECTRICITY

So far our development of thermodynamic relationships has been mainly restricted to simple systems, that is, to systems with only one reversible work mode. The exceptions were Sec. 6-6 where we included both surface and volume effects and Sec. 7-3 where we mentioned briefly the case of simultaneous electric and volume effects. Another illustration of a system with more than one reversible work mode is a *piezoelectric system*, in which there are simultaneous electric and elastic effects.

A *piezoelectric crystal* is a crystal in which electric polarity is produced by applied stress or, conversely, a crystal that becomes deformed when in an electric field. Piezoelectric effects are of great technical importance. For example, quartz plates are used to control and stabilize the frequency of oscillators, and rochelle-salt crystal plates are used in microphones, telephone receivers, etc., to make devices superior to their electromagnetic predecessors.

Piezoelectric crystals are generally highly anisotropic and their analysis is a complicated matter.[1] We will consider only the simple case where an elastic dielectric crystal is under stress in one direction in a uniform electric field in that direction. For simplicity in notation, we will write all the thermodynamic equations for a unit volume of the material. According to Eqs. (6-4a) and (7-24), for a piezoelectric system the work done is given by

$$dW = -\sigma \, d\epsilon - \mathbf{E} \, d\mathbf{P} \tag{7-31}$$

and the combined first and second laws then becomes

$$T \, ds = du - \sigma \, d\epsilon - \mathbf{E} \, d\mathbf{P} \tag{7-32}$$

Since there are two reversible work modes in a piezoelectric system, then according to the state principle, three independent properties are required to specify an equilibrium state of the system. Any one of the following sets: $(T, \sigma, \mathbf{E})$, $(T, \sigma, \mathbf{P})$, $(T, \epsilon, \mathbf{E})$, or $(T, \epsilon, \mathbf{P})$ would be adequate to describe the system. When the first set is chosen, the total differential of the entropy is given by

$$ds = \left(\frac{\partial s}{\partial T}\right)_{\sigma, \mathbf{E}} dT + \left(\frac{\partial s}{\partial \sigma}\right)_{T, \mathbf{E}} d\sigma + \left(\frac{\partial s}{\partial \mathbf{E}}\right)_{T, \sigma} d\mathbf{E}$$

or

$$T \, ds = T\left(\frac{\partial s}{\partial T}\right)_{\sigma, \mathbf{E}} dT + T\left(\frac{\partial s}{\partial \sigma}\right)_{T, \mathbf{E}} d\sigma + T\left(\frac{\partial s}{\partial \mathbf{E}}\right)_{T, \sigma} d\mathbf{E}$$

[1] See, for example, W. G. Cady, "Piezoelectricity," Dover Publications, New York, 1964.

where

$$T\left(\frac{\partial s}{\partial T}\right)_{\sigma,\,\mathbf{E}} = c_{\sigma,\,\mathbf{E}}$$

which is the specific heat at constant stress and constant electric field. In order to transform the other two partial derivatives it is helpful to define a piezo-electric Gibbs function by the equation

$$g = u - \sigma\epsilon - \mathbf{E}\mathbf{P} - T\,s$$

Upon differentiation,

$$dg = du - \sigma d\epsilon - \epsilon d\sigma - \mathbf{E}\,d\mathbf{P} - \mathbf{P}\,d\mathbf{E} - T\,ds - sdT$$

and when Eq. (7-32) is substituted,

$$dg = -\epsilon d\sigma - \mathbf{P}\,d\mathbf{E} - sdT$$

This equation yields directly three Maxwell relations:

$$\left(\frac{\partial \mathbf{P}}{\partial T}\right)_{\sigma,\,\mathbf{E}} = \left(\frac{\partial s}{\partial \mathbf{E}}\right)_{\sigma,\,T} \qquad \left(\frac{\partial \epsilon}{\partial T}\right)_{\mathbf{E},\,\sigma} = \left(\frac{\partial s}{\partial \sigma}\right)_{\mathbf{E},\,T} \qquad \left(\frac{\partial \epsilon}{\partial \mathbf{E}}\right)_{T,\,\sigma} = \left(\frac{\partial \mathbf{P}}{\partial \sigma}\right)_{T,\,\mathbf{E}}$$

Upon substituting the first two of the preceding Maxwell relations into the expression for $T\,ds$, we get

$$T\,ds = c_{\sigma,\,\mathbf{E}}\,dT + T\left(\frac{\partial \epsilon}{\partial T}\right)_{\mathbf{E},\,\sigma}\,d\sigma + T\left(\frac{\partial \mathbf{P}}{\partial T}\right)_{\sigma,\,\mathbf{E}}\,d\mathbf{E} \qquad (7\text{-}33)$$

When entropy s is considered as a function of T, σ, and $\mathbf{P}$, we have

$$T\,ds = T\left(\frac{\partial s}{\partial T}\right)_{\sigma,\,\mathbf{P}}\,dT + T\left(\frac{\partial s}{\partial \sigma}\right)_{T,\,\mathbf{P}}\,d\sigma + T\left(\frac{\partial s}{\partial \mathbf{P}}\right)_{T,\,\sigma}\,d\mathbf{P}$$

where

$$T\left(\frac{\partial s}{\partial T}\right)_{\sigma,\,\mathbf{P}} = c_{\sigma,\,\mathbf{P}}$$

which is the specific heat at constant stress and constant polarization. Now define a new function by the equation

$$g' = u - \sigma\epsilon - Ts$$

Then

$$dg' = du - \sigma d\epsilon - \epsilon d\sigma - T\,ds - s\,dT = \mathbf{E}\,d\mathbf{P} - \epsilon d\sigma - s\,dT$$

Whereupon

$$\left(\frac{\partial \epsilon}{\partial T}\right)_{\mathbf{P}, \sigma} = \left(\frac{\partial s}{\partial \sigma}\right)_{\mathbf{P}, T} \qquad \left(\frac{\partial \mathbf{E}}{\partial T}\right)_{\sigma, \mathbf{P}} = -\left(\frac{\partial s}{\partial \mathbf{P}}\right)_{\sigma, T} \qquad \left(\frac{\partial \mathbf{E}}{\partial \sigma}\right)_{T, \mathbf{P}} = -\left(\frac{\partial \epsilon}{\partial \mathbf{P}}\right)_{T, \sigma}$$

Thus we get

$$T \, ds = c_{\sigma, \mathbf{P}} \, dT + T\left(\frac{\partial \epsilon}{\partial T}\right)_{\mathbf{P}, \sigma} d\sigma - T\left(\frac{\partial \mathbf{E}}{\partial T}\right)_{\sigma, \mathbf{P}} d\mathbf{P} \tag{7-34}$$

When s is considered as a function of T, ϵ, and $\mathbf{E}$, we have

$$T \, ds = T\left(\frac{\partial s}{\partial T}\right)_{\epsilon, \mathbf{E}} dT + T\left(\frac{\partial s}{\partial \epsilon}\right)_{T, \mathbf{E}} d\epsilon + T\left(\frac{\partial s}{\partial \mathbf{E}}\right)_{T, \epsilon} d\mathbf{E}$$

where

$$T\left(\frac{\partial s}{\partial T}\right)_{\epsilon, \mathbf{E}} = c_{\epsilon, \mathbf{E}}$$

which is the specific heat at constant strain and constant electric field. Now define a new function by the equation

$$g'' = u - \mathbf{EP} - Ts$$

Then

$$dg'' = du - \mathbf{E} \, d\mathbf{P} - \mathbf{P} \, d\mathbf{E} - T \, ds - s \, dT = \sigma d\epsilon - \mathbf{P} \, d\mathbf{E} - s \, dT$$

Whereupon

$$\left(\frac{\partial \mathbf{P}}{\partial T}\right)_{\epsilon, \mathbf{E}} = \left(\frac{\partial s}{\partial \mathbf{E}}\right)_{\epsilon, T} \qquad \left(\frac{\partial \sigma}{\partial T}\right)_{\mathbf{E}, \epsilon} = -\left(\frac{\partial s}{\partial \epsilon}\right)_{\mathbf{E}, T} \qquad \left(\frac{\partial \sigma}{\partial \mathbf{E}}\right)_{T, \epsilon} = -\left(\frac{\partial \mathbf{P}}{\partial \epsilon}\right)_{T, \mathbf{E}}$$

Thus we get

$$T \, ds = c_{\epsilon, \mathbf{E}} \, dT - T\left(\frac{\partial \sigma}{\partial T}\right)_{\mathbf{E}, \epsilon} d\epsilon + T\left(\frac{\partial \mathbf{P}}{\partial T}\right)_{\epsilon, \mathbf{E}} d\mathbf{E} \tag{7-35}$$

When s is considered as a function of T, ϵ, and $\mathbf{P}$ we have

$$T \, ds = T\left(\frac{\partial s}{\partial T}\right)_{\epsilon, \mathbf{P}} dT + T\left(\frac{\partial s}{\partial \epsilon}\right)_{T, \mathbf{P}} d\epsilon + T\left(\frac{\partial s}{\partial \mathbf{P}}\right)_{T, \epsilon} d\mathbf{P}$$

where

$$T\left(\frac{\partial s}{\partial T}\right)_{\epsilon, \mathbf{P}} = c_{\epsilon, \mathbf{P}}$$

which is the specific heat at constant strain and constant polarization. Now from the definition of Helmhotz function,

$$a = u - Ts$$

we have

$$da = du - T \, ds - s \, dT = \sigma d\epsilon + \mathbf{E} \, d\mathbf{P} - s \, dT$$

Whereupon

$$\left(\frac{\partial \mathbf{E}}{\partial T}\right)_{\epsilon, \mathbf{P}} = -\left(\frac{\partial s}{\partial \mathbf{P}}\right)_{\epsilon, T} \qquad \left(\frac{\partial \sigma}{\partial T}\right)_{\mathbf{P}, \epsilon} = -\left(\frac{\partial s}{\partial \epsilon}\right)_{\mathbf{P}, T} \qquad \left(\frac{\partial \sigma}{\partial \mathbf{P}}\right)_{T, \epsilon} = \left(\frac{\partial \mathbf{E}}{\partial \epsilon}\right)_{T, \mathbf{P}}$$

Thus we get

$$T \, ds = c_{\epsilon, \mathbf{P}} \, dT - T \left(\frac{\partial \sigma}{\partial T}\right)_{\mathbf{P}, \epsilon} d\epsilon - T \left(\frac{\partial \mathbf{E}}{\partial T}\right)_{\epsilon, \mathbf{P}} d\mathbf{P} \qquad (7\text{-}36)$$

7-5 THERMODYNAMIC RELATIONS IN A GRAVITATIONAL FIELD

In our treatment of compressible systems we have assumed that the system is either entirely homogeneous in its properties or it is a composite of a limited number of homogeneous phases. However, when a system is situated in a gravitational field, it must be considered inhomogeneous. It is observed empirically that two portions of a system of constant temperature and composition, in general, will differ in their pressure and density due to the presence of a gravitational field. So far we have ignored inhomogeneity because its effect in most cases is negligible. But there are cases, such as in the study of the earth's atmosphere, where we can no longer neglect inhomogeneity. We now extend the treatment to include the effect of the earth's gravitational field.

The gravitational field is characterized by a gravitational potential Ψ. When the mass of the system under consideration is small compared with that of the earth, the gravitational potential at each point in space has a definite value and is independent of the presence or state of any material there. In this case, the *gravitational potential* can be written as

$$\Psi = gz \qquad (7\text{-}37)$$

where g is the acceleration due to gravity and z is the distance from the system to the center of the earth.

Let us imagine the system under study to be divided into small volume elements of infinitesimal thickness in the direction of the field, thus forming a collection of open phases in contact with one another. The gravitational potential at each volume element can be considered constant. The work done against the field in increasing the mass of component i of a volume element by the amount dm_i is given by

$$\Psi\, dm_i = M_i\, \Psi\, dn_i$$

where Ψ is the gravitational potential at the volume element, n_i is the number of moles of component i, and M_i is its molecular weight. When Eq. (4-3) is extended to include the gravitational work, we have for each open phase the fundamental equation

$$dU^{(\alpha)} = T^{(\alpha)}\, dS^{(\alpha)} - p^{(\alpha)}\, dV^{(\alpha)} + \sum_{i=1}^{r} (\mu_i^{(\alpha)} + M_i\, \Psi^{(\alpha)}) dn_i^{(\alpha)}, \quad \alpha = 1, 2, \ldots \quad (7\text{-}38)$$

where the subscript i denotes the component and the superscript (α) denotes the phase; and $\mu_i^{(\alpha)}$ is the chemical potential of component i in the α phase in the absence of the gravitational field.

Similarly, from Eqs. (4-4) to (4-6) we obtain

$$dH^{(\alpha)} = T^{(\alpha)}\, dS^{(\alpha)} + V^{(\alpha)}\, dp^{(\alpha)} + \sum_{i=1}^{r} (\mu_i^{(\alpha)} + M_i\, \Psi^{(\alpha)}) dn_i^{(\alpha)} \qquad (7\text{-}39)$$

$$dA^{(\alpha)} = -S^{(\alpha)}\, dT^{(\alpha)} - p^{(\alpha)}\, dV^{(\alpha)} + \sum_{i=1}^{r} (\mu_i^{(\alpha)} + M_i\, \Psi^{(\alpha)}) dn_i^{(\alpha)} \qquad (7\text{-}40)$$

$$dG^{(\alpha)} = -S^{(\alpha)}\, dT^{(\alpha)} + V^{(\alpha)}\, dp^{(\alpha)} + \sum_{i=1}^{r} (\mu_i^{(\alpha)} + M_i\, \Psi^{(\alpha)}) dn_i^{(\alpha)} \qquad (7\text{-}41)$$

From the preceding equations it is seen that when a gravitational field is included in the analysis of a compressible system, a total potential as defined by the expression $(\mu_i^{(\alpha)} + M_i\, \Psi^{(\alpha)})$ takes the place of the chemical potential $\mu_i^{(\alpha)}$. Thus, in an analogy with Eq. (5-24), the condition for phase equilibrium, say between α and β phases, is given by the equation

$$\mu_i^{(\alpha)} + M_i\, \Psi^{(\alpha)} = \mu_i^{(\beta)} + M_i\, \Psi^{(\beta)} \qquad (7\text{-}42)$$

for each component.

In addition to satisfying the condition of uniformity of total potential as expressed by Eq. (7-42), a compressible system in a gravitational field must have uniformity of temperature to insure thermal equilibrium. From these two general conditions, an equation for the condition of hydrostatic equilibrium can be derived. Let us consider the case of a single-component system for the purpose of illustration. According to Eq. (7-42), for a single-component system, by considering the thermodynamic variables to be a function of z, we have

$$\frac{d\mu}{dz} + M \frac{d\Psi}{dz} = 0 \tag{7-43}$$

But applying Eq. (7-41) to a single-component system, since the molecular weight M and the gravitational potential Ψ are independent of p and T, gives us by cross-differentiation,

$$\left(\frac{\partial \mu}{\partial p}\right)_{T,n} = \left(\frac{\partial V}{\partial n}\right)_{T,p} = v$$

where v is the molal volume of the single-component system. Hence, for a unit mole of the system at constant temperature, we have

$$\frac{d\mu}{dz} = v \frac{dp}{dz}$$

Substituting this equation into Eq. (7-43) leads to

$$v \frac{dp}{dz} + M \frac{d\Psi}{dz} = 0$$

But the density ρ of the system is given by

$$\rho = \frac{M}{v}$$

Finally we obtain

$$\frac{dp}{dz} = -\rho \frac{d\Psi}{dz} \tag{7-44}$$

This is the condition of hydrostatic equilibrium in a gravitational field. With the gravitational potential as given by Eq. (7-37), the preceding equation can be rewritten as

$$dp = -\rho \, g \, dz \tag{7-45}$$

assuming g to be a constant.

For a single-component ideal gas, since $\rho = p \, M/RT$, Eq. (7-45) becomes

$$RT \frac{dp}{p} = -M \, g \, dz$$

which may be integrated for an isothermal atmosphere to yield

$$p = p_0 e^{-(Mg/RT)(z-z_0)} \tag{7-46}$$

or

$$\rho = \rho_0 e^{-(Mg/RT)(z-z_0)} \tag{7-46a}$$

where p_0 and ρ_0 are the respective pressure and density at the earth's surface which is at a distance z_0 from the center of the earth, and p and ρ are the respective pressure and density at an elevation $(z - z_0)$ above the earth's surface. Equation (7-46) is the well-known *barometric equation* which describes approximately the isothermal atmosphere over a limited-altitude range. According to this equation, the higher the altitude the smaller the barometric pressure and density.

> *EXAMPLE 7-5.* When you are going up in an elevator in a skyscraper your ears pop at the 50th floor, about 180 meters above the ground. The temperature is 295°K throughout the elevator shaft. What is the pressure change that causes your ears to pop?
>
> *Solution.*
> The average molecular weight of air is $M = 29$ g/g mole. Substituting the given values into the barometric equation, we obtain

$$\frac{p}{p_0} = \exp\left[-\frac{Mg}{RT}(z - z_0)\right]$$

$$= \exp\left[-\frac{\left(29\ \frac{g}{g\ mole}\right)\left(9.80\ \frac{m}{sec^2}\right)(180\ m)}{\left(1000\ \frac{g}{kg}\right)\left(8.314\ \frac{(newtons)(m)}{(g\ mole)(°K)}\right)(295°K)}\right] = 0.9794$$

When $p_0 = 1$ atm at the ground floor, we have at the 50th floor

$$p = 0.9794\ p_0 = 0.9794\ atm$$

Therefore, the pressure change is

$$p - p_0 = 0.9794 - 1 = -0.0206\ atm$$

7-6 CENTRIFUGAL FIELDS

The thermodynamic treatment of a compressible system subject to the influence of a centrifugal field is identical to a similar case involving a gravitational field. The *centrifugal potential* Ψ characterizing a centrifugal field is defined so that

$$d\Psi = -\omega^2 z\ dz$$

where ω is the angular velocity and z is the distance from the axis of rotation. Thus, $\omega^2 z$ is the centrifugal acceleration. The centrifugal potential at any given values of ω and z is then

$$\Psi = -\omega^2 \int_0^z z \, dz = -\frac{1}{2} \omega^2 z^2 \tag{7-47}$$

All thermodynamic relations derived in the preceding section for a gravitational field can be applied to problems involving a centrifugal field if the potential is defined by Eq. (7-47) instead of Eq. (7-37). Seeing that there are many direct applications involving centrifugal fields, such as the separation of components in multicomponent systems and the determination of molecular weight of large molecules, we will now develop a few equations in terms appropriate to centrifugal problems. Note that all equations derived in this section can also be applied to gravitational problems if the potential is defined by Eq. (7-37).

For component i of a multicomponent system in equilibrium in a centrifugal field, by considering the thermodynamic variables to be functions of z, from Eq. (7-42) we have

$$\frac{d\mu_i}{dz} + M_i \frac{d\Psi}{dz} = 0 \tag{7-48}$$

But since μ_i is the chemical potential of component i in the absence of the field, from Eq. (4-21a) we have

$$\frac{d\mu_i}{dz} = -\bar{S}_i \frac{dT}{dz} + \bar{V}_i \frac{dp}{dz} + \sum_{k=2}^{r} \left(\frac{\partial \mu_i}{\partial x_k}\right)_{T,p} \frac{dx_k}{dz}$$

Substituting this equation into Eq. (7-48) yields

$$-\bar{S}_i \frac{dT}{dz} + \bar{V}_i \frac{dp}{dz} + \sum_{k=2}^{r} \left(\frac{\partial \mu_i}{\partial x_k}\right)_{T,p} \frac{dx_k}{dz} + M_i \frac{d\Psi}{dz} = 0$$

where $\bar{S}_i$ and $\bar{V}_i$ are the partial molal entropy and volume, respectively, of component i. At constant temperature this equation reduces to

$$\bar{V}_i \frac{dp}{dz} + M_i \frac{d\Psi}{dz} + \sum_{k=2}^{r} \left(\frac{\partial \mu_i}{\partial x_k}\right)_{T,p} \frac{dx_k}{dz} = 0 \tag{7-49}$$

On the other hand, at constant temperature the Gibbs-Duhem equation (Eq. 4-11) gives

$$\sum_{i=1}^{r} n_i \, d\mu_i = V \, dp \tag{7-50}$$

From Eqs. (7-48) and (7-50) we can write

$$\sum_{i=1}^{r} n_i \frac{d\mu_i}{dz} = - \sum_{i=1}^{r} M_i \, n_i \frac{d\Psi}{dz} = V \frac{dp}{dz} \tag{7-51}$$

But

$$\sum_{i=1}^{r} M_i \, n_i = m = \text{mass of the system}$$

and

$$\frac{m}{V} = \rho = \text{mean density of the system}$$

Thus Eq. (7-51) can be rewritten as

$$\frac{dp}{dz} = -\rho \frac{d\Psi}{dz} \tag{7-52}$$

which is the condition for hydrostatic equilibrium and is the same as Eq. (7-44).
When Eq. (7-52) is substituted into Eq. (7-49) we obtain

$$(M_i - \rho \bar{V}_i) \frac{d\Psi}{dz} + \sum_{k=2}^{r} \left(\frac{\partial \mu_i}{\partial x_k} \right)_{T,\,p} \frac{dx_k}{dz} = 0 \tag{7-53}$$

This equation can be integrated only in certain simple cases. We now give the solution for the case of a binary ideal solution.
For a binary system Eq. (7-53) becomes

$$(M_2 - \rho \bar{V}_2) \frac{d\Psi}{dz} + \left(\frac{\partial \mu_2}{\partial x_2} \right)_{T,\,p} \frac{dx_2}{dz} = 0 \tag{7-54}$$

where M_2, $\bar{V}_2$, μ_2, and x_2 are the molecular weight, partial molal volume, chemical potential, and mole fraction respectively of component 2. If the system is an ideal solution, from Eq. (4-78) we have

$$\left(\frac{\partial \mu_2}{\partial x_2} \right)_{T,\,p} = \frac{RT}{x_2}$$

and from Eq. (4-73) we have

$$\bar{V}_2 = v_2^0$$

where v_2^0 is the molal volume of pure component 2 at T and p of the solution. Thus Eq. (7-54) may be rewritten as

$$(M_2 - \rho v_2^0) \frac{d\Psi}{dz} + \frac{RT}{x_2} \frac{dx_2}{dz} = 0$$

or

$$\frac{d(\ln x_2)}{dz} = -\frac{M_2 - \rho v_2^0}{RT} \frac{d\Psi}{dz} \tag{7-55}$$

For a centrifugal field, substituting Eq. (7-47) into Eq. (7-55) leads to

$$d(\ln x_2) = \frac{M_2 - \rho v_2^0}{RT} \omega^2 z \, dz$$

Assuming the product ρv_2^0 to be independent of z, we obtain upon integration of the preceding equation at constant temperature,

$$\ln \frac{x_2^{(\beta)}}{x_2^{(\alpha)}} = \frac{M_2 - \rho v_2^0}{2RT} \omega^2 \left[(z^{(\beta)})^2 - (z^{(\alpha)})^2 \right] \tag{7-56}$$

where $x_2^{(\alpha)}$ and $x_2^{(\beta)}$ are the mole fractions of component 2 at radii $z^{(\alpha)}$ and $z^{(\beta)}$ respectively. This equation can be used in molecular weight determinations.

For the case of an ideal gas mixture, it is convenient to deal with the partial pressure p_i of component i which is related to the total pressure p by the equation $p_i = x_i p$. Since in an ideal gas mixture the presence of other components has no effect on the behavior of a given component, equivalent to Eq. (7-44) we may write the condition of hydrostatic equilibrium of component i as

$$\frac{dp_i}{dz} = -\rho_i \frac{d\Psi}{dz} = -\frac{p_i M_i}{RT} \frac{d\Psi}{dz} \tag{7-57}$$

In a centrifugal field, $d\Psi/dz = -\omega^2 z$, thus

$$\frac{dp_i}{dz} = \frac{p_i M_i}{RT} \omega^2 z \qquad \text{or} \qquad \frac{dp_i}{p_i} = \frac{M_i \omega^2}{RT} z \, dz$$

Integrating at constant temperature,

$$\ln \frac{p_i^{(\beta)}}{p_i^{(\alpha)}} = \frac{M_i \omega^2}{2RT} \left[(z^{(\beta)})^2 - (z^{(\alpha)})^2 \right] \tag{7-58}$$

where $p_i^{(\alpha)}$ and $p_i^{(\beta)}$ are the partial pressures of component i at radii $z^{(\alpha)}$ and $z^{(\beta)}$ respectively.

Applying Eq. (7-58) to two different components 1 and 2 and taking the difference between the two resulting equations, we have

$$\ln \frac{p_2^{(\beta)}/p_1^{(\beta)}}{p_2^{(\alpha)}/p_1^{(\alpha)}} = \frac{(M_2 - M_1)\omega^2}{2RT} [(z^{(\beta)})^2 - (z^{(\alpha)})^2] \tag{7-59}$$

This equation can be used to investigate the separation of two gaseous components by a centrifuge.

EXAMPLE 7-6. Natural uranium is 99.3% of U^{238} and 0.7% of U^{235}. In order to obtain enriched U^{235}, the natural uranium is first converted into uranium hexafluoride (UF_6), which is a compound volatile enough to be handled as a vapor. In the centrifugal separation of uranium isotopes at $100°C$, let us assume that UF_6 can be treated as an ideal gas. At an angular velocity of 10^4 rad/sec, what will be the abundance of U^{235} in percentage at a radius of 2.5 cm, assuming the isotopes in natural proportion to prevail at a radius of 10 cm?

Solution
We denote UF_6 from U^{235} by the subscript 1 and UF_6 from U^{238} by the subscript 2. We also denote the mixture at the radii of 2.5 cm and 10 cm by the superscripts (α) and (β) respectively. Thus

$M_2 - M_1 = 3$ g/g mole

$z^{(\alpha)} = 0.025$ m $\qquad z^{(\beta)} = 0.1$ m

$x_1^{(\beta)} = 0.007$

Since for an ideal gas mixture the partial pressure p_i of component i is related to the total pressure p by the equation $p_i = x_i p$, we have then

$$\frac{p_2^{(\beta)}}{p_1^{(\beta)}} = \frac{x_2^{(\beta)}}{x_1^{(\beta)}} = \frac{1 - x_1^{(\beta)}}{x_1^{(\beta)}} = \frac{1 - 0.007}{0.007} = 141.9$$

and

$$\frac{p_2^{(\alpha)}}{p_1^{(\alpha)}} = \frac{x_2^{(\alpha)}}{x_1^{(\alpha)}} = \frac{1 - x_1^{(\alpha)}}{x_1^{(\alpha)}}$$

Substituting the given values into Eq. (7-59), we have

$$\ln \frac{141.9}{\left(\frac{1 - x_1^{(\alpha)}}{x_1^{(\alpha)}}\right)} = \frac{(3)(10^8)}{(1000)(2)(8.314)(373)} (0.10^2 - 0.025^2) = 0.4535$$

Thus

$$\frac{1 - x_1^{(\alpha)}}{x_1^{(\alpha)}} = \frac{141.9}{e^{0.4535}} = 90.15$$

Therefore

$$x_1^{(\alpha)} = 1.1\%$$

which is the abundance of U^{235} at a radius of 2.5 cm.

PROBLEMS

7-1. A toroidal coil ring has 100 turns per meter and carries a current of 8 amp.

(a) Determine the magnetic field intensity $\mathbf{H}$.

(b) When there is no core material inside the ring, determine the magnetic induction $\mathbf{B}_{no\ core}$.

(c) When there is an iron core, determine the magnetic induction $\mathbf{B}_{core}$ and the magnetization $\mathbf{M}$, assuming the relative permeability of iron as defined by the ratio $\mathbf{B}_{core}/\mathbf{B}_{no\ core}$ to be 2,000.

7-2. For a paramagnetic solid if the Curie equation (Example 7-1) holds, show that the heat capacity at constant magnetization C_M is independent of the magnetic field and equal to the zero-field heat capacity.

7-3. For a simple magnetic system that obeys Curie equation (Example 7-1), show that for a reversible adiabatic process $HM^{-\gamma}$ = constant where $\gamma = C_H/C_M$ is assumed to be a constant.

7-4. Iron at $1200^\circ K$ is assumed to obey the Curie-Weiss equation (Example 7-2), and the characteristic constants are $C_{cw} = 0.0155$ $(m^3)(^\circ K)/kg$ mole and $\Theta = 1093^\circ K$.

(a) Calculate the entropy change when one kg of iron is placed in a magnetic field of 1×10^6 amp/m at a constant temperature of $1,200^\circ K$.

(b) Calculate the change in temperature when the iron is then adiabatically demagnetized, using $C_M = 22.2$ kJ/(kg mole)($^\circ K$).

7-5. For a system subject to both magnetic and volume effects, the basic equation of combined first and second laws can be written as

$$T\,dS = dU + p\,dV - \mu_0\,\mathbf{H}\,d\mathbf{I}$$

where V and $\mathbf{I}$ are the total volume and total magnetic moment respectively of the entire system. Prove the following relationship between the volume magnetostriction and the pressure coefficient of magnetization:

$$\left(\frac{\partial V}{\partial \mathbf{H}}\right)_{p,\,T} = -\mu_0 \left(\frac{\partial \mathbf{I}}{\partial p}\right)_{\mathbf{H},\,T}$$

Furthermore, prove that when the magnetic field is increased at constant temperature and pressure from zero to **H**, the small change in volume is given by

$$\frac{\Delta V}{V} = \frac{1}{2} \mu_0 \mathbf{H}^2 \left[\kappa \chi - \left(\frac{\partial \chi}{\partial p} \right)_{\mathbf{H}, T} \right]$$

where

$$\kappa = - \frac{1}{V} \left(\frac{\partial V}{\partial p} \right)_{\mathbf{H}, T} = \text{isothermal compressibility}$$

and

$\chi = \mathbf{I}/\mathbf{H}V = $ magnetic susceptibility

7-6. For a simple dielectric system,

(a) define the properties enthalpy H, Helmholtz function A, and Gibbs function G

(b) write the basic property relations of dU, dH, dA and dG

(c) write the four Maxwell relations.

7-7. The heat capacities of a simple dielectric system at constant electric field and at constant electric polarization are defined as follows

$$C_{\mathbf{E}} = T \left(\frac{\partial S}{\partial T} \right)_{\mathbf{E}} \quad \text{and} \quad C_{\mathbf{P}} = T \left(\frac{\partial S}{\partial T} \right)_{\mathbf{P}}$$

Assuming the thermal equation of state of the system to be expressed by the equation

$$\mathbf{P} = \frac{\mathbf{P}'}{V} = K \frac{\mathbf{E}}{T}$$

where K is a constant and V is the constant volume of the system, show that

(a) $C_{\mathbf{E}} - C_{\mathbf{P}} = KV \dfrac{\mathbf{E}^2}{T^2}$

(b) $dU = C_{\mathbf{P}} \, dT$

(c) $dH = C_{\mathbf{E}} \, dT - \dfrac{2KV}{T} \mathbf{E} \, d\mathbf{E}$

where $H = U - V \, \mathbf{EP} = U - \mathbf{EP}'$

7-8. The thermal equation of state of a simple dielectric system is given by

$$\mathbf{P} = \frac{\mathbf{P}'}{V} = \left(a + \frac{b}{T} \right) \mathbf{E}$$

where a and b are constants. Show that:

(a) In a reversible isothermal change of electric field, the heat transfer is given by

$$Q = -\frac{bV}{2T} (E_2^2 - E_1^2)$$

(b) In a reversible adiabatic change of electric field, the small temperature change is given by

$$\Delta T = \frac{bV}{2TC_E} (E_2^2 - E_1^2)$$

where C_E is the heat capacity at constant electric field.

7-9. For a compressible dielectric system,
   (a) define the properties enthalpy H, Helmholtz function A, and Gibbs function G
   (b) write the basic property relations of dU, dH, dA and dG
   (c) write all the Maxwell relations pertaining to this system.

7-10. The static value of the dielectric constant of liquid iodomethane (CH_3I) as a function of temperature may be expressed by the equation

$$\kappa = \frac{2160}{T} - 0.39$$

where T is in $°K$. Calculate the work done per cm^3 of iodomethane when the electric field is increased from zero to 10^4 volts/m while the temperature is increased from $-20°C$ to $20°C$, assuming that the temperature varies linearly with the electric field during the process.

7-11. A balloon containing an ideal gas and having a volume of $1 m^3$ at the surface of the earth rises slowly to a height of 10 km in an isothermal atmosphere at $10°C$. Calculate the work done by the balloon on the atmosphere, assuming the acceleration of gravity to be a constant at its sea level value.

7-12. Solve the preceding problem with the following data: volume $1 ft^3$, height 5 miles, and temperature $40°F$.

7-13. Atmospheric air at sea level may be considered to contain 20.16 mole percent of oxygen, 78.72 mole percent of nitrogen, and 1.12 mole percent of minor components. The total pressure due to the two main components is 0.995 atm at sea level. Neglecting the effect of the minor components in the air, and considering the air to be at a constant temperature of $300°K$, estimate the total pressure and the density of air at an altitude of 4000 m. What are the partial pressures of oxygen and nitrogen at that altitude? Assume that the acceleration of gravity is a constant at its sea level value.

7-14. Atmospheric air near the earth's surface is composed mostly of oxygen and nitrogen with a small amount of other gases. Assuming that there is 0.1

mole percent of hydrogen in the atmosphere at the earth surface, estimate the altitude at which the concentration of hydrogen reaches 50 mole percent of the total in an isothermal atmosphere at $0°C$. Assume that the acceleration of gravity is a constant at its sea level value.

7-15. A long horizontal cylinder, closed at one end and open to a large reservoir at the other end, rotates at (a) 2×10^3 rpm and (b) 2×10^4 rpm about a vertical shaft at the open end. The reservoir contains a gaseous mixture of 50 mole percent helium and 50 mole percent krypton at $300°K$. Estimate the composition of the mixture at a radius of 1 m from the axis of rotation.

Chapter 8
LOW TEMPERATURES

8-1 PRODUCTION OF LOW TEMPERATURES

As a result of the attempt to liquefy and to solidify the so-called permanent gases, such as oxygen, nitrogen, hydrogen, and helium, a fascinating and useful branch of thermodynamics known as *cryogenics* has grown steadily to near maturity through the work of such pioneers as Cailletet, Dewar, Kamerlingh-Onnes, Giauque, Simon, and Kurti. Now microdegree temperatures approaching absolute zero can be achieved experimentally. In this and the next two chapters, we will study the methods of producing very low temperatures and investigate various low-temperature phenomena.

There are a few common methods of producing cryogenic temperatures. These are: (a) evaporation under reduced pressure, (b) Joule-Thomson expansion, (c) adiabatic expansion against a restraining force, and (d) adiabatic demagnetization of paramagnetic materials. We will discuss the first three methods in sections 1 and 2 of this chapter, leaving the description of the last method for later sections.

When a gas at a temperature lower than its critical temperature and higher than its triple-point temperature is compressed isothermally, the gas will liquify when the pressure reaches the vapor pressure at the given temperature. Once in liquid form, it can be allowed to evaporate in an insulated container under reduced pressure; the vapor formed is constantly pumped away, carrying with it the latent heat of vaporization. When the pumping of the vapor is rapid enough, the triple-point temperature of the substance can be achieved and the substance solidified. Since the vapor pressure of a solid falls off rapidly as the temperature goes below the triple point, this method fairly soon becomes impractical.

If different substances having consecutively lower triple-point temperatures are used in a series of compression-evaporation processes, very low temperatures can be achieved. In such a series of processes, the gas with higher triple point is liquefied first; this liquid is then allowed to evaporate under reduced pressure, thus cooling itself. This cooled liquid is then used as the cooling fluid during isothermal compression of a second gas, thus liquefying the second gas.

The "cascade" process described above is a rather simple method of liquefying low-boiling-point gases. The first gas in the series should be one with critical temperature above room temperature. A possible cascade process in producing liquid nitrogen is to use the following four substances in a sequence: ammonia, ethylene, methane, and nitrogen. The feasibility of this sequence can be seen from Table 8-1, where data on critical temperature and triple-point temperature for a number of substances are shown. Note that although methane is included in the sequence to make the operating pressures more convenient, it is not strictly necessary.

Inspection of the data shown in Table 8-1 reveals that a gap exists between the triple-point temperature of nitrogen and the critical temperature of hydrogen. Therefore the cascade sequence mentioned above cannot be extended to include the liquefaction of hydrogen. In order to liquefy hydrogen or helium, either Joule-Thomson free expansion or adiabatic expansion against a restraining force, or a combination of these two methods, can be used.

The thermodynamic process of Joule-Thomson expansion has been analyzed in Sec. 2-6. When the Joule-Thomson expansion is used to produce a cooling effect, the initial temperature of the gas must be below its maximum inversion temperature. Table 8-2 gives the maximum inversion temperatures for a number of substances. Since the maximum inversion temperatures of hydrogen and helium are below the normal room temperature, precooling is necessary before these gases can be throttled to produce a cooling effect.

TABLE 8-1. Critical Temperature and Triple-point Temperature of a Number of Substances

Substance	Critical temperature, $°K$	Triple-point temperature, $°K$
Ammonia	405.5	195.42
Carbon dioxide	304.2	216.55
Ethylene	283.1	104.00
Methane	190.7	90.67
Oxygen	154.6	54.35
Argon	151.	83.78
Carbon monoxide	133.	68.14
Nitrogen	126.2	63.15
Hydrogen (normal)	33.3	13.84
Helium (He^4)	5.20	—
Helium (He^3)	3.32	—

TABLE 8-2. Maximum Inversion Temperatures

Gas	Maximum inversion temperature, °K
Carbon dioxide	1,500
Oxygen	761
Argon	723
Nitrogen	621
Air	603
Hydrogen	202
Helium	40

Of course, a single expansion is, in general, not enough to produce liquefaction. In practical operations, the gas that has been cooled by throttling is sent back to a heat exchanger to cool the incoming gas which is on the way to the throttling device. Thus the gas arriving at the throttling device gets cooler and cooler. A number of successive coolings will eventually produce partial liquefaction. The liquid portion is then withdrawn from the system. Further cooling can be accomplished by allowing the liquid to evaporate under reduced pressure.

In contrast to a Joule-Thomson expansion, an adiabatic expansion against a restraining force always produces a cooling, regardless of the initial temperature. In the ideal case, the expansion would be reversible and adiabatic, and the drop in temperature corresponding to a drop in pressure is given by the relation:

$$\frac{T_1}{T_2} = \left(\frac{p_1}{p_2}\right)^{k-1/k} \tag{2-61}$$

where $k = c_p/c_v$ and is considered to be a constant in a given temperature range.

From a thermodynamic point of view, a reversible adiabatic expansion is superior to a Joule-Thomson expansion. Between any two given pressures, the former always gives a greater temperature drop than the latter. However, as may be seen from the equation

$$T_1 - T_2 = T_1 \left[1 - \left(\frac{p_2}{p_1}\right)^{k-1/k} \right]$$

which is written directly from Eq. (2-61), the temperature drop due to a reversible adiabatic expansion decreases as the initial temperature decreases for a given pressure drop. Whereas in the case of a Joule-Thomson expansion the reverse is true. Furthermore, at low temperatures there are lubrication difficulties for a moving expansion mechanism such as an engine. On the contrary, in a

Joule-Thomson expansion device, there are no moving parts and so lubrication is no problem. On account of these reasons, a combination of these two methods, namely the use of an adiabatic expansion against a restraining force in the higher-temperature range followed by a Joule-Thomson expansion in the lower-temperature range, is in common use to take advantage of both methods. We will illustrate the actual procedures in the next section.

8-2 SOME GAS-LIQUEFACTION SYSTEMS

In order to illustrate the use of Joule-Thomson and reversible adiabatic expansions in cryogenic applications we proceed now to discuss a few basic systems in gas liquefaction. Variations and modifications of these basic systems shall not be included in our discussions.

The simplest liquefaction system is the Linde-Hampson system, shown schematically in Fig. 8-1(a). The idealized cycle of this system in temperature-entropy plane for the liquefaction of air is shown in Fig. 8-1(b). Here the gas is compressed from state 1 to state 2 reversibly and isothermally. In actual

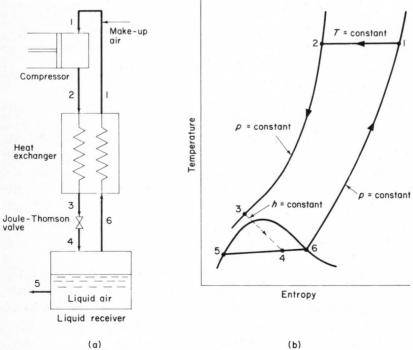

(a)   (b)

FIG. 8-1. Linde-Hampson system for air liquefaction.

operations, the compression process could be carried out in multistages with intercooling and aftercooling. The gas is then cooled in a counterflow heat exchanger, where it gives heat to the returning low-pressure stream of gas. Upon throttling through an expansion valve, the gas cools itself and then returns to the heat exchanger to cool the incoming gas so that the gas arriving at the expansion valve gets cooler and cooler. After many repetitions, eventually a low-temperature state such as indicated by point 3 is reached, from which a Joule-Thomson expansion to state 4 results in partial liquefaction of the gas. At state 4 the gas is in the two-phase region, consisting of saturated liquid at state 5 and saturated vapor at state 6. The liquid is withdrawn from the liquid receiver and the vapor leaves the liquid receiver and returns through the heat exchanger to the compressor.

As mentioned in the preceding section, there are advantages in the use of an adiabatic expansion against a restraining force in a higher temperature range followed by a Joule-Thomson expansion in a lower temperature range. This idea is utilized in the Claude system shown in Fig. 8-2. Here the gas is first compressed and then cooled in passing through the first heat exchanger. The gas is then divided into two streams. A major part of the gas stream is diverted to an expander, where the gas gets cooler by expanding against a restraining force. The remainder of the gas stream is sent to the second heat exchanger on its way to the third heat exchanger and the throttling valve to be partially liquefied. The liquid portion is withdrawn from the liquid receiver, and the vapor portion leaves the liquid receiver and joins the cooled gas from the expander to serve as the cooling agency on its return trip to the compressor.

Since the maximum inversion temperatures of neon, hydrogen, and helium are below normal room temperature, these gases must be precooled before a Joule-Thomson expansion can be used to produce a cooling effect. The precoolant must have a triple-point temperature below that of the maximum inversion temperature of the gas to be liquefied. From a practical standpoint, nitrogen would be a good choice as the precoolant for the liquefaction of hydrogen or neon. Figure 8-3(a) shows a schematic diagram of a precooled Linde-Hampson system for hydrogen or neon, using liquid nitrogen as the precoolant. Figure 8-3(b) shows the idealized cycle in T-s plane for the main gas only.

Since an adiabatic expansion against a restraining force always produced a cooling regardless of the initial temperature, a Claude system as shown in Fig. 8-2 without modification can be used to liquefy hydrogen or neon. However, if a liquid-nitrogen precooling is used with a Claude system the preformance of the system can certainly be improved.

For the liquefaction of helium, S. C. Collins in 1946 succeeded in developing a compact system known as the Collins helium liquefier, as shown in Fig. 8-4. It

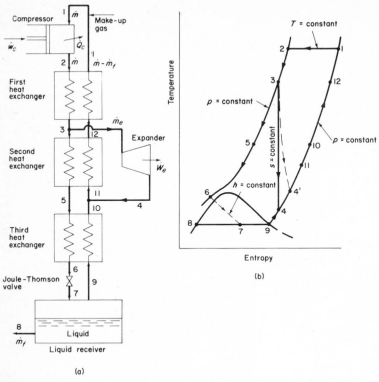

FIG. 8-2. Claude system.

is essentially an extension of the Claude system with two or more expansion engines. The availability of this compact helium liquefier has resulted in a tremendous increase in scientific and technological work at very low temperatures since 1946.

EXAMPLE 8-1. In a Claude liquefaction system as depicted in Fig. 8-2, nitrogen gas at $25°C$ and 1 atm is compressed reversibly and isothermally to 50 atm. The gas is then cooled to $0°C$ by passing through a heat exchanger. At the outlet of this heat exchanger, 60% of the total flow is diverted and is expanded adiabatically to 1 atm through an expander which has an adiabatic efficiency of 75%. The expander work is utilized as part of the compressor work requirement. All the heat exchangers are working under ideal conditions. Determine the liquid yield per unit mass of gas compressed and the net work required to liquefy a unit mass of gas.

Solution.
In Fig. 8-2, $\dot{m}$ and $\dot{m}_e$ denote the mass flow rates through the compressor

and the expander, respectively, and $\dot{m}_f$ denotes the liquid discharge rate from the liquid receiver.

From the data on nitrogen in the appendix we obtain the following values:

$$h_1 = 460 \text{ joules/g} \qquad s_1 = 4.414 \text{ joules/(g)(}^{\circ}\text{K)} \qquad \text{at} \quad \begin{matrix} p_1 = 1 \text{ atm} \\ T_1 = 298^{\circ}\text{K} \end{matrix}$$

$$h_2 = 450 \text{ joules/g} \qquad s_2 = 3.222 \text{ joules/(g)(}^{\circ}\text{K)} \qquad \text{at} \quad \begin{matrix} p_2 = 50 \text{ atm} \\ T_2 = 298^{\circ}\text{K} \end{matrix}$$

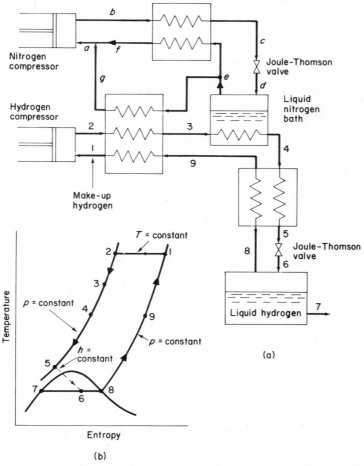

FIG. 8-3. Precooled Linde-Hampson system for hydrogen (or neon) liquefaction, using liquid nitrogen as the precoolant.

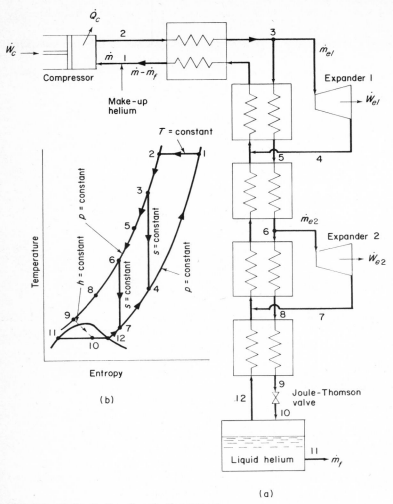

FIG. 8-4. Collins helium-liquefaction system.

$$h_3 = 422 \text{ joules/g} \quad s_3 = 3.123 \text{ joules/(g)(}^\circ\text{K)} \quad \text{at} \quad \begin{array}{l} p_3 = 50 \text{ atm} \\ T_3 = 273^\circ\text{K} \end{array}$$

$$h_4 = 239 \text{ joules/g} \quad \text{at} \quad \begin{array}{l} p_4 = 1 \text{ atm} \\ s_4 = s_3 = 3.123 \text{ joules/(g)(}^\circ\text{K)} \end{array}$$

$$h_8 = 29.4 \text{ joules/g} \quad \text{at} \quad \begin{array}{l} p_8 = 1 \text{ atm} \\ \text{saturated liquid} \end{array}$$

The adiabatic efficiency of the expander is defined as

$$\eta_{exp} = \frac{h_3 - h_{4'}}{h_3 - h_4}$$

Whence

$$h_{4'} = h_3 - \eta_{exp}\,(h_3 - h_4) = 422 - 0.75\,(422 - 239) = 285 \text{ joules/g}$$

The work done by the gas during expansion is

$$\frac{\dot{W}_e}{\dot{m}} = \frac{\dot{W}_{34'}}{\dot{m}} = \frac{\dot{m}_e}{\dot{m}}\,(h_3 - h_{4'}) = 0.60\,(422 - 285)$$

$$= 82.2 \text{ joules/g compressed}$$

Since the compression process is reversible and isothermal, the heat rejected from the gas during compression is

$$\frac{\dot{Q}_c}{\dot{m}} = \frac{-\dot{Q}_{12}}{\dot{m}} = T_1\,(s_1 - s_2) = 298\,(4.414 - 3.222)$$

$$= 355 \text{ joules/g compressed}$$

From the first law for steady flow the work required to compress the gas is

$$\frac{\dot{W}_c}{\dot{m}} = \frac{-\dot{W}_{12}}{\dot{m}} = \frac{\dot{Q}_c}{\dot{m}} + h_2 - h_1 = 355 + 450 - 460$$

$$= 354 \text{ joules/g compressed}$$

The net work required is then

$$\frac{\dot{W}_{net}}{\dot{m}} = \frac{\dot{W}_c}{\dot{m}} - \frac{\dot{W}_e}{\dot{m}} = 345 - 82.2 = 263 \text{ joules/g compressed}$$

Applying the first law for steady flow to the system as a whole we have

$$\dot{W}_c - \dot{W}_e - \dot{Q}_c + \dot{m}_f h_1 - \dot{m}_f h_8 = 0$$

assuming that the condition of the make-up gas is the same as that of state 1. Dividing through by $\dot{m}$ and rearranging, we have

$$\frac{\dot{m}_f}{\dot{m}} = \frac{-\dot{W}_c/\dot{m} + \dot{W}_e/\dot{m} + \dot{Q}_c/\dot{m}}{h_1 - h_8} = \frac{-345 + 82.2 + 355}{460 - 29.4} = 0.214$$

This is the liquid yield per unit mass of gas compressed.

The net work required to liquefy a unit mass of gas is then

$$\frac{\dot{W}_{net}}{\dot{m}_f} = \frac{\dot{W}_{net}/\dot{m}}{\dot{m}_f/\dot{m}} = \frac{263}{0.214} = 1,230 \text{ joules/g liquefied}$$

EXAMPLE 8-2. For the ideal two-expander Collins helium-liquefaction system as depicted in Fig. 8-4, show that the liquid yield per unit mass of gas compressed can be expressed by the equation

$$\frac{\dot{m}_f}{\dot{m}} = \frac{1}{h_1 - h_{11}} \left[(h_1 - h_2) + \frac{\dot{m}_{e1}}{\dot{m}} (h_3 - h_4) + \frac{\dot{m}_{e2}}{\dot{m}} (h_6 - h_7) \right] \tag{8-1}$$

wherein $\dot{m}$, $\dot{m}_{e1}$, and $\dot{m}_{e2}$ denote the mass-flow rates through the compressor, expander 1, and expander 2, respectively; and $\dot{m}_f$ denotes the liquid discharge rate from the liquid receiver.

For such a system, let $p_1 = 1$ atm, $p_2 = 15$ atm, $T_1 = 300°$K, $T_3 = 60°$K, $T_6 = 15°$K, $\dot{m}_{e1}/\dot{m} = 0.25$, and $\dot{m}_{e2}/\dot{m} = 0.50$. Calculate the liquid helium yielded per unit mass of gaseous helium compressed.

Solution.
Applying the first law for steady flow to the system as a whole except the helium compressor, we have

$$\dot{W}_{e1} + \dot{W}_{e2} + \dot{m}_f h_{11} + (\dot{m} - \dot{m}_f) h_1 - \dot{m} h_2 = 0$$

But

$$\dot{W}_{e1} = \dot{m}_{e1} (h_3 - h_4) \qquad \text{and} \qquad \dot{W}_{e2} = \dot{m}_{e2} (h_6 - h_7)$$

Whence

$$\dot{m}_{e1} (h_3 - h_4) + \dot{m}_{e2} (h_6 - h_7) + \dot{m}_f (h_{11} - h_1) + \dot{m} (h_1 - h_2) = 0$$

Dividing the preceding equation by $\dot{m}$ and rearranging we obtain Eq. (8-1).

Now from the data on helium in the appendix we obtain the following values:

$$h_1 = 1,574 \text{ joules/g} \quad \text{at} \quad \begin{array}{l} p_1 = 1 \text{ atm} \\ T_1 = 300°\text{K} \end{array}$$

$$h_2 = 1,579 \text{ joules/g} \quad \text{at} \quad \begin{array}{l} p_2 = 15 \text{ atm} \\ T_2 = 300°\text{K} \end{array}$$

$$h_3 = 328.5 \text{ joules/g} \quad s_3 = 17.41 \text{ joules/(g)(}°\text{K)} \quad \text{at} \quad \begin{array}{l} p_3 = 15 \text{ atm} \\ T_3 = 60°\text{K} \end{array}$$

$$h_6 = 80.5 \text{ joules/g} \quad s_6 = 9.62 \text{ joules/(g)(}°\text{K)} \quad \text{at} \quad \begin{array}{l} p_6 = 15 \text{ atm} \\ T_6 = 15°\text{K} \end{array}$$

$$h_4 = 119.6 \text{ joules/g} \quad \text{at} \quad \begin{array}{l} p_4 = 1 \text{ atm} \\ s_4 = s_3 = 17.41 \text{ joules/(g)(}°\text{K)} \end{array}$$

$$h_7 = 36.2 \text{ joules/g} \quad \text{at} \quad \begin{array}{l} p_7 = 1 \text{ atm} \\ \\ s_7 = s_6 = 9.62 \text{ joules/(g)($^\circ$K)} \end{array}$$

$$h_{11} = 9.76 \text{ joules/g} \quad \text{at} \quad \begin{array}{l} p_{11} = \text{ atm} \\ \\ \text{saturated liquid} \end{array}$$

Substituting numerical values in Eq. (8-1) leads to

$$\frac{\dot{m}_f}{\dot{m}} = \frac{1}{1574 - 9.76} \left[(1574 - 1579) + 0.25 \, (328.5 - 119.6) \right.$$

$$\left. + 0.50 \, (80.5 - 36.2) \right] = 0.044 \text{ g/g compressed}$$

This is the liquid yield per unit mass of gas compressed.

8-3 PHASE BEHAVIOR OF HELIUM ISOTOPES

Helium is the only known substance that remains in the liquid phase down to the absolute zero of temperature. This and other unique properties have led to extensive experimental and theoretical research on helium. We will examine the phase behavior of helium in this section and study its properties in greater detail in the next chapter.

Helium has two stable isotopes, namely *helium four* (He^4) and *helium three* (He^3). The helium four atom has two protons and two neutrons in the nucleus, with an atomic mass of 4.0026; the helium three atom has two protons and only one neutron in the nucleus, with an atomic mass of 3.0160. The heavier He^4 is the predominant isotope. Ordinary helium contains He^3 and He^4 in the ratio of

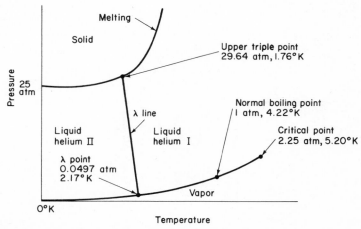

FIG. 8-5. Pressure-temperature phase diagram for He^4.

about one part to a million. The phase behaviors of the two helium isotopes are quite different and require separate studies.

The p-T phase diagram for He^4 is shown in Fig. 8-5. This phase diagram differs in form from that of any other known substance in many respects. First of all, as mentioned previously, helium remains as liquid right down to the absolute zero of temperature. Thus, liquid He^4 does not freeze under its own vapor pressure even if the temperature is reduced to absolute zero, and there is no triple point in which all three states of aggregation coexist. It can be solidified only through the use of pressure as well as low temperature. Even down to absolute zero temperature, solid He^4 cannot exist at pressures lower than 25 atm, which is considerably higher than its critical pressure of 2.25 atm. Thus no sublimation can ever occur.

Another unique feature in the phase behavior of He^4 is the existence of two different liquid phases: *liquid helium I* and *liquid helium II*. These two liquid phases have markedly different properties. The transition curve separating the two liquid phases is called the λ *line*. The point at which the λ line intersects the vapor-pressure curve is called the λ *point*; and the point at which the λ line

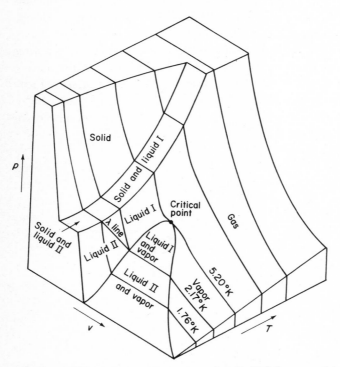

FIG. 8-6. Pressure-volume-temperature surface for He^4 (not to scale).

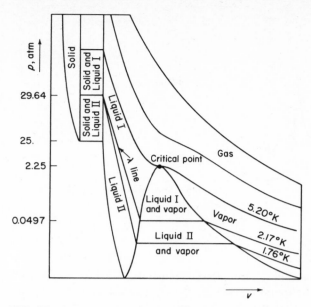

FIG. 8-7. Pressure-volume diagram for He⁴ (not to scale).

intersects the melting curve is called the *upper triple point*. These two points are the two triple points in which the two liquid phases coexist with vapor and solid respectively.

Figure 8-6 shows the three-dimensional p-v-T surface, and Fig. 8-7 shows the p-v diagram for He⁴. Neither diagram is drawn to scale.

The transition from liquid helium I to liquid helium II at the upper triple point and along the λ line appears to be a normal first-order phase transition with finite latent heat. However, at the λ point the transition between the two liquid phases is of higher order. The volume and entropy show no discontinuities, whereas the specific heat shows a huge anomaly at the λ point, as illustrated in Fig. 8-8. As a result of these observations the phase change at λ point appears to be a λ transition. As a matter of fact, the designation of λ point and λ line, just as the designation of λ transition, were originated from the resemblance of the c_p vs. T curve to the Greek letter λ.

The p-T phase diagram for He³ is shown in Fig. 8-9. It can be seen that He³ has only one liquid phase extending right down to the absolute zero of temperature, and there is no triple point of any kind. Although a superfluid phase at extremely low temperatures has been predicted from theoretical considerations, this new phase has not yet been experimentally observed.

When liquid helium is forced to evaporate by pumping its vapor away rapidly, temperatures lower than 1°K can be reached. Since the normal boiling

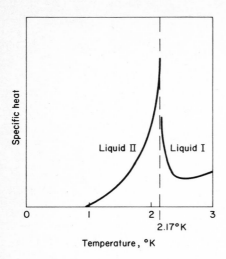

FIG. 8-8. Specific heat of liquid He^4.

point of He^3 is lower than that of He^4, lower temperatures can be reached by evaporating He^3 than by evaporating He^4 for the same pressure drop beginning from the atmospheric condition. For instance, when the vapor pressure of He^4 is lowered to 0.12 mm Hg, the temperature reached is 1°K; however, when the vapor pressure of He^3 is lowered to about the same value (0.14 mm Hg to be precise), the temperature reached is 0.5°K. To reach a temperature of 0.5°K with He^4 would require a vapor pressure as low as 1.6×10^{-5} mm Hg. As we will explain later, due to the creeping effect of liquid helium II, 0.7°K is about the lowest temperature which can be reached by evaporating He^4. With He^3 it is possible to reach about 0.3°K by rapid evaporation using a high vacuum pump.

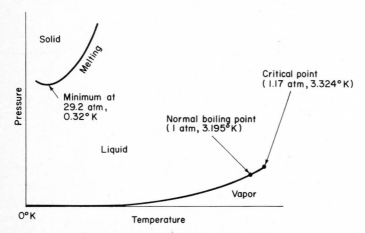

FIG. 8-9. Pressure-temperature phase diagram for He^3.

8-4 PARAMAGNETIC SOLIDS

Of the four methods of producing cryogenic temperatures mentioned at the beginning of this chapter, adiabatic demagnetization of paramagnetic solids is the ultimate method that shall enable us to achieve very, very low temperatures approaching the absolute zero. Before discussing this method, a study of the properties of the working material—paramagnetic solids—is in order.

As described in Sec. 7-1, a paramagnetic material is not a magnet if no external magnetic field is applied to it. In the presence of an external field it becomes slightly magnetized. The processes of magnetization and demagnetization of a paramagnetic material are reversible, and the state of the material can be described in terms of a few thermodynamic variables.

Most experiments on paramagnetic solids are performed at constant atmospheric pressure with negligible volume changes in the solids. Thus, magnetization would be the only work mode and the system may be considered as a simple paramagnetic system. In Sec. 7-2, we have derived the essential equations, including the first and second laws, the Maxwell relations, and specific heat relations, for simple magnetic systems. These equations can be directly applied to a paramagnetic solid. One additional bit of information we need now is some appropriate form of equation of state for a paramagnetic solid.

One simple form of equation of state for a paramagnetic solid was established from the experimental fact that the magnetic moment of a paramagnetic solid is directly proportional to the external magnetic field H and inversely proportional to the temperature T. For small values of the ratio H/T, this is expressed in the simple equation, known as *Curie's equation*:

$$I = V\mathbf{M} = C_c \frac{\mathbf{H}}{T} \tag{8-2}$$

where I is the total magnetic moment, $\mathbf{M}$ is the magnetization or magnetic moment per unit volume, V is the volume of the specimen, and C_c is a proportional constant, called the *Curie constant*. The Curie constants for a number of paramagnetic salts are given in Table 8-3. Curie's equation may also be rewritten in an alternate form,

$$\chi T = C_c \tag{8-3}$$

where $\chi = I/H$, the paramagnetic susceptibility, is a function of temperature.

Curie's equation predicts that the magnetization of a paramagnetic solid would increase indefinitely with increasing external field and decreasing temperature. This is obviously not true, because when all atomic magnets in the material are substantially aligned with the applied field, the magnetization would

TABLE 8-3. Data on Paramagnetic Salts*

Paramagnetic salt	$m^\dagger$ g/g ion	J	g_L	$C_c^\ddagger$ (cm³)(°K)/g ion	A_c (joule)(°K)/g ion
$Cr_2(SO_4)_3 \cdot K_2SO_4 \cdot 24H_2O$ (chromium potassium alum)	499	$\frac{3}{2}$	2	1.84	0.150
$Fe_2(SO_4)_3 \cdot (NH_4)_2SO_4 \cdot 24H_2O$ (iron ammonium alum)	482	$\frac{5}{2}$	2	4.39	0.108
$Gd_2(SO_4)_3 \cdot 8H_2O$ (gadolinium sulfate)	373	$\frac{7}{2}$	2	7.80	2.91
$2Ce(NO_3)_3 \cdot 3Mg(NO_3)_2 \cdot 24H_2O$ (cerium magnesium nitrate)	765	$\frac{5}{2}$	1.84	0.317	5.07×10^{-5}

*M. W. Zemansky, "Heat and Thermodynamics," 5th ed., pp. 445 and 455, McGraw-Hill Book Company, New York, 1968.

†Note that m is the gram ionic mass of crystal. A mass of crystal containing exactly N_A (Avogadro's number) magnetic ions is known as 1 gram ion.

‡To obtain the value of C_c in rationalized mks system of units, the tabulated value is multiplied by $4\pi \times 10^{-3}$, and the unit of C_c then becomes (m³)(°K)/kg ion.

become saturated. As a matter of fact, Curie's equation breaks down long before a state of saturation occurs. The breakdown temperature for many paramagnetic salts is usually below 1°K.

A more general equation of state for paramagnetic salts was developed by L. Brillouin based on quantum-mechanical concepts. This equation, known as *Brillouin's equation*, reads:

$$I = VM = Ng_L\mu_B JB_J(\varsigma) \tag{8-4}$$

where

I = total magnetic moment

M = magnetization or magnetic moment per unit volume

V = volume of the specimen

N = number of paramagnetic ions

g_L = Landé splitting factor, which is an atomic constant (see Table 8-3 for its values)

J = total angular momentum quantum number, which is another atomic constant (see Table 8-3 for its values)

μ_B = Bohr magneton = 9.27×10^{-21} erg/oersted

and

$$B_J(\varsigma) = \frac{1}{J}\left[\left(J + \frac{1}{2}\right)\coth\left(J + \frac{1}{2}\right)\varsigma - \frac{1}{2}\coth\frac{1}{2}\varsigma\right]$$

in which

$$\zeta = \frac{g_L \mu_B H}{k_B T}$$

and k_B = Boltzmann constant = 1.38×10^{-16} erg/°K. The function $B_J(\zeta)$ is called the *Brillouin function*, a plot of which is given in Fig. 8-10.

Brillouin's equation was derived on the condition that paramagnetic ions in crystalline paramagnetic salts are so dilute that they interact with one another only very weakly, just like molecules of an ideal gas do. This condition is well satisfied, because in the paramagnetic crystals commonly used in low-temperature work, the paramagnetic ions, such as chromium, iron, cerium, and gadolinium ions, are surrounded by a very large number of nonmagnetic particles. For example, each iron ion in $Fe_2(SO_4)_3 \cdot (NH_4)_2 \, SO_4 \cdot 24H_2O$ is surrounded by 1 nitrogen atom, 2 sulfur atoms, 20 oxygen atoms, and 28 hydrogen atoms—a total of 51 nonmagnetic particles.

For small values of the ratio H/T, Brillouin's equation reduces to Curie's equation, with the Curie constant expressed by the following equation:

$$C_c = \frac{N g_L^2 \mu_B^2 J(J+1)}{3 k_B} \tag{8-5}$$

As a consequence of the H-M-T relation in the forms of Brillouin's equation and Curie's equation, there are some important relations involving the specific

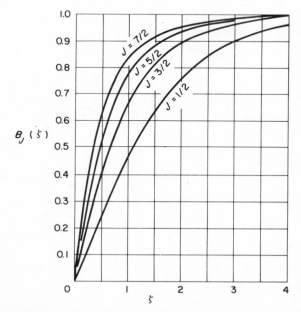

FIG. 8-10. The Brillouin function.

heats. From Eq. (7-8) we can write

$$\left(\frac{\partial U}{\partial \mathbf{H}}\right)_T = T\left(\frac{\partial S}{\partial \mathbf{H}}\right)_T + \mu_0 V \mathbf{H}\left(\frac{\partial \mathbf{M}}{\partial \mathbf{H}}\right)_T$$

Using the Maxwell equation (7-14d), this equation becomes

$$\left(\frac{\partial U}{\partial \mathbf{H}}\right)_T = \mu_0 V T\left(\frac{\partial \mathbf{M}}{\partial T}\right)_\mathbf{H} + \mu_0 V \mathbf{H}\left(\frac{\partial \mathbf{M}}{\partial \mathbf{H}}\right)_T$$

Now since Brillouin's equation is in the form $\mathbf{M} = f(\mathbf{H}/T)$, we can write

$$\left(\frac{\partial \mathbf{M}}{\partial T}\right)_\mathbf{H} = -\frac{\mathbf{H}}{T^2} f'\left(\frac{\mathbf{H}}{T}\right) \quad \text{and} \quad \left(\frac{\partial \mathbf{M}}{\partial \mathbf{H}}\right)_T = \frac{1}{T} f'\left(\frac{\mathbf{H}}{T}\right) \tag{8-6}$$

where $f'(\mathbf{H}/T)$ denotes the first derivative of $f(\mathbf{H}/T)$. Consequently,

$$\left(\frac{\partial U}{\partial \mathbf{H}}\right)_T = \mu_0 V T\left[-\frac{\mathbf{H}}{T^2} f'\left(\frac{\mathbf{H}}{T}\right)\right] + \mu_0 V \mathbf{H}\left[\frac{1}{T} f'\left(\frac{\mathbf{H}}{T}\right)\right] = 0$$

Thus, we conclude that internal energy is a function of temperature only for a paramagnetic salt obeying Brillouin's equation (and, of course, Curie's equation). Furthermore, according to Eq. (7-15),

$$C_\mathbf{M} = \left(\frac{\partial U}{\partial T}\right)_\mathbf{M}$$

We conclude also that $C_\mathbf{M}$ is a function of temperature only.

If we substitute Eq. (8-6) into Eq. (7-17), we obtain

$$C_\mathbf{H} - C_\mathbf{M} = -\mu_0 V \mathbf{H}\left(\frac{\partial \mathbf{M}}{\partial T}\right)_\mathbf{H} \tag{8-7}$$

This equation is valid for a paramagnetic solid obeying Brillouin's equation. In addition, if Curie's equation is also obeyed, then since

$$V\left(\frac{\partial \mathbf{M}}{\partial T}\right)_\mathbf{H} = -C_c \frac{\mathbf{H}}{T^2}$$

we have

$$C_\mathbf{H} - C_\mathbf{M} = \mu_0 C_c \frac{\mathbf{H}^2}{T^2} \tag{8-8}$$

For paramagnetic salts which conform to Curie's equation there is a useful specific heat-vs.-temperature relation of the form

$$C_M = \frac{A_c}{T^2} \tag{8-9}$$

where A_c is a constant, the values of which are listed in Table 8-3. Substituting Eq. (8-9) into Eq. (8-8) leads to

$$C_H = \frac{A_c}{T^2} + \mu_0 C_c \frac{H^2}{T^2} \tag{8-10}$$

EXAMPLE 8-3. Determine the Curie constant of gadolinium sulfate for which the atomic constants are $J = 7/2$ and $g_L = 2$.

Solution.
For a gram ion of the salt crystal, the number of paramagnetic ions is $N = N_A = 6.02 \times 10^{23}$. Substituting numerical values in Eq. (8-5) and noting that the unit oersted^2 has the same dimensions as erg/cm^3, we obtain

$$C_c = \frac{N\mu_B^2}{3k_B} \, g_L^2 \, J(J + 1)$$

$$= \frac{(6.02 \times 10^{23} \, 1/\text{g ion})(9.27^2 \times 10^{-42} \, \text{erg}^2/\text{oersted}^2)}{3(1.38 \times 10^{-16} \text{erg}/^\circ\text{K})} \times 2^2 \times \frac{7}{2} \times \frac{9}{2}$$

$$= 7.88 \, \frac{(\text{cm}^3)(^\circ\text{K})}{\text{g ion}}$$

This calculated value agrees very well with the measured value of 7.80 $(\text{cm}^3)(^\circ\text{K})/\text{g ion}$ as given in Table 8-3.

EXAMPLE 8-4. Determine the magnetic moment per gram of gadolinium sulfate at a temperature of 0.05°K and in a magnetic field of 1,000 oersteds, assuming that the Brillouin's equation is valid at these conditions. What error would be introduced if Curie's equation is used? The atomic constants for the salt are $J = 7/2$ and $g_L = 2$, and the gram ionic mass is 373.

Solution.
We first calculate the parameter ζ in the Brillouin's equation. Thus

$$\zeta = \frac{g_L \mu_B H}{k_B T} = \frac{2(9.27 \times 10^{-21} \, \text{erg/oersted})(1,000 \, \text{oersteds})}{(1.38 \times 10^{-16} \text{erg}/^\circ\text{K})(0.05^\circ\text{K})} = 2.69$$

The function $B_J(\zeta)$ can be obtained directly from Fig. 8-10, or calculated from the equation

$$B_J(\zeta) = \frac{1}{J}\left[\left(J + \frac{1}{2}\right)\coth\left(J + \frac{1}{2}\right)\zeta - \frac{1}{2}\coth\frac{1}{2}\zeta\right]$$

Thus,

$$B_J(2.69) = \frac{1}{3.5} \left[4 \coth (4 \times 2.69) - 0.5 \coth (0.5 \times 2.69) \right] = 0.979$$

The number of paramagnetic ions per gram of the salt is

$$N = \frac{N_A}{m} = \frac{6.02 \times 10^{23} \, \text{g ion}^{-1}}{373 \, \text{g/g ion}} = 1.61 \times 10^{21} \, \text{g}^{-1}$$

Substituting numerical values in Brillouin's equation and noting that the unit oersted^2 has the same dimensions as erg/cm^3, we obtain

Magnetic moment per gram

$$= N \mu_B g_L J B_J(\zeta)$$
$$= (1.61 \times 10^{21} \, \text{g}^{-1})(9.27 \times 10^{-21} \, \text{ergs/oersted})(2)(3.5)(0.979)$$
$$= 102 (\text{oersteds})(\text{cm}^3)/\text{g}$$

From the previous example we have

$$C_c = 7.88 \, (\text{cm}^3)(^\circ \text{K})/\text{g ion}$$

Then, if Curie's equation is used we obtain

$$\text{Magnetic moment per gram} = C_c \frac{\text{H}}{T}$$
$$= \frac{7.88 \, (\text{cm}^3)(^\circ \text{K})/\text{g ion}}{373 \, \text{g/g ion}} \cdot \frac{1000 \, \text{oersteds}}{0.05^\circ \text{K}} = 423 \, (\text{oersteds})(\text{cm}^3)/\text{g}$$

The error introduced by the use of Curie's equation is

$$\frac{423 - 102}{102} = 315\%$$

8-5 IONIC DEMAGNETIZATION

The method of *adiabatic ionic demagnetization*, suggested independently by Giauque[1] and by Debye[2] in 1926 and first performed experimentally by Giauque and MacDougall in 1933, is based on the fact that when a thermally isolated paramagnetic material is demagnetized its temperature drops. This fact can be verified analytically from the second dS equation for a simple paramagnetic system,

$$dS = \frac{C_H}{T} \, dT + \mu_0 V \left(\frac{\partial M}{\partial T} \right)_H dH \tag{7-19}$$

[1] W. F. Giauque, *J. Am. Chem. Soc.*, 49:1864, 1870 (1927).
[2] P. Debye, *Ann. Physik*, 81:1154 (1926).

Since at constant magnetic field an increase in temperature of a paramagnetic solid always causes a decrease in its magnetization, the derivative $(\partial M/\partial T)_H$ is always negative. It follows from Eq. (7-19) that for a reversible adiabatic demagnetization process, we have

$$\frac{dT}{T} = -\frac{\mu_0 V}{C_H}\left(\frac{\partial M}{\partial T}\right)_H dH \qquad (8\text{-}11)$$

indicating that a negative dH would give a negative dT, thus a drop in temperature. This ideal process is depicted by the vertical line 23 in the T-S diagram of Fig. 8-11.

The paramagnetic material must be magnetized before the demagnetization process. An ideal isothermal magnetization process is depicted by the horizontal line 12 in Fig. 8-11. According to Eq. (7-19), for this process we have

$$T\,dS = \mu_0 VT\left(\frac{\partial M}{\partial T}\right)_H dH \qquad (8\text{-}12)$$

Since $(\partial M/\partial T)_H$ is always negative, an increase in magnetic field results in an outflow of heat and a corresponding decrease in entropy. Since entropy is physically an index of disorder, it is apparent that an unmagnetized state in a paramagnetic material necessarily has a greater entropy or disorder than a magnetized state at the same temperature. The line of zero magnetic field must lie on the right of the line representing constant magnetic field H on the T-S diagram. This confirms that heat must be rejected from the material while being isothermally magnetized.

In order to carry out an experiment of magnetic cooling a sample of paramagnetic salt crystals is cooled first to about $1°K$. This is done by mounting the

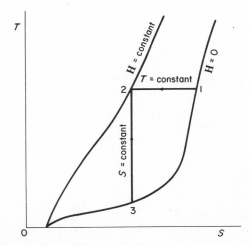

FIG. 8-11. Temperature-entropy diagram of a paramagnetic salt showing reversible processes of isothermal magnetization and adiabatic demagnetization.

salt sample inside a chamber which is completely immersed in a bath of liquid helium that boils at reduced pressure, as shown diagrammatically in Fig. 8-12. Surrounding the liquid helium is liquid nitrogen, also boiling at reduced pressure. The intervening space between the two liquids is evacuated. Inside the sample chamber is some helium gas, called exchange gas, serving as a heat-exchanging agent. When the salt sample is cooled to a temperature as low as possible, a strong magnetic field is then applied either by a conventional electromagnet or a superconducting magnet. The heat evolved due to magnetization is transported by the helium exchange gas to the surrounding liquid helium, causing some of the liquid to evaporate. The temperature of the salt sample is thus maintained essentially constant during magnetization. At the end of the magnetization process, the helium exchange gas in the sample chamber is pumped out, thereby breaking the thermal connection between the salt and its surroundings. The magnetic field is then reduced to zero and the temperature of the salt falls under adiabatic conditions. In such a cooling process, temperatures down to a few thousandths of a degree above absolute zero can be achieved without major difficulties.

EXAMPLE 8-5. A gadolinium sulfate crystal at $1.5°K$ is magnetized reversibly and isothermally from zero field to 1×10^6 amp/meter. For two g ions

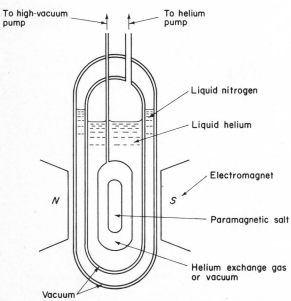

FIG. 8-12. Cryostat for ionic demagnetization (schematic).

of the salt, what are the heat, work, and change in entropy for the process of magnetization? Assume that Curie's equation is obeyed. Compute the final temperature of the salt when it is then demagnetized reversibly and adiabatically.

Solution.

From Eq. (7-19) we have

$$dS = \frac{C_H}{T} dT + \mu_0 \left(\frac{\partial I}{\partial T}\right)_H dH$$

But according to Curie's equation

$$I = C_c \frac{H}{T}$$

whence

$$\left(\frac{\partial I}{\partial T}\right)_H = -C_c \frac{H}{T^2}$$

Thus

$$dS = \frac{C_H}{T} dT - \mu_0 C_c \frac{H}{T^2} dH \qquad (8\text{-}13)$$

For the process of isothermal magnetization, the preceding equation reduces to

$$dS = -\mu_0 C_c \frac{H}{T^2} dH$$

which yields on integration (See Fig. 8-11).

$$\Delta S_{12} = S_2 - S_1 = -\mu_0 C_c \frac{H_2^2 - H_1^2}{2T_1^2} \qquad (8\text{-}14)$$

Now in SI units we have

$$\mu_0 = 4\pi \times 10^{-7} \text{ newton/amp}^2$$

$$C_c = \left(4\pi \times 7.80 \times 10^{-6} \frac{(m^3)(^\circ K)}{g \text{ ion}}\right) (2 \text{ g ions})$$

$$= 4\pi \times 15.6 \times 10^{-6} (m^3)(^\circ K)$$

$$H_2 = 1 \times 10^6 \text{ amp/m}$$

$$H_1 = 0$$

and

$$T_1 = 1.5^\circ K$$

Upon substituting into Eq. (8-14), we obtain

$$\Delta S_{12} = S_2 - S_1$$
$$= - \frac{(4\pi \times 10^{-7})(4\pi \times 15.6 \times 10^{-6})}{2 \times 1.5^2} [(1 \times 10^6)^2 - 0]$$
$$= - 54.7 \text{ (newton)(m)}/^\circ K = - 54.7 \text{ joules}/^\circ K$$

Whence

$$Q_{12} = T_1 (\Delta S_{12}) = 1.5 (- 54.7) = - 82.1 \text{ joules}$$

The minus sign indicates that heat is removed from the salt during the process of isothermal magnetization.

Now, since the internal energy is a function of temperature only for a paramagnetic salt obeying Curie's equation, we have $\Delta U_{12} = 0$. Therefore, from the first law we obtain

$$W_{12} = Q_{12} = -82.1 \text{ joules}$$

Here the minus sign indicates that work is done on the salt during the process of isothermal magnetization.

For the process of reversible adiabatic demagnetization, Eq. (8-13) reduces to

$$C_H \frac{dT}{T} - \mu_0 C_c \frac{H}{T^2} dH = 0 \qquad\qquad\qquad\qquad (8-15)$$

But from Eq. (8-10),

$$C_H = \frac{1}{T^2} (A_c + \mu_0 C_c H^2)$$

Thus we have

$$(A_c + \mu_0 C_c H^2) \frac{dT}{T} = \mu_0 C_c H \, dH \quad \text{or} \quad \frac{dT}{T} = \frac{\mu_0 C_c H \, dH}{A_c + \mu_0 C_c H^2}$$

Integrating between states 2 and 3 yields

$$\ln \frac{T_3}{T_2} = \frac{1}{2} \ln \frac{A_c + \mu_0 C_c H_3^2}{A_c + \mu_0 C_c H_2^2}$$

Thus

$$\frac{T_3}{T_2} = \left(\frac{A_c + \mu_0 C_c H_3^2}{A_c + \mu_0 C_c H_2^2} \right)^{1/2}$$

From Table 8-3 we have $A_c = 2.91$ (joule)($^\circ$K)/g ion for gadolinium sulfate. Therefore,

$$\frac{T_3}{T_2} = \left[\frac{2.91}{2.91 + (4\pi \times 10^{-7})(4\pi \times 7.80 \times 10^{-6})(1 \times 10^6)^2}\right]^{1/2} = 0.152$$

Hence

$$T_3 = 0.152\,T_2 = 0.152 \times 1.5 = 0.23°\text{K}$$

8-6 MAGNETIC REFRIGERATOR

In a low-temperature experiment using magnetic cooling, observations on the behavior of the paramagnetic salt or any substance cooled down by it commence immediately after the demagnetization process. However, because of the un-avoidable influx of heat into the salt either from the surroundings or due to energy dissipation involved in the experiment, the duration of time in which observations can be made at the very low temperatures is certainly limited. To remedy the situation, Daunt and Heer[3] have developed a *magnetic refrigerator* operating on a cycle to maintain temperatures well below 1°K. A commercial model of the machine has been constructed by A. D. Little Co. of Massachusetts.

The magnetic refrigerator consists essentially of a paramagnetic working salt, a thermal reservoir, and two superconducting thermal valves, as shown dia-grammatically in Fig. 8-13. These components are mounted in a vacuum chamber which is immersed in a liquid helium bath maintained at about 1°K. The upper thermal valve connects the working salt with the helium bath, while the lower thermal valve connects the working salt with the thermal reservoir. The thermal reservoir and the experimental space near it are maintained con-tinuously at the desired temperature below 1°K. Surrounding the liquid helium bath are the control magnets for the working salt and the two thermal valves.

The thermal valves are made of lead ribbons and work on the principles of superconductivity, which we will study in the next chapter. For the present, it will suffice to note that lead at a temperature below 7.22°K in the absence of an external magnetic field is a superconductor of electricity and a very poor con-ductor of heat. However, the superconductivity of lead can be destroyed by the application of a magnetic field, thus changing it back to a normal conductor of electricity and a good conductor of heat.

With the preceding brief mention on the principle of a thermal valve, we are now ready to present the sequence of operations for a magnetic refrigerator. The idealized processes of a magnetic refrigerator in the form of a Carnot cycle are shown in Fig. 8-14. The processes are as follows:

[3] C. V. Heer, C. B. Barnes, and J. G. Daunt, *Rev. Sci. Instr.*, 25:1088 (1954).

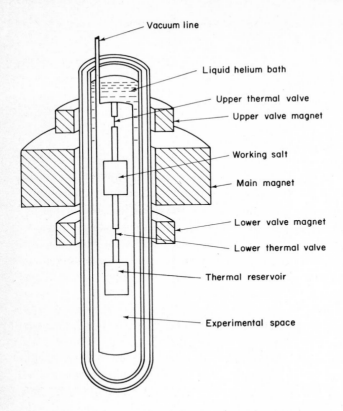

FIG. 8-13. Magnetic refrigerator of Daunt and Heer (schematic).

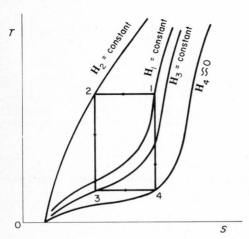

FIG. 8-14. Ideal thermodynamic cycle for the paramagnetic working salt in a magnetic refrigerator.

Process 12—The magnetic field around the working salt as applied by the main magnet is increased from H_1 to H_2. During this process the upper thermal valve is kept open by applying magnetic field from the upper valve magnet. This means that the upper thermal valve is in a state of normal conductivity and is a good heat conductor, capable of transferring heat from the working salt to the liquid helium bath. During this process, however, the lower thermal valve is kept closed, meaning that it is in a state of superconductivity and therefore is a poor heat conductor, incapable of transferring heat from the working salt to the thermal reservoir.

Process 23—Both thermal valves are kept closed, meaning that both are poor heat conductors. The magnetic field around the working salt is reduced from H_2 to H_3, causing the temperature of the working salt to fall adiabatically.

Process 34—The magnetic field around the working salt is reduced to zero (or near zero), while the lower thermal valve is kept open but the upper thermal valve is kept closed. Being in a state of normal conductivity, the lower thermal valve is capable of transferring heat from the thermal reservoir to the working salt, thus cooling the thermal reservoir and the experimental space to a temperature well below $1°K$ (as low as $0.2°K$ can be maintained continuously).

Process 41—Both thermal valves are kept closed. The magnetic field around the working salt is increased to its initial value, causing the temperature of the working salt to increase adiabatically to its initial value. This completes a cycle.

EXAMPLE 8-6.   Iron ammonium alum is used as the working substance in a Carnot magnetic refrigerating cycle operating between two reservoirs at $1°K$ and $0.2°K$. The maximum and minimum magnetic fields are 0.6×10^6 amp/meter and 0. Referring to Fig. 8-14, compute the magnetic field and magnetic moment per g ion at each of the four end states, assuming that Curie's equation is valid. For one g ion of the salt, find the heat and work interactions for each of the four processes. Also find the coefficient of performance for the refrigerator.

Solution.
Referring to Fig. 8-14, the given conditions are

$$T_1 = T_2 = 1°K \qquad T_3 = T_4 = 0.2°K$$

$$H_2 = 0.6 \times 10^6 \, \text{amp/m} \qquad H_4 = 0$$

In SI units, for iron ammonium alum we have

$$C_c = 4\pi \times 4.39 \times 10^{-6} (\text{m}^3)(°K)/\text{g ion}$$

$$A_c = 0.108 \, (\text{joule})(°K)/\text{g ion}$$

$$\mu_0 = 4\pi \times 10^{-7} \, \text{newton/amp}^2$$

According to Curie's equation

$$I = C_c \frac{H}{T}$$

we obtain

$$I_4 = C_c \frac{H_4}{T_4} = 0$$

$$I_2 = C_c \frac{H_2}{T_2} = \frac{(4\pi \times 4.39 \times 10^{-6})(0.6 \times 10^6)}{1.0} = 33.1 \text{ (amp)}(m^2)/\text{g ion}$$

From Eq. (7-18) we have

$$dS = \frac{C_M}{T} dT - \mu_0 \left(\frac{\partial H}{\partial T}\right)_I dI$$

But according to Curie's equation we have

$$\left(\frac{\partial H}{\partial T}\right)_I = \frac{I}{C_c}$$

Thus

$$dS = \frac{C_M}{T} dT - \frac{\mu_0}{C_c} I \, dI$$

Furthermore, from Eq. (8-9),

$$C_M = \frac{A_c}{T^2}$$

it follows that

$$dS = \frac{A_c}{T^3} dT - \frac{\mu_0}{C_c} I \, dI \tag{8-16}$$

Applying the preceding equation to the reversible adiabatic process 23, we have

$$A_c \int_{T_2}^{T_3} \frac{dT}{T^3} = \frac{\mu_0}{C_c} \int_{I_2}^{I_3} I \, dI \quad \text{or} \quad -\frac{A_c}{2}\left(\frac{1}{T_3^2} - \frac{1}{T_2^2}\right) = \frac{\mu_0}{2C_c}(I_3^2 - I_2^2)$$

Hence

$$I_3 = \left[I_2^2 - \frac{A_c C_c}{\mu_0}\left(\frac{1}{T_3^2} - \frac{1}{T_2^2}\right)\right]^{1/2}$$

$$= \left[(33.1)^2 - \frac{(0.108)(4\pi \times 4.39 \times 10^{-6})}{(4\pi \times 10^{-7})}\left(\frac{1}{0.2^2} - \frac{1}{1.0^2}\right)\right]^{1/2}$$

$$= 31.3 \text{ (amp)}(m^2)/\text{g ion}$$

Similarly,

$$
\begin{aligned}
\mathbf{I}_1 &= \left[\mathbf{I}_4^2 - \frac{A_c C_c}{\mu_0} \left(\frac{1}{T_1^2} - \frac{1}{T_4^2} \right) \right]^{1/2} \\
&= \left[0 - \frac{(0.108)(4\pi \times 4.39 \times 10^{-6})}{(4\pi \times 10^{-7})} \left(\frac{1}{1.0^2} - \frac{1}{0.2^2} \right) \right]^{1/2}
\end{aligned}
$$

$$
= 10.7 \ (\text{amp})(\text{m}^2)/\text{g ion}
$$

Substituting the values of $\mathbf{I}_1$ and $\mathbf{I}_3$ in Curie's equation, we obtain

$$
\mathbf{H}_1 = \frac{T_1 \mathbf{I}_1}{C_c} = \frac{(1.0)(10.7)}{4\pi \times 4.39 \times 10^{-6}} = 0.194 \times 10^6 \ \text{amp/m}
$$

$$
\mathbf{H}_3 = \frac{T_3 \mathbf{I}_3}{C_c} = \frac{(0.2)(31.3)}{4\pi \times 4.39 \times 10^{-6}} = 0.113 \times 10^6 \ \text{amp/m}
$$

The values of the magnetic field and magnetic moment at each of the four end points are tabulated as follows:

Points	1	2	3	4
H, amp/m	0.194×10^6	0.6×10^6	0.113×10^6	0
I, (amp)(m²)/g ion	10.7	33.1	31.3	0

We now calculate the amounts of heat transfer for each of the four processes. Since processes 23 and 41 are reversible and adiabatic, we have $Q_{23} = 0$ and $Q_{41} = 0$. For the reversible isothermal processes 12 and 34, Eq. (8-16) becomes

$$
\mathit{d}Q = T \, dS = - \frac{\mu_0}{C_c} T \mathbf{I} \, d\mathbf{I}
$$

Whence,

$$
Q_{12} = - \frac{\mu_0}{C_c} T_1 \int_{\mathbf{I}_1}^{\mathbf{I}_2} \mathbf{I} \, d\mathbf{I} = - \frac{\mu_0 T_1}{2 C_c} (\mathbf{I}_2^2 - \mathbf{I}_1^2)
$$

$$
= - \frac{(4\pi \times 10^{-7})(1.0)}{2(4\pi \times 4.39 \times 10^{-6})} (33.1^2 - 10.7^2) = - 11.2 \ \text{joules/g ion}
$$

and

$$
Q_{34} = - \frac{\mu_0 T_3}{2 C_c} (\mathbf{I}_4^2 - \mathbf{I}_3^2) = - \frac{(4\pi \times 10^{-7})(0.2)}{2(4\pi \times 4.39 \times 10^{-6})} (0 - 31.3^2)
$$

$$
= 2.23 \ \text{joules/g ion}
$$

For the sake of evaluating the work interactions, we first note that the internal energy of a paramagnetic salt which follows the Curie's equation is a function of temperature only. Thus from the first law we conclude that

$$W_{12} = Q_{12} = -11.2 \text{ joules/g ion}$$

and

$$W_{34} = Q_{34} = 2.23 \text{ joules/g ion}$$

For an adiabatic process the first law gives

$$đQ = dU + đW = 0 \quad \text{or} \quad đW = -dU$$

But from Eqs. (7-15) and (8-9), for a paramagnetic salt which obeys Curie's equation we have

$$dU = C_M dT = \frac{A_c}{T^2} dT$$

Whence

$$đW = -\frac{A_c}{T^2} dT$$

Therefore

$$W_{23} = -A_c \int_{T_2}^{T_3} \frac{dT}{T^2}$$

$$= A_c \left(\frac{1}{T_3} - \frac{1}{T_2} \right) = 0.108 \left(\frac{1}{0.2} - \frac{1}{1.0} \right) = 0.432 \text{ joules/g ion}$$

and

$$W_{41} = A_c \left(\frac{1}{T_1} - \frac{1}{T_4} \right) = 0.108 \left(\frac{1}{1.0} - \frac{1}{0.2} \right) = -0.432 \text{ joules/g ion}$$

The coefficient of performance of the refrigerator is defined as

$$\text{COP} = \frac{Q_{34}}{\text{net work input}} = \frac{Q_{34}}{-(W_{12} + W_{23} + W_{34} + W_{41})}$$

$$= \frac{2.23}{-(-11.2 + 0.432 + 2.23 - 0.432)} = 24.9\%$$

Of course, the COP of this Carnot refrigerator can also be calculated by the equation

$$\text{COP} = \frac{T_3}{T_1 - T_3} = \frac{0.2}{1.0 - 0.2} = 25\%$$

Note that the value 24.9% is in error due to slide rule calculations.

8-7 NUCLEAR DEMAGNETIZATION

Just as electrons, so the nuclear particles (protons and neutrons) also have spins that tend to orientate in certain senses of rotation about their own axes in relation to the atomic nucleus. In the absence of a magnetic field, the axes of nuclear spins orientate at random and no directional effect will be observed. In the presence of a magnetic field, the nuclear spins can be aligned. However, since the mass of a proton or a neutron is 1,836 times the mass of an electron, a nuclear particle can maintain the same angular momentum as does an electron by spinning much slower. Thus a nuclear magnetic moment is much smaller than that of an atom. The alignment of nuclear magnetic moments requires a very strong magnetic field even at very low temperatures. In fact, at 0.01°K an external magnetic field of 50,000 to 100,000 oersteds is needed to produce nuclear orientation. If the polarized nuclei are made to undergo a reversible adiabatic demagnetization, microdegrees can be reached. The method of *adiabatic nuclear demagnetization* was first suggested by Gorter[4] in 1934 and by Kurti and Simon[5] in 1935, and was first performed experimentally by Kurti and Simon in 1956.

Since nuclear demagnetization does not work effectively until the temperature is as low as 0.01°K, a two-state cooling is a necessity. The first stage consists of an ionic demagnetization of an ordinary paramagnetic salt to bring the temperature down to 0.01°K. The second state consists of a nuclear demagnetization of a nuclear paramagnetic material to bring the temperature farther down, possibly to one-millionth of a degree above absolute zero. A schematic illustration of a nuclear demagnetization cryostat is shown in Fig. 8-15. The sequence of operations would be as follows:

First step—With both the electronic and nuclear stages cooled down to below 1°K by the liquid helium bath, the electronic stage is first magnetized isothermally, giving up heat of magnetization to the liquid helium by way of the helium exchange gas.

Second step—The helium exchange gas is pumped out and the electronic state is demagnetized adiabatically, thus cooling itself to 0.01°K. With the thermal valve open, heat is transferred from the nuclear stage to the electronic stage, causing the temperature of the nuclear stage to drop to 0.01°K.

Third step—The nuclear stage is magnetized isothermally, giving up heat of magnetization to the electronic stage through the thermal valve.

---

[4] C. J. Gorter, *Physik. Z.*, **35**:923 (1934).
[5] N. Kurti and F. E. Simon, *Proc. Roy. Soc. (London)*, **A149**:152 (1935).

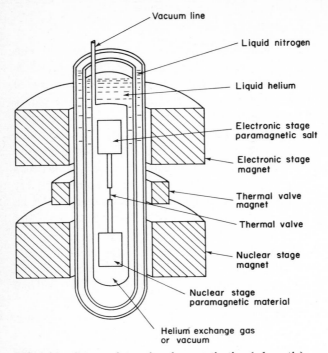

FIG. 8-15. Cryostat for nuclear demagnetization (schematic).

Fourth step—The thermal contact between the two stages is broken up by closing the thermal valve. Then the nuclear stage is demagnetized adiabatically, thus cooling itself to a temperature, say 10^{-6} °K.

In the pioneer experiment on nuclear demagnetization by Kurti and Simon in 1956, chromium potassium alum was used in the electronic stage and metallic copper was used in the nuclear stage, with copper wires connecting the two stages instead of a thermal valve. The temperature recorded was 0.000016 °K. However, this extremely low temperature was realized only for an instant.

It is apparent that in order to improve the technique of nuclear cooling, an efficient thermal valve must be incorporated. With the introduction of a thermal valve, temperatures in the order of 10^{-6} °K have been achieved. However, there are other obstacles much deeper in nature that must be overcome before any further progress in nuclear cooling can be made. Since in the material used for a nuclear stage, besides nuclear spins, there are crystal lattice vibrations and electron gas kinetic energy which are also measures of temperatures. In the temperature range encountered in nuclear cooling the energy exchange of the nuclear spins with the lattice vibrations and the electron gas is very slow. Thus in a nuclear cooling process one is faced with the problem of having the nuclear spins in a state corresponding to a very low temperature and the lattice vibrations and the electron gas in another state corresponding to a much higher temperature. With stronger magnetic fields, lower starting temperatures, and a very effective thermal valve, one may hope eventually to cool the material in the nuclear stage as a whole to very low temperatures.

8-8 MEASUREMENT OF TEMPERATURES BELOW 1°K

In the measurement of low temperatures above 1°K, the helium gas thermometer and the hydrogen or helium vapor-pressure thermometer are the two basic thermometers in common use. A schematic diagram of a constant-volume gas thermometer has been given in Sec. 1-4. A gas thermometer is, of course, based on the validity of the ideal gas equation $pv = RT$. A vapor-pressure thermometer, on the other hand, is based on the principle that the pressure exerted by a saturated vapor in thermal equilibrium with the saturated liquid or solid is a definite function of temperature. Thus in using a vapor-pressure thermometer, the vapor-pressure vs. temperature relationship for the substance must be exactly known, A He^3 vapor-pressure thermometer may be used to measure temperatures from its critical point of 3.32°K down to 0.2°K.

For the measurement of temperatures below 1°K, besides the limited range offered by a He^3 vapor-pressure thermometer, the only available thermometer is the magnetic thermometer. As we pointed out in Sec. 8-4, at low values of the ratio H/T many paramagnetic salts obey Curie's equation. In the temperature range where Curie's equation is obeyed, the temperature of the salt on the Kelvin scale is obtained by simply measuring the magnetic susceptibility and calculating from Eq. (8-3). However, at very low temperatures, usually below 1°K, deviations from Curie's equation do occur and therefore calibration is required for a temperature scale based on this equation. The temperature scale based on Curie's equation is defined by

$$T^* = \frac{C_c}{x} = \frac{C_c}{I/H} \tag{8-17}$$

where T^* has been given the name *magnetic temperature*. When Curie's equation is obeyed, the magnetic temperature T^* and the real Kelvin temperature T are the same; when Curie's equation is not obeyed, these two temperatures are different.

A schematic diagram illustrating in principle one method of measuring the magnetic susceptibility of a paramagnetic material is shown in Fig. 8-16. It consists of a primary coil and a secondary coil. The secondary coil is divided into two equal parts, the measuring coil and the compensating coil, which are wound

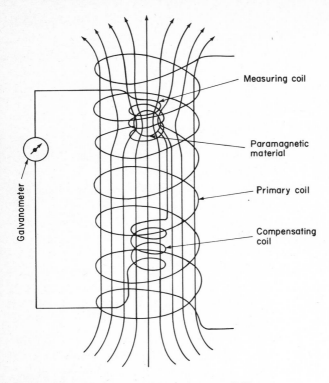

FIG. 8-16. Measurement of magnetic susceptibility.

in opposite directions. The coils are placed around the cryostat in such a position that the test sample of paramagnetic material is at the center of the measuring coil. When a current is switched on in the primary coil, a magnetic field is established inside it. Since the paramagnetic material has a magnetic susceptibility, magnetic lines of force are drawn into the measuring coil, resulting in an increased number of magnetic lines passing through the measuring coil as compared to the compensating coil. Thus a larger current is induced in the measuring coil than in the compensating coil. The current recorded by a galvanometer is then a direct measure of the magnetic susceptibility of the paramagnetic material.

In order to establish the relationship between the magnetic temperature T^* and the real Kelvin temperature T at very low temperatures where Curie's equation is not obeyed, we proceed to follow a number of isothermal magnetizations and adiabatic demagnetizations as depicted in Fig. 8-17. The magnetic susceptibility at points $a, b, c, d,$ and e can be determined by the method outlined in the last paragraph. With the known value of Curie's constant as determined at higher temperatures for the paramagnetic material, the magnetic temperatures at these points can be obtained from Eq. (8-17).

Now the entropy changes during the isothermal magnetizations, such as from a to e', may be calculated from Eq. (7-19).

$$dS = \frac{C_H}{T} dT + \mu_0 V \left(\frac{\partial M}{\partial T}\right)_H dH$$

Integrating this equation for the isothermal process ae' leads to

$$S_a - S_{e'} = -\mu_0 V \int_0^{H_e} \left(\frac{\partial M}{\partial T}\right)_H dH$$

At temperature T_a, even though Curie's equation is on the verge of breaking down, Brillouin's equation is still valid. The integral in the preceding equation can be evaluated through the use of Brillouin's equation. Thus we have the value of $(S_a - S_{e'})$, which is equal to $(S_a - S_e)$. Similarly, the values of $(S_a - S_d)$, $(S_a - S_c)$, and $(S_a - S_b)$ may be obtained. Thus we have the values of $(S_d - S_e)$, $(S_c - S_e)$ and $(S_b - S_e)$. A graph relating the entropy change $(S - S_e)$ and the magnetic temperature T^* at the various points along the line of zero magnetic field can be plotted, such as shown in Fig. 8-18(a).

The next step in the calibration procedure is to add and to measure the amount of heat required to change the magnetic material from the minimum temperature state e to higher temperatures, such as $d, c,$ etc., at zero magnetic

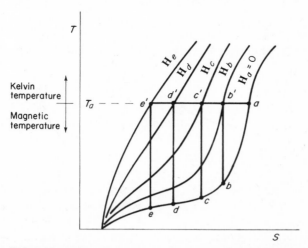

FIG. 8-17. Calibration processes for a magnetic thermometer.

field. Since the heating processes are along the line of zero magnetic field, no work is done. Hence, according to the first law, the heat added is equal to the change in internal energy. Thus, for instance, $Q_{ed} = U_d - U_e$. A graph relating the internal energy change $(U - U_e)$ and the magnetic temperature T^* at the various points along the line of zero magnetic field can be plotted, such as shown in Fig. 8-18(b).

Finally, when we combine the two graphs in Fig. 8-18(a) and (b), we obtain the graph as shown in Fig. 8-18(c). Since point e is fixed throughout the present calibration, U_e and S_e are constants, and therefore the slope of the curve in Fig. 8-18(c) at any point is $(\partial U/\partial S)_{\mathbf{H} = 0}$. Accordingly, from Eq. (7-8),

$$dU = T\,dS + \mu_0 V \mathbf{H}\,d\mathbf{M}$$

we have, for zero magnetic field

$$\left(\frac{\partial U}{\partial S}\right)_{\mathbf{H}=0} = T$$

Values of the real Kelvin temperature T at various points, such as a, b, c, etc., can now be calculated. These temperatures are then tabulated or plotted against the corresponding magnetic temperature T^* for calibration purposes. Figure 8-19 is such a graph.

PROBLEMS

8-1. With the help of some schematic flow diagrams and a minimum required thermophysical data, describe a cascade compression-evaporation process in the liquefaction of nitrogen.

8-2. For a gas obeying the van der Waals equation of state, find the ratio of the maximum inversion temperature to the critical temperature. On the basis of

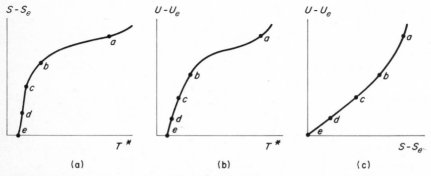

FIG. 8-18. Evaluation of Kelvin temperatures from data on magnetic temperatures.

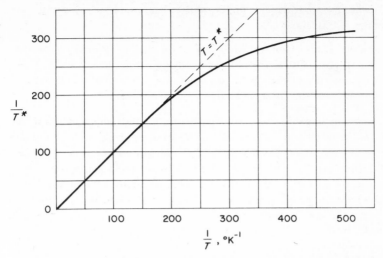

FIG. 8-19. T vs. T^* correlation for a sphere of cerium magnesium nitrate. (*Data from R. B. Frankel, D. A. Shirley, and N. J. Stone*, Phys. Rev., **140:3A:** *A1020* [*1965*].)

this information, explain that much less preliminary cooling is required to liquefy a gas by a Joule-Thomson expansion than by a cascade compression-evaporation process. You may use the expressions obtained in Example 2-5.

8-3. For a gas obeying the Redlich-Kwong equation of state, derive the expression for the Joule-Thomson coefficient and find the ratio of the maximum inversion temperature to the critical temperature.

8-4. A conventional vapor-compression refrigerating system consists of a compressor, a condenser, an expansion valve, and an evaporator. This refrigerating system is used to produce solid carbon dioxide (dry ice). Precooled carbon dioxide vapor at 4 atm is compressed reversibly and adiabatically to 50 atm. The vapor then flows into the condenser where heat is removed to a suitable coolant to condense the vapor to a saturated liquid at constant pressure. It finally expands through the expansion valve into the evaporator (called the snow chamber in this case), which is maintained at 4 atm. Solid carbon dioxide is removed from the snow chamber at a rate of 100 kg/hr, and vapor carbon dioxide is returned to the compressor. Make-up vapor is at $0°C$ and 4 atm. Determine the power requirement.

8-5. Air initially at 1 atm and $300°K$ is liquefied in an ideal simple Linde-Hampson system (see Fig. 8-1). A liquid yield of 0.09 kg liquid per kg gas compressed is desired. Determine the pressure to which the air must be compressed, and find the net work required per kg of air compressed. For the given initial condition, is there any upper limit in the liquid yield per unit mass of gas compressed? Explain.

8-6. An ideal Linde-Hampson air-liquefaction system compresses air isothermally from 1 atm, 300°K to 200 atm. Calculate the liquid yield per unit mass of air compressed and the kW-hr work per unit mass of air liquefied. What is the air temperature at the entrance to the expansion valve?

8-7. In order to improve the performance of a Linde-Hampson air liquefaction system, a Freon-12 vapor-compression refrigeration cycle is superimposed on it, as shown in Fig. P8-7. The air portion of the system operates between 1 atm and 200 atm, and the compression process is reversible and isothermal at 300°K. The Freon-12 portion of the system operates between 2 atm and 5 atm; saturated liquid enters the expansion valve; vapor at 2 atm and 300°K enters the compressor; and the compression process is reversible and adiabatic. The mass flow rate of Freon-12 is one tenth of the mass flow rate of air. All heat exchangers are perfectly effective. Show the state points of air on a *T-s* diagram and those of Freon-12 on a *T-s* and an *h-s* diagrams. Calculate the liquid air yielded per unit mass of air compressed and the kW-hr work per unit mass of air liquefied. What are the temperatures at points 3 and 4? Compare the results obtained in this problem with those obtained in Problem 8-6.

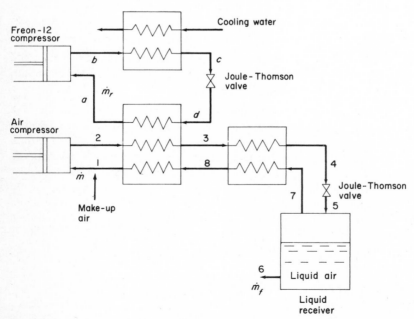

FIG. P8-7.

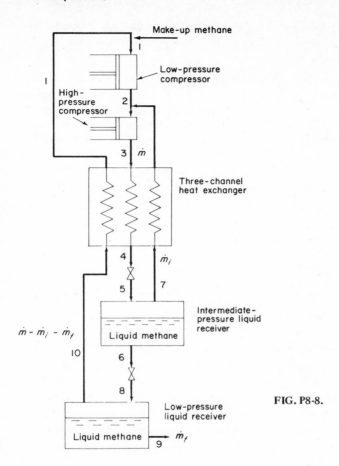

FIG. P8-8.

8-8. In order to reduce the total work requirement per unit mass of liquid yield, a dual-pressure Linde system as shown schematically in Fig. P8-8 can be used. Methane initially at 1 atm and $300°K$ is compressed reversibly and isothermally first to an intermediate pressure of 20 atm. After joining a return stream from the intermediate-pressure liquid receiver it is compressed to a high pressure of 160 atm. The high-pressure stream is then cooled in passing through the three-channel heat exchanger on its way to be throttled to the intermediate-pressure liquid receiver. From this receiver the saturated vapor is returned to the high-pressure compressor through the three-channel heat exchanger, and the saturated liquid is further throttled to the low-pressure liquid receiver. The mass-flow rate returning from the intermediate-pressure liquid receiver to the high-pressure compressor is 0.8 of the total mass-flow rate handled by this compressor. All processes are ideal. Show the processes on a T-s diagram. Calculate the liquid yield per unit mass compressed in the high-pressure compressor, and the work requirement per unit mass liquefied. If the gas is compressed directly from 1 atm to 160 atm in a single compressor, what is the work requirement per unit mass liquefied?

8-9. Oxygen gas initially at 14.7 psia and $60°F$ is liquefied in a Claude system. It is compressed reversibly and isothermally to 900 psia and then cooled to $10°F$ in passing through a heat exchanger. At the outlet of this heat exchanger, 60 percent of the total flow is diverted and expanded adiabatically to 14.7 psia through an expander which has an adiabatic efficiency of 75 percent. The expander work is utilized as part of the compressor work requirement. All the heat exchangers are working under ideal conditions. Calculate the liquid yield per unit mass of gas compressed and the net work required to liquefy a unit mass of gas.

8-10. Neon initially at 1 bar and $15°C$ is liquefied in an ideal precooled Linde-Hampson system as depicted in Fig. 8-3, except that the required liquid nitrogen as the precoolant is taken directly from a storage tank containing saturated nitrogen liquid at a temperature corresponding to a saturation pressure of 0.3 bar. The liquid nitrogen bath is maintained at 0.3 bar. After passing through the three-channel heat exchanger, the nitrogen gas is pumped away. The high-pressure side of the neon compressor is at 50 bars. Calculate the neon liquid yielded per unit mass compressed, the work required per unit mass of neon liquefied, and the mass of nitrogen boiled away per unit mass of neon liquefied.

8-11. Helium gas initially at 1 bar and $290°K$ is liquefied in a two-expander Collins liquefaction system as depicted in the flow diagram of Fig. 8-4. The compression process is reversible and isothermal with the high-pressure side at 15 bars. The temperatures at the intakes of the two expanders are $55°K$ and $14°K$ respectively. The respective mass flow rates through the two expanders are 0.25 and 0.50 times the mass flow rate through the compressor. Both expanders are adiabatic with an efficiency of 80 percent. The expander work outputs are utilized as parts of the compressor work requirement. All the heat exchangers are perfectly effective. Calculate the liquid yield per unit mass of gas compressed and the net work requirement per unit mass liquefied.

8-12. Determine the Curie constant of chromium potassium alum for which the atomic constants are $J = 3/2$ and $g_L = 2$. Give your answer in cgs-Gaussian system of units and in rationalized mks system of units.

8-13. A crystal of 2×10^{-3} kg ions of gadolinium sulfate is initially at $10°K$ and in a magnetic field of 1×10^6 amp/meter.

(a) When the field is reduced reversibly and isothermally to zero, calculate the heat transferred, the work done, and the change in entropy.

(b) When the field is reduced reversibly and adiabatically to zero, calculate the final temperature reached. Assume that Curie's law is valid.

8-14. One kilogram of iron ammonium alum at $1.2°K$ is magnetized reversibly and isothermally from zero field to 1.5×10^6 amp/meter. What are the heat, work, and change in entropy for the process of magnetization? Assume that Curie's law is valid. Compute the final temperature of the salt when it is then demagnetized reversibly and adiabatically.

8-15. Gadolinium sulfate is used as the working substance in a Carnot magnetic refrigerating cycle operating between two reservoirs at $1.5°K$ and $0.5°K$. The maximum and minimum magnetic fields are 2×10^6 amp/meter and 0, respectively. Referring to Fig. 8-14, compute the magnetic field and magnetic moment per g ion at each of the four end states, assuming that Curie's equation is valid. For one g ion of the salt, find the heat and work interactions for each of the four processes. Also find the coefficient of performance for the refrigerator.

Chapter 9

SUPERCONDUCTIVITY
AND SUPERFLUIDITY

9-1 SUPERCONDUCTIVITY—TYPE I SUPERCONDUCTORS

A metal has a crystalline structure with the atoms lying on a regular repetitive lattice. When the crystal lattice is perfect, the free electrons in the metal are able to pass through it without difficulty. However, there are two factors that generally ruin the perfect arrangement of a crystal lattice and thus give rise to electrical resistance. These are the thermal vibrations of the atoms and the impurities or imperfections of the material. As the temperature falls, the thermal vibrations of the lattice atoms decrease, and hence the electrical resistance of the metal decreases. But any real specimen of metal cannot be perfectly pure and will inevitably contain some impurities. The effect of impurity on electrical resistance is more or less independent of temperature. Thus we can see that impurities and lattice imperfections are mainly responsible for the small constant residual resistivity of a metal at very low temperatures.

There are many metals, however, that exhibit a most extraordinary behavior. After the residual resistivity of a metal has been reached, when the temperature is further reduced its electrical resistance suddenly disappears completely. Once a current is introduced in the metal at such low temperatures, the current will continue to flow undiminished for an indefinite period of time. This phenomenon was discovered in 1911 by Kamerlingh Onnes and was given the name *superconductivity*. A metal which shows superconductivity at low temperatures is called a *superconductor*. Since its discovery, the extraordinary behavior of superconductivity has been found in many metallic elements and in a very large number of alloys and compounds, some of them containing nonmetallic elements.

The unusual phenomenon of superconductivity has many applications. For instance, it can be applied in journal bearings to eliminate friction, in electric motors to reduce internal losses, in electromagnets to obtain very high magnetic fields, and in high-speed computers to form the so-called cryotrons to be used as logic, memory, and comparison elements. Owing to the great usefulness of the phenomenon of superconductivity, we now proceed to study it in some detail. We will emphasize the general nature of the phenomenon without regard to any particular application.

There are two kinds of superconductivity, known as type I and type II. Most of those elements which exhibit superconductivity belong to type I, while alloys generally belong to type II. These two types of superconductivity have many properties in common, but there are considerable differences in their magnetic behavior. We will discuss type I superconductivity in this and the next three sections, leaving the study of type II superconductivity to Section 9-5.

The temperature at which the transition from normal to superconductivity occurs in the absence of a magnetic field is called the *transition temperature* T_0. This temperature varies from metal to metal and depends not only on pressure but also on the thickness of the metal sample. The value of T_0 of a thin film is different from that of the bulk metal. For a strain-free pure bulk metal, the transition temperature is well-defined and can be measured accurately.

Superconductivity can be destroyed by a magnetic field. A magnetic field strength required to destroy superconductivity in a metal is called a *critical or threshold field* H_c. The uniqueness of the critical field at a given temperature depends on the shape and orientation of the superconductor as well as on any impurities and strains in it. In the ideal case, when a strain-free pure type I superconductor in the shape of a long thin cylinder is placed longitudinally in a uniform magnetic field, the transition between normal and superconductivity is sharp and a unique value of the critical field can be obtained at a given

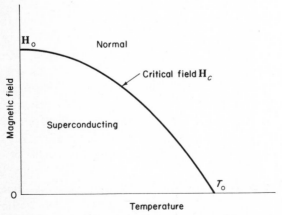

FIG. 9-1. Temperature dependence of the critical magnetic field of a type I superconductor.

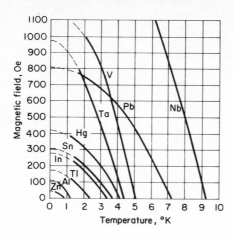

FIG. 9-2. Critical field curves for several type I superconducting elements.

temperature. This critical field is a function of temperature only. Figure 9-1 shows the temperature dependence of the critical field of a type I super-conductor. It is seen that the critical field curve forms the boundary of super-conducting states. The area enclosed by the critical field curve is the region in which the metal is superconducting. Whereas beyond the confines of the curve it is normal. Figure 9-2 shows the critical field curves for several type I super-conducting elements. Values of the transition temperature T_0 and the critical field at absolute zero temperature, H_0, for a number of elements are given in Table 9-1.

An inspection of Fig. 9-2 reveals that every critical field curve has a negative slope which increases in magnitude from zero at $0°K$ to a finite value at T_0. These curves may be approximated by a parabolic equation of the form

$$H_c = H_0 \left(1 - \frac{T^2}{T_0^2}\right) \tag{9-1}$$

In addition to the disappearance of electrical resistivity, a superconducting metal also shows a magnetic effect—the disappearance of magnetic induction. The phenomenon of zero magnetic induction was discovered by Meissner and Ochsenfeld[1] in 1933 and is now commonly known as the *Meissner effect*. In order to understand its meaning, let us consider a type I superconducting metal specimen going through a few processes as depicted in Fig. 9-3(*a*). The specimen is first cooled at zero magnetic field from a normal state *a* to a superconducting state *b*, and then is magnetized at constant temperature from state *b* to state *c*, which is well below the critical field curve. When the magnetic field is applied to the specimen in superconducting states, persistent currents induced on the

[1] W. Meissner and R. Ochsenfeld, *Naturwissenschaften*, **21**:787 (1933).

TABLE 9-1. The Superconducting Elements*

Element	Symbol	T_0 (°K)	H_0 (amp/m)
Aluminum	Al	1.2	0.79×10^4
Cadmium	Cd	0.5	0.24×10^4
Gallium	Ga	1.1	0.41×10^4
Indium	In	3.4	2.2×10^4
Iridium	Ir	0.1	$\sim 0.16 \times 10^4$
Lanthanum	La $\begin{cases} \alpha \\ \beta \end{cases}$	4.8 4.9	
Lead	Pb	7.2	6.4×10^4
Mercury	Hg $\begin{cases} \alpha \\ \beta \end{cases}$	4.2 4.0	3.3×10^4 2.7×10^4
Molybdenum	Mo	0.9	
Niobium	Nb	9.3	Type II
Osmium	Os	0.7	$\sim 0.5 \times 10^4$
Rhenium	Re	1.7	1.6×10^4
Ruthenium	Ru	0.5	0.53×10^4
Tantalum	Ta	4.5	6.6×10^4
Technetium	Tc	8.2	
Thallium	Tl	2.4	1.4×10^4
Thorium	Th	1.4	1.3×10^4
Tin	Sn	3.7	2.4×10^4
Titanium	Ti	0.4	
Tungsten	W	0.01	
Uranium	U $\begin{cases} \alpha \\ \beta \end{cases}$	0.6 1.8	
Vanadium	V	5.3	Type II
Zinc	Zn	0.9	0.42×10^4
Zirconium	Zr	0.8	0.37×10^4

*Data from A. C. Rose-Innes and E. H. Rhoderick, "Introduction to Superconductivity," p. 46, Pergamon Press, London, 1969.

surface of the specimen prevent the field from penetrating the metal. Thus, a illustrated in Fig. 9-3(*b*), at state *c* the magnetic lines of forces are bulgin around the specimen. When the specimen is heated in a constant magnetic fiel from state *c* to state *d*, as the critical field curve is passed the persistent current on its surface die out and magnetic flux penetrates into it. Thus, as illustrated i Fig. 9-3(*b*), at state *d* the magnetic field has uniformly penetrated the meta since the metal is now in a normal state and is virtually nonmagnetic.

The magnetic pattern during the cyclic processes *a b c d*, as explained in th preceding paragraph, is entirely expected because of the property of zer electrical resistance of superconductivity. However, when the specimen is made to go through the reversed cycle *a d c b*, the magnetic pattern cannot be explained by reason of zero electrical resistance alone. Of course, when a magnetic field is applied to the specimen at constant temperature from state *a* to state *d*, the magnetic field will uniformly penetrate the metal. But on cooling the specimen in a constant magnetic field from state *d* to state *c*, since the external field does not vary, no current would be induced on the specimen and the field would be undisturbed if infinite electrical conductivity is the only special property of

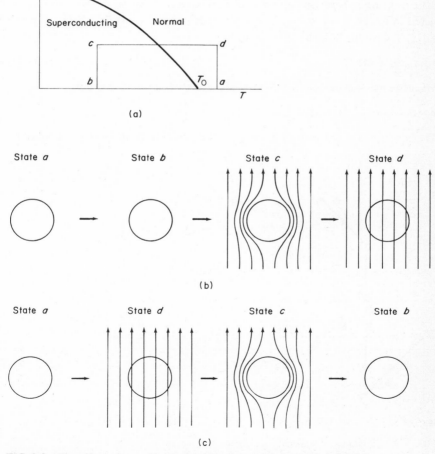

FIG. 9-3. The effect of zero electrical resistance and zero magnetic induction.

superconductivity. Nevertheless, during the process $d\,c$ the magnetic flux does not remain in the metal but is completely expelled when state c is reached. The magnetic pattern at state c is depicted in Fig. 9-3(c). The fact that the magnetic field is expelled from the metal when it becomes superconducting implies not only infinite electrical conductivity, but also perfect diamagnetism. This is the essence of the Meissner effect.

As a further illustration of the magnetic nature of superconductivity, let us consider the variations of the magnetic induction **B** and the magnetization **M** of a type **I** superconductor as the applied magnetic field **H** is increased isothermally across the critical field, as depicted by the process $a\,b\,d$ in Fig. 9-4. As mentioned before, in the ideal case when a pure strain-free type I superconductor in the form of a long thin cylinder is placed longitudinally in a magnetic field, the phase transition between normal and superconductivity is abrupt, such as depicted in the **H**-T plot of Fig. 9-4. On the **B-H** and **M-H** plots the phase transition is depicted by the line bb'. In general,

$$\mathbf{B} = \mu_0(\mathbf{H} + \mathbf{M}) \tag{7-1}$$

When the metal is in superconducting phase, between points a and b, we have

$$\mathbf{B} = 0 \quad \text{and} \quad \mathbf{M} = -\mathbf{H} \tag{9-2}$$

Whereas, beyond the critical field $\mathbf{H}_c$ corresponding to the given temperature,

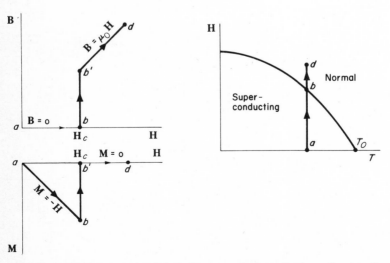

FIG. 9-4. Variations of magnetic induction and magnetization of a type I superconductor as the magnetic field is increased isothermally.

between points b' and d, the metal is normal. Since normal metals (excluding ferromagnetic metals, such as iron) are virtually nonmagnetic, it follows that

$$\mathbf{M} = 0 \quad \text{and} \quad \mathbf{B} = \mu_0 \mathbf{H} \tag{9-3}$$

There is another property of a metal which changes abruptly during transitions from normal to superconductivity. As we know, conduction heat transfer in a metal is mainly due to the mobility of free electrons. However, at a superconducting state the free electrons of a metal no longer interact with the lattice in such a way that the electrons can pick up heat energy from one part of the metal and deliver it to another part. Therefore when a metal becomes superconducting, its thermal conductivity in general decreases. At temperatures well below the transition temperature, the decrease in thermal conductivity is abrupt at the crossing of the critical field curve. Since superconductivity can be destroyed by the application of a magnetic field, the thermal conductivity of a superconductor can be easily controlled by means of a magnetic field. This is the basic principle for a thermal valve.

EXAMPLE 9-1. Calculate the critical field of lead (Pb) at $3°$K from data on T_0 and $\mathbf{H}_0$.

Solution.
From Table 9-1 we have for lead $T_0 = 7.2°$K and $\mathbf{H}_0 = 6.4 \times 10^4$ amp/m. If the transition curve follows the parabolic rule of Eq. (9-1), the critical field is given by

$$\mathbf{H}_c = \mathbf{H}_0 \left(1 - \frac{T^2}{T_0^2} \right) = (6.4 \times 10^4) \left(1 - \frac{3^2}{7.2^2} \right)$$

$$= 5.29 \times 10^4 \text{ amp/m}$$

$$= (5.29 \times 10^4 \text{ amp/m}) \left(4\pi \times 10^{-3} \frac{\text{oersted}}{\text{amp/m}} \right)$$

$$= 665 \text{ oersteds}$$

The experimental value as read from Fig. 9-2 is $\mathbf{H}_c = 680$ oersteds. Thus the value of $\mathbf{H}_c$ calculated by Eq. (9-1) has an error of

$$\frac{680 - 665}{680} \times 100 = 2.2\%$$

A more accurate value of $\mathbf{H}_c$ can be obtained by the use of an equation of the form

$$\mathbf{H}_c = \mathbf{H}_0 \left[1 + a \left(\frac{T}{T_0} \right)^2 + b \left(\frac{T}{T_0} \right)^3 + c \left(\frac{T}{T_0} \right)^4 + \cdots \right]$$

For lead, $a = -0.91$, $c = -0.09$, and $b = d = \cdots = 0$. Thus

$$\mathbf{H}_c = (6.4 \times 10^4)\left[1 - 0.91\left(\frac{3}{7.2}\right)^2 - 0.09\left(\frac{3}{7.2}\right)^4\right] = 5.37 \times 10^4 \text{ amp/m}$$

$$= (5.37 \times 10^4)(4\pi \times 10^{-3}) = 675 \text{ oersteds}$$

which has an error of

$$\frac{680 - 675}{680} \times 100 = 0.7\%$$

9-2 THERMODYNAMICS OF TYPE I SUPERCONDUCTING SYSTEMS

As demonstrated in Fig. 9-3 of the preceding section, a type I superconducting system at any values of **H** and T within the superconducting region has a state that is strictly fixed by the given values of **H** and T and is independent of how the system got there. Hence a type I superconductor may be considered as a thermodynamic system whose equilibrium states can be described by a few thermodynamic properties, and the transitions between normal and super-conducting states are reversible.

As mentioned previously, for the ideal case of a strain-free pure type I superconductor in the shape of a long thin cylinder, placed longitudinally in a magnetic field, there is a unique single critical field curve which divides the **H**-T plane into two regions. Below the critical field curve the system is in the super-conducting phase, and above the critical field curve it is in the normal phase. The critical field curve itself is the equilibrium line for coexisting phases.

In general, the transition between normal and superconducting phases, taking place at constant temperature T and constant critical field $\mathbf{H}_c$, involves a finite latent heat. It is a phase transition of the first order. We now derive the equation for the latent heat in relation to the temperature and the critical field curve. According to Eq. (7-13), the differential of the magnetic Gibbs function is given by

$$dG = -S\,dT - \mu_0 V \mathbf{M}\,d\mathbf{H}$$

At T = constant and $\mathbf{H} = \mathbf{H}_c$ = constant, we must have $dG = 0$, or

$$G^{(n)} = G^{(s)}$$

where the superscripts (n) and (s) denote respectively normal and super-conducting phases. When the equilibrium temperature and critical field are increased to $T + dT$ and $\mathbf{H}_c + d\mathbf{H}_c$ respectively, the new Gibbs functions of the two phases must also be equal, or

$$G^{(n)} + dG^{(n)} = G^{(s)} + dG^{(s)}$$

Then

$$dG^{(n)} = dG^{(s)}$$

Applying Eq. (7-13) we obtain

$$-S^{(n)}dT - \mu_0 V \mathbf{M}^{(n)}d\mathbf{H}_c = -S^{(s)}dT - \mu_0 V\mathbf{M}^{(s)}d\mathbf{H}_c$$

Therefore

$$-\frac{d\mathbf{H}_c}{dT} = \frac{S^{(n)} - S^{(s)}}{\mu_0 V(\mathbf{M}^{(n)} - \mathbf{M}^{(s)})}$$

But from Eqs. (9-2) and (9-3),

$$\mathbf{M}^{(n)} = 0 \quad \text{and} \quad \mathbf{M}^{(s)} = -\mathbf{H}_c$$

It follows that

$$S^{(n)} - S^{(s)} = -\mu_0 V \mathbf{H}_c \frac{d\mathbf{H}_c}{dT} \tag{9-4}$$

Now since $L = T(S^{(n)} - S^{(s)}) = $ latent heat, we obtain finally

$$L = -\mu_0 VT \mathbf{H}_c \frac{d\mathbf{H}_c}{dT} \tag{9-5}$$

Since $d\mathbf{H}_c/dT$ is always negative, we see from Eq. (9-4) that $S^{(n)}$ is greater than $S^{(s)}$. Since entropy is physically an index of orderliness, we conclude that more order exists in the superconducting than in the normal phase. From Eq. (9-5) we see that $L = 0$ at the two extremes of the critical field curve, i.e., at $T = 0$ where $d\mathbf{H}_c/dT = 0$, and at $T = T_0$ where $\mathbf{H}_c = 0$. Between these two extreme conditions, L is positive, indicating that heat addition is required in changing from superconducting to normal phase.

Since in the absence of a magnetic field, the transition between normal and superconducting phases takes place at T_0 without latent heat evolution, but measurements of specific heats show a discontinuity at the transition point, this phase transition is obviously of the second order.

The general patterns of variation of the specific heats of a type I superconductor in the superconducting phase as well as in the normal phase are shown in Fig. 9-5. Except for the transition from the normal to the superconducting phase, the specific heats are independent of the magnetic field. Experimental

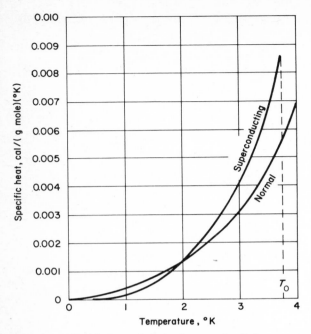

FIG. 9-5. Specific heat of tin in the normal and the superconduct-ing states. (*From measurements by W. H. Keesom and P. H. Van Laer*, Physica, 5:*193 (1938).*)

measurements of the specific heat for the superconducting phase can be made in the absence of any magnetic field, while measurements for the normal phase are made with the application of a magnetic field strong enough to destroy the superconductivity. It is seen from Fig. 9-5 that at the transition temperature T_0 there is an abrupt jump in the specific heats. At lower temperatures the specific heat of the superconducting phase falls more rapidly than the normal phase. As the absolute zero temperature is approached, the specific heat of the super-conducting phase tends toward zero less rapidly than the normal phase.

There are some useful relationships concerning the specific heat and the critical magnetic field. Thus by differentiating Eq. (9-4) with respect to T and multiplying the resulting equation by T, we obtain

$$T \frac{dS^{(n)}}{dT} - T \frac{dS^{(s)}}{dT} = -\mu_0 VT \frac{d}{dT}\left(\mathbf{H}_c \frac{d\mathbf{H}_c}{dT}\right)$$

Now since $c = T(ds/dT)$, the difference in specific heats is then

$$c^{(s)} - c^{(n)} = \mu_0 vT \frac{d}{dT}\left(\mathbf{H}_c \frac{d\mathbf{H}_c}{dT}\right) \tag{9-6}$$

or

$$c^{(s)} - c^{(n)} = \mu_0 v T \left(\frac{d\mathbf{H}_c}{dT}\right)^2 + \mu_0 v T \mathbf{H}_c \frac{d^2 \mathbf{H}_c}{dT^2}$$

Equation (9-6) can be used to calculate the difference in specific heats of the superconducting and normal states from measurements of magnetic properties for an ideal type I superconductor.

At the transition temperature $T = T_0$ and $\mathbf{H}_c = 0$. Equation (9-6) reduces to

$$\left(\frac{d\mathbf{H}_c}{dT}\right)^2_{T=T_0} = \frac{1}{\mu_0 v T_0} (c^{(s)} - c^{(n)})_{T=T_0} \tag{9-7}$$

This equation is known as the *Rutgers equation*. It may be used to calculate the slope of the critical field curve at the transition temperature T_0 from measurements on specific heats. Conversely, it may be used to calculate the magnitude of the specific heat jump at T_0 from measurements of the slope of the critical field curve.

Another useful equation may be obtained by integrating Eq. (9-6) from $T = 0°\text{K}$, $\mathbf{H}_c = \mathbf{H}_0$ to $T = T_0$, $\mathbf{H}_c = 0$ along the critical field curve. Thus

$$\int_0^{T_0} (c^{(s)} - c^{(n)}) dT = \mu_0 v \int_{\mathbf{H}_c}^0 T d\left(\mathbf{H}_c \frac{d\mathbf{H}_c}{dT}\right)$$

$$= \mu_0 v T \mathbf{H}_c \frac{d\mathbf{H}_c}{dT}\bigg]_0^{T_0} - \mu_0 v \int_{\mathbf{H}_0}^0 \mathbf{H}_c \frac{d\mathbf{H}_c}{dT} dT$$

where the method of integration by parts has been used. Now since at $T = T_0$, $\mathbf{H}_c = 0$, the first term in the last expression is zero, and therefore

$$\int_0^{T_0} (c^{(s)} - c^{(n)}) dT = \frac{1}{2} \mu_0 v \mathbf{H}_0^2$$

or

$$\mathbf{H}_0^2 = \frac{2}{\mu_0 v} \int_0^{T_0} (c^{(s)} - c^{(n)}) dT \tag{9-8}$$

which can be used to calculate $\mathbf{H}_0$ from measurements on specific heats.

EXAMPLE 9-2. For a type I superconductor in the form of a long thin rod, let $g^{(s)}(T,\mathbf{H})$ and $g^{(s)}(T,0)$ denote the magnetic Gibbs functions per unit volume of the superconducting state at a temperature T and in an applied magnetic field $\mathbf{H} = \mathbf{H}$ (parallel to the rod) and $\mathbf{H} = 0$ respectively; and $g^{(n)}(T, 0)$ denote the magnetic Gibbs function per unit volume of the normal state at a temperature T and in the absence of an applied magnetic field. Show that

$$g^{(s)}(T, \mathbf{H}) - g^{(s)}(T, 0) = \frac{1}{2} \mu_0 \mathbf{H}^2$$

and

$$\mathbf{H}_c(T) = \left\{ \frac{2}{\mu_0} \left[g^{(n)}(T, 0) - g^{(s)}(T, 0) \right] \right\}^{1/2}$$

where $\mathbf{H}_c(T)$ is the critical field at temperature T.

Solution.
According to Eq. (7-13), the magnetic Gibbs function per unit volume of a superconductor is given by

$$dg = -s\, dT - \mu_0 \mathbf{M}\, d\mathbf{H}$$

Applying this equation to the superconducting state and integrating it at constant temperature T from $\mathbf{H} = 0$ to $\mathbf{H} = \mathbf{H}$ yields

$$g^{(s)}(T, \mathbf{H}) - g^{(s)}(T, 0) = -\mu_0 \int_0^{\mathbf{H}} \mathbf{M}\, d\mathbf{H}$$

But from Eq. (9-2), $\mathbf{M} = -\mathbf{H}$. Therefore

$$g^{(s)}(T, \mathbf{H}) - g^{(s)}(T, 0) = \frac{1}{2} \mu_0 \mathbf{H}^2 \qquad\qquad (9\text{-}9)$$

Since the normal state is virtually nonmagnetic, the application of a magnetic field does not change its magnetic Gibbs function. However, due to the diamagnetic nature of the superconducting state, its magnetic Gibbs function will increase according to Eq. (9-9). When the applied field is increased to a value such that the magnetic Gibbs function of the superconducting state becomes higher than that of the normal state, or $g^{(s)}(T, \mathbf{H}) > g^{(n)}(T, 0)$, the transition from superconducting to normal state will occur. Accordingly, for such a case we have from Eq. (9-9)

$$g^{(n)}(T, 0) < g^{(s)}(T, 0) + \frac{1}{2} \mu_0 \mathbf{H}^2 \qquad \text{or} \qquad \frac{1}{2} \mu_0 \mathbf{H}^2 > g^{(n)}(T, 0) - g^{(s)}(T, 0)$$

Thus, at a given temperature there is a maximum magnetic field that can be applied to a superconducting metal without destroying its superconductivity. This maximum field is the critical field $\mathbf{H}_c(T)$ and is given by the equation

$$\mathbf{H}_c(T) = \left\{ \frac{2}{\mu_0} \left[g^{(n)}(T, 0) - g^{(s)}(T, 0) \right] \right\}^{1/2} \qquad\qquad (9\text{-}10)$$

EXAMPLE 9-3. Determine the latent heat and the entropy change during the transition from superconducting to normal phase for 0.1 m^3 of tin (Sn) at 3°K, assuming that the transition curve follows the parabolic rule of Eq. (9-1).

Solution.
From Table 9-1 we have for tin $T_0 = 3.7°$K and $H_0 = 2.4 \times 10^4$ amp/m. Differentiating Eq. (9-1) with respect to T and substituting into Eq. (9-5) lead to the equation for the latent heat

$$L = -\mu_0 \, VT \left[H_0 \left(1 - \frac{T^2}{T_0^2} \right) \right] \left(-\frac{H_0}{T_0^2} \, 2T \right) = 2\mu_0 \, VH_0^2 \left(\frac{T}{T_0} \right)^2 \left[1 - \left(\frac{T}{T_0} \right)^2 \right]$$

Inserting numerical values we obtain

$$L = 2 \left(4\pi \times 10^{-7} \, \frac{\text{newton}}{\text{amp}^2} \right) (0.1 \text{ m}^3) \left(2.4 \times 10^4 \, \frac{\text{amp}}{\text{m}} \right)^2$$

$$\cdot \left(\frac{3°\text{K}}{3.7°\text{K}} \right)^2 \left[1 - \left(\frac{3°\text{K}}{3.7°\text{K}} \right)^2 \right] = 32.6 \text{ joules}$$

The entropy change is given by

$$S^{(n)} - S^{(s)} = \frac{L}{T} = \frac{32.6 \text{ joules}}{3°\text{K}} = 10.9 \text{ joules/}°\text{K}$$

EXAMPLE 9-4. Calculate the difference in specific heats between superconducting and normal states of tin at 3°K, assuming that the transition curve follows the parabolic rule of Eq. (9-1).

Solution.
According to Eq. (9-1), the first and second derivatives of H_c with respect to T are

$$\frac{dH_c}{dT} = -\frac{H_0}{T_0^2} \, 2T \quad \text{and} \quad \frac{d^2H_c}{dT^2} = -\frac{2H_0}{T_0^2}$$

Substituting these expressions and Eq. (9-1) into Eq. (9-6) yields

$$c^{(s)} - c^{(n)} = \mu_0 vT \left[\left(\frac{dH_c}{dT} \right)^2 + H_c \left(\frac{d^2H_c}{dT^2} \right) \right]$$

$$= \mu_0 vT \left[\left(-\frac{2H_0}{T_0^2} \, T \right)^2 + H_0 \left(1 - \frac{T^2}{T_0^2} \right) \left(-\frac{2H_0}{T_0^2} \right) \right]$$

$$= -\frac{2}{T_0^2} \, \mu_0 vTH_0^2 \left[1 - 3 \left(\frac{T}{T_0} \right)^2 \right]$$

Now for tin, $v = 16.1 \times 10^{-6}$ m^3/g mole. From Table 9-1, we have $T_0 = 3.7°$K and $H_0 = 2.4 \times 10^4$ amp/m. Hence at $T = 3°$K we obtain

$$c^{(s)} - c^{(n)} = -\frac{2}{(3.7°K)^2}\left(4\pi \times 10^{-7}\ \frac{newton}{amp^2}\right)\left(16.1 \times 10^{-6}\ \frac{m^3}{g\ mole}\right)$$

$$\cdot (3°K)\left(2.4 \times 10^4\ \frac{amp}{m}\right)^2\left[1 - 3\left(\frac{3°K}{3.7°K}\right)^2\right]$$

$$= 4.97 \times 10^{-3}\ \frac{joules}{(g\ mole)(°K)} = 1.19 \times 10^{-3}\ \frac{cal}{(g\ mole)(°K)}$$

EXAMPLE 9-5. According to the data of Fig. 9-5, the specific heats of tin in the superconducting state, $c^{(s)}$, and in the normal state, $c^{(n)}$, are listed in the following table as a function of temperature up to the transition temperature of $3.72°$K.

$T, °$K	$c^{(s)}$ cal (g mole)(°K)	$c^{(n)}$ cal (g mole)(°K)	$c^{(s)} - c^{(n)}$ cal (g mole)(°K)
0	0	0	0
1.0	0.000175	0.000456	− 0.000281
1.5	0.000607	0.000848	− 0.000241
2.0	0.00146	0.00140	0.00006
2.5	0.00264	0.00214	0.00050
3.0	0.00426	0.00319	0.00107
3.5	0.00680	0.00471	0.00209
3.72	0.00852	0.00562	0.00290

Estimate for tin the critical field at absolute zero temperature H_0 and the slope of critical field curve at the transition temperature, $(dH_0/dT)_{T = T_0}$.

Solution.
From the given data a graph of $(c^{(s)} - c^{(n)})$ versus T is prepared, from which by graphical integration we obtain

$$\int_{0°K}^{T_0 = 3.72°K} (c^{(s)} - c^{(n)})dT = 1.53 \times 10^{-3}\ cal/g\ mole$$

Substituting this value and $v = 16.1 \times 10^{-6}$ m^3/g mole into Eq. (9-8), we have

$$H_0^2 = \frac{2}{\mu_0 v} \int_0^{T_0} (c^{(s)} - c^{(n)}) \, dT$$

$$= \frac{2(1.53 \times 10^{-3} \text{ cal/g mole})(4.187 \text{ joules/cal})}{(4\pi \times 10^{-7} \text{ newton/amp}^2)(16.1 \times 10^{-6} \text{ m}^3/\text{g mole})}$$

$$= 6.33 \times 10^8 \text{ amp}^2/\text{m}^2$$

Therefore

$$H_0 = 2.52 \times 10^4 \text{ amp/m}$$

We now use the Rutgers equation to estimate the slope of the critical field curve at the transition temperature. Thus

$$\left(\frac{d\mathbf{H}_c}{dT}\right)_{T=T_0}^2 = \frac{1}{\mu_0 v T_0} (c^{(s)} - c^{(n)})_{T=T_0}$$

$$= \frac{(0.00290 \text{ cal/(g mole)}(^\circ\text{K}))(4.187 \text{ joules/cal})}{(4\pi \times 10^{-7} \text{ N/amp}^2)(16.1 \times 10^{-6} \text{ m}^3/\text{g mole})(3.72^\circ\text{K})} = 1.61 \times 10^8$$

Therefore

$$\left(\frac{d\mathbf{H}_c}{dT}\right)_{T=T_0} = -1.27 \times 10^4 \text{ amp/(m)}(^\circ\text{K})$$

where the minus sign indicates that the slope is negative.

It should be noted that the values of H_0 and $(d\mathbf{H}_c/dT)_{T=T_0}$ obtained in this example are not very accurate. More accurate data on the specific heats, particularly at the transition temperature, would yield more accurate answers.

9-3 THE INTERMEDIATE STATE

In the last two sections when we studied the transitions between normal and superconducting phases of type I superconductors we referred to a specimen in the shape of a long thin cylinder placed longitudinally in a uniform external magnetic field. The reason for this restriction is to minimize the end effects so that the phase transition may occur simultaneously and completely throughout the specimen. It is known from electromagnetism that, with the exception of a toroid or a long thin body in a longitudinal magnetic field, the field inside the body differs from the applied field. For a paramagnetic or ferromagnetic body the internal field is less than the applied field, while for a diamagnetic body such as a superconductor the internal field exceeds the applied field.

The magnetic field distribution inside a body in general is complicated. However, for an ellipsoid of revolution in an applied magnetic field **H** parallel to the axis of revolution, the internal field $\mathbf{H}_i$ and magnetization **M** are constant and parallel to the applied field. The relationship between these quantities is given by

$$\mathbf{H}_i = \mathbf{H} - D_m \mathbf{M} \tag{9-11}$$

where D_m is the demagnetizing factor of the body. For a prolate ellipsoid this is given by

$$D_m = \left(\frac{1}{e^2} - 1\right)\left(\frac{1}{2e} \ln \frac{1+e}{1-e} - 1\right)$$

where e is the eccentricity and is equal to $(1 - b^2/a^2)^{1/2}$, a and b being the semimajor and semiminor axes respectively. For a sphere $D_m = 1/3$. For the case of an infinite cylinder it can be shown that $D_m = 0$ if the axis is parallel to the applied field, and $D_m = 1/2$ if the axis is perpendicular to the applied field.

For a superconductor, since $\mathbf{M} = -\mathbf{H}_i$, Eq. (9-11) becomes

$$\mathbf{H}_i = \frac{\mathbf{H}}{1 - D_m} \tag{9-12}$$

In the case of an infinite cylinder with axis parallel to the applied field, $D_m = 0$ and $\mathbf{H}_i = \mathbf{H}$. The magnetic field is, therefore, everywhere the same as the applied field. The cylinder will remain entirely superconducting until the applied field reaches the critical value $\mathbf{H}_c$. At $\mathbf{H}_c$ the entire cylinder then becomes normal. This is the ideal situation we referred to in the last two sections.

In the case of other ellipsoidal shapes, D_m is nonzero. For illustrative purposes, let us consider a superconducting sphere for which $D_m = 1/3$ and $\mathbf{H}_i = 3\mathbf{H}/2$. The inequality in field strength between the internal and applied fields raises the question of what happens when the applied field **H** is gradually increased to a value $\mathbf{H}_c' = 2\mathbf{H}_c/3$, so that the internal field $\mathbf{H}_i$ becomes equal to the critical field $\mathbf{H}_c$. One might expect that the sphere would be driven into the normal phase. But this would lead to a contradiction: If the sphere were to become normal we should have $\mathbf{M} = 0$ and $\mathbf{H}_i = \mathbf{H} = \mathbf{H}_c'$, which is smaller than $\mathbf{H}_c$. This means that we would have a superconductor in a normal state but in a field less than the critical field $\mathbf{H}_c$. Instead one must postulate that once the applied field reaches $\mathbf{H}_c' = 2\mathbf{H}_c/3$, the sphere will subdivide into small alternating normal and superconducting coexisting regions with boundaries parallel to the applied field. Such a state of a superconductor is known as the *intermediate state*.

The intermediate state is not an intrinsic feature of a superconducting material, but depends on the shape of the specimen. It happens when a type I superconducting body with a nonzero demagnetizing factor D_m is in an applied field between $H'_c = (1 - D_m)H_c$ and H_c. When the applied field is raised beyond H'_c, the normal regions will grow larger and larger, and finally when the applied field reaches H_c the material becomes entirely normal.

Since in the intermediate state there is magnetic flux passing through the normal regions, let us define an average or effective flux density $\overline{B}$ as the total flux passing through the specimen divided by its maximum cross-sectional area. The average magnetization is then defined as

$$\overline{M} = \frac{\overline{B}}{\mu_0} - H_i$$

The magnetic properties of a type I superconductor in the intermediate state is depicted in Fig. 9-6, in which the variations of $\overline{B}$, $\overline{M}$, and H_i as functions of the applied field H are shown. Note that the intermediate state with the normal and superconducting phases coexisting can happen only if the internal field H_i is equal to the critical field H_c.

9-4 THEORETICAL DEVELOPMENT OF SUPERCONDUCTIVITY

Many attempts have been made to seek a fundamental understanding of the phenomenon of superconductivity since its discovery in 1911. It was not until 1957 that an acceptable fundamental theory of superconductivity was formulated on a firm basis. The complete treatment of the theory of superconductivity is extremely complicated. It requires an advanced knowledge of quantum mechanics and is beyond the scope of this book. It is our intention here to give only a brief descriptive introduction of the theories involved.

Shortly after the discovery of the Meissner effect, two theories of phenomenological nature were advanced with considerable success. In 1934 Gorter and Casimir[2] first proposed a two-fluid model, which suggested that the free electrons in a superconducting metal below the transition temperature were composed of normal electrons and highly ordered superelectrons. All entropy effects are assumed to be associated with the normal electrons. Superelectrons can pass through the superconducting metal without resistance, while normal electrons still encounter resistances just like conduction electrons in a normal metal. At the absolute zero temperature all conduction electrons behave like

[2] C. J. Gorter and H. B. G. Casimir, *Physica*, 1:306 (1934).

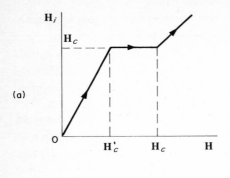

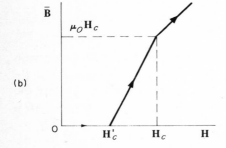

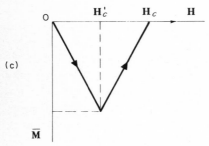

FIG. 9-6. The intermediate state of a type I superconductor. Graphs showing the variations of internal magnetic field H_i, average magnetic flux density $\bar{B}$, and average magnetization $\bar{M}$ as functions of the applied magnetic field H.

superelectrons. The fraction of superelectrons is to decrease as the temperature is raised towards the transition temperature. Eventually, at the transition temperature, all the conduction electrons become normal electrons and the metal returns to normal behavior. The principal virtue of the two-fluid model is to provide a conceptual tool of primarily qualitative nature. It can be used at best only semiquantitatively.

About the same time when Gorter and Casimir introduced the two-fluid model, two brothers Fritz and Heinz London[3] succeeded in formulating a new

[3] F. London and H. London, *Proc. Roy. Soc.* (*London*), **A**149:71 (1935); *Physica*, **2:** 341 (1935).

electrodynamic theory by adding new relations to Maxwell's well-known electromagnetic equations in order to treat the electromagnetic properties of superconductors. In ordinary electrodynamics a steady current is associated with a steady magnetic field, but a steady magnetic field does not result in a current. In order to cause a current in a normal conductor, the magnetic field must vary with time. On the other hand, in London's new electrodynamics a steady magnetic field will also cause a steady current in a superconductor. These are the persistent currents of the Meissner effect when a superconductor is cooled to superconducting states in a steady magnetic field. This new theory permits a complete phenomenological description of superconductivity, but it cannot be regarded as a physical theory of fundamental nature.

Owing to the Meissner effect, a metal in a superconducting state never permits a magnetic flux to exist in its interior. As a result, electric currents cannot pass through the body of a superconducting metal, but flow only on the surface. Nevertheless, currents cannot be confined entirely to the mathematical surface of zero thickness, they must flow within a layer of the surface, however thin it may be. Consequently, when a superconducting metal is in an applied magnetic field, the magnetic flux does not fall to zero abruptly at the boundary of the metal, but penetrates the metal and dies away within the surface layer where the screening currents are flowing. The term *penetration depth* is defined as the depth of the surface layer within which if the magnetic flux of the applied field remains constant there would be the same amount of flux inside the metal as the actual penetration. For an infinitely thick specimen, as shown in Fig. 9-7, the penetration depth λ can be defined by

$$\lambda = \frac{1}{B_a} \int_0^\infty B(x)\,dx \qquad (9\text{-}13)$$

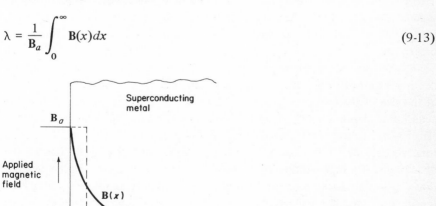

FIG. 9-7. Penetration of magnetic flux into the surface of a superconducting metal.

where $\mathbf{B}_a$ is the magnetic flux of the applied field at the surface, and $\mathbf{B}(x)$ is the magnetic flux at a distance x inside the metal. According to the London theory, the magnetic flux decays exponentially inside a superconducting metal; whence the Landon penetration depth λ_L is defined by the relation

$$\mathbf{B}(x) = \mathbf{B}_a \exp\left(\frac{-x}{\lambda_L}\right) \tag{9-14}$$

Thus the London penetration depth is the distance at which the magnetic flux inside a superconducting metal falls to $1/e = 1/2.718$ of its value at the surface.

The penetration depth is found to vary with temperature according to the equation

$$\lambda = \lambda_0 \left[1 - \left(\frac{T}{T_0}\right)^4\right]^{-1/2} \tag{9-15}$$

where λ_0 is the penetration depth at $0°K$, and T_0 is the transition temperature. The values of λ_0 are, for example, 5.0×10^{-6} cm for aluminum and 3.9×10^{-6} cm for lead.

In the study of the intermediate state of a type I superconductor, we are led to the belief that the phase boundary between normal and superconducting regions cannot be sharp and the changes of properties must be gradual. In terms of the two-fluid model, it implies that the concentration of superelectrons cannot drop abruptly to zero at the boundary, but rather decreases gradually over a certain distance. This distance was called by Pippard[4] the *coherence length* ξ, the value of which was found to be of the order of 10^{-4} cm for pure superconductors. Within the coherence length, the superelectrons do not behave completely independently of each other, but exhibit some sort of "long-range order."

In the classical treatment of superconductivity by London, the penetration depth is assumed to be independent of the strength of the applied field and also of the size of the specimen. However, it has been found experimentally that the penetration depth actually increases with applied magnetic field, although the effect is not large except near the transition temperature for bulk superconductors. As regards to the size effect, it has been found that when one of the dimensions of a superconducting specimen becomes comparable to the penetration depth, its critical magnetic field becomes much higher than that of a bulk specimen of the same material at the same temperature. In 1950 Ginzburg and Landau[5] introduced a new phenomenological theory through the use of

---

[4] A. B. Pippard, *Proc. Roy. Soc.* (*London*), **A203**:210 (1950).
[5] V. L. Ginzburg and L. D. Landau, *J.E.T.P.* (*USSR*), **20**:1064 (1950).

quantum mechanics. In this theory the penetration depth is assumed to vary with the applied magnetic field and the size of specimen in addition to the temperature. One of the contributions of the Ginzburg-Landau theory is its correct prediction that for very thin films the transition between superconducting and normal phases is of the second order, without latent heat or change in entropy but with a discontinuity in specific heat. For thicker films and bulk specimens, the transition in an external field, as discussed in Sec. 9-2, is always of the first order.

All the theories mentioned so far in this section are essentially phenomenological theories. They cannot be regarded as physical theories of fundamental nature. It was not until 1957 that an acceptable fundamental theory was successfully developed when quantum mechanics was applied to the free electrons in a crystal lattice. The first step in establishing the microscopic theory was suggested independently by Fröhlich and Bardeen in 1950 while the successful development of the theory, including the solution of the intricate mathematical problems involved, was accomplished by Bardeen, Cooper, and Schrieffer[6] in 1957.

According to the Bardeen-Cooper-Schrieffer theory, the mechanism leading to superconductivity is due to the manner in which the lattice vibrations are influenced by the presence of free electrons. A metal crystal consists of a lattice of positive ions and a "gas" of free electrons moving at random through the crystal lattice. When one negative electron appears in a lattice of positive ions, the attraction between them gives rise to a slight local contraction of the crystal lattice. This slightly crowded positive ions then serve to attract another electron. The attractive force produced by lattice interaction can take place over a distance of as much as 10^{-4} cm, which is the distance previously postulated by Pippard as the coherence length ξ. The right kind of attractive interaction that causes superconductivity is the one in which pairs of electrons having opposite spins and opposite velocities are involved. Such pairs of electrons, known as *Cooper pairs*, can move as a unit without exchanging energy with the vibrating lattice ions, thus promoting superconductivity.

9-5 TYPE II SUPERCONDUCTORS

In the preceding section we defined two special properties of superconductors, the penetration depth λ and the coherence length ξ. These two properties play an important role in determining the types of superconductivity. In general, λ and ξ are different in length. In most metallic elements λ is shorter than ξ, as

---

[6] J. Bardeen, L. N. Cooper, and J. R. Schrieffer, *Phys. Rev.*, **108**:1175 (1957).

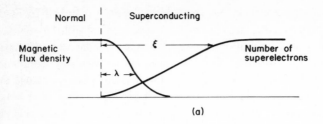

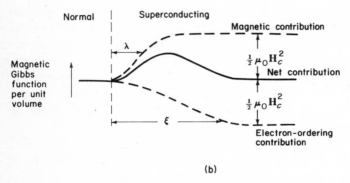

(b)

FIG. 9-8. Type I superconductor. (*a*) Penetration depth λ is shorter than coherence length ξ. (*b*) Surface energy is positive.

shown schematically in Fig. 9-8*a*. In this case, there is a positive surface energy at the boundary between normal and superconducting phases, meaning that extra energy is required to form the boundary.

The origin of positive surface energy at an interphase boundary when $\lambda < \xi$ can be explained as follows. Since the normal phase is virtually nonmagnetic, the application of a magnetic field does not change its magnetic Gibbs function. However, due to the diamagnetic nature the magnetic Gibbs function of the superconducting phase will increase by the amount $\frac{1}{2}\mu_0 H_c^2$ per unit volume under the critical field H_c (Example 9-2). This positive contribution to the magnetic Gibbs function is well inside the superconducting phase and it just cancels the negative contribution due to the presence of ordered superelectrons. As a result, the magnetic Gibbs function per unit volume well inside the superconducting phase will be the same as that of the neighboring normal phase, thus satisfying the condition of equilibrium between phases. However, due to the difference in length between λ and ξ, the positive and negative contributions to the magnetic Gibbs function of the superconducting phase do not cancel near the boundary, as shown schematically by the dotted lines in Fig. 9-8*b*. Thus the total magnetic Gibbs function of the superconducting phase, shown by the solid line in Fig. 9-8b, is increased at the boundary. This is what we called the positive surface energy and is a characteristic of type I superconductors.

On the other hand, in many alloys and a few metallic elements, the coherence length ξ is shorter than the penetration depth λ, as shown schematically in Fig. 9-9. By the same reasoning as described in the preceding paragraph we may conclude that in such a case there is a negative surface energy, meaning that energy is released on forming the interphase boundary. This is a characteristic of *type II superconductors*. If the negative surface energy is sufficiently large, a type II superconducting metal may contain a large number of phase boundaries in order to have a minimum magnetic Gibbs function. Thus at a certain range of temperature and magnetic field the metal would become a mixture of small-scale superconducting and normal regions. The normal regions thread through the superconducting metal with their boundaries lying parallel to the applied magnetic field such that the ratio of surface to volume of normal material is a maximum. Such a state of a type II superconductor is called the *mixed state* or *mixed phase*.

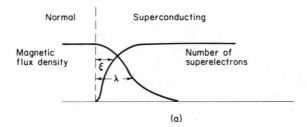

(a)

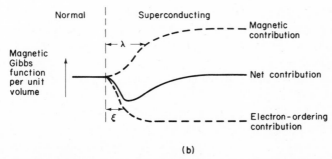

(b)

FIG. 9-9. Type II superconductor. (*a*) Penetration depth λ is longer than coherence length ξ. (*b*) Surface energy is negative.

Since $\lambda < \xi$ for type I superconductors and $\lambda > \xi$ for type II superconductors, it is convenient to define the ratio of λ and ξ by the parameter

$$\kappa_{G-L} = \frac{\lambda}{\xi} \tag{9-16}$$

known as the *Ginzburg-Landau constant*. Thus a superconductor is type I or type II depending approximately on whether its value of κ_{G-L} is less or greater than unity. However, more detailed studies reveal that the distinction between type I and type II superconductors depends more exactly on whether its value of κ_{G-L} is less or greater than $1/\sqrt{2} = 0.71$, instead of unity.

It is important to note that the mixed state defined in this section is different from the intermediate state defined in Sec. 9-3. The intermediate state occurs only in those type I superconductors which have nonzero demagnetizing factor. It is not an intrinsic feature of a type I superconducting material, but depends on the shape of the specimen. On the other hand, a mixed state is an intrinsic feature of a type II superconducting material, and its appearance does not depend on the shape of the specimen. Furthermore, with a positive surface energy a type I superconductor is energetically unfavorable to permit the coexistence of normal and superconducting phases. When the demagnetizing factor is nonzero, coexistent phases do occur in a type I superconductor, but the structure of the intermediate state is relatively coarse with the lowest possible interphase boundary area in order to have a minimum magnetic Gibbs function. However, the structure of a mixed state in a type II superconductor is on a much finer scale, comprising a large number of interphase regions.

The mixed state of a type II superconductor in general covers a wide range of magnetic field from the transition temperature down to the absolute zero. Inside this range the metal remains electrically superconductive, and the penetration of the metal by magnetic flux is a gradual process. The minimum magnetic field strength required to drive a type II superconductor into the mixed state is called the *lower critical field* H_{c1}, and the maximum magnetic field strength up to which the mixed state can persist is called the *upper critical field* H_{c2}. Both the lower and upper critical fields are temperature dependent, as shown schematically in the H-T diagram in Fig. 9-10a.

In Fig. 9-10a, the so-called thermodynamic critical field H_c is also shown. It is defined by the equation

$$H_c(T) = \left\{ \frac{2}{\mu_0} \ [g^{(n)}(T, 0) - g^{(s)}(T, 0)] \right\}^{1/2}$$

which is the same equation as applied to a type I superconductor (Example 9-2). The term $[g^{(n)}(T, 0) - g^{(s)}(T, 0)]$ is the characteristic difference in magnetic

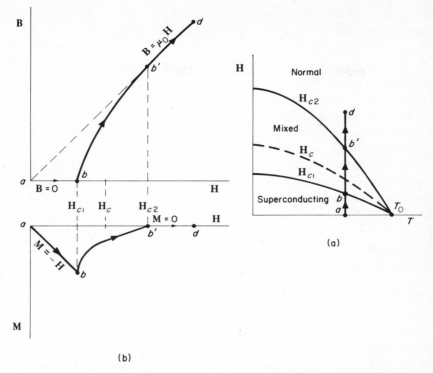

FIG. 9-10. Type II Superconductor. (a) Phase diagram showing the superconducting, mixed, and normal phases. (b) Variations of magnetic induction and magnetization.

Gibbs functions per unit volume for the normal and superconducting states at a temperature T and in the absence of an applied magnetic field. The thermo-dynamic critical field for a type II superconductor may be loosely considered as the critical field of an equivalent type I superconductor with the same transition temperature.

The values of the lower critical field $\mathbf{H}_{c1}$ and the upper critical field $\mathbf{H}_{c2}$ can be expressed approximately in relation to the thermodynamic critical field $\mathbf{H}_c$ by the following equations:

$$\mathbf{H}_{c1} = \frac{\mathbf{H}_c}{(\sqrt{2}\kappa_{G-L})^{0.65}} \tag{9-17}$$

and

$$\mathbf{H}_{c2} = \sqrt{2}\kappa_{G-L}\mathbf{H}_c \tag{9-18}$$

The magnetic properties of a type II superconductor is depicted in Fig.

9-10b by the **B-H** and **M-H** plots for an isothermal process $a\ b\ b'\ d$, in which the applied magnetic field is increased across the lower and upper critical fields. At applied field $H < H_{c1}$ between points a and b, the metal is in superconducting phase, exhibiting perfect diamagnetism just like a type I superconductor. When the applied magnetic field reaches H_{c1} at point b, the metal goes from the completely superconducting phase to the mixed phase, and normal cores are being formed. The magnetic flux in the metal is no longer zero and the magnitude of the magnetization suddenly decreases. When the applied magnetic field is increased beyond H_{c1} between points b and b', as the normal cores pack closer together, the average magnetic flux in the metal increases and the magnitude of the magnetization decreases gradually. As the applied magnetic field reaches H_{c2} at point b' the normal cores merge together and the metal goes to the normal phase. Beyond the upper critical field the metal is normal, thus $B = \mu_0 H$ and $M = 0$. Note that Fig. 9-10 for a type II superconductor is equivalent to Fig. 9-4 for a type I superconductor.

For many years the anomalous behavior of a mixed state was thought to be primarily due to impurities. It was not until 1957 that the existence of the mixed state was finally recognized as an intrinsic feature of a group of super-conductors of what we called type II when Abrikosov[7] published a full theory of this type of superconductivity. Since the range of a mixed state generally extends to very high magnetic fields, a type II superconductor can remain super-conducting in a very large magnetic field or can carry a large current without becoming normal. As a result, type II superconductors are quite useful in many applications, especially in the winding of superconducting magnets to generate strong magnetic fields.

9-6 SUPERFLUIDITY

Of all known materials, helium at very low temperatures possesses the most spectacular and unique properties. For instance, when liquid He^4 is cooled by boiling under reduced pressure, in the liquid helium I region it boils violently with a very turbulent liquid surface. But as soon as the λ point of $2.17°K$ is passed and the liquid helium II region is entered, the violent ebullition suddenly stops and the liquid becomes very quiet. The reason for this radical change is that the heat transport in liquid helium II is so rapid that it is virtually impossible to set up appreciable temperature gradients within the liquid, and hence all the evaporation takes place from the liquid surface without any bubbling.

[7] A. A. Abrikosov, *J.E.T.P.* (*USSR*), **32**:1442 (1957).

An unusual and seemingly unbelievable phenomenon to watch is the creeping film of liquid helium II. If an empty glass beaker is partially immersed in a reservoir of liquid helium II, as illustrated in Fig. 9-11(a), a liquid film (of the order of 10^{-6} cm thick) will creep up the outside wall of the glass beaker, go over the top, and then run down the inside wall; thus gradually filling the glass beaker until the surface levels inside and outside are equal. If the glass beaker is lifted to a position as illustrated in Fig. 9-11(b), the creeping film will now flow in the opposite direction until the surface levels again are equal. Even if the glass beaker is listed entirely out of the reservoir, as illustrated in Fig. 9-11(c), the liquid in the glass beaker will gradually creep out and drip off the bottom at a definite rate.

As mentioned briefly in Sec. 8-3, the creeping film is the phenomenon that set the lower temperature limit which can be reached by evaporating He^4. When liquid helium II creeps up the container walls and reaches the warmer region on the top it evaporates. Part of this vapor then recondenses, thus transferring heat to the liquid helium and preventing further decrease in temperature.

The most spectacular and strange phenomenon of liquid helium II is the *thermomechanical effect*, commonly known as the *fountain effect*. To demonstrate this phenomenon, an open glass tube with a capillary on the upper end and a fine powder package on the lower end, is immersed in a bath of liquid helium II, as shown in Fig. 9-12. When heat is supplied to the liquid helium between the powders by thermal radiation or by a small electric heater, a jet of the liquid will eject from the top opening of the tube. Liquid helium fountains as high as 30 cm have been observed.

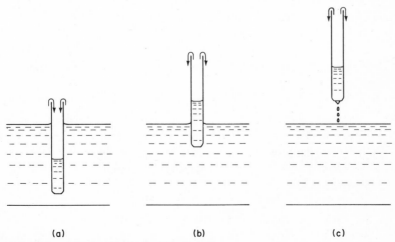

(a)	(b)	(c)

FIG. 9-11. Creeping film in liquid helium II.

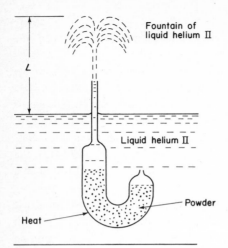

L

Fountain of
liquid helium II

Liquid helium II

Powder

Heat

FIG. 9-12. Apparatus for showing the thermomechanical (fountain) effect.

All the unusual phenomena exhibited by liquid helium II are somehow related to a basic property, namely the viscosity of the liquid. Measurements on
the flow of liquid helium through very fine capillaries or narrow cracks reveal
that as the temperature is lowered to the λ point, the rate of flow increases
abruptly and rapidly. This indicates that liquid helium II has vanishingly low
viscosity. This rather peculiar behavior of liquid helium II has led to the name
superfluidity for it.

However, there is a paradox in the measurements of viscosity of liquid
helium II. If viscous drag is measured by the rotating disk method, liquid helium
II behaves like a normal liquid with a well-defined coefficient of viscosity. It was
this viscosity paradox that initially stimulated the concept of two-fluid model of
liquid helium II, which we will now introduce.

Based on quantum statistical mechanics two theories were advanced independently by London[8] and Landau.[9] London's idea was immediately used by
Tisza[10] to formulate a microscopic phenomenological theory called the *two-
fluid model* of liquid helium II. According to this theory liquid helium II is
postulated to contain two freely intermingled fluids: the normal fluid with
normal viscosity, and the superfluid with zero viscosity and zero-point energy
and entropy. For simplicity we will assign zero entropy to the superfluid. If ρ is
the density of liquid helium II, ρ_n the density of normal fluid, and ρ_s the
density of superfluid, then

[8] F. London, *Nature*, **141**:643 (1938); *Phys. Rev.*, 54: 947 (1938).

[9] L. D. Landau, *J. Phys. (USSR)*, 5:71 (1941).

[10] L. Tisza, *Nature*, **141**:913 (1938); *Phys. Rev.*, 72:838 (1947).

$$\rho = \rho_n + \rho_s \tag{9-19}$$

At the λ point, the entire fluid is normal with $\rho_n/\rho = 1$, whereas at $0°K$ the entire fluid is superfluid with $\rho_n/\rho = 0$. Figure 9-13 shows the normal fluid concentration ρ_n/ρ versus temperature. The normal fluid concentration can be measured directly by employing an oscillating disk technique. It can also be evaluated indirectly from measurements of the velocity of "second sound," which we will study in Sec. 9-8.

The two-fluid model as postulated in the preceding paragraph is capable of explaining most the striking phenomena concerning liquid helium II. To resolve the paradox in the measurements of viscosity of liquid helium II, we note that in the flow type of measurement the motion of the normal fluid is restricted by its viscosity, whereas the superfluid moves freely without friction. On the other hand, in the rotating disk measurement, since superfluid has no viscosity and transmits no torque, the measured viscosity is solely that of the normal fluid. Hence the two types of measurements actually deal with two different fluids and there is no contradiction.

To explain the creeping film, as illustrated in Fig. 9-11, we note that the surfaces of the beaker which are exposed to the liquid are in equilibrium with the saturated vapor of the liquid and therefore will be covered with a thin film of adsorbed helium. This film acts as a form of capillary siphon through which the superfluid can move without friction.

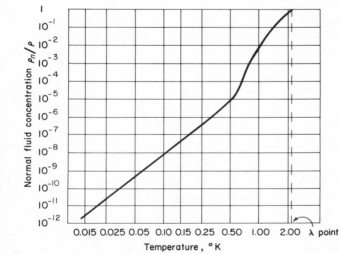

FIG. 9-13. Normal fluid concentration ρ_n/ρ versus temperature. (*Data from D. de Klerk, R. P. Hudson, and J. R. Pellam,* Phys. Rev., *93:28 (1954).*)

In the next section we will use the two-fluid model to explain the fountain effect and to derive equations pertaining to this phenomenon.

9-7 THERMOMECHANICAL EFFECT

The two-fluid model offers a very simple explanation for the *thermomechanical* or *fountain effect*. When heat is supplied to the liquid helium II in the glass tube, shown in Fig. 9-12, some of the superfluid in it is transformed into normal. Since normal fluid cannot pass through the narrow passages in the fine powder package because of its viscosity, but superfluid with no viscosity can pass through freely, more superfluid will enter the tube to equilize its concentrations on both sides of the narrow passages. As a result of this inflow of superfluid, the amount of liquid helium in the tube quickly builds up. This liquid finally escapes from the tube through the open end of the capillary in the form of a fountain.

To derive the equations pertaining to the thermomechanical effect, let us consider two vessels A and B, each equipped with a piston and connected with a fine capillary, as shown in Fig. 9-14. Contained in the two vessels are two quantities of liquid helium II, which are maintained at different temperatures T_0 and T by two heat reservoirs. Since the fine capillary is impassable to the normal fluid, but passable to the superfluid, it acts as a form of semipermeable membrane. Suppose both pistons are moved slowly to the right so that the pressure p_0 in A and p in B are kept constant. During this process an infinitesimal mass dm of superfluid is transferred from A to B. Since the superfluid carries no entropy, the specific entropy tends to increase in A and decrease in B; thus the temperature tends to increase in A and decrease in B. To prevent these changes a quantity of heat

$$dQ_0 = T_0 s_0 dm$$

must be transferred from A to the reservoir at T_0, while a quantity of heat

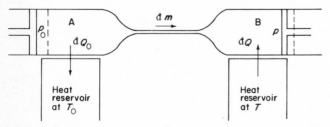

FIG. 9-14. Reversible transfer of superfluid mass dm through a capillary tube.

$$dQ = Ts\,dm$$

must be transferred to B from the reservoir at T, where s_0 and s are the specific entropies of liquid helium II in A (at constant T_0 and p_0) and in B (at constant T and p). The net heat transfer to the liquid helium in both vessels is then

$$dQ - dQ_0 = (Ts - T_0 s_0)\,dm$$

Now from the first law of thermodynamics we have

$$(Ts - T_0 s_0)\,dm = (u - u_0)\,dm + (pv - p_0 v_0)\,dm$$

from which we obtain

$$g_0(T_0, p_0) = g(T, p) \qquad (9\text{-}20)$$

where u_0, v_0, and g_0 are the specific internal energy, specific volume, and specific Gibbs function respectively of liquid helium II at T_0 and p_0; while u, v, and g are the corresponding properties at T and p.

From Eq. (9-20) we see that during a reversible transfer of superfluid between two vessels, each maintained at constant temperature and pressure, there is no change in specific Gibbs function; thus $dg = 0$ for the process. Therefore from Eq. (2-11) we have

$$-s\,dT + v\,dp = 0$$

or

$$\left(\frac{\partial p}{\partial T}\right)_g = \frac{s}{v} \qquad (9\text{-}21)$$

It follows that there is a pressure difference associated with a temperature difference between the two vessels. This is what we call the thermomechanical effect.

EXAMPLE 9-6. Liquid helium II at a temperature of $1.6°$K is contained in an apparatus for showing the thermomechanical effect (see Fig. 9-12). If the temperature increase is $0.005°$K, what would be the height of the fountain? Calculate also the heat required per gram of helium in the fountain. The specific entropy of liquid helium between $1.2°$K and $2.17°$K can be represented by the empirical formula $s = 0.005\,T^{5.6}$, where s is in cal/(g)($°$K) and T in $°$K.

Solution.
Let L denote the level difference or the height of the helium fountain column in cm, ρ denote the helium density in g/cm^3, and g denote the gravitational acceleration in cm/sec^2. The pressure of the helium column in dynes/cm^2 is then given by

$$p = L\rho\text{g}$$

Substituting this equation into Eq. (9-21) leads to

$$\left(\frac{\partial p}{\partial T}\right)_g = \rho g \left(\frac{\partial L}{\partial T}\right)_g = \frac{s}{v}$$

Whence

$$\left(\frac{\partial L}{\partial T}\right)_g = \frac{s}{g}$$

Now for $T = 1.6°K$, we have

$$s = 0.005(1.6)^{5.6} = 0.0695 \text{ cal/(g)(°K)}$$

Hence

$$\left(\frac{\partial L}{\partial T}\right)_g = \frac{(0.0695 \text{ cal/(g)(°K))}(4.187 \times 10^7 \text{ erg/cal})}{980 \text{ cm/sec}^2} = 2.97 \times 10^3 \text{ cm/°K}$$

Since the temperature increase is $0.005°K$, the height of the fountain column is then

$$L = (2.97 \times 10^3 \text{ cm/°K})(5 \times 10^{-3}°K) = 14.9 \text{ cm}$$

The heat required to produce the fountain effect per gram of helium in the fountain is given by

$$Q = Ts = (1.6°K)(0.0695 \text{ cal/(g)(°K))} = 0.111 \text{ cal/g}$$

9-8 SECOND SOUND

There is still another unique and interesting property of liquid helium II—its ability to transmit thermal waves. We will explain this phenomenon according to the two-fluid theory. When heat is supplied locally for a short period at a point in a bath of liquid helium II, the local concentration of normal fluid will become larger. This local variation in the relative densities of the normal and superfluids will be propagated away from the source of temperature pulse, causing a temperature increase while it is passing a point. The temperature drops again as soon as the wave passes the point. Such a wave propagation is analogous to the acoustic wave of ordinary sound. But instead of the propagation of fluctuations in the density ρ of the whole medium as in ordinary sound, it is the propagation of local variations in the ratio ρ_n/ρ_s without altering the sum $\rho_n + \rho_s = \rho$. The name *second sound* has been given to the thermal waves in contrast with the name first sound (ordinary sound) for the acoustic waves. Of course, the second sound cannot be heard, it can be detected only with sensitive thermometers.

In order to derive the equation for the velocity of second sound we will follow the simple, though not quite rigorous, derivation given by Gogate and

Pathak[11] and restrict ourselves to the one-dimensional case. Let ρ_n, v_n and ρ_s, v_s denote the densities and velocities of motion for the normal and superfluids, respectively. The density ρ of the entire fluid is given by

$$\rho = \rho_n + \rho_s \tag{9-19}$$

Since the normal and superfluids possess independent motions and there is no net mass flow, we may write

$$\rho_n v_n + \rho_s v_s = 0 \tag{9-22}$$

Inasmuch as the superfluid contributes no entropy, the entropy of the entire fluid is attributed to the normal fluid only. Thus we may write

$$\rho s = \rho_n s_n \tag{9-23}$$

where s and s_n refer to the specific entropies of the entire fluid and the normal fluid respectively.

The kinetic energy per unit volume of the fluid is given by

$$K = \frac{1}{2} (\rho_n v_n^2 + \rho_s v_s^2)$$

The use of Eq. (9-19) permits the preceding equation to be written as

$$K = \frac{1}{2\rho} (\rho_n v_n + \rho_s v_s)^2 + \frac{1}{2\rho} \rho_n \rho_s (v_n - v_s)^2$$

The first term on the right of this equation represents the kinetic energy of the actual mass transfer when the normal and superfluids move together; the second term represents the kinetic energy of internal convection when the normal and superfluids move in opposite directions. In the propagation of second sound, the velocity v_n moves in the direction of propagation and the velocity v_s moves in the opposite direction, but there is no net mass motion. Hence in this case the kinetic energy per unit volume of the fluid is then

$$K = \frac{1}{2\rho} \rho_n \rho_s (v_n - v_s)^2$$

Using Eqs. (9-19) and (9-22), the preceding equation can be rewritten to read

$$K = \frac{1}{2\rho_s} \rho \rho_n v_n^2 \tag{9-24}$$

Now consider a layer of liquid helium II in the form of a parallelopiped of

[11] D. V. Gogate and P. D. Pathak, *Proc. Phys. Soc. (London)*, 59:457 (1947).

unit cross-section and thickness dx, across which the temperature varies by dT. The amount of entropy entering the volume element per unit time through one unit-area face (which is perpendicular to the direction of wave propagation) is

$$\rho_n v_n s_n = \rho v_n s$$

The amount of entropy leaving the volume element per unit time through the opposite parallel face is

$$\rho v_n s + \frac{\partial}{\partial x} (\rho v_n s) dx$$

The net outflow of entropy per unit time per unit volume is then $(\partial/\partial x)(\rho v_n s)$. This accounts for the decrease in entropy in the volume element per unit time per unit volume, which can be expressed as $(- \partial/\partial \tau)(\rho s)$ where τ denotes time. Thus

$$\frac{\partial}{\partial x} (\rho v_n s) = - \frac{\partial}{\partial \tau} (\rho s) \tag{9-25}$$

This is the continuity equation for the entropy in the volume element. This equation expresses the fact that the entropy flow follows the movement of the normal fluid in the direction of wave propagation.

We ought to mention now that in our derivation we are to consider terms linear to the following items as small of first order and will neglect products in any two of these quantities: v_n, v_s, and the gradients and the rates of changes of T, s, ρ_n and ρ_s. In addition, if we assume ρ to be sensibly constant we can rewrite Eq. (9-25) as

$$\frac{\partial s}{\partial \tau} = -s \frac{\partial v_n}{\partial x} \tag{9-26}$$

From the second law of thermodynamics we know that when an amount of heat Q is transferred reversibly from T to $T - dT$, the work done is given by

$$\text{Work} = Q \frac{dT}{T}$$

Now the heat current across the layer of thickness dx is $\rho_n v_n s_n T = \rho s T v_n$. This heat current carries the amount of heat

$$Q = \rho s T v_n d\tau \tag{9-27}$$

through the layer of unit cross-section in time $d\tau$. Therefore the reversible work done is

$$\text{Work} = (\rho s T v_n d\tau) \frac{dT}{T} = \rho s v_n d\tau dT$$

This work brings about a change in kinetic energy of internal convection within the layer of unit cross-section and thickness dx. We accordingly write

$$dK \, dx = -\rho s v_n d\tau dT \quad \text{or} \quad \frac{\partial K}{\partial \tau} = -\rho s v_n \frac{\partial T}{\partial x} \tag{9-28}$$

The derivative of the kinetic energy per unit volume with respect to time can be obtained by differentiating Eq. (9-24). When the second-order terms are neglected, we have

$$\frac{\partial K}{\partial \tau} = \frac{\rho \rho_n}{\rho_s} v_n \frac{\partial v_n}{\partial \tau}$$

Substituting this equation in Eq. (9-28) we obtain

$$\frac{\partial v_n}{\partial \tau} = -\frac{s \rho_s}{\rho_n} \frac{\partial T}{\partial x} \tag{9-29}$$

Since the specific heat for liquid helium is

$$c = T \left(\frac{\partial s}{\partial T} \right)_v \approx T \left(\frac{\partial s}{\partial T} \right)_p$$

we have

$$\frac{\partial T}{\partial s} = \frac{T}{c} = \frac{\partial T}{\partial x} \frac{\partial x}{\partial s} \quad \text{or} \quad \frac{\partial T}{\partial x} = \frac{T}{c} \frac{\partial s}{\partial x}$$

Substituting this equation in Eq. (9-29) gives

$$\frac{\partial v_n}{\partial \tau} = -\frac{s \rho_s}{\rho_n} \frac{T}{c} \frac{\partial s}{\partial x} \tag{9-30}$$

Differentiating Eq. (9-26) with respect to τ and Eq. (9-30) with respect to x and combining them results in

$$\frac{\partial^2 s}{\partial \tau^2} = \frac{\rho_s}{\rho_n} \frac{s^2 T}{c} \frac{\partial^2 s}{\partial x^2}$$

or

$$v_2 = \frac{\partial x}{\partial \tau} = \left(\frac{\rho_s}{\rho_n} \frac{s^2 T}{c} \right)^{1/2} \tag{9-31}$$

where v_2 denotes the velocity of second sound.

Figure 9-15 shows the experimentally determined temperature dependence of the velocity of second sound in liquid helium II. At the λ point the velocity is zero; between $1°K$ to almost $2°K$ it remains fairly constant; but below $1°K$ it

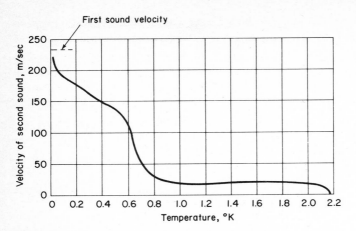

FIG. 9-15. Temperature dependence of the velocity of second sound.
(*Data below 1°K from D. de Klerk, R. P. Hudson, and J. R. Pellam,*
Phys. Rev., **93**:*28 (1954); and data above 1°K from V. Peshkov, J.*
Phys. (USSR), **10**:*389 (1946).*)

rises rapidly to very high values, approaching the velocity of ordinary sound near
$0°K$.

   EXAMPLE 9-7. Calculate the normal fluid concentration ρ_n/ρ and the
density ratio ρ_s/ρ_n for liquid helium II at $1°K$ from the following data:[12]

   specific heat $c = 104.2$ joules/(kg)($°K$)
   specific entropy $s = 16.8$ joules/(kg)($°K$)
   velocity of second sound $v_2 = 19.35$ m/sec

Solution.
From Eq. (9-31) we have

$$\frac{\rho_s}{\rho_n} = \frac{cv_2^2}{s^2 T}$$

   But

$$\rho_n + \rho_s = \rho$$

   Thus

$$\frac{\rho}{\rho_n} - 1 = \frac{cv_2^2}{s^2 T}$$

---

[12] Data for c and s from H. C. Kramers, J. D. Wasscher, and C. J. Gorter, *Physica*, **18**:
329 (1952); datum for v_2 from D. de Klerk, R. P. Hudson, and J. R. Pellam, *Phys. Rev.*,
93:28 (1954).

or

$$\frac{\rho_n}{\rho} = \left[1 + \frac{cv_2^2}{s^2 T}\right]^{-1}$$

Inserting numerical values gives

$$\frac{\rho_n}{\rho} = \left[1 + \frac{(104.2)(19.35)^2}{(16.8)^2(1.0)}\right]^{-1} = 7.18 \times 10^{-3}$$

Whence

$$\frac{\rho_s}{\rho} = 1 - \frac{\rho_n}{\rho} = 1 - 0.00718 = 0.993$$

and

$$\frac{\rho_s}{\rho_n} = \frac{\rho_s/\rho}{\rho_n/\rho} = \frac{0.993}{0.00718} = 138$$

PROBLEMS

9-1. From a study of literature on superconductivity, give a brief description on the following superconductive devices:

(a) superconducting bearings
(b) superconducting motors
(c) superconducting magnets
(d) any other superconductive devices of interest.

9-2. Discuss the current status of the use of superconducting computer elements.

9-3. Starting from the equation

$$dG = -S\, dT + V\, dp - \mu_0 V \mathbf{M}\, d\mathbf{H}$$

derive the following three equations for superconducting phase transitions:

$$\left(\frac{\partial p}{\partial T}\right)_{\mathbf{H}} = \frac{S^{(n)} - S^{(s)}}{V^{(n)} - V^{(s)}}$$

$$\left(\frac{\partial \mathbf{H}_c}{\partial p}\right)_T = \frac{V^{(n)} - V^{(s)}}{\mu_0 V^{(s)} \mathbf{H}_c}$$

$$\left(\frac{\partial \mathbf{H}_c}{\partial T}\right)_p = -\frac{S^{(n)} - S^{(s)}}{\mu_0 V^{(s)} \mathbf{H}_c}$$

Note that the last of the preceding three equations is for the most common case of superconducting phase change.

9-4. The transition from normal to superconductivity of a type I superconductor in the absence of a magnetic field is a second-order phase transition. Starting from the equations obtained in the preceding problem, derive the two Ehrenfest's equations (Sec. 5-9) for this transition.

9-5. Calculate the heat absorbed and the work done by 100 cm^3 of lead during the transition from the superconducting to the normal phase at 3°K, assuming that the transition curve follows the parabolic rule of Eq. (9-1).

9-6. Solve the preceding problem using the data given in Fig. 9-2.

9-7. For a superconductor whose transition curve follows the parabolic rule of Eq. (9-1), at what value of T/T_0 will the latent heat of transition from superconducting to normal phase be a maximum? Determine this maximum value of latent heat for lead. The specific volume of lead is $v = 18.3$ cm^3/g mole.

9-8. The specific heats of tin (Sn) in the superconducting state and normal state can be represented approximately up to its transition temperature of 3.72°K by the respective equations

$$c^{(s)} = 464.5 \left(\frac{T}{140}\right)^3$$

and

$$c^{(n)} = 464.5 \left(\frac{T}{185}\right)^3 + 4 \times 10^{-4} T$$

where c is in cal/(g mole)(°K) and T in °K. Evaluate for tin the critical field at absolute zero temperature, H_0, and the slope of critical field curve at the transition temperature, $(dH_c/dT)_{T = T_0}$. The specific volume of tin is $v = 16.1$ cm^3/g mole.

9-9. For aluminum, $T_0 = 1.2$°K, $H_0 = 0.79 \times 10^4$ amp/m, and $v = 9.93$ cm^3/g mole. Calculate the values of $(c^{(s)} - c^{(n)})$ at various temperatures up to T_0 and plot these values against temperature.

9-10. The ratio between the thermal conductivities of a metal in the normal state $k^{(n)}$ and in the superconducting state $k^{(s)}$, both at the same temperature T, can be represented approximately by the equation

$$\frac{k^{(n)}}{k^{(s)}} = \left(\frac{T_0}{T}\right)^2$$

where T_0 is the transition temperature. Calculate this ratio for lead at 0.5°K. With the answer obtained in this problem as a guide, explain the working procedure of a thermal valve in a magnetic refrigerator.

9-11. The critical field of lead (a type I superconductor) at 4.2°K is 4.4 $\times$ 10^4 amp/m. If this is the thermodynamic critical field of a certain type II superconductor, estimate its upper and lower critical fields, assuming that the Ginzburg-Landau constant of this type II superconductor is 30.

9-12. With the data in the preceding problem as a guide, discuss the importance of type II superconductors in modern technology.

9-13. For liquid helium II at $0.7°K$, $c = 0.0098$ joule/(g)($°K$), $s = 0.00276$ joule/(g)($°K$), and $\rho_n/\rho = 2.00 \times 10^{-4}$. Calculate ρ_s/ρ, ρ_s/ρ_n, and the velocity of second sound at $0.7°K$.

9-14. Between $1°$ to $1.8°K$, the velocity of second sound is approximately 20 m/sec. What is the variation of ρ_n/ρ with temperature in this range of temperature? Assume that the specific entropy of liquid helium II between $1°$ to $1.8°K$ can be represented by the equation

$$s = (1.847 - 2.955\ T + 1.238\ T^2) \times 10^3$$

where s is in joule/(kg)($°K$) and T is in $°K$.

Chapter 10
THE THIRD LAW OF THERMODYNAMICS

10-1 THE THIRD LAW

A basic law of thermodynamics was born from the attempt to calculate equilibrium constants of chemical reactions entirely from thermal data (i.e., enthapies and heat capacities). What has come to be known as the third law of thermodynamics had its origin in the Nernst heat theorem, enunciated by Nernst in 1906. It stated that there is no charge in entropy if a chemical change takes place between pure crystalline solids at absolute zero. This theorem was later modified and generalized by Planck, Simon, Lewis, Guggenheim, and others. One well-known version of the modified theorem, sometimes known as the *Nernst-Simon statement of the third law of thermodynamics* is as follows: The entropy change associated with any isothermal reversible process of a condensed system (i.e., a solid or a liquid) approaches zero as the temperature approaches absolute zero. Symbolically we may write

$$\lim_{T \to 0} (\Delta S)_T = 0 \tag{10-1}$$

In view of the empirical fact that in any cooling process the lower the temperature achieved, the more difficult it is to go lower, another form of the third law has been advanced. It is stated as follows: It is impossible by means of any process, no matter how idealized, to reduce the temperature of a system to the absolute zero in a finite number of steps. This is known as the *unattainability statement of the third law of thermodynamics.*

The Nernst-Simon and the unattainability statements of the third law are entirely equivalent to each other in their consequences. To demonstrate this equivalence, it is convenient to use a paramagnetic system in typical magnetic cooling processes, such as shown in Fig. 10-1. According to the Nernst-Simon statement, the curves for constant magnetic fields must come together at the absolute zero temperature in order to satisfy the condition that $(\Delta S)_T \to 0$ as

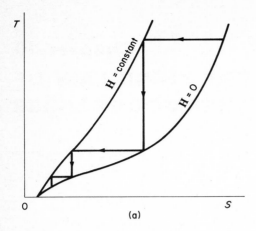

(a)

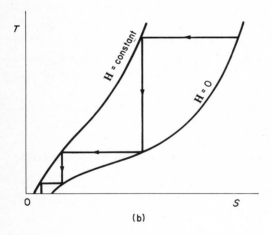

FIG. 10-1. Schematic temperature-entropy diagrams for magnetic cooling processes. (a) In
agreement with the third law.
(b) In violation of the third law.

(b)

$T \to 0$. Figure 10-1a shows such correct curves. It is clear that the temperature
drops in successive demagnetizations gradually decrease and an infinite number
of demagnetization steps would be required to attain the absolute zero of temperature, thus confirming the unattainability statement. On the other hand, if
the Nernst-Simon statement is not obeyed and the curves for constant magnetic
fields do not meet at the absolute zero, as depicted in Fig. 10-1b, then finite
steps could suffice to lower the temperature down to the absolute zero. This is,
of course, a violation of the unattainability statement.

The above argument concerning the equivalence of the two statements of the
third law may be expressed more formally by writing the equation for $(\Delta S)_T$ as
$T \to 0$. We are to show that the violation of either statement can always be made

to result in a violation of the other one. Let us suppose that the paramagnetic system passes reversibly and adiabatically from state i (with temperature T_i and entropy S_i) to state f (with temperature T_f and entropy S_f), as depicted in Fig. 10-2. The entropy change between state a (with $T = 0$ and $\mathbf{H} = \mathbf{H}_i$) and state i is

$$S_i - S_a = \int_0^{T_i} \frac{C_{H=H_i}}{T} \, dT$$

where C_H is the heat capacity at constant magnetic field. Similarly, the entropy change between state b (with $T = 0$ and $\mathbf{H} = 0$) and state f is

$$S_f - S_b = \int_0^{T_f} \frac{C_{H=0}}{T} \, dT$$

Since $S_i = S_f$, it follows that

$$S_a - S_b = \int_0^{T_f} \frac{C_{H=0}}{T} \, dT - \int_0^{T_i} \frac{C_{H=H_i}}{T} \, dT \tag{10-2}$$

Now suppose that there were a certain value of T_i that would lead to $T_f = 0$, thereby violating the unattainability statement. Accordingly, Eq. (10-2) would reduce to

$$S_a - S_b = -\int_0^{T_i} \frac{C_{H=H_i}}{T} \, dT \tag{10-3}$$

As C_H is a positive quantity, we would then have a negative value for $(S_a - S_b)$, a clear violation of the Nernst-Simon statement. On the other hand, suppose that

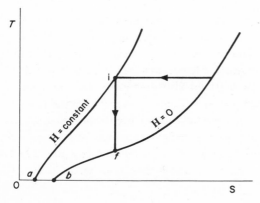

FIG. 10-2. A temperature-entropy diagram in reference to Eq. (10-2).

$(S_a - S_b)$ had a negative value, thereby violating the Nernst-Simon Statement. Then there would be a certain value of T_i that would render Eq. (10-2) to reduce to Eq. (10-3), thus making the first integral in Eq. (10-2) to vanish and $T_f = 0$, thereby violating the unattainability statement.

There are many experimental processes which support the validity of the third law. One of the most convincing evidences comes from the solid-liquid phase transition of ordinary helium (He^4) at extremely low temperatures. As shown in Fig. 10-3a (see also Fig. 8-5), the melting curve changes its gradient abruptly in the neighborhood of the upper triple point, and the gradient dp/dT vanishes rapidly as absolute zero temperature is approached, as shown in Fig. 10-3b. Since this transition is of the first order, from the Clapeyron equation we have

$$\frac{dp}{dT} = \frac{S^{(\alpha)} - S^{(\beta)}}{V^{(\alpha)} - V^{(\beta)}}$$

where the superscripts (α) and (β) denote liquid and solid phases respectively. The vanishing gradient of the melting curve corresponds to the vanishing entropy difference between the liquid and the solid as absolute zero temperature is approached. Figure 10-3c shows how the entropy curves of the liquid and the solid converge as temperature approaches absolute zero.

Another striking verification of the third law is seen from the transition between normal and superconductivity of a metal at the absolute zero temperature. As shown in Fig. 9-1, the gradient of the critical magnetic field curve approaches zero as the absolute zero temperature is approached. Now we recall that the gradient of the critical field curve of a metal is related to the entropy difference between normal and superconducting phases by the equation

$$S^{(n)} - S^{(s)} = -\mu_0 V H_c \frac{dH_c}{dT} \tag{9-4}$$

The vanishing gradient of the critical field curve corresponds to the vanishing entropy difference as absolute zero temperature is approached.

To conclude the general discussion on the third law of thermodynamics, we need to mention still another statement of it postulated by Planck. We realize that both the Nernst-Simon and the unattainability statements call for the disappearance of the entropy differences between different states of a system at the absolute zero temperature. It is not necessary that the entropies themselves vanish. Planck in 1911 extended the Nernst theorem by making the additional postulate that the absolute value of the entropy of a pure solid or a pure liquid approaches zero as the temperature approaches absolute zero or symbolically,

$$\lim_{T \to 0} S = 0 \tag{10-4}$$

This postulate is reasonable only when there is no isotope effect and when no net nuclear or electronic moments are possessed by the atoms of the material. A material composed of atoms with nuclei having nonvanishing nuclear spins will possess a so-called zero-point entropy when it is at the absolute zero temperature. Nevertheless, Planck's postulate is often used to establish a conventional scale of "absolute" entropies for tabulational purposes.

It is imperative to note that the statement as expressed in Eq. (10-4) applies only to systems in thermodynamic equilibrium including metastable equilibrium.

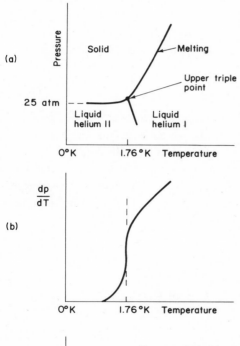

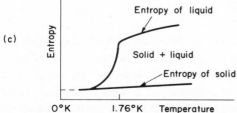

FIG. 10-3. (*a*) Melting curve of He^4. (*b*) The gradient dp/dT of the melting curve of He^4. (*c*) The entropies of liquid and solid He^4.

It cannot apply to systems in nonequilibrium states. To be more precise we may restate it as follows: The contribution to the entropy by each aspect which is in thermodynamic equilibrium tends to zero at absolute zero temperature.

10-2 THERMODYNAMIC PROPERTIES AS TEMPERATURE APPROACHES ABSOLUTE ZERO

The third law of thermodynamics is concerned with the limiting behavior of systems as the temperature approaches absolute zero. Based on this law we are now to derive a number of limiting relationships for thermodynamic properties applicable at the absolute zero of temperature.

It is well known experimentally that the specific heats of all substances tend to zero as the absolute zero of temperature is approached. For instance, for a dielectric solid, the specific heat arises solely from the vibrations of the lattice. Its value at low temperatures vanishes as the cube of the temperature, as given by the *Debye equation*,

$$c_p \approx c_v = 464.5 \, T^3/\Theta^3 \text{ cal/(g mole)(°K)} \tag{10-5}$$

where Θ is the Debye temperature, which is a characteristic of a given substance. For a metal, however, a linear term of T must be added to the Debye equation to account for the specific heat of the electrons. At the absolute zero of temperature, the lattice and electronic contributions to the specific heat vanish independently. The vanishing of specific heat, at the absolute zero temperature can be verified analytically from the third law. Thus, for instance, for the case of a system at constant pressure, from Eq. (2-27) we have

$$C_p = T\left(\frac{\partial S}{\partial T}\right)_p$$

Integrating this equation between $0°K$ and $T°K$ yields

$$S_T - S_0 = \int_0^T \frac{C_p dT}{T} \tag{10-6}$$

where S_T and S_0 are the entropies of the system at temperature T and absolute zero respectively, and C_p is the heat capacity at constant pressure. Now since the third law asserts that the entropy of any substance must be finite or zero at absolute zero, and finite values are observed for ΔS at higher temperatures, it follows that the entropy of a substance must be finite at any finite temperature. Consequently, from Eq. (10-6) we must conclude that

$$\lim_{T \to 0} C_p = 0 \tag{10-7}$$

Indeed, if C_p remained constant down to the absolute zero of temperature, the integral in Eq. (10-6) would diverge at its lower limit, and hence the entropy would not be finite. By a similar argument, we can show that

$$\lim_{T \to 0} C_v = 0 \tag{10-8}$$

In fact, we can make the general conclusion that all heat capacities tend to zero as the temperature approaches absolute zero.

The derivatives of many thermodynamic properties are related to the derivatives of the entropy through the Maxwell relations. The limiting behavior of these properties can be written with the help of the third law. For example, for a simple compressible system, we have

$$\left(\frac{\partial S}{\partial V}\right)_T = \left(\frac{\partial P}{\partial T}\right)_V \tag{2-13c}$$

and

$$\left(\frac{\partial S}{\partial p}\right)_T = -\left(\frac{\partial V}{\partial T}\right)_p \tag{2-13d}$$

But by the third law,

$$\lim_{T \to 0} \left(\frac{\partial S}{\partial V}\right)_T = 0 \quad \text{and} \quad \lim_{T \to 0} \left(\frac{\partial S}{\partial p}\right)_T = 0$$

Consequently,

$$\lim_{T \to 0} \left(\frac{\partial p}{\partial T}\right)_V = 0 \tag{10-9}$$

and

$$\lim_{T \to 0} \left(\frac{\partial V}{\partial T}\right)_p = 0 \tag{10-10}$$

Since the coefficient of thermal expansion α is defined as

$$\alpha = \frac{1}{V} \left(\frac{\partial V}{\partial T}\right)_p \tag{2-17}$$

it follows that the coefficients of thermal expansion of solids and liquids must vanish at the absolute zero of temperature. This is confirmed by experiment.

EXAMPLE 10-1. For a simple paramagnetic system, prove that

$$\lim_{T \to 0} \left(\frac{\partial \mathbf{M}}{\partial T}\right)_{\mathbf{H}} = 0 \quad \text{and} \quad \lim_{T \to 0} \left(\frac{\partial \chi}{\partial T}\right)_{\mathbf{H}} = 0$$

Using the third law, explain the limitation of Curie's equation.

Solution.

For a simple paramagnetic system, a Maxwell relation (Eq. 7-14d) shows that

$$\left(\frac{\partial S}{\partial \mathbf{H}}\right)_{T} = \mu_0 V \left(\frac{\partial \mathbf{M}}{\partial T}\right)_{\mathbf{H}}$$

By the third law, $(\partial S/\partial \mathbf{H})_T$ must vanish as $T \to 0$. It follows that

$$\lim_{T \to 0} \left(\frac{\partial \mathbf{M}}{\partial T}\right)_{\mathbf{H}} = 0$$

Since the magnetic susceptibility χ is defined by the equation

$$\chi = \frac{V\mathbf{M}}{\mathbf{H}}$$

we have

$$\left(\frac{\partial \chi}{\partial T}\right)_{\mathbf{H}} = \frac{V}{\mathbf{H}} \left(\frac{\partial \mathbf{M}}{\partial T}\right)_{\mathbf{H}}$$

Whence

$$\lim_{T \to 0} \left(\frac{\partial \chi}{\partial T}\right)_{\mathbf{H}} = 0$$

If the Curie equation $V\mathbf{M} = C_c \mathbf{H}/T$ is obeyed we have

$$\left(\frac{\partial \mathbf{M}}{\partial T}\right)_{\mathbf{H}} = - \frac{C_c}{V} \frac{\mathbf{H}}{T^2}$$

As $T \to 0$, $(\partial \mathbf{M}/\partial T)_{\mathbf{H}} \to \infty$ and does not vanish. Thus the third law requires Curie's equation to fail at low enough temperatures. This is indeed found experimentally.

10-3 ALLOTROPIC TRANSFORMATIONS

A pure substance may exist in more than one crystalline solid phase. A transition from one solid phase to another is known as an *allotropic transformation*. A pressure-volume-temperature surface showing different solid phases of water is depicted in Fig. 1-19. A pressure-temperature phase diagram showing the two different crystalline forms of solid sulfur is depicted in Fig. 10-4.

In the case of phosphine (PH_3) there are four crystalline solid phases, not counting the phase change in a λ transition. Figure 10-5 shows the specific heat

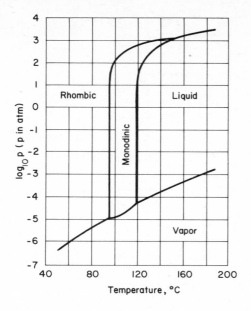

FIG. 10-4. Phase diagram of sulfur showing the two crystalline solids.

of phosphine at 1 atm as a function of temperature and Table 10-1 gives data on the temperatures and latent heats of the phase transitions. At $88.10°K$ there is a first order phase transition from the δ solid to the α solid. This transition is rapid and the δ solid can only be supercooled by a few tenths of a degree below $88.10°K$. When the α solid is cooled slowly to $49.43°K$, it will undergo a first order phase transformation to the β solid. The β solid below $49.43°K$ is stable and will persist to the absolute zero of temperature. The transition from the α solid to the β solid is a slow process, requiring many hours or several days to insure complete conversion. If cooled with sufficient rapidity, the α solid may be supercooled far below $49.43°K$ with practically no conversion. This metastable form of the α solid upon cooling will undergo a λ transition at $35.66°K$ without latent heat evolution. Further cooling of the α solid below the λ transition temperature causes it to undergo a first-order phase transition at $30.29°K$ to the γ solid. The γ solid will then persist to the absolute zero of temperature.

Experimental measurements of the properties of different allotropic solids of a pure substance at very low temperatures provide a very accurate verification of the third law of thermodynamics. Thus, if at atmospheric pressure, the changes of entropy from $0°K$ to the transition temperature for two allotropic solids are determined separately, and the change of entropy due to the allotropic transformation is also determined, we can check as to whether or not $\Delta S = 0$ for the

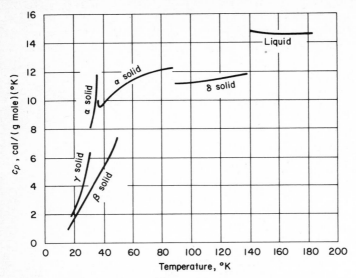

FIG. 10-5. Specific heat of phosphine at 1 atm. [*Data from C. C. Stephenson and W. F. Giauque*, J. Chem. Phys., 5:*149* (*1937*).]

two solids at the absolute zero of temperature. The following example serves to illustrate the validity of the third law.

EXAMPLE 10-2. From the data given in Fig. 10-5 and Table 10-1 on phosphine, demonstrate the validity of the third law by calculating the entropy of the α solid at 49.43°K and 1 atm by two different paths: (a) use the specific heat data on the γ and α solids and the latent heat at 30.29°K, and (b) use the specific heat data on the β solid and the latent heat at 49.43°K.

TABLE 10-1. Data on Phase Transitions for Phosphine at 1 atm*

Phase transition	Temperature, °K	Latent heat cal/g mole
γ solid to α solid	30.29	19.6
λ transition	35.66	0
β solid to α solid	49.43	185.7
α solid to δ solid	88.10	115.8
δ solid to liquid	139.35	270.4
Liquid to vapor	185.38	3,489

*See reference quoted in Fig. 10-5.

Solution.

The numerical values shown in this example are mainly those given by Stephenson and Giauque.* The methods of obtaining these values are as follows.

The entropy of the α solid at $49.43^\circ K$ may be calculated by either of the following two equations according to the two paths specified in the problem statement:

$$s(\alpha, \text{ at } 49.43^\circ K) = s(\gamma, \text{ at } 0^\circ K) + \Delta s(\gamma, 0 \text{ to } 15^\circ K)$$

$$+ \Delta s(\gamma, 15 \text{ to } 30.29^\circ K) + \Delta s(\gamma \text{ to } \alpha, \text{ at } 30.29^\circ K)$$

$$+ \Delta s(\alpha, 30.29 \text{ to } 49.43^\circ K) \quad (10\text{-}11)$$

and

$$s(\alpha, \text{ at } 49.43^\circ K) = s(\beta, \text{ at } 0^\circ K) + \Delta s(\beta, 0 \text{ to } 15^\circ K)$$

$$+ \Delta s(\beta, 15 \text{ to } 49.43^\circ K) + \Delta s(\beta \text{ to } \alpha, \text{ at } 49.43^\circ K) \quad (10\text{-}12)$$

Now from the given latent heat data, we have

$$\Delta s(\gamma \text{ to } \alpha, \text{ at } 30.29^\circ K) = \frac{19.6}{30.29} = 0.647 \text{ cal/(g mole)}(^\circ K)$$

$$\Delta s(\beta \text{ to } \alpha, \text{ at } 49.43^\circ K) = \frac{185.7}{49.43} = 3.757 \text{ cal/(g mole)}(^\circ K)$$

In order to obtain the values of $\Delta s(\gamma, 15 \text{ to } 30.29^\circ K)$, $\Delta s(\alpha, 30.29 \text{ to } 49.43^\circ K)$ and $\Delta s(\beta, 15 \text{ to } 49.43^\circ K)$, graphs of c_p/T vs. T for the γ, α, and β solids were prepared based on the given specific heat data of Fig. 10-5. From these graphs by graphical integration for $\int (c_p/T)dT$ we have the following values:

$$\Delta s(\gamma, 15 \text{ to } 30.29^\circ K) = 2.185 \text{ cal/(g mole)}(^\circ K)$$

$$\Delta s(\alpha, 30.29 \text{ to } 49.43^\circ K) = 4.800 \text{ cal/(g mole)}(^\circ K)$$

$$\Delta s(\beta, 15 \text{ to } 49.43^\circ K) = 4.041 \text{ cal/(g mole)}(^\circ K)$$

The values of $\Delta s(\gamma, 0 \text{ to } 15^\circ K)$ and $\Delta s(\beta, 0 \text{ to } 15^\circ K)$ can be obtained by the extrapolation of the low-temperature specific heat data by means of the Debye function. When Eq. (10-5) is used, the Debye temperature Θ as determined from the low temperature specific heat data are $\Theta = 101.8$ for the γ solid and $\Theta = 115.6$ for the β solid. Thus in general at low temperatures we

*See reference quoted in Fig. 10-5.

have

$$\Delta s(0 \text{ to } T^\circ \text{K}) = \int_0^T \frac{c_p dT}{T} = \frac{464.5}{\Theta^3} \int_0^T T^2 \, dT = \frac{464.5}{\Theta^3} \frac{T^3}{3}$$

From which we have

$$\Delta s(\gamma, 0 \text{ to } 15^\circ \text{K}) = \frac{464.5}{101.8^3} \frac{15^3}{3} = 0.495$$

$$\Delta s(\beta, 0 \text{ to } 15^\circ \text{K}) = \frac{464.5}{115.6^3} \frac{15^3}{3} = 0.338$$

Substituting numerical values in Eqs. (10-11) and (10-12), we have

$$s(\alpha, \text{ at } 49.43^\circ \text{K}) - s(\gamma, \text{ at } 0^\circ \text{K}) = 0.495 + 2.185 + 0.647 + 4.800$$

$$= 8.13 \, \frac{\text{cal}}{(\text{g mole})(^\circ \text{K})}$$

$$s(\alpha, \text{ at } 49.43^\circ \text{K}) - s(\beta, \text{ at } 0^\circ \text{K}) = 0.338 + 4.041 + 3.757 = 8.14 \, \frac{\text{cal}}{(\text{g mole})(^\circ \text{K})}$$

The close agreement of the results as obtained by the two paths is an excellent verification of the third law of thermodynamics, indicating that

$$s(\gamma, \text{ at } 0^\circ \text{K}) = s(\beta, \text{ at } 0^\circ \text{K})$$

10-4 GLASSES AND THE THIRD LAW

When a material is in the liquid state, its atoms or molecules have a certain degree of mobility and do not remain in definite positions relative to each other. But when a material is in the crystalline solid state, its atoms or molecules remain in definite positions relative to each other in a structure known as a crystal lattice. For most materials at atmospheric pressure, when the temperature of the liquid is lowered to its normal freezing point, transition to the solid state with structural readjustment will occur promptly. In some instance it is possible to supercool the liquid to temperatures below its freezing point without crystallization. Such a supercooled liquid is in a metastable equilibrium state. On further cooling it eventually passes over spontaneously into the more stable solid state. However, for some materials the viscosity of the supercooled liquid at reduced temperatures is very high, and therefore the process of crystallization is very sluggish. When cooled below a certain range of temperature, such a supercooled liquid becomes so viscous that practically all further crystallization will be prevented. This condition of the material is referred to as the *glassy state*. It is a continuation of the liquid state without discontinuous change in structural arrangement. Nevertheless, a glass is relatively rigid due to its high viscosity. We may therefore say that a glass is a rigid liquid.

When a supercooled liquid becomes glassy it practically ceases to undergo internal transformations appropriate to its actual temperature, and all the spatial disorder characteristics of the liquid are abruptly "frozen-in." There are differences in orderliness, and therefore differences in entropies, between the glassy and crystalline states at the same temperature. The entropy difference corresponding to the configurational entropy of the glass which is frozen-in will persist down to the absolute zero of temperature. However, the persistence of an entropy difference down to absolute zero in this case is not a violation of the third law of thermodynamics because the glassy state is not in thermodynamic equilibrium. Although the process of forming a crystal lattice in a glass practically ceases it does not really stop. But the time for relaxation of the structure to the equilibrium configuration becomes so lengthy that equilibrium may never be reached, especially at lower temperatures. In the absence of equilibrium the third law cannot apply at the absolute zero temperature.

EXAMPLE 10-3. From the data given in Fig. 10-6 on sulfuric acid trihydrate ($H_2SO_4 \cdot 3H_2O$), calculate the entropy difference between the glassy and crystalline states at the absolute zero temperature and 1 atm. What is the absolute entropy of the liquid at $298.16°K$ and 1 atm when the effects of nuclear spin and isotopes are not included?

Solution.
The numerical values shown in this example are those given by Kunzler and Giaugue.* The methods of obtaining these values are as follows.

s(liquid, at $298.16°K$) $- s$(glass, at $0°K$) $= \Delta s$(glass, 0 to $15°K$)

$$+ \Delta s(\text{glass and liquid, 15 to } 298.16°K) \quad (10\text{-}13)$$

s(liquid, at $298.16°K$) $- s$(crystal, at $0°K$) $= \Delta s$(crystal, 0 to $15°K$)

$$+ \Delta s(\text{crystal, 15 to } 236.82°K) + \Delta s(\text{crystal to liquid, at } 236.82°K)$$

$$+ \Delta s(\text{liquid, 236.82 to } 298.16°K) \quad (10\text{-}14)$$

Now from the given latent heat data, we have

$$\Delta s(\text{crystal to liquid, at } 236.82°K) = \frac{5786}{236.82} = 24.432 \ \frac{\text{cal}}{(\text{g mole})(°K)}$$

By graphical integration for $\int c_p d \ln T$ based on the given specific heat data, we have

*See reference quoted in Fig. 10-6.

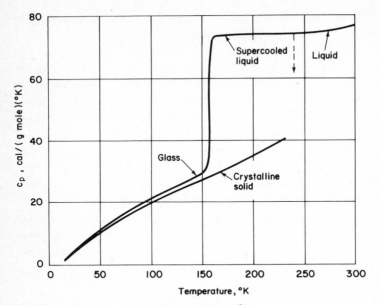

Note: Melting point temperature = 236.82° K
       Heat of fusion = 5,786 cal/g mole

FIG. 10-6. Specific heat of sulfuric acid trihydrate glass and crystal at 1 atm. [*Data from J. E. Kunzler and W. F. Giauque, J. Am. Chem. Soc., 74: 797 (1952).*]

Δs(glass and liquid, 15 to 298.16°K) = 76.54 cal/(g mole)(°K)

Δs(crystal, 15 to 236.82°K) = 40.744 cal/(g mole)(°K)

Δs(liquid, 236.82 to 298.16°K) = 17.301 cal/(g mole)(°K)

By extrapolating the low-temperature specific heat data by means of the Debye function we have

Δs(glass, 0 to 15°K) = 0.49 $\dfrac{\text{cal}}{(\text{g mole})(°\text{K})}$

Δs(crystal, 0 to 15°K) = 0.42 $\dfrac{\text{cal}}{(\text{g mole})(°\text{K})}$

Substituting numerical values in Eqs. (10-13) and (10-14) we have

s(liquid, at 298.16°K) − s(glass, at 0°K) = 0.49 + 76.54 = 77.0 cal/(g mole)(°K)

s(liquid, at 298.16°K) − s(crystal, at 0°K) = 0.42 + 40.744 + 24.432 + 17.301

$$= 82.90 \text{ cal/(g mole)(°K)}$$

From which we have

s(glass, at $0°$K) $- s$(crystal, at $0°$K) $= 82.90 - 77.0 = 5.90$ cal/(g mole)($°$K)

Note that since the glassy state is not in thermodynamic equilibrium, the measured values of c_p for the glass are not equilibrium values. This is why the third law fails to apply.

When the effects of nuclear spin and isotopes are not included in entropy calculations, we may assign s(crystal, at $0°$K) $= 0$. Thus

s(liquid, at $298.16°$K) $= 82.90$ cal/(g mole)($°$K)

10-5 NEGATIVE KELVIN TEMPERATURES

Experimental studies in 1951 by Pound, Purcell, and Ramsey[1] on the magnetic resonance of the nuclei of a pure lithium fluoride crystal have revealed the existence of negative temperatures on the Kelvin scale. At this very outset in our discussion of negative temperatures, it is worth mentioning that the possibility of reaching a negative Kelvin temperature does not imply that one can reach absolute zero. We now discuss the meaning of negative temperatures from the standpoint of thermodynamics, with some helpful remarks from statistical mechanics.

From the combined first and second laws for a simple system in general, we have the relation

$$dU = T\,dS + Y\,dX$$

where Y is the generalized force and X the generalized displacement. From this it follows that

$$T = \left(\frac{\partial U}{\partial S}\right)_X = \frac{1}{\left(\dfrac{\partial S}{\partial U}\right)_X} \tag{10-15}$$

In thermodynamics there is no explicit statement that S must increase monotonically with U to always give positive values for T; indeed such a restriction is not necessary for the derivation of many thermodynamic relations. If and when $(\partial S/\partial U)_X$ becomes negative, we then have a negative temperature.

In order to understand the meaning of negative temperatures, it is best to make use of the Boltzmann distribution equation from statistical mechanics,

$$N_i = A'e^{-\epsilon_i/k_B T}$$

[1] N. F. Ramsey, *Phys. Rev.*, **103**:1:20 (1956).

where N_i is the number of particles in an energy level with energy ϵ_i, k_B the Boltzmann constant, and A' a characteristic constant. According to this equation, the number of particles per level decreases exponentially with the energy associated with the level. Low-energy levels will be more populated than high-energy levels. Since most systems encountered in practice have an infinite number of higher-energy levels, an increase of temperature will produce increased populations of higher-energy levels, but no energy level will get populated more than the one below it. However, if the particles of a system may assume only a finite number of energy levels, there would be an upper limit to the possible energy of the allowed states. For the purpose of argument, let us consider a system whose particles may assume only four energy levels. The exponential curve in Fig. 10-7a depicts the distribution of the particles at a certain positive temperature. When energy is supplied to the system, causing some of the particles to move up to higher levels, the distribution curve is again exponential but with a lesser slope, such as shown in Fig. 10-7b. If the supply of energy is continued, eventually there would be more particles in the upper levels than in the lower ones, as shown in Fig. 10-7c. In this last case, in order to satisfy the Boltzmann distribution equation with an exponential curve, the system must have a negative T.

As a further illustration, for a system whose particles may assume only a finite number of energy levels, let us assume that at the lowest energy state of the system, all the particles are in the lowest energy level. This is clearly a highly ordered state of the system and corresponds to zero entropy. Likewise, if at the

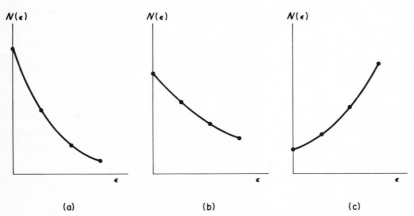

FIG. 10-7. Possible distributions of particles obeying Boltzmann equation for a system of particles that can exist in only four energy levels. (a) Positive temperature, lower energy. (b) Positive temperature, higher energy. (c) Negative temperature, very high energy.

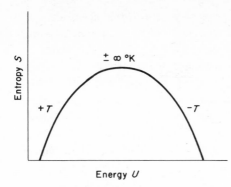

FIG. 10-8. Entropy-energy diagram of a system whose elements may assume only a limited number of discrete energy levels.

highest energy state of the system, all the particles are in the highest energy level, the system is again highly ordered with zero entropy. At intermediate energies, when the particles are distributed among the various levels, there are greater disorders and correspondingly greater entropies. Between the lowest and the highest energy states of the system, the entropy must pass through a maximum value and then decreases with increasing energy, as illustrated in Fig. 10-8. To the left of the maximum $(\partial S/\partial U)_X$ is positive, and hence the temperature of the system also is positive. To the right of the maximum $(\partial S/\partial U)_X$ is negative, and hence the temperature of the system also is negative. At the maximum point $(\partial S/\partial U)_X = 0$ and hence the temperature of the system is infinite.

From Fig. 10-8 it is clear that as the energy decreases and cooling takes place from a negative to a positive temperature, the system in question will pass through $\infty°$K instead of through $0°$K. In other words, negative temperatures are not colder than $0°$K but rather they are hotter than $\infty°$K. Here the words "hot" and "cold" are used in the sense that when two bodies are brought into thermal contact heat will flow from the hotter to the colder body. Thus when we write the Kelvin temperatures in order, such that those to the right are hotter than those to the left, the sequence is as follows:

$$+0°K, \ldots, +300°K, \ldots, \pm\infty°K, \ldots, -300°K, \ldots, -0°K$$

We see in this sequence that $+0°$K is the coldest and $-0°$K is the hottest. It would be psychologically more appealing to have the coldest temperature represented by $-\infty$ and the hottest temperature by $+\infty$. This can be done by defining a new temperature scale with numerical values equal to the negative reciprocal of the Kelvin temperatures. On this new scale the preceding temperature sequence then becomes

$$-\infty°, \ldots, -0.0033°, \ldots, \mp 0°, \ldots, +0.0033°, \ldots, +\infty°$$

Here the algebraic order agrees with the order from cold to hot, and $-\infty°$

appears truly unattainable. However, a major objection to this new scale is that the temperatures of all thermodynamic systems normally encountered in practice are negative.

With the understanding that systems with negative temperatures are hot and possess high energies, we can now describe briefly the experiment performed by Pound, Purcell, and Ramsey to realize negative temperatures. The chief requirement for a thermodynamic system to be capable of negative temperatures is that the elements of the system may assume only a limited number of discrete energy levels; thus there must be an upper limit to the possible energy of the allowed states. A nuclear-spin system is an example of such a system. Experimental studies of a nuclear-spin system require that the spin system be well isolated from the lattice system. In a pure lithium fluoride (LiF) crystal the nuclear spins take an unusually long time to come into equilibrium with the lattice. In fact the spin-lattice relaxation time characterizing the approach of equilibrium is as large as 5 minutes at room temperature, while the spin-spin relaxation time is less than 10^{-5} seconds. Thus within a short moment the nuclear spins behave essentially as an isolated system, amenable to experimental measurement.

In essence, the procedure of the experiment of Pound, Purcell, and Ramsey is as follows. A crystal of lithium fluoride is placed in a strong magnetic field, thus forcing some of the nuclei to have their magnetic dipole moments oriented in the direction of the applied field. Then the applied field is suddenly reversed in direction, but the nuclear spins are unable to follow the field and remain in their original orientation. Since the field and the spins are now in opposite directions, the spins which formerly had the least energy now have the most, and vice versa. In other words, the sudden reversal of the applied field has caused a population inversion; thus more of the nuclei now have high rather than low energies. Such a distribution of energy is the characteristic of a negative temperature.

10-6 RESTATEMENT OF THE LAWS OF THERMODYNAMICS
FOR NEGATIVE TEMPERATURES

Since the restriction of a monotonic increase of entropy with internal energy is not essential to the development of thermodynamics, the laws of thermodynamics and all relations derived from them should apply equally well to systems at either positive or negative temperatures. However, in order to be consistent in both temperatures ranges, some restatement of the laws of thermodynamics are in order. In his treatment on this subject Ramsey* pro-

*See the footnote which appears in the preceding section.

posed to retain the usual definitions of work and heat and preserve the conventional formulations of the first law of thermodynamics for both positive and negative temperatures. In the case of the second law of thermodynamics, although the Clausius statement, the Carathéodory statement, and the principle of increase of entropy are applicable without modification for negative temperatures, the Kelvin-Planck statement, however, must be modified. The statements of the third law of thermodynamics also require rewording.

In order to illustrate the necessity of revising the Kelvin-Planck statement of the second law, let us use a specific numerical example for easy visualization. Suppose there are two heat reservoirs, one at $-100°K$ and the other at $-200°K$. Of course, we understand that $-100°K$ is a higher temperature than $-200°K$. If Q_1 units of heat is transferred from the hot reservoir to an engine and Q_2 units of heat is transferred from the engine to the cold reservoir, then by the definition of Kelvin temperature scale based on a Carnot engine, we have

$$\frac{Q_1}{Q_2} = \frac{T_1}{T_2} = \frac{-100}{-200} = \frac{1}{2}$$

This implies that for each unit of heat leaving the hot reservoir, twice as much heat must enter the cold reservoir. Thus there is a net heat rejection from the engine. Accordingly, work must be done on the engine instead of by the engine in order to concur with the first law of thermodynamics, as depicted in Fig. 10-9a. If work output is desired, a reversed cycle as depicted in Fig. 10-9b is appropriate where heat is supplied to the engine from the cold reservoir and a

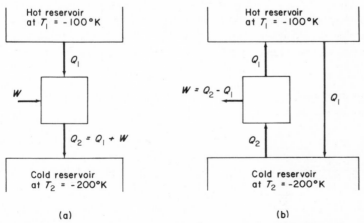

(a)   (b)

FIG. 10-9. (a) Work must be done on a heat engine which absorbs heat from a hot reservoir and rejects heat to a cold reservoir. (b) A heat engine that can convert heat from a cold reservoir completely into work.

smaller quantity of heat is rejected from the engine to the hot reservoir. If the heat transferred to the hot reservoir in this reverse cycle is allowed to flow back to the cold reservoir by a direct heat-transfer process, then there exists an engine that operates in a cycle and produce no other effect than the absorption of heat from a reservoir and the conversion of this heat into work. This is a violation of the conventional form of the Kelvin-Planck statement of the second law.

The truth of the matter is that the conventional Kelvin-Planck statement is designed for positive temperatures only. To include negative temperatures this statement should be modified to read: No process is possible whose sole result is either (1) the absorption of heat from a reservoir at a positive temperature and the conversion of this heat into work or (2) the conversion of work into heat and the rejection of this heat into a reservoir at a negative temperature. We will call this the Kelvin-Planck-Ramsey statement of the second law. With this modification, the proposed reversed cycle shown in Fig. 10-9b is no longer a violation and may in fact exist.

Using the foregoing numerical example on negative temperatures, it is a simple matter to demonstrate that a direct heat-transfer process is irreversible and must result in an increase in entropy. Thus, for the direct heat-transfer process shown in Fig. 10-9b, the total entropy change is

$$\frac{-Q_1}{-100} + \frac{Q_1}{-200} = \frac{Q_1}{200}$$

which is positive, indicating an increase in entropy.

It is interesting to note that the conventional Clausius statement without modification is applicable for both positive and negative temperatures, but the physical reasons are different. To illustrate this difference, let us examine the device shown in Fig. 10-10, attempting to transfer heat from a cold reservoir to a hot reservoir in two steps. Step A is to extract heat from the cold reservoir and

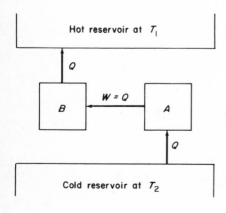

FIG. 10-10. It is impossible to transfer heat from one body to another body at a higher temperature with no other effect being produced.

convert it into work with no other effect being produced. Step B is to convert this work into heat and deliver the heat to the hot reservoir with no other effect being produced. At positive temperatures, step B is possible but step A is prohibited by the Kelvin-Planck statement. At negative temperatures, step A is possible but step B is prohibited by the Kelvin-Planck-Ramsey statement.

Since a system at $-0°K$ is in its highest-energy state where it can absorb no more energy, and a system at $+0°K$ is in its lowest-energy state where it can give up no more energy, it is apparent that the difficulty of heating a hot system at negative temperatures is analogous to the difficulty of cooling a cold system at positive temperatures. Accordingly, the third law of thermodynamics ought to be reworded to include both $+0°K$ and $-0°K$. Thus the unattainability statement of the third law should be reworded to read: It is impossible by means of any process, no matter how idealized, in a finite number of steps to reduce the positive temperature of a system to $+0°K$, or to raise the negative temperature of a system to $-0°K$.

It is significant to note that since the conversion of heat into work is easy at negative temperatures, systems at such temperatures could be useful for some applications. In addition, since at negative temperatures most resistances are negative, negative-temperature systems are intrinsically amplifiers just as ordinary resistance networks at positive temperatures are attenuators. However, systems having negative temperatures cannot maintain equilibrium with ordinary systems having positive temperatures. A high-energy negative-temperature system will inevitably lose energy to its surroundings and cool down to positive temperatures.

PROBLEMS

10.1. Use the third law to prove that
(a) For a surface film of He^4 or He^3,

$$\lim_{T \to 0} \left(\frac{d\gamma}{dT} \right) = 0$$

(b) For a reversible cell for which $- E d\mathbf{Z}$ is the only work term,

$$\lim_{T \to 0} \left(\frac{dE}{dT} \right) = 0$$

10-2. For a simple compressible system, show that

$$\lim_{T \to 0} \left(\frac{C_p}{T} \right) = 0 \quad \text{and} \quad \lim_{T \to 0} \left(\frac{C_v}{T} \right) = 0$$

10-3. For a simple compressible system, if the heat capacity at constant volume C_v is a known function of V and T, derive the expressions for the dependence of $(\partial S/\partial V)_T$ and of S on V and T.

10-4. The absolute entropy of methane in the ideal gas state at $298.15^\circ K$ is 44.49 cal/(g mole)($^\circ K$). Calculate the absolute entropy of methane at $450^\circ K$, assuming

$$c_p = 5.65 + 11.44 \times 10^{-3} \, T - 0.46 \times 10^5 \, T^{-2}$$

where c_p is in cal/(g mole)($^\circ K$), and T in $^\circ K$.

10-5. Cyclobutane exists in two different crystalline solid phases (Solids I and II). The transition between them at 1 atm occurs at about $145.7^\circ K$ with a heat of transition of 1363.5 cal/g mole. The normal melting point is at $182.43^\circ K$ with a heat of fusion of 260.1 cal/g mole. The normal boiling point is at $285.67^\circ K$ with a heat of vaporization of 5781 cal/g mole. The specific heat data are shown in Fig. P10-5. Determine the absolute entropy of the actual gas at its normal boiling point.

10-6. Nitrogen exists in two different crystalline solid phases (α solid and β solid). The transition between them at 1 atm occurs at $35.61^\circ K$ with a heat of transition of 54.71 cal/g mole. The normal melting point is at $63.14^\circ K$ with a heat of fusion of 172.3 cal/g mole. The normal boiling point is at $77.32^\circ K$ with a heat of vaporization of 1332.9 cal/g mole. The specific heat c_p in cal/(g mole) ($^\circ K$) for the α solid, β solid, and liquid nitrogen at 1 atm are as follows:

$T, ^\circ K$	15.82	19.51	24.85	28.32	31.29
c_p	3.124	4.577	6.380	7.540	8.643
$T, ^\circ K$	32.84	35.33	35.61		39.13
c_p	9.397	10.67	Transition		8.948
$T, ^\circ K$	48.07	55.88	61.41	63.14	
c_p	9.752	10.44	11.07	Melting	
$T, ^\circ K$	65.02	70.28	76.58	77.32	
c_p	13.33	13.45	13.68	Boiling	

Calculate the absolute entropy of the actual gas at its normal boiling point, knowing that for the α solid between 0° to $10^\circ K$ the change in entropy is 0.458 cal/(g mole)($^\circ K$).

10-7. Since absolute entropies calculated from calorimetric specific-heat data at 1 atm are values for such a pressure. For many real gases, however, the entropy at 1 atm pressure is not the same as the entropy of the gas in its ideal gas state. To obtain entropies in ideal gas state, it is necessary to apply a correction to the real gas entropy. Based on the entropy value obtained in the preceding problem for real gas nitrogen at $77.32^\circ K$ and 1 atm, calculate its value in the ideal gas state, using the Berthelot equation of state

$$pv = RT\left[1 + \frac{9}{128}\frac{p}{p_c}\frac{T_c}{T}\left(1 - 6\frac{T_c^2}{T^2}\right)\right]$$

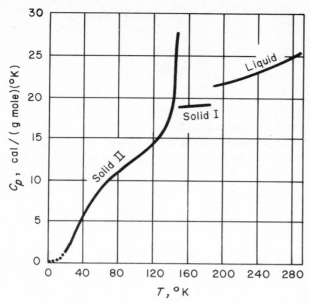

FIG. P10-5. Heat capacity of cyclobutane at 1 atm.

10-8. Below 220.4°K, methylammonium chloride exists in two different crystalline solid phases, the β solid and the γ solid, with a transition between them at 220.4°K. From the following data (at 1 atm), find the heat of phase transition in cal/g mole from the β solid to the γ solid at 220.4°K:

Δs(β solid, from 0° to 12.04°K) = 0.067 cal/(g mole)(°K)

Δs(β solid, from 12.04° to 220.4°K) = 22.326 cal/(g mole)(°K)

Δs(γ solid, from 220.4° to 264.5°K) = 3.690 cal/(g mole)(°K)

Δs(γ solid, from 0° to 19.55°K) = 0.442 cal/(g mole)(°K)

Δs(γ solid, from 19.55° to 264.5°K) = 27.610 cal/(g mole)(°K)

10-9. Benzothiophene exhibites a large λ transition from crystal I to crystal II between 250° and 261.6°K. From specific heat measurements of the stable crystal I and the supercooled crystal II down to 12°K, together with the extrapolation of the low-temperature specific heat data by means of the Debye function, the following entropy data were obtained:

Δs(crystal I, 0° to 12°K) = 0.338 cal/(g mole)(°K)

Δs(from crystal I at 12°K to crystal II at 261.6°K after the λ transition)

$$= 38.153 \text{ cal/(g mole)(°K)}$$

Δs (supercooled crystal II, $0°$ to $12°K$) = 0.528 cal/(g mole)($°K$)

Δs (supercooled crystal II, $12°$ to $261.6°K$) = 36.504 cal/(g mole)($°K$)

Calculate the value of: s(supercooled crystal II at $0°K$) $-$ s(stable crystal I at $0°K$). Does this entropy difference indicate an apparent contradiction of the third law?

10-10. Discuss the efficiency of a Carnot engine operating between

(a) two reservoirs at positive Kelvin temperatures

(b) two reservoirs at negative Kelvin temperatures

(c) a reservoir at a negative Kelvin temperature and a reservoir at a positive Kelvin temperature.

Chapter 11
REACTION EQUILIBRIUM

11-1 INTRODUCTION

All the preceding chapters have been concerned with nonreactive systems. We now turn our attention to systems involving chemical reactions. A chemical reaction is a process in which the interatomic bonds in the molecules of a collection of certain chemical constituents are broken, followed by the rearrangement of the atoms and electrons to form different constituents of new atomic combinations. Associated with every chemical reaction is a chemical equation. For example, the reaction between methane and oxygen to form carbon dioxide and water is expressed by the equation

$$CH_4 + 2O_2 \rightarrow CO_2 + 2H_2O$$

where the constituents on the left are called the reactants and those on the right the products. The numerical coefficients which precede the chemical symbols to give equal number of atoms of each chemical element on both sides of the equation are called *stoichiometric coefficients*.

From a thermodynamic point of view there are three major aspects of a chemical reaction: (1) a mass balance through the use of the principle of conservation of mass, thus establishing a chemical reaction equation; (2) an energy balance through the use of the first law of thermodynamics, thus determining the energy transfers and conversions, and (3) an equilibrium study through the use of the second law of thermodynamics, thus predicting the extent and direction of a given reaction. In addition, the third law of thermodynamics is used to compute entropies of different species involved in a reaction. In this section we will give two numerical examples to illustrate the mass-balance procedure. Analyses of chemical reactions on the basis of the three laws of thermodynamics will be taken up in the succeeding sections of this chapter.

Although the principles treated in this chapter apply to chemical reactions in general, one particular kind of reaction, namely, combustion is of prime importance in engineering applications. *Combustion* is a rapid reaction between fuels and oxygen in which chemical energy is liberated. In a combustion process of a hydrocarbon fuel, if all the carbon present is converted into CO_2 and all the hydrogen is converted into H_2O, the process is described as *complete combustion*. If CO appears in the products, the combustion is incomplete.

In most combustion processes the oxygen is supplied as air rather than pure oxygen. The ratio of nitrogen to oxygen on a molal basis in the atmosphere is approximately $79/21 = 3.76$. In the normal temperature range of many combustion processes, the nitrogen is not involved in the chemical reaction. At high temperatures, such as those achieved in certain internal-combustion engines, some direct reactions of nitrogen and oxygen may take place to form oxides of nitrogen. In our analyses, unless otherwise stated we will assume that the nitrogen is not involved in the chemical reaction. However, the temperature of nitrogen after the reaction process, in general, is not the same as before. This factor should be taken care of when the energy balance is made for the process.

The minimum amount of air which supplies sufficient oxygen for complete combustion of a fuel is called the *theoretical air*. The amount of air actually supplied may be less than or in excess of the theoretical air. If the actual air supply is less than the theoretical air, the reaction will be incomplete and CO and even unburned fuel will appear in the products. If excess air is supplied, the excess oxygen in a complete combustion process will appear in the products unchanged. However, even when some excess air is supplied, due to uneven mixing and some other factors, the reaction may still be incomplete and CO_2, CO, and O_2 appear together in the products. The amount of *excess air* is usually expressed as a percentage of the theoretical air required for complete combustion of the fuel. In practice, 20 percent excess air is usually regarded as a minimum in order to insure that no CO and unburned fuel are contained in the products.

The relationship between the fuel and the air supply can also be expressed in terms of *air-fuel ratio* which is defined as the mass of air supplied per unit mass of fuel. When theoretical air is supplied to a combustion process, this ratio is referred to as the theoretical or stoichiometric air-fuel ratio.

EXAMPLE 11-1. A gaseous mixture of 70% CH_4, 20% CO, 5% O_2 and 5% N_2 on a mole basis is burned completely at 1 atm with 20% excess air. Write the chemical equation and determine (a) the volume of air supplied per unit volume of fuel, both being measured at the same pressure and temperature, (b) the mass of air supplied per kg of fuel, and (c) the volumetric analysis of the products of combustion.

Solution.

Let the number of moles of oxygen supply needed for the complete combustion of one mole of fuel be denoted by a. Thus the theoretical air supplied is $(a\ O_2 + 3.76\ a\ N_2)$. With 20% excess air, the actual air supplied is then $1.2\ (a\ O_2 + 3.76\ a\ N_2)$ moles of air per mole of fuel. Therefore the chemical equation for complete combustion per mole of fuel can be written as

$$(0.70\ CH_4 + 0.20\ CO + 0.05\ O_2 + 0.05\ N_2) + 1.2\ (a\ O_2 + 3.76\ a\ N_2)$$

$$\rightarrow x\ CO_2 + y\ H_2O + 0.2\ a\ O_2 + z\ N_2$$

where x, y, and z are the unknown numbers of moles of CO_2, H_2O, and N_2 respectively in the products. The coefficient $0.2a$ for O_2 on the right-hand side of the equation represents the 20% excess O_2 that appears in the products unchanged.

The unknown quantities in the preceding equation can be determined by mass balances for each of the atomic species. Thus,

C balance : $x = 0.7 + 0.2 = 0.9$ moles CO_2/mole fuel

H_2 balance : $y = 2 \times 0.7 = 1.4$ moles H_2O/mole fuel

O_2 balance : $\dfrac{1}{2} \times 0.2 + 0.05 + 1.2a = x + \dfrac{1}{2}\ y + 0.2a$

$\qquad a = x + \dfrac{1}{2}\ y - 0.15\ = 1.45$ moles O_2/mole fuel (theoretical value)

N_2 balance : $z = 0.05 + 1.2 \times 3.76a = 6.59$ moles N_2/mole fuel

The chemical equation then becomes

$$(0.70\ CH_4 + 0.20\ CO + 0.05\ O_2 + 0.05\ N_2) + 1.2\ (1.45\ O_2 + 5.45\ N_2)$$

$$\rightarrow 0.9\ CO_2 + 1.4\ H_2O + 0.29\ O_2 + 6.59\ N_2$$

(a) According to the chemical equation, we have the ratio

$$\frac{\text{mole air}}{\text{mole fuel}} = 1.2\ (1.45 + 5.45) = 8.28$$

Since a mole of air and a mole of fuel at the same pressure and temperature occupy the same volume, the value 8.28 moles of air per mole of fuel also represents 8.28 m^3 of air per m^3 of fuel.

(b) The apparent or average molecular weight of air is 29. The apparent or average molecular weight of the fuel is

$0.70 \times 16 + 0.20 \times 28 + 0.05 \times 32 + 0.05 \times 28 = 19.8$ kg fuel/kg mole fuel

Thus the air-fuel mass ratio is

$$\left(8.28 \ \frac{\text{kg mole air}}{\text{kg mole fuel}}\right) \frac{(29 \text{ kg air/kg mole air})}{(19.8 \text{ kg fuel/kg mole fuel})} = 12.1 \text{ kg air/kg fuel}$$

(c) According to Eq. (4-42), for an ideal gas mixture the volume fraction of a constituent equals its mole fraction. There are a total of 0.9 + 1.4 + 0.29 + 6.59 = 9.18 moles of products. The volumetric analysis of the products is then

CO_2:   0.9/9.18 = 9.80%
H_2O: 1.4/9.18 = 15.25%
O_2:   0.29/9.18 = 3.16%
N_2:   6.59/9.18 = 71.79%

EXAMPLE 11-2. A gasoline has a mass analysis by chemical elements of 85% carbon and 15% hydrogen. In an automobile engine test, the dry products of combustion (exclusive of the water vapor in the products) show the following volumetric analysis: CO_2 12.5%, O_2 2.4%, CO 0.9%, and N_2 84.2%. The atmospheric air supplied for combustion contains 0.028 kg of water vapor per kg of dry air (i.e., O_2 and N_2). Write the chemical equation and determine per kg of gasoline (a) the amount of atmospheric air supplied, and (b) the amount of water in the products.

Solution.
For one kg of the fuel, there are 0.85/12 = 0.07083 kg mole of C, and 0.15/2 = 0.075 kg mole of H_2. Let a be the number of kg moles of oxygen supplied per kg of fuel and b be the number of kg moles of dry products per kg of fuel. For the combustion of one kg of fuel, the chemical equation when dry air is supplied for combustion may be written as

$$(0.07083 \text{ C} + 0.075 \text{ H}_2) + (a \text{ O}_2 + 3.76 a \text{ N}_2) \rightarrow (0.125 \ b \text{ CO}_2 + 0.024 \ b \text{ O}_2$$

$$+ \ 0.009 \ b \text{ CO} + 0.842 \ b \text{ N}_2) + x \text{ H}_2\text{O}$$

where x is the number of kg moles of H_2O in the products when dry air is supplied for combustion.

We now determine the unknown quantities by making mass balances. Thus,

C balance : 0.07083 = 0.125 b + 0.009 b

b = 0.5286 kg mole dry products/kg fuel

H_2 balance : x = 0.075 kg mole H_2O/kg fuel

O_2 balance : $a = 0.125 \ b + 0.024 \ b + \frac{1}{2} \times 0.009 \ b + \frac{1}{2} x = 0.1535 \ b + \frac{1}{2} x$

= 0.1186 kg mole O_2/kg fuel

As a check, a N_2 balance gives

$$a = \frac{0.842\ b}{3.76} = 0.1184 \text{ kg mole } O_2/\text{kg fuel}$$

The chemical equation then becomes

$$(0.07083 \text{ C} + 0.075 \text{ H}_2) + (0.1186 \text{ O}_2 + 0.445 \text{ N}_2) \rightarrow (0.0661 \text{ CO}_2$$
$$+ 0.0127 \text{ O}_2 + 0.00476 \text{ CO} + 0.445 \text{ N}_2) + 0.075 \text{ H}_2\text{O}$$

(a) According to the preceding chemical equation, the amount of dry air supplied is

$$\left(0.1186\ \frac{\text{kg mole } O_2}{\text{kg fuel}}\right)\left(32\ \frac{\text{kg } O_2}{\text{kg mole } O_2}\right) + \left(0.445\ \frac{\text{kg mole } N_2}{\text{kg fuel}}\right)\left(28\ \frac{\text{kg } N_2}{\text{kg mole } N_2}\right)$$
$$= 16.26 \text{ kg dry air/kg fuel}$$

Since the atmospheric air has 0.028 kg of water vapor per kg of dry air, the amount of water introduced by atmospheric moisture is then

$$\left(0.028\ \frac{\text{kg water}}{\text{kg dry air}}\right)\left(16.26\ \frac{\text{kg dry air}}{\text{kg fuel}}\right) = 0.455 \text{ kg water/kg fuel}$$

Therefore the amount of atmospheric air supplied for combustion is

$$\left(16.26\ \frac{\text{kg dry air}}{\text{kg fuel}}\right) + \left(0.455\ \frac{\text{kg water}}{\text{kg fuel}}\right) = 16.7\ \frac{\text{kg atmospheric air}}{\text{kg fuel}}$$

(b) The amount of water in the products is the sum of the water formed by combustion and the water introduced from atmospheric air. Its value is then

$$\left(0.075\ \frac{\text{kg mole } H_2O}{\text{kg fuel}}\right)\left(18\ \frac{\text{kg } H_2O}{\text{kg mole } H_2O}\right) + \left(0.455\ \frac{\text{kg } H_2O}{\text{kg fuel}}\right)$$
$$= 1.350 + 0.455 = 1.805 \text{ kg } H_2O/\text{kg fuel}$$

11-2 FIRST LAW ANALYSIS OF CHEMICAL REACTIONS

Chemical reactions are usually carried out either in a closed system at constant volume or constant pressure, or in an open system at steady-flow conditions. We now investigate the energy effects that accompany chemical changes in these processes on the basis of the first law of thermodynamics.

Consider first a constant-volume reaction in a closed simple compressible system. Since for this process $W = 0$, the first law becomes

$$Q = U_p - U_r$$

where the subscript r denotes the initial state of the reactants and the subscript p denotes the final state of the products.

Since the chemical aggregations of the reactants and the products are different, the state of zero internal energy cannot be arbitrarily chosen for both. Instead we must assign the zero-energy states so that for a constant volume reaction the change in internal energy between the reactants and the products equals the amount of heat transfer. The change in internal energy between the reactants and the products when both at the same temperature is called the *internal energy of reaction* at that temperature. The temperature variations of internal energy in proper relative values for the reactants and the products are depicted in Fig. 11-1. For convenience it is assumed in this plot that the internal energy of a chemical species is a function of temperature alone. This is, of course, exactly true only for ideal gases. In Fig. 11-1, the vertical distance at any temperature between the two curves is the internal energy of reaction at that temperature. In particular, the internal energy of reaction $\Delta U_R = U_{p_0} - U_{r_0}$ at a standard temperature T_0 is indicated in the figure.

Referring to Fig. 11-1, for a process which begins with the reactants at state r and ends with the products at state p, the transfer of heat is given by

$$Q = U_p - U_r = (U_p - U_{p_0}) + (U_{p_0} - U_{r_0}) + (U_{r_0} - U_r)$$

$$= (U_p - U_{p_0}) + \Delta U_R - (U_r - U_{r_0}) \tag{11-1}$$

or

$$Q = \sum_{\text{products}} n(u_p - u_{p_0}) + \Delta U_R - \sum_{\text{reactants}} n(u_r - u_{r_0}) \tag{11-1a}$$

where the u's are molal internal energies, and n is the number of moles.

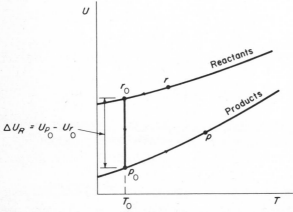

FIG. 11-1. Internal energies of reactants and products as functions of temperature.

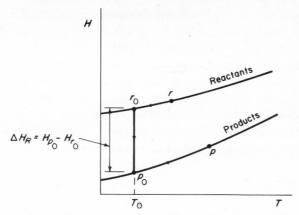

FIG. 11-2. Enthalpies of reactants and products as functions of temperature.

In the case of a steady-flow open system, the first law gives

$$Q = H_p - H_r$$

when changes in kinetic and potential energies are neglected and when there is no shaft work done. The preceding equation is also the expression of the first law for a constant pressure reaction in a closed system, because for such a process

$$Q = U_p - U_r + p(V_p - V_r) = H_p - H_r$$

When the enthalpy is a function of temperature alone, as in ideal gases, the H-T data in proper relative values for the reactants and the products can be shown as in Fig. 11-2. Referring to this figure, we can then write for the transfer of heat

$$Q = H_p - H_r = (H_p - H_{p_0}) + (H_{p_0} - H_{r_0}) + (H_{r_0} - H_r)$$
$$= (H_p - H_{p_0}) + \Delta H_R - (H_r - H_{r_0}) \tag{11-2}$$

or

$$Q = \sum_{products} n(h_p - h_{p_0}) + \Delta H_R - \sum_{reactants} n(h_r - h_{r_0}) \tag{11-2a}$$

where the h's are molal enthalpies, and $\Delta H_R = H_{p_0} - H_{r_0}$ is the *enthalpy of reaction* at the standard temperature T_0. Note that we will use $\Delta H_R(T, p)$ to denote the general form of the enthalpy of reaction at any temperature T and pressure p.

It should be noted that the subscript R in ΔH_R and ΔU_R serves as a reminder that these changes are due to chemical reactions instead of temperature changes. Values of ΔH_R and ΔU_R for various reactions are usually presented in the literature in a standard ideal gas condition at 1 atm and $T_0 = 298°K$. Table 11-1 gives the values of ΔH_R and ΔU_R for a number of simple reactions.

In order to establish an enthalpy scale for elements and compounds involved in chemical reactions, we now introduce the concept of an enthalpy of formation. The *enthalpy of formation* ΔH_F of a compound at the standard condition of 25°C and 1 atm is defined as the enthalpy of reaction at that condition for the formation of the compound from its elements which are originally in their most stable state at the standard condition. Thus we can write

$$\Delta H_F = h_{\text{compound}} - \sum_i (v_i h_i)_{\text{elements}}$$

where h_{compound} is the molal enthalpy of the compound and h_i is the molal enthalpy of the ith element, all at the standard condition; v_i is the stoichiometric coefficient of the ith element in forming a single mole of the compound. It is convenient to assign arbitrarily a value of zero to the enthalpy of all stable elements at 25°C and 1 atm. Accordingly we can write

$$\Delta H_F = h_{\text{compound}}$$

Therefore, the molal enthalpy of any compound at the standard condition is merely its enthalpy of formation at that condition.

Table 11-2 gives values of ΔH_F at 25°C and 1 atm for a number of substances. Values of ideal-gas enthalpies as a function of temperature for a number of substances are given in Appendix Tables A-15 through A-20. In these tables the enthalpy data are presented in the form of $(h - h_{298°K})$. An equivalent set of enthalpy data is given by the Keenan and Kaye gas tables (Appendix Table A-12). In the Keenan and Kaye tables, however, the zero of enthalpy for all substances is chosen to be at 0°K.

For any reaction the value of ΔH_R can be calculated from data on ΔH_F for each constituent involved in the reaction. Since by definition, the enthalpy of reaction at the standard condition of 25°C and 1 atm can be expressed symbolically as

$$\Delta H_R = \sum_{\text{products}} (vh) - \sum_{\text{reactants}} (vh)$$

where the v's are the stoichiometric coefficients of the individual constituents based on a balanced chemical equation. Now since the molal enthalpy h of any compound at the standard condition is merely its enthalpy of formation ΔH_F at that condition, we may substitute ΔH_F values for h values in the preceding

TABLE 11-1. Enthalpies and Internal Energies of Reaction at 25° C (77°F)

(Water appears as a liquid in the products of combustion)

Fuel	Formula	Enthalpy of vaporization (kcal/g mole of fuel)	Reaction	ΔH_R kcal/g mole of fuel	ΔU_R kcal/g mole of fuel
Carbon (graphite)	C		$C(s) + O_2(g) \rightarrow CO_2(g)$	– 94.054	– 94.054
Carbon (graphite)	C		$C(s) + 0.5 O_2(g) \rightarrow CO(g)$	– 26.417	– 26.713
Hydrogen	H_2		$H_2(g) + 0.5 O_2(g) \rightarrow H_2O(l)$	– 68.315	– 67.427
Carbon monoxide	CO		$CO(g) + 0.5 O_2(g) \rightarrow CO_2(g)$	– 67.637	– 67.341
Sulfur	S		$S(s) + O_2(g) \rightarrow SO_2(g)$	– 70.944	– 70.944
Methane	CH_4		$CH_4(g) + 2O_2(g) \rightarrow CO_2(g) + 2H_2O(l)$	– 212.789	– 211.605
Acetylene	C_2H_2		$C_2H_2(g) + 2.5 O_2(g) \rightarrow 2CO_2(g) + H_2O(l)$	– 310.613	– 309.725
Ethylene	C_2H_4		$C_2H_4(g) + 3O_2(g) \rightarrow 2CO_2(g) + 2H_2O(l)$	– 337.278	– 336.094
Ethane	C_2H_6		$C_2H_6(g) + 3.5 O_2(g) \rightarrow 2CO_2(g) + 3H_2O(l)$	– 372.813	– 371.333
Propane	C_3H_8	3.605	$C_3H_8(g) + 5O_2(g) \rightarrow 3CO_2(g) + 4H_2O(l)$	– 530.602	– 528.826
n-Butane	C_4H_{10}	5.035	$C_4H_{10}(g) + 6.5 O_2(g) \rightarrow 4CO_2(g) + 5H_2O(l)$	– 687.641	– 685.569
n-Pentane	C_5H_{12}	6.316	$C_5H_{12}(g) + 8O_2(g) \rightarrow 5CO_2(g) + 6H_2O(l)$	– 845.160	– 842.792
Benzene	C_6H_6	8.090	$C_6H_6(g) + 7.5 O_2(g) \rightarrow 6CO_2(g) + 3H_2O(l)$	– 789.089	– 787.609
n-Hexane	C_6H_{14}	7.540	$C_6H_{14}(g) + 9.5 O_2(g) \rightarrow 6CO_2(g) + 7H_2O(l)$	– 1002.569	– 999.905
n-Heptane	C_7H_{16}	8.735	$C_7H_{16}(g) + 11O_2(g) \rightarrow 7CO_2(g) + 8H_2O(l)$	– 1160.008	– 1157.048
n-Octane	C_8H_{18}	9.915	$C_8H_{18}(g) + 12.5 O_2(g) \rightarrow 8CO_2(g) + 9H_2O(l)$	– 1317.447	– 1314.191
Methyl alcohol	CH_3OH	9.08	$CH_3OH(g) + 1.5 O_2(g) \rightarrow CO_2(g) + 2H_2O(l)$	– 182.724	– 181.836
Ethyl alcohol	C_2H_5OH	10.18	$C_2H_5OH(g) + 3O_2(g) \rightarrow 2CO_2(g) + 3H_2O(l)$	– 336.863	– 335.679

Note: 1 cal/g mole = 1.8 Btu/lb mole

SOURCE: Calculated from the ΔH_F data given in Table 11-2.

TABLE 11-2. Enthalpies of Formation and Absolute Entropies of Substances at 25°C (77°F) and 1 atm

Substance	Formula	ΔH_F kcal/g mole	s cal/(g mole)(°K)
Acetylene	C_2H_2 (g)	54.190	48.004
Ammonia	NH_3 (g)	− 11.02	45.97
Argon	Ar (g)	0	36.9822
Benzene	C_6H_6 (g)	19.820	64.34
Benzene	C_6H_6 (1)	11.718	41.30
n-Butane	C_4H_{10} (g)	− 30.15	74.12
Carbon (graphite)	C (s)	0	1.359
Carbon	C (g)	170.886	37.761
Carbon dioxide	CO_2 (g)	− 94.054	51.072
Carbon monoxide	CO (g)	− 26.417	47.214
Carbon tetrachloride	CCl_4 (g)	− 22.940	74.019
Ethane	C_2H_6 (g)	− 20.24	54.85
Ethyl alcohol	C_2H_5OH (g)	− 56.19	67.54
Ethyl alcohol	C_2H_5OH (1)	− 66.37	38.4
Ethylene	C_2H_4 (g)	12.540	52.396
Helium	He (g)	0	30.1244
n-Heptane	C_7H_{16} (g)	− 44.89	102.24
n-Heptane	C_7H_{16} (1)	− 53.63	78.52
n-Hexane	C_6H_{14} (g)	− 39.96	92.83
n-Hexane	C_6H_{14} (1)	− 47.52	70.72
Hydrogen	H (g)	52.100	27.392
Hydrogen	H_2 (g)	0	31.208
Hydrogen sulfide	H_2S (g)	− 4.880	49.151
Hydroxyl	OH (g)	9.432	43.880
Krypton	Kr (g)	0	39.1905
Mercury	Hg (1)	0	18.17
Mercury	Hg (g)	14.652	41.794
Methane	CH_4 (g)	− 17.895	44.490
Methyl alcohol	CH_3OH (g)	− 47.96	57.29
Methyl alcohol	CH_3OH (1)	− 57.04	30.3
Neon	Ne (g)	0	34.9471
Nitric oxide	NO (g)	21.580	50.347
Nitrogen	N (g)	112.965	36.614
Nitrogen	N_2 (g)	0	45.770
Nitrogen dioxide	NO_2 (g)	7.910	57.343
n-Octane	C_8H_{18} (g)	− 49.82	111.55
n-Octane	C_8H_{18} (1)	− 59.74	86.23
Oxygen	O (g)	59.559	38.468
Oxygen	O_2 (g)	0	49.004
n-Pentane	C_5H_{12} (g)	− 35.00	83.40

TABLE 11-2. Enthalpies of Formation and Absolute Entropies of Substances at 25°C (77°F) and 1 atm (*Continued*)

Substance	Formula	ΔH_F kcal/g mole	s cal/(g mole)(°K)
n-Pentane	$C_5 H_{12}$ (1)	-41.36	62.92
Propane	$C_3 H_8$ (g)	-24.82	64.51
Propylene	$C_3 H_6$ (g)	4.879	63.80
Sulfur	S (s)	0	7.60
Sulfur	S (g)	66.636	40.084
Sulfur dioxide	SO_2 (g)	-70.944	59.30
Water	$H_2 O$ (g)	-57.798	45.106
Water	$H_2 O$ (1)	-68.315	16.71
Xenon	Xe (g)	0	40.5290

Note: 1 cal/g mole = 1.8 Btu/lb mole.

SOURCES: JANAF Thermochemical Tables, second edition, NSRDS-NBS 37, *Natl. Bur. Std.*, Washington, D.C., 1971; Selected Values of Chemical Thermodynamic Properties, Tech. Notes 270-3 and 4, *Natl. Bur. Std.*, Washington, D.C., 1968 and 1969; and Selected Values of Physical and Thermodynamic Properties of Hydrocarbons and Related Compounds, *API Res. Proj. 44*, Carnegie Press, Pittsburgh, Pa., 1953.

equation. Thus

$$\Delta H_R = \sum_{products} (\nu \Delta H_F) - \sum_{reactants} (\nu \Delta H_F) \qquad (11\text{-}3)$$

This equation applies both to reactions involving any mixture of reactants and to incomplete reactions.

EXAMPLE 11-3.   Liquid octane ($C_8 H_{18}$) at 77°F is sprayed into a low-pressure steady-flow combustion chamber and is burned with 400% theoretical air supplied at 260°F. The gaseous products leave the chamber at 980°F. Determine the amount of heat released per mole of fuel.

Solution.
For 400% theoretical air (or 300% excess air) the chemical equation for the combustion of octane is

$$C_8 H_{18}(1) + 4(12.5)O_2(g) + 4(12.5)(3.76)N_2(g) \rightarrow 8CO_2(g) + 9H_2O(g)$$

$$+ 3(12.5)O_2(g) + 4(12.5)(3.76)N_2(g)$$

or

$$C_8 H_{18}(1) + 50\,O_2(g) + 188\,N_2(g) \rightarrow 8CO_2(g) + 9H_2O(g)$$

$$+ 37.5\,O_2(g) + 188N_2(g)$$

At $980°F$ the H_2O in the products is in the gaseous state. We will assume ideal gas behavior for all the constituents except the liquid octane.

From Eq. (11-2a) the transfer of heat is given by

$$Q = \sum_{products} n(h_p - h_{p_0}) + \Delta H_R - \sum_{reactants} n(h_r - h_{r_0})$$

Referring to Fig. 11-2, liquid octane is supplied at point r_0 where $T_0 = 77°F = 537°R$, air is supplied at point r where $T = 260°F = 720°R$, and the products leave the combustion chamber at point p where $T = 980°F = 1440°R$.

The enthalpy of reaction ΔH_R at $77°F$ may be obtained from Table 11-1. However, the tabulated value of $-2,371,000$ Btu/lb mole is for gaseous octane and for water as a liquid product. In our present problem, octane is in the liquid state and water is in the vapor state. When a liquid fuel is used, the fuel is vaporized first, so that the energy released is less than that for a gaseous fuel at the same temperature. The enthalpy of vaporization of octane at $77°F$ is 17,860 Btu/lb mole. Therefore the corrected ΔH_R value for liquid octane is $-(2,371,000 - 17,860) = -2,353,000$ Btu/lb mole. Now to correct for gaseous water product, we must further lower the ΔH_R value by the enthalpy of vaporization of water, which at $77°F$ has a value of 1,050 Btu/lb water = 18,900 Btu/lb mole water. The final corrected value of ΔH_R is then $-(2,353,000 - 9 \times 18,900) = -2,183,000$ Btu/lb mole octane.

With the required enthalpy values taken from the JANAF tables (Tables A-15 through A-20), we have from Eq. (11-2a) the amount of heat transfer

$$Q = -2,183,000 + 8(9,815) + 9(7,740) + 37.5(6,815)$$

$$+ 188(6,473) - 50(1,303) - 188(1,278) = -868,000 \text{ Btu/lb mole octane}$$

Thus 868,000 Btu of heat is released per lb mole of octane burned.

11-3   SECOND LAW ANALYSIS OF CHEMICAL REACTIONS

In a first law analysis of a chemical reaction it is assumed that the reaction under consideration simply takes place, and no question is asked about the possibility of the given reaction. The answers to the questions about the extent and direction of a reaction can only be furnished by the second law of thermodynamics. We now discuss the restrictions placed by the second law on the occurrence of chemical reactions.

We have established in Section 5-1 the criteria of equilibrium based on the second law of thermodynamics. The criteria as expressed by Eqs. (5-2) through (5-11) can now be applied to determine the possibility of chemical reactions. For instance, Eq. (5-10) states that the Helmholtz function must decrease and seek its minimum value when a chemical reaction occurs at constant temperature and volume. Once the minimum of the Helmholtz function is attained, a condition of chemical equilibrium is reached and the chemical reaction ceases— meaning that the rates of reaction toward both directions are the same. In the case of a chemical reaction which occurs at constant temperature and pressure, Eq. (5-11) states that the Gibbs function must decrease and seek its minimum value, and once this minimum is attained the chemical reaction ceases.

Let us consider the equilibrium condition at constant temperature and pressure of a reactive mixture capable of undergoing a chemical reaction according to the following balanced equation:

$$\nu_1 C_1 + \nu_2 C_2 \rightleftarrows \nu_3 C_3 + \nu_4 C_4 \tag{11-4}$$

where $C_1, \ldots, C_4$ denote the chemical constituents, and $\nu_1, \ldots, \nu_4$ denote the stoichiometric coefficients. From Eq. (4-6) the infinitesimal change in Gibbs function at constant T and p is

$$dG_{T,p} = \mu_1 dn_1 + \mu_2 dn_2 + \mu_3 dn_3 + \mu_4 dn_4$$

where $n_1, \ldots, n_4$ denote the numbers of moles of each constituent present, and $\mu_1, \ldots, \mu_4$ denote the respective chemical potentials. In a reacting mixture the changes in the number of moles of the various constituents present are not independent of each other, but are proportional to the corresponding stoichiometric coefficients. Thus

$$-\frac{dn_1}{\nu_1} = -\frac{dn_2}{\nu_2} = \frac{dn_3}{\nu_3} = \frac{dn_4}{\nu_4}$$

Accordingly, based on Eq. (5-11) we must have

$$dG_{T,p} = (\mu_1 \nu_1 + \mu_2 \nu_2 - \mu_3 \nu_3 - \mu_4 \nu_4) \frac{dn_1}{\nu_1} \leqslant 0$$

As ν_1 is a positive number, we conclude that

(a) if $\mu_1 \nu_1 + \mu_2 \nu_2 > \mu_3 \nu_3 + \mu_4 \nu_4$ then $dn_1 < 0$, and the reaction proceeds to the right

(b) if $\mu_1 \nu_1 + \mu_2 \nu_2 < \mu_3 \nu_3 + \mu_4 \nu_4$ then $dn_1 > 0$, and the reaction proceeds to the left.

Therefore the condition of equilibrium at constant T and p is

$$\mu_1 \nu_1 + \mu_2 \nu_2 = \mu_3 \nu_3 + \mu_4 \nu_4 \tag{11-5}$$

This is called the *equation of reaction equilibrium*. It provides a means for determining the equilibrium composition of a reactive mixture at a given temperature and pressure.

11-4 IDEAL GAS REACTIONS

The reacting mixtures in most chemical reactions encountered in engineering applications can often be considered as ideal gas mixtures. We will use the equation of reaction equilibrium as derived in the last section to determine the equilibrium compositions of reacting ideal gas mixtures. To do this we need first to show how the chemical potential for an ideal gas is calculated. From Eqs. (4-62) and (4-63), for the ith component in an ideal gas mixture, we may write

$$(d\mu_i)_T = RTd\left[\ln(x_ip)\right] = RTd(\ln x_i + \ln p)$$

At a constant total pressure of the mixture, this equation becomes,

$$(d\mu_i)_T = RTd(\ln x_i)$$

Integrating isothermally at constant total pressure,

$$\int_{\mu_i=g_i}^{\mu_i} (d\mu_i)_T = RT \int_{x_i=1}^{x_i} d(\ln x_i)$$

from which

$$\mu_i = g_i(T,p) + RT \ln x_i \tag{11-6}$$

where μ_i is the chemical potential of component i with a mole fraction of x_i, and $g_i(T, p)$ is the molal Gibbs function of component i at T and p of the mixture.

Substituting Eq. (11-6) in Eq. (11-5) results in

$$[\nu_3 g_3(T,p) + \nu_4 g_4(T,p) - \nu_1 g_1(T,p) - \nu_2 g_2(T,p)]$$

$$+ RT(\nu_3 \ln x_3 + \nu_4 \ln x_4 - \nu_1 \ln x_1 - \nu_2 \ln x_2) = 0$$

or

$$\Delta G_R(T,p) + RT \ln \frac{x_3{}^{\nu_3} x_4{}^{\nu_4}}{x_1{}^{\nu_1} x_2{}^{\nu_2}} = 0 \tag{11-7}$$

where the Gibbs function change for complete reaction $\Delta G_R(T, p)$ is defined as

$$\Delta G_R(T,p) = \nu_3 g_3(T,p) + \nu_4 g_4(T,p) - \nu_1 g_1(T,p) - \nu_2 g_2(T,p) \tag{11-8}$$

It is convenient to express ΔG_R in Eq. (11-7) in terms of a reference pressure p_0, which is commonly taken as 1 atm. Thus, applying Eq. (2-54) to the definition of Gibbs function and noting that the enthalpy h and the variable ϕ are both functions of temperature only for an ideal gas, we obtain

$$g(T,p) = g(T,p_0) + RT \ln\left(\frac{p}{p_0}\right) \tag{11-9}$$

Substituting this equation into Eq. (11-8) results in

$$\Delta G_R(T,p) = \Delta G_R(T,p_0) + RT \ln\left(\frac{p}{p_0}\right)^{\nu_3+\nu_4-\nu_1-\nu_2}$$

where

$$\Delta G_R(T,p_0) = \nu_3 g_3(T,p_0) + \nu_4 g_4(T,p_0) - \nu_1 g_1(T,p_0) - \nu_2 g_2(T,p_0) \tag{11-10}$$

Accordingly, Eq. (11-7) can be written as

$$\Delta G_R(T,p_0) + RT \ln\left[\left(\frac{x_3{}^{\nu_3} x_4{}^{\nu_4}}{x_1{}^{\nu_1} x_2{}^{\nu_2}}\right)\left(\frac{p}{p_0}\right)^{\nu_3+\nu_4-\nu_1-\nu_2}\right]$$

or

$$\left(\frac{x_3{}^{\nu_3} x_4{}^{\nu_4}}{x_1{}^{\nu_1} x_2{}^{\nu_2}}\right)\left(\frac{p}{p_0}\right)^{\nu_3+\nu_4-\nu_1-\nu_2} = \exp\left[\frac{-\Delta G_R(T,p_0)}{RT}\right] \tag{11-11}$$

For a given chemical reaction and a given reference pressure p_0, the Gibbs function change $\Delta G_R(T,p_0)$ as defined by Eq. (11-10) is a function of the temperature T only. As the right side of Eq. (11-11) is a function of the temperature T only, the left side of this equation must depend only on the temperature T and not on the pressure p. For convenience let us define this temperature function as

$$K(T) = \frac{x_3{}^{\nu_3} x_4{}^{\nu_4}}{x_1{}^{\nu_1} x_2{}^{\nu_2}} \left(\frac{p}{p_0}\right)^{\nu_3+\nu_4-\nu_1-\nu_2} \tag{11-12}$$

and call it the *equilibrium constant*.

The equilibrium constant $K(T)$ can be expressed in terms of other variables. For instance, according to Eq. (4-40), the partial pressure of component i in an ideal gas mixture is given by

$$p_i = x_i p$$

Thus, in terms of partial pressures, Eq. (11-12) becomes

$$K(T) = \frac{p_3{}^{\nu_3} p_4{}^{\nu_4}}{p_1{}^{\nu_1} p_2{}^{\nu_2}} (p_0)^{\nu_1+\nu_2-\nu_3-\nu_4} \tag{11-13}$$

With the equilibrium constant so defined, Eq. (11-11) can be rewritten to give the following relation between $K(T)$ and $\Delta G_R(T,p_0)$:

$$\Delta G_R(T,p_0) = -RT \ln K(T) \tag{11-14}$$

It is worth drawing attention again to the fact that, for an ideal gas mixture capable of undergoing the reaction represented by Eq. (11-4), the equilibrium constant $K(T)$ is a function of temperature only and is a dimensionless property of the equilibrium mixture. Once the equilibrium values of the mole fractions or the partial pressures are determined, or if $\Delta G_R(T, p_0)$ is known, the equilibrium constant can be calculated at the desired temperature. Table 11-3 gives the values of $K(T)$ for several reactions.

The equilibrium composition of a reacting ideal gas mixture at a given temperature can be determined if the equilibrium constant is known. We will give a numerical example to illustrate the method employed in such a calculation.

The enthalpy of reaction $\Delta H_R(T, p_0)$ and the entropy change of reaction $\Delta S_R(T, p_0)$ for the complete reaction as expressed by Eq. (11-4) are defined as follows:

$$\Delta H_R(T, p_0) = \nu_3 h_3(T, p_0) + \nu_4 h_4(T, p_0) - \nu_1 h_1(T, p_0) - \nu_2 h_2(T, p_0) \quad (11\text{-}15)$$

$$\Delta S_R(T, p_0) = \nu_3 s_3(T, p_0) + \nu_4 s_4(T, p_0) - \nu_1 s_1(T, p_0) - \nu_2 s_2(T, p_0) \quad (11\text{-}16)$$

where $h(T, p_0)$ and $s(T, p_0)$ are the molal enthalpy and entropy of individual components at temperature T and pressure p_0. The three quantities $\Delta H_R(T, p_0)$, $\Delta S_R(T, p_0)$, and $\Delta G_R(T, p_0)$ are related by the usual thermodynamic equation

$$\Delta G_R(T, p_0) = \Delta H_R(T, p_0) - T\Delta S_R(T, p_0) \quad (11\text{-}17)$$

Through Eqs. (11-14) and (11-17) we can now derive a general expression for the variation of $K(T)$ with temperature. Substituting Eq. (11-17) in Eq. (11-14) leads to

$$-R \ln K(T) = \frac{1}{T} \Delta H_R(T, p_0) - \Delta S_R(T, p_0)$$

Differentiating with respect to temperature gives

$$-R \frac{d[\ln K(T)]}{dT} = -\frac{\Delta H_R(T, p_0)}{T^2} + \frac{1}{T} \frac{d[\Delta H_R(T, p_0)]}{dT} - \frac{d[\Delta S_R(T, p_0)]}{dT}$$

$$(11\text{-}18)$$

Now applying the basic equation $T\,dS = dH - V\,dp$ at constant pressure separately to the products and the reactants, and taking the difference between the two resulting equations, we obtain

$$Td(S_{\text{products}} - S_{\text{reactants}}) = d(H_{\text{products}} - H_{\text{reactants}})$$

or

$$Td[\Delta S_R(T, p_0)] = d[\Delta H_R(T, p_0)]$$

TABLE 11-3. Logarithms to the Base 10 of the Equilibrium Constant $K(T)$

$$K(T) = \frac{x_3^{\nu_3} x_4^{\nu_4}}{x_1^{\nu_1} x_2^{\nu_2}} \left(\frac{p}{p_0}\right)^{\nu_3 + \nu_4 - \nu_1 - \nu_2} \quad \text{and} \quad p_0 = 1 \text{ atm for the reaction } \nu_1 C_1 + \nu_2 C_2 \rightleftharpoons \nu_3 C_3 + \nu_4 C_4$$

T, °K	$H_2 \rightleftharpoons 2H$	$O_2 \rightleftharpoons 2O$	$N_2 \rightleftharpoons 2N$	$H_2O(g)$ $\rightleftharpoons H_2 + \frac{1}{2}O_2$	$H_2O(g)$ $\rightleftharpoons OH + \frac{1}{2}H_2$	CO_2 $\rightleftharpoons CO + \frac{1}{2}O_2$	$\frac{1}{2}O_2 + \frac{1}{2}N_2$ $\rightleftharpoons NO$	$CO_2 + H_2$ $\rightleftharpoons CO + H_2O(g)$	T, °K
298	-71.224	-81.208	-159.600	-40.048	-46.137	-45.066	-15.171	-5.018	298
500	-40.316	-45.880	-92.672	-22.886	-26.182	-25.025	-8.783	-2.139	500
1000	-17.292	-19.614	-43.056	-10.062	-11.309	-10.221	-4.062	-0.159	1000
1200	-13.414	-15.208	-34.754	-7.899	-8.811	-7.764	-3.275	$+0.135$	1200
1400	-10.630	-12.054	-28.812	-6.347	-7.021	-6.014	-2.712	$+0.333$	1400
1600	-8.532	-9.684	-24.350	-5.180	-5.677	-4.706	-2.290	$+0.474$	1600
1800	-6.896	-7.836	-20.874	-4.270	-4.631	-3.693	-1.962	$+0.577$	1800
2000	-5.580	-6.356	-18.092	-3.540	-3.793	-2.884	-1.699	$+0.656$	2000
2200	-4.502	-5.142	-15.810	-2.942	-3.107	-2.226	-1.484	$+0.716$	2200
2400	-3.600	-4.130	-13.908	-2.443	-2.535	-1.679	-1.305	$+0.764$	2400
2600	-2.834	-3.272	-12.298	-2.021	-2.052	-1.219	-1.154	$+0.802$	2600
2800	-2.178	-2.536	-10.914	-1.658	-1.637	-0.825	-1.025	$+0.833$	2800
3000	-1.606	-1.898	-9.716	-1.343	-1.278	-0.485	-0.913	$+0.858$	3000
3200	-1.106	-1.340	-8.664	-1.067	-0.963	-0.189	-0.815	$+0.878$	3200
3500	-0.462	-0.620	-7.312	-0.712	-0.559	$+0.190$	-0.690	$+0.902$	3500
4000	$+0.402$	$+0.340$	-5.504	-0.238	-0.022	$+0.692$	-0.524	$+0.930$	4000
4500	$+1.074$	$+1.086$	-4.094	$+0.133$	$+0.397$	$+1.079$	-0.397	$+0.946$	4500
5000	$+1.612$	$+1.686$	-2.962	$+0.430$	$+0.731$	$+1.386$	-0.296	$+0.956$	5000
5500	$+2.054$	$+2.176$	-2.032	$+0.675$	$+1.004$	$+1.635$	-0.214	$+0.960$	5500
6000	$+2.422$	$+2.584$	-1.250	$+0.880$	$+1.232$	$+1.841$	-0.147	$+0.961$	6000

SOURCE: JANAF Thermochemical Tables, second edition, NSRDS-NBS 37, *Natl. Bur. Std.*, Washington, D.C., 1971.

411

Upon substituting this equation into Eq. (11-18), we have

$$\frac{d\left[\ln K(T)\right]}{dT} = \frac{\Delta H_R(T, p_0)}{RT^2} \quad \text{or} \quad \frac{d\left[\ln K(T)\right]}{d(1/T)} = -\frac{\Delta H_R(T, p_0)}{R} \qquad (11\text{-}19)$$

Either of these equations is known as the *van't Hoff isobar equation*. Since $\Delta H_R(T, p_0)$ is approximately constant for a reaction over a wide temperature range, a plot of $\ln K(T)$ vs. $1/T$ according to Eq. (11-19) is almost a straight line. Such a linear plot is useful in estimating the values of $K(T)$ at other temperatures when its values at two temperatures are known. Furthermore, if the values of $K(T)$ are found by measuring the equilibrium compositions at two different temperatures, a linear plot of Eq. (11-19) can be used for evaluating the enthalpy of reaction at the standard temperature T_0.

> *EXAMPLE 11-4.* Gaseous propane (C_3H_8) is burned with 80% of theoretical air in a steady-flow process at 1 atm. Both the fuel and the air are supplied at $25°C$. The products which consist of CO_2, CO, H_2O, H_2, and N_2 in equilibrium, leave the combustion chamber at $1,500°K$. Determine the composition of the products and the amount of heat transfer in this process per kg of propane.
>
> *Solution.*
> Let y and z be the numbers of moles of CO_2 and H_2O in the products. The chemical equation can then be written as
>
> $C_3H_8 + 0.8(5\ O_2 + 3.76 \times 5\ N_2) \rightarrow y\ CO_2 + (3 - y)CO + z\ H_2O + (4 - z)H_2$
> $\qquad\qquad\qquad\qquad\qquad\qquad\qquad\qquad\qquad + 0.8 \times 3.76 \times 5N_2$
>
> where the coefficients for CO and H_2 are obtained by mass balances on C and H_2. A mass balance for O_2 then gives
>
> $0.8 \times 5 = y + \dfrac{1}{2}\ (3 - y) + \dfrac{1}{2}\ z$
>
> or
>
> $y + z = 5$ \hfill (11-20)
>
> Another relation between y and z can be obtained from the equilibrium condition for the reaction
>
> $CO_2 + H_2 \rightleftarrows CO + H_2O$ \hfill (11-21)
>
> According to Eq. (11-13), the equilibrium constant for this reaction is
>
> $K(T) = \dfrac{p_{CO}\, p_{H_2O}}{p_{CO_2}\, p_{H_2}}$
>
> where the p's are the partial pressures in the equilibrium mixture of

$y\ CO_2 + (3 - y)CO + z\ H_2O + (4 - z)H_2 + 15.04\ N_2$

For a total pressure of 1 atm, the values of partial pressures in atm are

$$p_{CO} = \frac{(3 - y)}{y + (3 - y) + z + (4 - z) + 15.04} = \frac{3 - y}{22.04}$$

$$p_{H_2O} = \frac{z}{22.04} \quad , \quad p_{CO_2} = \frac{y}{22.04} \quad , \quad p_{H_2} = \frac{4 - z}{22.04}$$

Thus

$$K(T) = \frac{(3 - y)z}{y(4 - z)}$$

From Table 11-3 for the reaction as expressed by Eq. (11-21), we have at $1,500°K$,

$$\log_{10}[K(T)] = 0.409 \quad \text{or} \quad K(T) = 2.56$$

Therefore

$$\frac{(3 - y)z}{y(4 - z)} = 2.56 \tag{11-22}$$

Solving Eqs. (11-20) and (11-22) simultaneously yields

$$y = 1.81 \quad \text{and} \quad z = 3.19$$

The chemical equation for the combustion process is then

$$C_3H_8 + 4\ O_2 + 15.04\ N_2 \rightarrow 1.81\ CO_2 + 1.19\ CO + 3.19\ H_2O$$

$$+ 0.81\ H_2 + 15.04\ N_2 \tag{11-23}$$

Thus the equilibrium mole fractions of the products are

$$x_{CO_2} = \frac{1.81}{1.81 + 1.19 + 3.19 + 0.81 + 15.04} = \frac{1.81}{22.04} = 0.0821 = 8.21\%$$

$$x_{CO} = \frac{1.19}{22.04} = 0.0540 = 5.40\%$$

$$x_{H_2O} = \frac{3.19}{22.04} = 0.1447 = 14.47\%$$

$$x_{H_2} = \frac{0.81}{22.04} = 0.0368 = 3.68\%$$

$$x_{N_2} = \frac{15.04}{22.04} = 0.6824 = 68.24\%$$

According to Eq. (11-3), the enthalpy of reaction ΔH_R at 25°C and 1 atm for the reaction expressed by Eq. (11-23) is

$$\Delta H_R = \sum_{\text{products}} (\nu \Delta H_F) - \sum_{\text{reactants}} (\nu \Delta H_F)$$

$$= 1.81\,\Delta H_{F,\,CO_2} + 1.19\,\Delta H_{F,\,CO} + 3.19\,\Delta H_{F,\,H_2O} - \Delta H_{F,\,C_3H_8}$$

in which the ΔH_F value of O_2, N_2, and H_2 has been assigned a value of zero. Upon substituting numerical values of ΔH_F from Table 11-2, we obtain

$$\Delta H_R = 1.81\,(-94.054) + 1.19(-26.417) + 3.19(-57.798) - 1(-24.82)$$

$$= -361.23 \text{ kcal/g mole } C_3H_8$$

Now applying Eq. (11-2a) to the reaction represented by Eq. (11-23), we have

$$Q = \Delta H_R + \sum_{\text{products}} n(h_p - h_{p_0}) - \sum_{\text{reactants}} n(h_r - h_{r_0})$$

$$= \Delta H_R + 1.81(h_p - h_{p_0})_{CO_2} + 1.19(h_p - h_{p_0})_{CO}$$

$$+ 3.19(h_p - h_{p_0})_{H_2O} + 0.81(h_p - h_{p_0})_{H_2} + 15.04(h_p - h_{p_0})_{N_2}$$

in which the terms for the reactants are zero because the reactants enter the process at the reference temperature of 25°C. Substituting numerical values from Appendix Tables A-15 through A-20, at the final products temperature of 1,500°K, we have

$$Q = -361,230 + 1.81(14,750) + 1.19(9,285)$$

$$+ 3.19(11,495) + 0.81(8,668) + 15.04(9,179)$$

$$= -141,740 \text{ cal/g mole } C_3H_8 = -3,221 \text{ kcal/kg } C_3H_8$$

11-5 APPLICATION OF THE THIRD LAW TO CHEMICAL REACTIONS

In Section 10-1 we studied Planck's statement of the third law which is: the absolute value of the entropy of a pure substance approaches zero as the temperature approaches absolute zero. This form of the third law permits the determination of the absolute entropy of all pure substances with a common base, namely $S = 0$ at $T = 0$.

By virtue of the third law, the entropy changes for processes involving chemical reactions can be obtained directly from experimentally measurable quantities. For a chemical reaction, the change of entropy at temperature T may be written as

$$S_p - S_r = (S_p - S_{p,0°K}) - (S_r - S_{r,0°K}) + (S_{p,0°K} - S_{r,0°K})$$

where the subscripts p and r denote the products and reactants and the subscript $0°K$ denotes the condition at the absolute zero of temperature. But according to Eq. (10-1), we must have

$$(S_{p,0°K} - S_{r,0°K}) = 0$$

Therefore

$$S_p - S_r = (S_p - S_{p,0°K}) - (S_r - S_{r,0°K})$$

The quantities $(S_p - S_{p,0°K})$ and $(S_r - S_{r,0°K})$ are the changes in entropy of the products and the reactants from $0°K$ to the temperature T. These quantities are experimentally measurable. For example, when the heat capacities at constant volume from $0°K$ to $T°K$ for the products and the reactants are known, we can evaluate these quantities by the relations

$$(S_p - S_{p,0°K}) = \int_0^T \frac{C_{v_p}}{T}\, dT \quad \text{and} \quad (S_r - S_{r,0°K}) = \int_0^T \frac{C_{v_r}}{T}\, dT$$

Now if the entropy $S_{p,0°K}$ and $S_{r,0°K}$ are assigned a value of zero according to Planck's postulate, then $S_p = (S_p - S_{p,0°K})$ and $S_r = (S_r - S_{r,0°K})$. They are then the absolute entropies of the products and the reactants.

The absolute entropies of a large number of pure substances at 25°C and 1 atm are listed in Table 11-2. In addition, Appendix Tables A-15 through A-20 give the absolute entropy at atmospheric pressure of a number of pure substances as a function of temperature. An equivalent set of data is given by the Keenan and Kaye gas tables (Table A-12), in which the tabulated ϕ values may be used directly as the absolute entropy values for ideal gases at 1 atm.

As outlined above, we can obtain the value of $\Delta S_R(T, p_0)$ for a chemical reaction at temperature T and a standard pressure p_0 from measurable thermal data. In addition, we can also obtain the value of $\Delta H_R(T, p_0)$ from ordinary calorimetry data. Then the value of $\Delta G_R(T, p_0)$ for the reaction is calculated from Eq. (11-17), and finally, the equilibrium constant $K(T)$ is calculated from Eq. (11-14). Thus, by virtue of the third law, the equilibrium constant of a reacting mixture is obtained entirely from thermal data.

EXAMPLE 11-5. Determine the equilibrium constant at $1,000°K$ for the reaction

$$NO + \frac{1}{2} O_2 \rightleftharpoons NO_2$$

Solution.

From Eqs. (11-15) and (11-16) we first determine the values of $\Delta H_R(T, p_0)$ and $\Delta S_R(T, p_0)$ at $T = 1,000°K$ and $p_0 = 1$ atm, using the enthalpy and absolute entropy data from Tables A-15 through A-20. Thus

$$\Delta H_R(T, p_0) = h_{NO_2} - h_{NO} - \frac{1}{2} h_{O_2} = (7,730 + 7,910) - (5,313 + 21,652)$$

$$- \frac{1}{2} (5,427) = -14,039 \text{ cal}$$

$$\Delta S_R(T, p_0) = s_{NO_2} - s_{NO} - \frac{1}{2} s_{O_2} = 70.215 - 59.377 - \frac{1}{2} (58.192)$$

$$= -18.258 \text{ cal/°K}$$

Now according to Eq. (11-17) we have

$$\Delta G_R(T, p_0) = \Delta H_R(T, p_0) - T \Delta S_R(T, p_0)$$

$$= -14,039 - 1,000 (-18.258) = 4,219 \text{ cal}$$

And from Eq. (11-14) we have

$$\ln K(T) = -\frac{\Delta G_R(T, p_0)}{RT} = -\frac{4,219}{1.986 \times 1,000} = -2.1244$$

Therefore

$$K(T) = 0.1195$$

11-6 FUEL CELL

As an application of a chemical reaction analysis we now give a brief description of a fuel cell. A *fuel cell* is an electrochemical device which converts chemical energy directly into electrical energy. A hydrogen-oxygen fuel cell is shown schematically in Fig. 11-3, in which two porous catalytic electrodes are separated from each other by an electrolyte (or an ion-conducting membrane). Hydrogen and oxygen gases under pressure are supplied to the anode and the cathode sides respectively. At the anode-electrolyte interface, hydrogen molecules dissociate into hydrogen ions and electrons according to the expression

$$2H_2(g) \rightarrow 4H^+ + 4e^- \tag{11-24}$$

These electrons flow through an external circuit where they do electrical work and then return to the cell at the cathode. Meanwhile, the hydrogen ions diffuse through the electrolyte to the electrolyte-cathode interface where they combine with the returning electrons and oxygen gas to form liquid water according to the expression

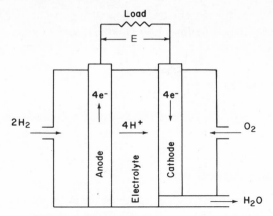

FIG. 11-3. Schematic of a hydrogen-oxygen fuel cell.

$$4H^+ + 4e^- + O_2(g) \rightarrow 2H_2O(l) \tag{11-25}$$

The overall chemical reaction is therefore

$$2H_2(g) + O_2(g) \rightarrow 2H_2O(l) \tag{11-26}$$

As a result of the reaction expressed by the preceding equation, a quantity of energy representing the enthalpy of reaction ΔH_R is released. Part of this energy is available for conversion to electrical energy, the remaining energy is transferred out as heat. Since in this reaction process there are four electrons passing through the external circuit per molecule of O_2 (or per two molecules of H_2), the electrical energy output per mole of O_2 (or per two moles of H_2) is then

$$W_e = 4 N_A eE \tag{11-27}$$

wherein N_A is the Avogadro's number, e is the electronic charge, and E is the terminal voltage developed.

In order to determine the maximum possible electrical power output, let us write the first and second laws for the cell, assuming steady-flow, steady-state conditions with negligible changes in kinetic and potential energies for the fluids. Furthermore, let us consider only the electrostatic potential energy of the electrons, and assume that there is no temperature difference between the flowing fluids and the cell. Writing the first law for an infinitesimal period of time, we have

$$(h \, dn)_{H_2O} - (h \, dn)_{H_2} - (h \, dn)_{O_2} + eE \, dN_e + dQ_{out} = 0 \tag{11-28}$$

where dN_e is the number of electrons and the dn's are the numbers of moles of fluids crossing the control volume of the system, and dQ_{out} is the amount of heat transferred out to the surroundings. Writing the second law for the same infinitesimal process, we have

$$\frac{dQ_{out}}{T} + (s\,dn)_{H_2O} - (s\,dn)_{H_2} - (s\,dn)_{O_2} \geqslant 0 \qquad (11\text{-}29)$$

where the equality holds for a reversible process and the inequality for an irreversible process. Combining Eqs. (11-28) and (11-29) and noting that $g = h - Ts$, we have then

$$eE\,dN_e \leqslant - [(g\,dn)_{H_2O} - (g\,dn)_{H_2} - (g\,dn)_{O_2}] \qquad (11\text{-}30)$$

in which the term $eE\,dN_e$ is the electrical energy output during the infinitesimal time period for which Eqs. (11-28) and (11-29) are written. Now since we are assuming a stoichiometric reaction of Eq. (11-26), the electrical energy output as expressed by Eq. (11-30) per mole of O_2 (or per two moles of H_2) can be written as

$$W_e \leqslant - \Delta G_R \qquad (11\text{-}31)$$

Thus, the maximum electrical power output is equal to the change in Gibbs function for the reaction.

From Eqs. (11-27) and (11-31) it follows that the maximum voltage which can be developed by the cell is given by

$$E = \frac{-\Delta G_R}{nN_A e} \qquad (11\text{-}32)$$

where $n = 4$ is the number of electrons released per molecule of O_2 (or per two molecules of H_2), and ΔG_R is the change in Gibbs function per mole of O_2 (or per two moles of H_2).

The maximum or ideal efficiency of a fuel cell is defined as

$$\eta_{max} = \frac{\Delta G_R}{\Delta H_R} = \frac{-nN_A eE}{\Delta H_R} \qquad (11\text{-}33)$$

Since $\Delta G_R = \Delta H_R - T\Delta S_R$ the maximum efficiency of a fuel cell is less than unity, even for a cell which operates reversibly as long as heat is being rejected to the surroundings. Unlike a heat engine, the maximum efficiency of a fuel cell is not limited by the Carnot principle. The maximum efficiency of a fuel cell may be as high as 90 percent. Owing to some undesirable effects, such as the Joule heating which occurs in the electrolyte, the actual voltage of a fuel cell will be less than that given by Eq. (11-32), and the actual efficiency will be less than that given by Eq. (11-33).

EXAMPLE 11-6. For an H_2-O_2 fuel cell working under steady-flow, steady-state conditions at $298°K$ and 1 atm, calculate the maximum voltage and the maximum efficiency. The water formed leaves the cell as a liquid.

Solution.

Since $g = h - Ts$, from Table 11-2 we obtain the following values:

For H_2, $g = 0 - (298)(31.208) = -9,300$ cal/g mole

For O_2, $g = 0 - (298)(49.004) = -14,603$ cal/g mole

For liquid H_2O, $g = -68,315 - (298)(16.71) = -73,295$ cal/g mole

For the reaction

$$2H_2(g) + O_2(g) \rightarrow 2H_2O(l)$$

the change in Gibbs function at $298°K$ and 1 atm is then

$$\Delta G_R = 2(-73,295) - 2(-9,300) - 1(-14,603) = -113,387 \text{ cal/g mole } O_2$$

$$= -474,751 \text{ joules/g mole } O_2$$

Now since $n = 4$ for the reaction, and

$$N_A e = (6.022 \times 10^{23})(1.602 \times 10^{-19}) = 96,470 \text{ coul} = 96,470 \text{ joules/volt}$$

then from Eq. (11-32) we obtain for the maximum voltage

$$E = -\frac{\Delta G_R}{nN_A e} = -\frac{-474,751}{4(96,470)} = 1.23 \text{ volts}$$

From Table 11-1, the ΔH_R value for the given reaction is

$$\Delta H_R = -68,315 \text{ cal/g mole } H_2O$$

$$= -136,630 \text{ cal/g mole } O_2$$

The maximum efficiency is then

$$\eta_{max} = \frac{\Delta G_R}{\Delta H_R} = \frac{-113,387}{-136,630} = 0.83$$

PROBLEMS

11-1. A fuel-gas mixture has the following volumetric analysis: 75.5% CH_4 and 24.5% C_2H_6. For complete combustion with 20% excess air, write the chemical equation and calculate the volumetric analysis of the dry products. The dew point of the products is defined as the saturation temperature corresponding to the partial pressure of water vapor in the products. Calculate the dew point of the products if the total pressure of the products is 1 atm.

11-2. A gaseous fuel mixture has the following volumetric analysis: CH_4 20%, C_2H_6 10%, CO 10%, CO_2 10%, H_2 10%, O_2 10%, and N_2 30%. For complete combustion with 30% excess air, write the chemical equation. Both the fuel and the air are supplied at 14.7 psia and $100°F$. Determine

(a) the volume of dry air per unit volume of fuel

(b) the volumetric analysis of the dry products

(c) the mass of the total products per unit mass of fuel

(d) the mass of water vapor in the products per unit mass of fuel

(e) the mass of dry air supplied per unit mass of fuel

(f) the dew point of the products if dry air is supplied for combustion (see preceding problem for definition of dew point)

(g) the dew point of the products if the air supplied for combustion contains 0.0295 lbm of water vapor per lbm of dry air.

11-3. An unknown hydrocarbon fuel (C_xH_y) undergoes a combustion process such that the dry products have the following volumetric analysis: CO_2 9.0%, CO 1.1%, O_2 8.8%, and N_2 81.1%. Determine on a mass basis

(a) the composition of the fuel

(b) the air-fuel ratio

(c) the percent theoretical air.

11.4. Determine the value of ΔH_R at $25°C$ for the following reaction:

$$CH_4 + C_2H_6 + 5O_2 \rightarrow 0.1CH_4 + 0.1C_2H_6 + 2.2CO_2$$

$$+ 0.5CO + 4.5H_2O(l) + 0.3O_2$$

11.5. Liquid octane (C_8H_{18}) at $25°C$ is burned completely with 400% theoretical air supplied at $25°C$ in a low-pressure steady-flow process. No work or changes in kinetic and potential energies are involved. If the combustion process is adiabatic, determine the temperature of the products. This temperature is known as the maximum adiabatic flame temperature.

11.6. Liquid octane is burned in an internal combustion engine with 20% excess air. Ninety percent of the carbon burns to CO_2 and the remainder burns to CO. The reactants enter the engine at $25°C$, and the products leave at $625°C$. The engine uses 0.2 kg of fuel per sec. Determine the power output of the engine, assuming the heat transfer from the engine and the work done by the engine being numerically equal.

11-7. Determine the equilibrium constant $K(T)$ at $298°K$ for the gaseous reaction $CO_2 + H_2 \rightleftharpoons CO + H_2O$ by using

(a) the enthalpy and entropy data from Table 11-2

(b) the $K(T)$ data from Table 11-3 for the simple reactions $CO_2 \rightleftharpoons CO + \frac{1}{2}O_2$ and $H_2O \rightleftharpoons H_2 + \frac{1}{2}O_2$.

Compare your answers with that read directly from Table 11-3 for the reaction $CO_2 + H_2 \rightleftharpoons CO + H_2O$.

11-8. Determine the equilibrium constant $K(T)$ at $3,000°K$ for the gaseous reaction $CO_2 + H_2 \rightleftharpoons CO + H_2O$ by the use of the data from Tables A-15 through A-20. Compare your answer with that given in Table 11-3.

11-9. Solve the preceding problem for the reaction $H_2 \rightleftharpoons 2H$.

11-10. For the gaseous reaction $H_2O(g) \rightleftharpoons OH + \frac{1}{2}H_2$, the equilibrium constant at $2,000°K$ is 0.0001611. Estimate the equilibrium constant at $3,000°K$, using the van't Hoff isobar equation.

11-11. Determine the mole fraction of OH in the product mixture when H_2O is heated to $5,000°K$ at 1 atm pressure.

11-12. For the reaction of CO with the stoichiometric amount of O_2 to form CO_2, determine the composition of the equilibrium mixture at $3,000°K$ and

(a) 1 atm

(b) 5 atm.

11-13. Methane (CH_4) is burned with the theoretical amount of air in a steady-flow process at 1 atm. Both the fuel and the air are supplied at $25°C$. The products which consist of CO_2, CO, H_2O, H_2, O_2, and N_2 in equilibrium, leave the combustion chamber at $2,200°K$. Determine the composition of the products and the amount of heat transfer in this process per mole of methane.

11-14. For a CH_4-O_2 fuel cell working under steady-flow, steady-state conditions at $298°K$ and 1 atm,

(a) write the chemical equations at the anode and at the cathode

(b) calculate the maximum voltage and the maximum efficiency.

The water formed leaves the cell as a liquid.

11-15. Solve Problem 11-14 for a carbon-oxygen fuel cell.

Appendix 1
LAGRANGE'S METHOD OF UNDETERMINED MULTIPLIERS

Lagrange's method of undetermined multipliers is a general procedure for determining an extremum point such as a maximum or minimum in a continuous function of many variables subject to a number of constraints. Consider a function $f(x_1, x_2, \ldots, x_n)$, where n denotes the number of variables. The variables $x_1, x_2, \ldots, x_n$ are not all independent but are related according to the following equations of constraint:

$$\phi_j(x_1, x_2, \ldots, x_n) = 0 \qquad j = 1, 2, \ldots, m \tag{A1-1}$$

where m is the number of equations of constraint and $m < n$.

At an extremum point of the function $f(x_1, x_2, \ldots, x_n)$ it is necessary that

$$df = \frac{\partial f}{\partial x_1} dx_1 + \frac{\partial f}{\partial x_2} dx_2 + \cdots + \frac{\partial f}{\partial x_n} dx_n = \sum_{i=1}^{n} \frac{\partial f}{\partial x_i} dx_i = 0 \tag{A1-2}$$

Of course, if all the variables are independent the following set of n independent equations

$$\frac{\partial f}{\partial x_i} = 0 \qquad i = 1, 2, \ldots, n \tag{A1-3}$$

will give the values of the n independent variables at the extremum condition. But since the variables are not all independent, Eq. (A1-3) is not valid, although an extremum point is still given by Eq. (A1-2). However, we now have the equations of constraint, Eq. (A1-1), in addition to Eq. (A1-2). A straightforward but pedestrian way to find the extremum point is to solve Eq. (A1-1) for m variables in terms of the other $n-m$ variables. These solutions when inserted in $f(x_1, x_2, \ldots, x_n)$ give a function of $n-m$ independent variables, and the extremum can then be found from the $n-m$ equations analogous to Eq. (A1-3).

A more elegant and useful method to solve this type of problem is to use Lagrange's method of undetermined multipliers. According to this method, one first writes the total differential of each constraining relation to yield

$$d\phi_j = 0 = \sum_{i=1}^{n} \frac{\partial \phi_j}{\partial x_i} dx_i \qquad j = 1, 2, \ldots, m \tag{A1-4}$$

Next let each $d\phi_j$ of the preceding equations be multiplied by some undetermined multiplier λ_j (the Lagrangian multiplier) and then added to Eq. (A1-2). The result is

$$\sum_{i=1}^{n} \frac{\partial f}{\partial x_i} dx_i + \sum_{j=1}^{m} \left[\lambda_j \sum_{i=1}^{n} \left(\frac{\partial \phi_j}{\partial x_i} dx_i \right) \right] = 0$$

or

$$\left(\frac{\partial f}{\partial x_1} + \lambda_1 \frac{\partial \phi_1}{\partial x_1} + \lambda_2 \frac{\partial \phi_2}{\partial x_1} + \cdots + \lambda_m \frac{\partial \phi_m}{\partial x_1} \right) dx_1$$

$$+ \left(\frac{\partial f}{\partial x_2} + \lambda_1 \frac{\partial \phi_1}{\partial x_2} + \lambda_2 \frac{\partial \phi_2}{\partial x_2} + \cdots + \lambda_m \frac{\partial \phi_m}{\partial x_2} \right) dx_2 + \cdots \tag{A1-5}$$

$$+ \left(\frac{\partial f}{\partial x_n} + \lambda_1 \frac{\partial \phi_1}{\partial x_n} + \lambda_2 \frac{\partial \phi_2}{\partial x_n} + \cdots + \lambda_m \frac{\partial \phi_m}{\partial x_n} \right) dx_n = 0$$

Now, if $\lambda_1, \lambda_2, \ldots, \lambda_m$ are so chosen that

$$\frac{\partial f}{\partial x_i} + \lambda_1 \frac{\partial \phi_1}{\partial x_i} + \lambda_2 \frac{\partial \phi_2}{\partial x_i} + \cdots + \lambda_m \frac{\partial \phi_m}{\partial x_i} = 0 \qquad i = 1, 2, \ldots, n \tag{A1-6}$$

then, together with the constraining equations

$$\phi_j(x_1, x_2, \ldots, x_n) = 0 \qquad j = 1, 2, \ldots, m \tag{A1-1}$$

the $n + m$ "unknowns" $x_1, x_2, \ldots, x_n, \lambda_1, \lambda_2, \ldots, \lambda_m$ can be determined to satisfy the necessary condition for an extremum of the function $f(x_1, x_2, \ldots, x_n)$.

Appendix 2
UNITS OF MEASUREMENT

TABLE A-1. System International Units (SI)

(a) Basic Units

Quantity	Name	Symbol
Length	meter	m
Mass	kilogram	kg
Time	second	sec
Electric current	ampere	amp
Temperature	kelvin	K
Luminous intensity	candela	cd

(b) Supplementary Units

Quantity	Name	Symbol
Plane angle	radian	rad
Solid angle	steradian	sr

(c) Derived Units with Special Names

Quantity	Name	Definition
Force	newton	$(kg)(m)/sec^2$
Energy	joule	$(kg)(m^2)/sec^2 = (newton)(m)$
Power	watt	$(kg)(m^2)/sec^3 = joule/sec$
Electric charge	coulomb	$(amp)(sec)$
Electric potential	volt	$(kg)(m^2)/(sec^3)(amp) = joule/(amp)(sec)$ $= joule/coul$
Electric resistance	ohm (Ω)	$(kg)(m^2)/(sec^3)(amp^2) = volt/amp$
Electric capacitance	farad	$(amp^2)(sec^4)/(kg)(m^2) = (amp)(sec)/volt$ $= coul/volt$
Magnetic flux	weber	$(kg)(m^2)/(sec^2)(amp) = (volt)(sec)$
Magnetic induction (Magnetic flux density)	tesla	$kg/(sec^2)(amp) = (volt)(sec)/m^2$ $= weber/m^2$
Inductance	henry	$(kg)(m^2)/(sec^2)(amp^2) = (volt)(sec)/amp$

TABLE A-1.   System International Units (SI)(*Continued*)

(d) Some Other Derived Units

Quantity	Unit	Symbol
Area	square meter	m^2
Volume	cubic meter	m^3
Pressure	newton per square meter	newton/m^2
Stress	newton per square meter	newton/m^2
Density	kilogram per cubic meter	kg/m^3
Specific weight	newton per cubic meter	newton/m^3
Velocity	meter per second	m/sec
Acceleration	meter per second squared	m/sec^2
Angular velocity	radian per second	rad/sec
Specific heat	joule per kilogram per kelvin	joule/(kg)($^{\circ}$K)
Specific entropy	joule per kilogram per kelvin	joule/(kg)($^{\circ}$K)
Specific enthalpy	joule per kilogram	joule/kg
Electric field intensity	volt per meter	volt/m
Magnetic field intensity	ampere per meter	amp/m

(e) Alternative Units

Quantity	Name	Symbol	Relation to basic units
Temperature	degree Celsius	$^{\circ}$C	t ($^{\circ}$C) = K $-$ 273.15
Volume	liter	liter	1 liter = 10^{-3} m^3 = 10^3 cm^3
Pressure	bar	bar	1 bar = 10^5 newton/m^2

(f) Decimal Multiples

Multiple or submultiple	Prefix	Symbol
10^{12}	tera	T
10^9	giga	G
10^6	mega	M
10^3	kilo	k
10^2	hecto	h
10	deka	da
10^{-1}	deci	d
10^{-2}	centi	c
10^{-3}	milli	m
10^{-6}	micro	μ
10^{-9}	nano	n
10^{-12}	pico	p
10^{-15}	femto	f
10^{-18}	atto	a

TABLE A-2. Conversion Factors

Length	1 m = 3.28084 ft = 39.3701 in.
	1 in. = 2.54 cm
	1 mile = 5280 ft = 1.60934 km
	1 micron (μ) = 10^{-6} m = 3.28084 $\times$ 10^{-6} ft
Volume	1 liter = 1000 cm^3 (exactly, as redefined in SI units) = 0.03531 ft^3
	= 0.26417 US gallon
	1 m^3 = 35.31 ft^3
	1 in.^{3} = 16.387 cm^3
Mass	1 kg = 2.20462 lbm = 6.85218 $\times$ 10^{-2} slug
	1 lbm = 453.592 g
	1 slug = 32.174 lbm
Density	1 g/cm^3 = 62.428 lbm/ft^3
	1 lbm/ft^3 = 0.0160185 g/cm^3
Force	1 newton = 1 kg m/sec^2 = 10^5 dynes
	1 dyne = 1 g cm/sec^2
	1 lbf = 1 slug-ft/sec^2 = 4.44822 newtons = 4.44822 $\times$ 10^5 dynes
Pressure	1 lbf/in.^{2} = 6,894.76 newtons/m^2
	1 bar = 10^5 newtons/m^2 = 10^6 dynes/cm^2 = 0.986923 atm
	1 atm = 14.6959 lbf/in.^{2} = 1.01325 bars = 760 mm Hg at 32°F
	= 29.9213 in. Hg at 32°F
Temperature	T (Rankine) = 1.8 T (Kelvin)
	t (Fahrenheit) = 1.8 t (Celsius) + 32
	T (Kelvin) = t (Celsius) + 273.15
	T (Rankine) = t (Fahrenheit) + 459.67
Energy	1 joule = 1 newton-meter = 10^7 ergs
	1 erg = 1 dyne-cm = 9.86923 $\times$ 10^{-7} atm-cm^3
	1 cal (International Table Calorie) = 4.1868 joules = 3.96832 $\times$ 10^{-3} Btu
	1 Btu = 778.169 ft-lbf
	1 kW-kr = 1.34102 hp-hr = 3.6 $\times$ 10^6 joules
Specific energy	1 Btu/lbm = 2.326 joules/g = 0.555556 cal/g
	1 joule/g = 334.553 ft-lbf/lbm = 0.429923 Btu/lbm
Power	1 watt = 1 joule/sec
	1 hp = 550 ft-lbf/sec = 2,544.43 Btu/hr
	1 kW = 1.34102 hp = 3,412.14 Btu/hr

Appendix 3
PHYSICAL CONSTANTS

TABLE A-3. Physical Constants

Avogadro's number*	6.022169×10^{26} (kg mole)$^{-1}$
Boltzmann constant	1.380622×10^{-23} joule/$^\circ$K
Planck's constant	6.626196×10^{-34} joule-sec
Speed of light	2.9979250×10^{8} m/sec
Electronic charge	$1.6021917 \times 10^{-19}$ coul
Bohr magneton	9.274096×10^{-24} amp-m^2
Permeability of free space	$4\pi \times 10^{-7}$ newton/amp^2
Permittivity of free space	8.85434×10^{-12} coul^2/(newton)(m^2)
Standard gravitational acceleration	9.80665 m/sec^2 = 32.174 ft/sec^2

*This value is based on the international atomic mass scale on which $M_{C^{12}} = 12$ units of mass per mole.
Source: Data mostly from B. N. Taylor, W. H. Parker, and D. N. Langenberg, *Rev. Mod. Phys.*, 41:375 (1969).

TABLE A-4. Universal Gas Constant in Various Units*

8.31429×10^{3} joule/(kg mole)($^\circ$K)
1.98583 kcal/(kg mole)($^\circ$K)
2.30953×10^{-3} (kW)(hr)/(kg mole)($^\circ$K)
8.20560×10^{-2} (atm)(m^3)/(kg mole)($^\circ$K)
1.54531×10^{3} (ft)(lbf)/(lb mole)($^\circ$R)
1.98583 Btu/(lb mole)($^\circ$R)
1.07314×10 (psia)(ft^3)/(lb mole)($^\circ$R)
7.30225×10^{-1} (atm)(ft^3)/(lb mole)($^\circ$R)

*The universal gas constant is based on the international atomic mass scale on which $M_{C^{12}} = 12$ units of mass per mole.
Source: J. Kestin, "A Course in Thermodynamics," vol. II, Blaisdell Publishing Co., Waltham, Massachusetts, 1968.

Appendix 4
PROPERTIES OF SUBSTANCES

TABLE A-5. Critical Constants

Substance	Formula	Molecular weight*	T_c °K	P_c atm	$\dfrac{v_c}{\text{cm}^3}$ (g mole)	Z_c
Acetylene	C_2H_2	26.038	309.5	61.6	113.	0.274
Ammonia	NH_3	17.031	405.5	111.3	72.5	0.243
Argon	Ar	39.948	151.	48.0	75.2	0.291
Benzene	C_6H_6	78.115	562.61	48.6	260.	0.274
n-Butane	C_4H_{10}	58.124	425.17	37.47	255.	0.274
i-Butane	C_4H_{10}	58.124	408.14	36.00	263.	0.283
1-Butene	C_4H_8	56.108	419.6	39.7	241.	0.278
Carbon dioxide	CO_2	44.010	304.2	72.9	94.0	0.275
Carbon disulfide	CS_2	76.139	552.	78.	170.	0.293
Carbon monoxide	CO	28.011	133.	34.5	93.1	0.294
Carbon tetrachloride	CCl_4	153.823	556.4	45.0	276.	0.272
Chlorine	Cl_2	70.906	417.	76.1	124.	0.276
n-Decane	$C_{10}H_{22}$	142.287	619.	20.8	602.	0.247
Ethane	C_2H_6	30.070	305.43	48.20	148.	0.285
Ethylene	C_2H_4	28.055	283.06	50.50	124.	0.270
Freon-12	CCl_2F_2	120.914	384.7	39.6	218.	0.273
Helium	He	4.003	5.20	2.25	57.5	0.303
n-Heptane	C_7H_{16}	100.206	540.16	27.01	426.	0.260
n-Hexane	C_6H_{14}	86.178	507.9	29.92	368.	0.264
Hydrogen	H_2	2.016	33.3	12.80	65.0	0.304
Hydrogen sulfide	H_2S	34.080	373.6	88.9	97.7	0.283
Krypton	Kr	83.80	209.4	54.3	92.2	0.291
Methane	CH_4	16.043	190.7	45.80	99.	0.290
Methyl chloride	CH_3Cl	50.488	416.3	65.9	143.	0.276
Neon	Ne	20.183	44.5	26.9	41.7	0.307
Nitric oxide	NO	30.006	180.	64.	58.	0.251
Nitrogen	N_2	28.013	126.2	33.5	90.1	0.291
Nitrous oxide	N_2O	44.013	309.7	71.7	96.3	0.272

TABLE A-5.   Critical Constants (*Continued*)

Substance	Formula	Molecular weight*	T_c °K	P_c atm	v_c cm³ (g mole)	Z_c
n-Nonane	C_9H_{20}	128.260	595.	22.5	543.	0.250
n-Octane	C_8H_{18}	114.233	569.4	24.64	486.	0.256
Oxygen	O_2	31.999	154.6	49.8	73.4	0.288
n-Pentane	C_5H_{12}	72.151	469.78	33.31	311.	0.269
i-Pentane	C_5H_{12}	72.151	461.0	32.9	308.	0.268
neo-Pentane	C_5H_{12}	72.151	433.76	31.57	303.	0.269
Propane	C_3H_8	44.097	369.97	42.01	200.	0.277
Propylene	C_3H_6	42.081	365.1	45.4	181.	0.274
Sulfur dioxide	SO_2	64.063	430.7	77.8	122.	0.269
Sulfur trioxide	SO_3	80.062	491.4	83.8	126.	0.262
Toluene	C_7H_8	92.142	594.0	40.	320.	0.263
Water	H_2O	18.015	647.4	218.3	56.	0.230
Xenon	Xe	131.30	289.75	58.0	118.8	0.290

*The values of molecular weight are based on the international atomic mass scale on which $M_{C^{12}} = 12$ units of mass per mole.

Source: Hydrocarbon data from F. D. Rossini, K. S. Pitzer, R. L. Arnett, R. M. Braun, and G. C. Pimentel, "Selected Values of Physical and Thermodynamic Properties of Hydrocarbons and Related Compounds," Carnegie Press, Carnegie Institute of Technology, Pittsburgh, Pa., 1953. Other data mostly from K. A. Kobe and R. E. Lynn, *Chem. Rev.*, **52**:117 (1953).

TABLE A-6. Properties of Water-saturated Liquid and Vapor (Temperature Table, Metric Units)

Temp. °C	Press. Bars	Specific volume cm³/g		Internal energy joule/g			Enthalpy joule/g			Entropy joule/(g)(°K)		
		Sat. liquid	Sat. vapor	Sat. liquid	Evap.	Sat. vapor	Sat. liquid	Evap.	Sat. vapor	Sat. liquid	Evap.	Sat. vapor
t	p	v_f	v_g	u_f	u_{fg}	u_g	h_f	h_{fg}	h_g	s_f	s_{fg}	s_g
0.01	0.006113	1.0002	206,136	0.00	2,375.3	2,375.3	0.01	2,501.3	2,501.4	0.0000	9.1562	9.1562
5	0.008721	1.0001	147,120	20.97	2,361.3	2,382.3	20.98	2,489.6	2,510.6	0.0761	8.9496	9.0257
10	0.012276	1.0004	106,379	42.00	2,347.2	2,389.2	42.01	2,477.7	2,519.8	0.1510	8.7498	8.9008
15	0.017051	1.0009	77,926	62.99	2,333.1	2,396.1	62.99	2,465.9	2,528.9	0.2245	8.5569	8.7814
20	0.02339	1.0018	57,791	83.95	2,319.0	2,402.9	83.96	2,454.1	2,538.1	0.2966	8.3706	8.6672
25	0.03169	1.0029	43,360	104.88	2,304.9	2,409.8	104.89	2,442.3	2,547.2	0.3674	8.1905	8.5580
30	0.04246	1.0043	32,894	125.78	2,290.8	2,416.6	125.79	2,430.5	2,556.3	0.4369	8.0164	8.4533
35	0.05628	1.0060	25,216	146.67	2,276.7	2,423.4	146.68	2,418.6	2,565.3	0.5053	7.8478	8.3531
40	0.07384	1.0078	19,523	167.56	2,262.6	2,430.1	167.57	2,406.7	2,574.3	0.5725	7.6845	8.2570
45	0.09593	1.0099	15,258	188.44	2,248.4	2,436.8	188.45	2,394.8	2,583.2	0.6387	7.5261	8.1648
50	0.12349	1.0121	12,032	209.32	2,234.2	2,443.5	209.33	2,382.7	2,592.1	0.7038	7.3725	8.0763
55	0.15758	1.0146	9,568	230.21	2,219.9	2,450.1	230.23	2,370.7	2,600.9	0.7679	7.2234	7.9913
60	0.19940	1.0172	7,671	251.11	2,205.5	2,456.6	251.13	2,358.5	2,609.6	0.8312	7.0784	7.9096
65	0.2503	1.0199	6,197	272.02	2,191.1	2,463.1	272.06	2,346.2	2,618.3	0.8935	6.9375	7.8310
70	0.3119	1.0228	5,042	292.95	2,176.6	2,469.6	292.98	2,333.8	2,626.8	0.9549	6.8004	7.7553
75	0.3858	1.0259	4,131	313.90	2,162.0	2,475.9	313.93	2,321.4	2,635.3	1.0155	6.6669	7.6824
80	0.4739	1.0291	3,407	334.86	2,147.4	2,482.2	334.91	2,308.8	2,643.7	1.0753	6.5369	7.6122
85	0.5783	1.0325	2,828	355.84	2,132.6	2,488.4	355.90	2,296.0	2,651.9	1.1343	6.4102	7.5445
90	0.7014	1.0360	2,361	376.85	2,117.7	2,494.5	376.92	2,283.2	2,660.1	1.1925	6.2866	7.4791
95	0.8455	1.0397	1,981.9	397.88	2,102.7	2,500.6	397.96	2,270.2	2,668.1	1.2500	6.1659	7.4159
100	1.0135	1.0435	1,672.9	418.94	2,087.6	2,506.5	419.04	2,257.0	2,676.1	1.3069	6.0480	7.3549
105	1.2082	1.0475	1,419.4	440.02	2,072.3	2,512.4	440.15	2,243.7	2,683.8	1.3630	5.9328	7.2958
110	1.4327	1.0516	1,210.2	461.14	2,057.0	2,518.1	461.30	2,230.2	2,691.5	1.4185	5.8202	7.2387

TABLE A-6. Properties of Water-saturated Liquid and Vapor (Temperature Table, Metric Units) (Continued)

Temp. °C	Press. Bars	Specific volume cm³/g		Internal energy joule/g			Enthalpy joule/g			Entropy joule/(g)(°K)		
		Sat. liquid	Sat. vapor	Sat. liquid	Evap.	Sat. vapor	Sat. liquid	Evap.	Sat. vapor	Sat. liquid	Evap.	Sat. vapor
t	p	v_f	v_g	u_f	u_{fg}	u_g	h_f	h_{fg}	h_g	s_f	s_{fg}	s_g
115	1.6906	1.0559	1,036.6	482.30	2,041.4	2,523.7	482.48	2,216.5	2,699.0	1.4734	5.7100	7.1833
120	1.9853	1.0603	891.9	503.50	2,025.8	2,529.3	503.71	2,202.6	2,706.3	1.5276	5.6020	7.1296
125	2.321	1.0649	770.6	524.74	2,009.9	2,534.6	524.99	2,188.5	2,713.5	1.5813	5.4962	7.0775
130	2.701	1.0697	668.5	546.02	1,993.9	2,539.9	546.31	2,174.2	2,720.5	1.6344	5.3925	7.0269
135	3.130	1.0746	582.2	567.35	1,977.7	2,545.0	567.69	2,159.6	2,727.3	1.6870	5.2907	6.9777
140	3.613	1.0797	508.9	588.74	1,961.3	2,550.0	589.13	2,144.7	2,733.9	1.7391	5.1908	6.9299
145	4.154	1.0850	446.3	610.18	1,944.7	2,554.9	610.63	2,129.6	2,740.3	1.7907	5.0926	6.8833
150	4.758	1.0905	392.8	631.68	1,927.9	2,559.5	632.20	2,114.3	2,746.5	1.8418	4.9960	6.8379
155	5.431	1.0961	346.8	653.24	1,910.8	2,564.1	653.84	2,098.6	2,752.4	1.8925	4.9010	6.7935
160	6.178	1.1020	307.1	674.87	1,893.5	2,568.4	675.55	2,082.6	2,758.1	1.9427	4.8075	6.7502
165	7.005	1.1080	272.7	696.56	1,876.0	2,572.5	697.34	2,066.2	2,763.5	1.9925	4.7153	6.7078
170	7.917	1.1143	242.8	718.33	1,858.1	2,576.5	719.21	2,049.5	2,768.7	2.0419	4.6244	6.6663
175	8.920	1.1207	216.8	740.17	1,840.0	2,580.2	741.17	2,032.4	2,773.6	2.0909	4.5347	6.6256
180	10.021	1.1274	194.05	762.09	1,821.6	2,583.7	763.22	2,015.0	2,778.2	2.1396	4.4461	6.5857
185	11.227	1.1343	174.09	784.10	1,802.9	2,587.0	785.37	1,997.1	2,782.4	2.1879	4.3586	6.5465
190	12.544	1.1414	156.54	806.19	1,783.8	2,590.0	807.62	1,978.8	2,786.4	2.2359	4.2720	6.5079
195	13.978	1.1488	141.05	828.37	1,764.4	2,592.8	829.98	1,960.0	2,790.0	2.2835	4.1863	6.4698
200	15.538	1.1565	127.36	850.65	1,744.7	2,595.3	852.45	1,940.7	2,793.2	2.3309	4.1014	6.4323
205	17.230	1.1644	115.21	873.04	1,724.5	2,597.5	875.04	1,921.0	2,796.0	2.3780	4.0172	6.3952
210	19.062	1.1726	104.41	895.53	1,703.9	2,599.5	897.76	1,900.7	2,798.5	2.4248	3.9337	6.3585
215	21.04	1.1812	94.79	918.14	1,682.9	2,601.1	920.62	1,879.9	2,800.5	2.4714	3.8507	6.3221
220	23.18	1.1900	86.19	940.87	1,661.5	2,602.4	943.62	1,858.5	2,802.1	2.5178	3.7683	6.2861
225	25.48	1.1992	78.49	963.73	1,639.6	2,603.3	966.78	1,836.5	2,803.3	2.5639	3.6863	6.2503

230	27.95	1.2088	71.58	986.74	1,617.2	2,603.9	990.12	1,813.8	2,804.0	2.6099	3.6047	6.2146
235	30.60	1.2187	65.37	1,009.89	1,594.2	2,604.1	1,013.62	1,790.5	2,804.2	2.6558	3.5233	6.1791
240	33.44	1.2291	59.76	1,033.21	1,570.8	2,604.0	1,037.32	1,766.5	2,803.8	2.7015	3.4422	6.1437
245	36.48	1.2399	54.71	1,056.71	1,546.7	2,603.4	1,061.23	1,741.7	2,803.0	2.7472	3.3612	6.1083
250	39.73	1.2512	50.13	1,080.39	1,522.0	2,602.4	1,085.36	1,716.2	2,801.5	2.7927	3.2802	6.0730
255	43.19	1.2631	45.98	1,104.28	1,496.7	2,600.9	1,109.73	1,689.8	2,799.5	2.8383	3.1992	6.0375
260	46.88	1.2755	42.21	1,128.39	1,470.6	2,599.0	1,134.37	1,662.5	2,796.9	2.8838	3.1181	6.0019
265	50.81	1.2886	38.77	1,152.74	1,443.9	2,596.6	1,159.28	1,634.4	2,793.6	2.9294	3.0368	5.9662
270	54.99	1.3023	35.64	1,177.36	1,416.3	2,593.7	1,184.51	1,605.2	2,789.7	2.9751	2.9551	5.9301
275	59.42	1.3168	32.79	1,202.25	1,387.9	2,590.2	1,210.07	1,574.9	2,785.0	3.0208	2.8730	5.8938
280	64.12	1.3321	30.17	1,227.46	1,358.7	2,586.1	1,235.99	1,543.6	2,779.6	3.0668	2.7903	5.8571
285	69.09	1.3483	27.77	1,253.00	1,328.4	2,581.4	1,262.31	1,511.0	2,773.3	3.1130	2.7070	5.8199
290	74.36	1.3656	25.57	1,278.92	1,297.1	2,576.0	1,289.07	1,477.1	2,766.2	3.1594	2.6227	5.7821
295	79.93	1.3839	23.54	1,305.2	1,264.7	2,569.9	1,316.3	1,441.8	2,758.1	3.2062	2.5375	5.7437
300	85.81	1.4036	21.67	1,332.0	1,231.0	2,563.0	1,344.0	1,404.9	2,749.0	3.2534	2.4511	5.7045
305	92.02	1.4247	19.948	1,359.3	1,195.9	2,555.2	1,372.4	1,366.4	2,738.7	3.3010	2.3633	5.6643
310	98.56	1.4474	18.350	1,387.1	1,159.4	2,546.4	1,401.3	1,326.0	2,727.3	3.3493	2.2737	5.6230
315	105.47	1.4720	16.867	1,415.5	1,121.1	2,536.6	1,431.0	1,283.5	2,714.5	3.3982	2.1821	5.5804
320	112.74	1.4988	15.488	1,444.6	1,080.9	2,525.5	1,461.5	1,238.6	2,700.1	3.4480	2.0882	5.5362
330	128.45	1.5607	12.996	1,505.3	993.7	2,498.9	1,525.3	1,140.6	2,665.9	3.5507	1.8909	5.4417
340	145.86	1.6379	10.797	1,570.3	894.3	2,464.6	1,594.2	1,027.9	2,622.0	3.6594	1.6763	5.3357
350	165.13	1.7403	8.813	1,641.9	776.6	2,418.4	1,670.6	893.4	2,563.9	3.7777	1.4335	5.2112
360	186.51	1.8925	6.945	1,725.2	626.3	2,351.5	1,760.5	720.5	2,481.0	3.9147	1.1379	5.0526
370	210.3	2.213	4.925	1,844.0	384.5	2,228.5	1,890.5	441.6	2,332.1	4.1106	0.6865	4.7971
374.136	220.9	3.155	3.155	2,029.6	0	2,029.6	2,099.3	0	2,099.3	4.4298	0	4.4298

Source: Abridged from J. H. Keenan, F. G. Keyes, P. G. Hill, and J. G. Moore, "Steam Tables," International Edition, John Wiley & Sons, Inc., New York, 1969.

TABLE A-7. Properties of Water-saturated Liquid and Vapor (Pressure Table, Metric Units)

Press. Bars	Temp. °C	Specific volume cm³/g		Internal energy joule/g			Enthalpy joule/g			Entropy joule/(g)(°K)		
		Sat. liquid	Sat. vapor	Sat. liquid	Evap.	Sat. vapor	Sat. liquid	Evap.	Sat. vapor	Sat. liquid	Evap.	Sat. vapor
p	t	v_f	v_g	u_f	u_{fg}	u_g	h_f	h_{fg}	h_g	s_f	s_{fg}	s_g
0.01	6.98	1.0002	129,208	29.30	2,355.7	2,385.0	29.30	2,484.9	2,514.2	0.1059	8.8697	8.9756
0.02	17.50	1.0013	67,004	73.48	2,326.0	2,399.5	73.48	2,460.0	2,533.5	0.2607	8.4629	8.7237
0.03	24.08	1.0027	45,665	101.04	2,307.5	2,408.5	101.05	2,444.5	2,545.5	0.3545	8.2231	8.5776
0.04	28.96	1.0040	34,800	121.45	2,293.7	2,415.2	121.46	2,432.9	2,554.4	0.4226	8.0520	8.4746
0.05	32.88	1.0053	28,192	137.81	2,282.7	2,420.5	137.82	2,423.7	2,561.5	0.4764	7.9187	8.3951
0.10	45.81	1.0102	14,674	191.82	2,246.1	2,437.9	191.83	2,392.8	2,584.7	0.6493	7.5009	8.1502
0.15	53.97	1.0141	10,022	225.92	2,222.8	2,448.7	225.94	2,373.1	2,599.1	0.7549	7.2536	8.0085
0.20	60.06	1.0172	7,649	251.38	2,205.4	2,456.7	251.40	2,358.3	2,609.7	0.8320	7.0766	7.9085
0.25	64.97	1.0199	6,204	271.90	2,191.2	2,463.1	271.93	2,346.3	2,618.2	0.8931	6.9383	7.8314
0.30	69.10	1.0223	5,229	289.20	2,179.2	2,468.4	289.23	2,336.1	2,625.3	0.9439	6.8247	7.7686
0.40	75.87	1.0265	3,993	317.53	2,159.5	2,477.0	317.58	2,319.2	2,636.8	1.0259	6.6441	7.6700
0.50	81.33	1.0300	3,240	340.44	2,143.4	2,483.9	340.49	2,305.4	2,645.9	1.0910	6.5029	7.5939
0.60	85.94	1.0331	2,732	359.79	2,129.8	2,489.6	359.86	2,293.6	2,653.5	1.1453	6.3867	7.5320
0.70	89.95	1.0360	2,365	376.63	2,117.8	2,494.5	376.70	2,283.3	2,660.0	1.1919	6.2878	7.4797
0.80	93.50	1.0386	2,087	391.58	2,107.2	2,498.8	391.66	2,274.1	2,665.8	1.2329	6.2017	7.4346
1.0	99.63	1.0432	1,694.0	417.36	2,088.7	2,506.1	417.46	2,258.0	2,675.5	1.3026	6.0568	7.3594
1.2	104.80	1.0473	1,428.4	439.20	2,072.9	2,512.1	439.32	2,244.2	2,683.5	1.3608	5.9373	7.2981
1.4	109.31	1.0510	1,236.6	458.24	2,059.1	2,517.3	458.39	2,232.1	2,690.4	1.4109	5.8355	7.2464
1.6	113.32	1.0544	1,091.4	475.19	2,046.7	2,521.9	475.36	2,221.1	2,696.5	1.4550	5.7467	7.2017
1.8	116.93	1.0576	977.5	490.49	2,035.4	2,525.9	490.68	2,211.2	2,701.8	1.4944	5.6679	7.1623
2.0	120.23	1.0605	885.7	504.49	2,025.0	2,529.5	504.70	2,201.9	2,706.7	1.5301	5.5970	7.1271
2.2	123.27	1.0633	810.1	517.40	2,015.4	2,532.8	517.63	2,193.4	2,711.0	1.5628	5.5325	7.0953

2.4	7.0663	5.4733	1.5930	2,715.0	2,185.4	529.65	2,535.8	2,006.4	529.39	746.7	1.0660	126.10
2.6	7.0396	5.4186	1.6210	2,718.7	2,177.8	540.90	2,538.6	1,998.0	540.62	692.8	1.0685	128.73
2.8	7.0149	5.3677	1.6472	2,722.1	2,170.7	551.48	2,541.2	1,990.0	551.18	646.3	1.0709	131.21
3.0	6.9919	5.3201	1.6718	2,725.3	2,163.8	561.47	2,543.6	1,982.4	561.15	605.8	1.0732	133.55
3.5	6.9405	5.2130	1.7275	2,732.4	2,148.1	584.33	2,548.9	1,965.0	583.95	524.3	1.0786	138.88
4.0	6.8959	5.1193	1.7766	2,738.6	2,133.8	604.74	2,553.6	1,949.3	604.31	462.5	1.0836	143.63
4.5	6.8565	5.0359	1.8207	2,743.9	2,120.7	623.25	2,557.6	1,934.9	622.77	414.0	1.0882	147.93
5.0	6.8213	4.9606	1.8607	2,748.7	2,108.5	640.23	2,561.2	1,921.6	639.68	374.9	1.0926	151.86
6.0	6.7600	4.8288	1.9312	2,756.8	2,086.3	670.56	2,567.4	1,897.5	669.90	315.7	1.1006	158.85
7.0	6.7080	4.7158	1.9922	2,763.5	2,066.3	697.22	2,572.5	1,876.1	696.44	272.9	1.1080	164.97
8.0	6.6628	4.6166	2.0462	2,769.1	2,048.0	721.11	2,576.8	1,856.6	720.22	240.4	1.1148	170.43
9.0	6.6226	4.5280	2.0946	2,773.9	2,031.1	742.83	2,580.5	1,838.6	741.83	215.0	1.1212	175.38
10.0	6.5865	4.4478	2.1387	2,778.1	2,015.3	762.81	2,583.6	1,822.0	761.68	194.44	1.1273	179.91
11	6.5536	4.3744	2.1792	2,781.7	2,000.4	781.34	2,586.4	1,806.3	780.09	177.53	1.1330	184.09
12	6.5233	4.3067	2.2166	2,784.8	1,986.2	798.65	2,588.8	1,791.5	797.29	163.33	1.1385	187.99
13	6.4953	4.2438	2.2515	2,787.6	1,972.7	814.93	2,591.0	1,777.5	813.44	151.25	1.1438	191.64
14	6.4693	4.1850	2.2842	2,790.0	1,959.7	830.30	2,592.8	1,764.1	828.70	140.84	1.1489	195.07
15	6.4448	4.1298	2.3150	2,792.2	1,947.3	844.89	2,594.5	1,751.3	843.16	131.77	1.1539	198.32
20	6.3409	3.8935	2.4474	2,799.5	1,890.7	908.79	2,600.3	1,693.8	906.44	99.63	1.1767	212.42
25	6.2575	3.7028	2.5547	2,803.1	1,841.0	962.11	2,603.1	1,644.0	959.11	79.98	1.1973	223.99
30	6.1869	3.5412	2.6457	2,804.2	1,795.7	1,008.42	2,604.1	1,599.3	1,004.78	66.68	1.2165	233.90
35	6.1253	3.4000	2.7253	2,803.4	1,753.7	1,049.75	2,603.7	1,558.3	1,045.43	57.07	1.2347	242.60
40	6.0701	3.2737	2.7964	2,801.4	1,714.1	1,087.31	2,602.3	1,520.0	1,082.31	49.78	1.2522	250.40
50	5.9734	3.0532	2.9202	2,794.3	1,640.1	1,154.23	2,597.1	1,449.3	1,147.81	39.44	1.2859	263.99
60	5.8892	2.8625	3.0267	2,784.3	1,571.0	1,213.35	2,589.7	1,384.3	1,205.44	32.44	1.3187	275.64
70	5.8133	2.6922	3.1211	2,772.1	1,505.1	1,267.00	2,580.5	1,323.0	1,257.55	27.37	1.3513	285.88
80	5.7432	2.5364	3.2068	2,758.0	1,441.3	1,316.64	2,569.8	1,264.2	1,305.57	23.52	1.3842	295.06
90	5.6772	2.3915	3.2858	2,742.1	1,378.9	1,363.26	2,557.8	1,207.3	1,350.51	20.48	1.4178	303.40
100	5.6141	2.2544	3.3596	2,724.7	1,317.1	1,407.56	2,544.4	1,151.4	1,393.04	18.026	1.4524	311.06

TABLE A-7. Properties of Water-saturated Liquid and Vapor (Pressure Table, Metric Units) (Continued)

Press. Bars p	Temp. °C t	Specific volume cm³/g Sat. liquid v_f	Sat. vapor v_g	Internal energy joule/g Sat. liquid u_f	Evap. u_{fg}	Sat. vapor u_g	Enthalpy joule/g Sat. liquid h_f	Evap. h_{fg}	Sat. vapor h_g	Entropy joule/(g)(°K) Sat. liquid s_f	Evap. s_{fg}	Sat. vapor s_g
110	318.15	1.4886	15.987	1,433.7	1,096.0	2,529.8	1,450.1	1,255.5	2,705.6	3.4295	2.1233	5.5527
120	324.75	1.5267	14.263	1,473.0	1,040.7	2,513.7	1,491.3	1,193.6	2,684.9	3.4962	1.9962	5.4924
130	330.93	1.5671	12.780	1,511.1	985.0	2,496.1	1,531.5	1,130.7	2,662.2	3.5606	1.8718	5.4323
140	336.75	1.6107	11.485	1,548.6	928.2	2,476.8	1,571.1	1,066.5	2,637.6	3.6232	1.7485	5.3717
150	342.24	1.6581	10.337	1,585.6	869.8	2,455.5	1,610.5	1,000.0	2,610.5	3.6848	1.6249	5.3098
160	347.44	1.7107	9.306	1,622.7	809.0	2,431.7	1,650.1	930.6	2,580.6	3.7461	1.4994	5.2455
170	352.37	1.7702	8.364	1,660.2	744.8	2,405.0	1,690.3	856.9	2,547.2	3.8079	1.3698	5.1777
180	357.06	1.8397	7.489	1,698.9	675.4	2,374.3	1,732.0	777.1	2,509.1	3.8715	1.2329	5.1044
190	361.54	1.9243	6.657	1,739.9	598.1	2,338.1	1,776.5	688.0	2,464.5	3.9388	1.0839	5.0228
200	365.81	2.036	5.834	1,785.6	507.5	2,293.0	1,826.3	583.4	2,409.7	4.0139	0.9130	4.9269
210	369.89	2.207	4.952	1,842.1	388.5	2,230.6	1,888.4	446.2	2,334.6	4.1075	0.6938	4.8013
220	373.80	2.742	3.568	1,961.9	125.2	2,087.1	2,022.2	143.4	2,165.6	4.3110	0.2216	4.5327
220.9	374.14	3.155	3.155	2,029.6	0	2,029.6	2,099.3	0	2,099.3	4.4298	0	4.4298

Source: Abridged from J. H. Keenan, F. G. Keyes, P. G. Hill, and J. G. Moore, "Steam Tables," International Edition, John Wiley & Sons, Inc., New York, 1969.

TABLE A-8. Properties of Water-superheated Vapor (Metric Units)

p in bars
t in °C
v in cm³/g
u in joule/g
h in joule/g
s in joule/(g)(°K)

t	p = 0.1 (t_{sat} = 45.81)				p = 0.5 (t_{sat} = 81.33)				p = 1.0 (t_{sat} = 99.63)			
	v	u	h	s	v	u	h	s	v	u	h	s
Sat.	14,674	2,437.9	2,584.7	8.1502	3,240	2,483.9	2,645.9	7.5939	1,694.0	2,506.1	2,675.5	7.3594
100	17,196	2,515.5	2,687.5	8.4479	3,418	2,511.6	2,682.5	7.6947	1,695.8	2,506.7	2,676.2	7.3614
150	19,512	2,587.9	2,783.0	8.6882	3,889	2,585.6	2,780.1	7.9401	1,936.4	2,582.8	2,776.4	7.6134
200	21,825	2,661.3	2,879.5	8.9038	4,356	2,659.9	2,877.7	8.1580	2,172	2,658.1	2,875.3	7.8343
250	24,136	2,736.0	2,977.3	9.1002	4,820	2,735.0	2,976.0	8.3556	2,406	2,733.7	2,974.3	8.0333
300	26,445	2,812.1	3,076.5	9.2813	5,284	2,811.3	3,075.5	8.5373	2,639	2,810.4	3,074.3	8.2158
400	31,063	2,968.9	3,279.6	9.6077	6,209	2,968.5	3,278.9	8.8642	3,103	2,967.9	3,278.2	8.5435
500	35,679	3,132.3	3,489.1	9.8978	7,134	3,132.0	3,488.7	9.1546	3,565	3,131.6	3,488.1	8.8342
600	40,295	3,302.5	3,705.4	10.1608	8,057	3,302.2	3,705.1	9.4178	4,028	3,301.9	3,704.7	9.0976
700	44,911	3,479.6	3,928.7	10.4028	8,981	3,479.4	3,928.5	9.6599	4,490	3,479.2	3,928.2	9.3398
800	49,526	3,663.8	4,159.0	10.6281	9,904	3,663.6	4,158.9	9.8852	4,952	3,663.5	4,158.6	9.5652
900	54,141	3,855.0	4,396.4	10.8396	10,828	3,854.9	4,396.3	10.0967	5,414	3,854.8	4,396.1	9.7767
1,000	58,757	4,053.0	4,640.6	11.0393	11,751	4,052.9	4,640.5	10.2964	5,875	4,052.8	4,640.3	9.9764
1,200	67,987	4,467.9	5,147.8	11.4091	13,597	4,467.8	5,147.7	10.6662	6,799	4,467.7	5,147.6	10.3463

t	p = 2 (t_{sat} = 120.23)				p = 4 (t_{sat} = 143.63)				p = 6 (t_{sat} = 158.85)			
	v	u	h	s	v	u	h	s	v	u	h	s
Sat.	885.7	2,529.5	2,706.7	7.1272	462.5	2,553.6	2,738.6	6.8959	315.7	2,567.4	2,756.8	6.7600
150	959.6	2,576.9	2,768.8	7.2795	470.8	2,564.5	2,752.8	6.9299				

TABLE A-8. Properties of Water-superheated Vapor (Metric Units) (*Continued*)

p in bars
t in °C
v in cm³/g
u in joule/g
h in joule/g
s in joule/(g)(°K)

t	$p = 2$ ($t_{sat} = 120.23$)				$p = 4$ ($t_{sat} = 143.63$)				$p = 6$ ($t_{sat} = 158.85$)			
	v	u	h	s	v	u	h	s	v	u	h	s
200	1,080.3	2,654.4	2,870.5	7.5066	534.2	2,646.8	2,860.5	7.1706	352.0	2,638.9	2,850.1	6.9665
250	1,198.8	2,731.2	2,971.0	7.7086	595.1	2,726.1	2,964.2	7.3789	393.8	2,720.9	2,957.2	7.1816
300	1,316.2	2,808.6	3,071.8	7.8926	654.8	2,804.8	3,066.8	7.5662	434.4	2,801.0	3,061.6	7.3724
400	1,549.3	2,966.7	3,276.6	8.2218	772.6	2,964.4	3,273.4	7.8985	513.7	2,962.1	3,270.3	7.7079
500	1,781.4	3,130.8	3,487.1	8.5133	889.3	3,129.2	3,484.9	8.1913	592.0	3,127.6	3,482.8	8.0021
600	2,013	3,301.4	3,704.0	8.7770	1,005.5	3,300.2	3,702.4	8.4558	669.7	3,299.1	3,700.9	8.2674
700	2,244	3,478.8	3,927.6	9.0194	1,121.5	3,477.9	3,926.5	8.6987	747.2	3,477.0	3,925.3	8.5107
800	2,475	3,663.1	4,158.2	9.2449	1,237.2	3,662.4	4,157.3	8.9244	824.5	3,661.8	4,156.5	8.7367
900	2,706	3,854.5	4,395.8	9.4566	1,352.9	3,853.9	4,395.1	9.1362	901.7	3,853.4	4,394.4	8.9486
1,000	2,937	4,052.5	4,640.0	9.6563	1,468.5	4,052.0	4,639.4	9.3360	978.8	4,051.5	4,638.8	9.1485
1,200	3,399	4,467.5	5,147.3	10.0262	1,699.6	4,467.0	5,146.8	9.7060	1,133.0	4,466.5	5,146.3	9.5185

t	$p = 8$ ($t_{sat} = 170.43$)				$p = 10$ ($t_{sat} = 179.91$)				$p = 15$ ($t_{sat} = 198.32$)			
	v	u	h	s	v	u	h	s	v	u	h	s
Sat.	240.4	2,576.8	2,769.1	6.6628	194.44	2,583.6	2,778.1	6.5865	131.77	2,594.5	2,792.2	6.4448
200	260.8	2,630.6	2,839.3	6.8158	206.0	2,621.9	2,827.9	6.6940	132.48	2,598.1	2,796.8	6.4546
250	293.1	2,715.5	2,950.0	7.0384	232.7	2,709.9	2,942.6	6.9247	151.95	2,695.3	2,923.3	6.7090
300	324.1	2,797.2	3,056.5	7.2328	257.9	2,793.2	3,051.2	7.1229	169.66	2,783.1	3,037.6	6.9179

Panel 1 (continued)

T	v	u	h	s
350	354.4	2,878.2	3,161.7	7.4089
400	384.3	2,959.7	3,267.1	7.5716
500	443.3	3,126.0	3,480.6	7.8673
600	501.8	3,297.9	3,699.4	8.1333
700	560.1	3,476.2	3,924.2	8.3770
800	618.1	3,661.1	4,155.6	8.6033
900	676.1	3,852.8	4,393.7	8.8153
1,000	734.0	4,051.0	4,638.2	9.0153
1,200	849.7	4,466.1	5,145.9	9.3855

$p = 20\ (t_{sat} = 212.42)$

T	v	u	h	s
Sat.	99.63	2,600.3	2,799.5	6.3409
250	111.44	2,679.6	2,902.5	6.5453
300	125.47	2,772.6	3,023.5	6.7664
350	138.57	2,859.8	3,137.0	6.9563
400	151.20	2,945.2	3,247.6	7.1271
450	163.53	3,030.5	3,357.5	7.2846
500	175.68	3,116.2	3,467.6	7.4317
600	199.60	3,290.9	3,690.1	7.7024
700	223.2	3,470.9	3,917.4	7.9487
800	246.7	3,657.0	4,150.3	8.1765
900	270.0	3,849.3	4,389.4	8.3895
1,000	293.3	4,048.0	4,634.6	8.5901
1,200	339.8	4,463.3	5,142.9	8.9607

$p = 40\ (t_{sat} = 250.40)$

T	v	u	h	s
Sat.	49.78	2,602.3	2,801.4	6.0701
300	58.84	2,725.3	2,960.7	6.3615

Panel 2 (continued)

T	v	u	h	s
350	282.5	2,875.2	3,157.7	7.3011
400	306.6	2,957.3	3,263.9	7.4651
500	354.1	3,124.4	3,478.5	7.7622
600	401.1	3,296.8	3,697.9	8.0290
700	447.8	3,475.3	3,923.1	8.2731
800	494.3	3,660.4	4,154.7	8.4996
900	540.7	3,852.2	4,392.9	8.7118
1,000	587.1	4,050.5	4,637.6	8.9119
1,200	679.8	4,465.6	5,145.4	9.2822

$p = 25\ (t_{sat} = 223.99)$

T	v	u	h	s
Sat.	79.98	2,603.1	2,803.1	6.2575
250	87.00	2,662.6	2,880.1	6.4085
300	98.90	2,761.6	3,008.8	6.6438
350	109.76	2,851.9	3,126.3	6.8403
400	120.10	2,939.1	3,239.3	7.0148
450	130.14	3,025.5	3,350.8	7.1746
500	139.98	3,112.1	3,462.1	7.3234
600	159.30	3,288.0	3,686.3	7.5960
700	178.32	3,468.7	3,914.5	7.8435
800	197.16	3,655.3	4,148.2	8.0720
900	215.90	3,847.9	4,387.6	8.2853
1,000	234.6	4,046.7	4,633.1	8.4861
1,200	271.8	4,462.1	5,141.7	8.8569

$p = 50\ (t_{sat} = 263.99)$

T	v	u	h	s
Sat.	39.44	2,597.1	2,794.3	5.9734
300	45.32	2,698.0	2,924.5	6.2084

Panel 3 (continued)

T	v	u	h	s
350	186.56	2,867.6	3,147.5	7.1017
400	203.0	2,951.3	3,255.8	7.2690
500	235.2	3,120.3	3,473.1	7.5698
600	266.8	3,293.9	3,694.0	7.8385
700	298.1	3,473.1	3,920.2	8.0837
800	329.2	3,658.7	4,152.5	8.3109
900	360.3	3,850.8	4,391.2	8.5235
1,000	391.3	4,049.2	4,636.1	8.7238
1,200	453.1	4,464.4	5,144.1	9.0942

$p = 30\ (t_{sat} = 233.90)$

T	v	u	h	s
Sat.	66.68	2,604.1	2,804.2	6.1869
250	70.58	2,644.0	2,855.8	6.2872
300	81.14	2,750.1	2,993.5	6.5390
350	90.53	2,843.7	3,115.3	6.7428
400	99.36	2,932.8	3,230.9	6.9212
450	107.87	3,020.4	3,344.0	7.0834
500	116.19	3,108.0	3,456.5	7.2338
600	132.43	3,285.0	3,682.3	7.5085
700	148.38	3,466.5	3,911.7	7.7571
800	164.14	3,653.5	4,145.9	7.9862
900	179.80	3,846.5	4,385.9	8.1999
1,000	195.41	4,045.4	4,631.6	8.4009
1,200	226.52	4,460.9	5,140.5	8.7720

$p = 60\ (t_{sat} = 275.64)$

T	v	u	h	s
Sat.	32.44	2,589.7	2,784.3	5.8892
300	36.16	2,667.2	2,884.2	6.0674

TABLE A-8. Properties of Water-superheated Vapor (Metric Units) (Continued)

p in bars
t in °C
v in cm³/g
u in joule/g
h in joule/g
s in joule/(g)(°K)

t	\(p = 40\) (\(t_{sat} = 250.40\))				\(p = 50\) (\(t_{sat} = 263.99\))				\(p = 60\) (\(t_{sat} = 275.64\))			
	v	u	h	s	v	u	h	s	v	u	h	s
350	66.45	2,826.7	3,092.5	6.5821	51.94	2,808.7	3,068.4	6.4493	42.23	2,789.6	3,043.0	6.3335
400	73.41	2,919.9	3,213.6	6.7690	57.81	2,906.6	3,195.7	6.6459	47.39	2,892.9	3,177.2	6.5408
450	80.02	3,010.2	3,330.3	6.9363	63.30	2,999.7	3,316.2	6.8186	52.14	2,988.9	3,301.8	6.7193
500	86.43	3,099.5	3,445.3	7.0901	68.57	3,091.0	3,433.8	6.9759	56.65	3,082.2	3,422.2	6.8803
550					73.68	3,181.8	3,550.3	7.1218	61.01	3,174.6	3,540.6	7.0288
600	98.85	3,279.1	3,674.4	7.3688	78.69	3,273.0	3,666.5	7.2589	65.25	3,266.9	3,658.4	7.1677
700	110.95	3,462.1	3,905.9	7.6198	88.49	3,457.6	3,900.1	7.5122	73.52	3,453.1	3,894.2	7.4234
800	122.87	3,650.0	4,141.5	7.8502	98.11	3,646.6	4,137.1	7.7440	81.60	3,643.1	4,132.7	7.6566
900	134.69	3,843.6	4,382.3	8.0647	107.62	3,840.7	4,378.8	7.9593	89.58	3,837.8	4,375.3	7.8727
1,000	146.45	4,042.9	4,628.7	8.2662	117.07	4,040.4	4,625.7	8.1612	97.49	4,037.8	4,622.7	8.0751
1,200	169.87	4,458.6	5,138.1	8.6376	135.87	4,456.3	5,135.7	8.5331	113.21	4,454.0	5,133.3	8.4474

t	\(p = 70\) (\(t_{sat} = 285.88\))				\(p = 80\) (\(t_{sat} = 295.06\))				\(p = 90\) (\(t_{sat} = 303.40\))			
	v	u	h	s	v	u	h	s	v	u	h	s
Sat.	27.37	2,580.5	2,772.1	5.8133	23.52	2,569.8	2,758.0	5.7432	20.48	2,557.8	2,742.1	5.6772
300	29.47	2,632.2	2,838.4	5.9305	24.26	2,590.9	2,785.0	5.7906				
350	35.24	2,769.4	3,016.0	6.2283	29.95	2,747.7	2,987.3	6.1301	25.80	2,724.4	2,956.6	6.0361
400	39.93	2,878.6	3,158.1	6.4478	34.32	2,863.8	3,138.3	6.3634	29.93	2,848.4	3,117.8	6.2854
450	44.16	2,978.0	3,287.1	6.6327	38.17	2,966.7	3,272.0	6.5551	33.50	2,955.2	3,256.6	6.4844
500	48.14	3,073.4	3,410.3	6.7975	41.75	3,064.3	3,398.3	6.7240	36.77	3,055.2	3,386.1	6.6576

550	51.95	3,167.2	3,530.9	6.9486	45.16	3,159.8	3,521.0	6.8778	39.87	3,152.2	3,511.0	6.8142
600	55.65	3,260.7	3,650.3	7.0894	48.45	3,254.4	3,642.0	7.0206	42.85	3,248.1	3,633.7	6.9589
650					51.66	3,349.0	3,762.3	7.1545	45.74	3,343.6	3,755.3	7.0943
700	62.83	3,448.5	3,888.3	7.3476	54.81	3,443.9	3,882.4	7.2812	48.57	3,439.3	3,876.5	7.2221
800	69.81	3,639.5	4,128.2	7.5822	60.97	3,636.0	4,123.8	7.5173	54.09	3,632.5	4,119.3	7.4596
900	76.69	3,835.0	4,371.8	7.7991	67.02	3,832.1	4,368.3	7.7351	59.50	3,829.2	4,364.8	7.6783
1,000	83.50	4,035.3	4,619.8	8.0020	73.01	4,032.8	4,616.9	7.9384	64.85	4,030.3	4,614.0	7.8821
1,200	97.03	4,451.7	5,130.9	8.3747	84.89	4,449.5	5,128.5	8.3115	75.44	4,447.2	5,126.2	8.2556
	$p = 100$ ($t_{sat} = 311.06$)				$p = 125$ ($t_{sat} = 327.89$)				$p = 150$ ($t_{sat} = 342.24$)			
Sat.	18.026	2,544.4	2,724.7	5.6141	13.495	2,505.1	2,673.8	5.4624	10.337	2,455.5	2,610.5	5.3098
350	22.42	2,699.2	2,923.4	5.9443	16.126	2,624.6	2,826.2	5.7118	11.470	2,520.4	2,692.4	5.4421
400	26.41	2,832.4	3,096.5	6.2120	20.00	2,789.3	3,039.3	6.0417	15.649	2,740.7	2,975.5	5.8811
450	29.75	2,943.4	3,240.9	6.4190	22.99	2,912.5	3,199.8	6.2719	18.445	2,879.5	3,156.2	6.1404
500	32.79	3,045.8	3,373.7	6.5966	25.60	3,021.7	3,341.8	6.4618	20.80	2,996.6	3,308.6	6.3443
550	35.64	3,144.6	3,500.9	6.7561	28.01	3,125.0	3,475.2	6.6290	22.93	3,104.7	3,448.6	6.5199
600	38.37	3,241.7	3,625.3	6.9029	30.29	3,225.4	3,604.0	6.7810	24.91	3,208.6	3,582.3	6.6776
650	41.01	3,338.2	3,748.2	7.0398	32.48	3,324.4	3,730.4	6.9218	26.80	3,310.3	3,712.3	6.8224
700	43.58	3,434.7	3,870.5	7.1687	34.60	3,422.9	3,855.3	7.0536	28.61	3,410.9	3,840.1	6.9572
800	48.59	3,628.9	4,114.8	7.4077	38.69	3,620.0	4,103.6	7.2965	32.10	3,610.9	4,092.4	7.2040
900	53.49	3,826.3	4,361.2	7.6272	42.67	3,819.1	4,352.5	7.5182	35.46	3,811.9	4,343.8	7.4279
1,000	58.32	4,027.8	4,611.0	7.8315	46.58	4,021.6	4,603.8	7.7237	38.75	4,015.4	4,596.6	7.6348
1,200	67.89	4,444.9	5,123.8	8.2055	54.30	4,439.3	5,118.0	8.0987	45.23	4,433.8	5,112.3	8.0108
	$p = 175$ ($t_{sat} = 354.75$)				$p = 200$ ($t_{sat} = 365.81$)				$p = 250$			
Sat.	7.920	2,390.2	2,528.8	5.1419	5.834	2,293.0	2,409.7	4.9269				
375	10.554	2,568.5	2,753.2	5.4944	7.666	2,449.2	2,602.5	5.2269	1.9731	1,798.7	1848.0	4.0320
400	12.447	2,685.0	2,902.9	5.7213	9.942	2,619.3	2,818.1	5.5540	6.004	2,430.1	2,580.2	5.1418

TABLE A-8. Properties of Water-superheated Vapor (Metric Units) (Continued)

p in bars
t in °C
v in cm³/g
u in joule/g
h in joule/g
s in joule/(g)(°K)

t	p = 175 (t_sat = 354.75)				p = 200 (t_sat = 365.81)				p = 250			
	v	u	h	s	v	u	h	s	v	u	h	s
425	13.914	2,771.4	3,014.9	5.8848	11.458	2,723.7	2,952.9	5.7507	7.881	2,609.2	2,806.3	5.4723
450	15.174	2,844.2	3,109.7	6.0184	12.695	2,806.2	3,060.1	5.9017	9.162	2,720.7	2,949.7	5.6744
500	17.358	2,970.3	3,274.1	6.2383	14.768	2,942.9	3,238.2	6.1401	11.123	2,884.3	3,162.4	5.9592
550	19.288	3,083.9	3,421.4	6.4230	16.555	3,062.4	3,393.5	6.3348	12.724	3,017.5	3,335.6	6.1765
600	21.06	3,191.5	3,560.1	6.5866	18.178	3,174.0	3,537.6	6.5048	14.137	3,137.9	3,491.4	6.3602
650	22.74	3,296.0	3,693.9	6.7357	19.693	3,281.4	3,675.3	6.6582	15.433	3,251.6	3,637.4	6.5229
700	24.34	3,398.7	3,824.6	6.8736	21.13	3,386.4	3,809.0	6.7993	16.646	3,361.3	3,777.5	6.6707
800	27.38	3,601.8	4,081.1	7.1244	23.85	3,592.7	4,069.7	7.0544	18.912	3,574.3	4,047.1	6.9345
900	30.31	3,804.7	4,335.1	7.3507	26.45	3,797.5	4,326.4	7.2830	21.045	3,783.0	4,309.1	7.1680
1,000	33.16	4,009.3	4,589.5	7.5589	28.97	4,003.1	4,582.5	7.4925	23.10	3,990.9	4,568.5	7.3802
1,200	38.76	4,428.3	5,106.6	7.9360	33.91	4,422.8	5,101.0	7.8707	27.11	4,412.0	5,089.9	7.7605

t	p = 300				p = 350				p = 400			
	v	u	h	s	v	u	h	s	v	u	h	s
375	1.7892	1,737.8	1,791.5	3.9305	1.7003	1,702.9	1,762.4	3.8722	1.6407	1,677.1	1,742.8	3.8290
400	2.790	2,067.4	2,151.1	4.4728	2.100	1,914.1	1,987.6	4.2126	1.9077	1,854.6	1,930.9	4.1135
425	5.303	2,455.1	2,614.2	5.1504	3.428	2,253.4	2,373.4	4.7747	2.532	2,096.9	2,198.1	4.5029
450	6.735	2,619.3	2,821.4	5.4424	4.961	2,498.7	2,672.4	5.1962	3.693	2,365.1	2,512.8	4.9459
500	8.678	2,820.7	3,081.1	5.7905	6.927	2,751.9	2,994.4	5.6282	5.622	2,678.4	2,903.3	5.4700

n	p = 500				p = 600				p = 700			
550	10.168	2,970.3	3,275.4	6.0342	8.345	2,921.0	3,213.0	5.9026	6.984	2,869.7	3,149.1	5.7785
600	11.446	3,100.5	3,443.9	6.2331	9.527	3,062.0	3,395.5	6.1179	8.094	3,022.6	3,346.4	6.0114
650	12.596	3,221.0	3,598.9	6.4058	10.575	3,189.8	3,559.9	6.3010	9.063	3,158.0	3,520.6	6.2054
700	13.661	3,335.8	3,745.6	6.5606	11.533	3,309.8	3,713.5	6.4631	9.941	3,283.6	3,681.2	6.3750
800	15.623	3,555.5	4,024.2	6.8332	13.278	3,536.7	4,001.5	6.7450	11.523	3,517.8	3,978.7	6.6662
900	17.448	3,768.5	4,291.9	7.0718	14.883	3,754.0	4,274.9	6.9886	12.962	3,739.4	4,257.9	6.9150
1,000	19.196	3,978.8	4,554.7	7.2867	16.410	3,966.7	4,541.1	7.2064	14.324	3,954.6	4,527.6	7.1356
1,200	22.589	4,401.3	5,079.0	7.6692	19.360	4,390.7	5,068.3	7.5910	16.940	4,380.1	5,057.7	7.5224

n	p = 800				p = 900				p = 1,000			
375	1.5594	1,638.6	1,716.6	3.7639	1.5028	1,609.4	1,699.5	3.7141	1.4592	1,585.6	1,687.7	3.6730
400	1.7309	1,788.1	1,874.6	4.0031	1.6335	1,745.4	1,843.4	3.9318	1.5664	1,713.2	1,822.8	3.8775
425	2.007	1,959.7	2,060.0	4.2734	1.8165	1,892.7	2,001.7	4.1626	1.7063	1,847.8	1,967.2	4.0881
450	2.486	2,159.6	2,284.0	4.5884	2.085	2,053.9	2,179.0	4.4121	1.8931	1,990.2	2,122.7	4.3068
500	3.892	2,525.5	2,720.1	5.1726	2.956	2,390.6	2,567.9	4.9321	2.466	2,290.6	2,463.2	4.7619
550	5.118	2,763.6	3,019.5	5.5485	3.956	2,658.8	2,896.2	5.3441	3.227	2,563.9	2,789.7	5.1715
600	6.112	2,942.0	3,247.6	5.8178	4.834	2,861.1	3,151.2	5.6452	3.976	2,783.4	3,061.7	5.4925
650	6.966	3,093.5	3,441.8	6.0342	5.595	3,028.8	3,364.5	5.8829	4.650	2,965.3	3,290.8	5.7479
700	7.727	3,230.5	3,616.8	6.2189	6.272	3,177.2	3,553.5	6.0824	5.256	3,124.5	3,492.4	5.9606
800	9.076	3,479.8	3,933.6	6.5290	7.459	3,441.5	3,889.1	6.4109	6.318	3,403.4	3,845.7	6.3066
900	10.283	3,710.3	4,224.4	6.7882	8.508	3,681.0	4,191.5	6.6805	7.252	3,651.6	4,159.2	6.5862
1,000	11.411	3,930.5	4,501.1	7.0146	9.480	3,906.4	4,475.2	6.9127	8.110	3,882.1	4,449.8	6.8240
1,200	13.561	4,359.1	5,037.2	7.4058	11.317	4,338.2	5,017.2	7.3083	9.722	4,317.3	4,997.8	7.2240

n	p = 800				p = 900				p = 1,000			
375	1.4239	1,565.5	1,679.4	3.6379	1.3941	1,548.0	1,673.4	3.6070	1.3685	1,532.5	1,669.4	3.5794
400	1.5154	1,687.1	1,808.3	3.8330	1.4743	1,665.1	1,797.7	3.7952	1.4399	1,646.0	1,790.0	3.7620

TABLE A-8. Properties of Water-superheated Vapor (Metric Units) (Continued)

p in bars
t in °C
v in cm³/g

u in joule/g
h in joule/g
s in joule/(g)(°K)

t	p = 800				p = 900				p = 1,000			
	v	u	h	s	v	u	h	s	v	u	h	s
425	1.6299	1,813.6	1,943.9	4.0309	1.5717	1,785.8	1,927.2	3.9840	1.5251	1,762.3	1,914.8	3.9441
450	1.7746	1,945.0	2,086.9	4.2321	1.6908	1,909.9	2,062.0	4.1737	1.6266	1,881.1	2,043.8	4.1256
500	2.188	2,218.9	2,394.0	4.6425	2.013	2,165.6	2,346.7	4.5542	1.8903	2,123.8	2,312.8	4.4852
550	2.763	2,483.9	2,704.9	5.0323	2.458	2,418.6	2,639.8	4.9216	2.246	2,365.7	2,590.3	4.8329
600	3.386	2,711.8	2,982.7	5.3601	2.971	2,648.2	2,915.6	5.2469	2.671	2,592.7	2,859.8	5.1509
650	3.976	2,904.6	3,222.7	5.6276	3.483	2,848.0	3,161.6	5.5211	3.115	2,796.2	3,107.7	5.4271
700	4.518	3,073.2	3,434.6	5.8512	3.967	3,024.0	3,381.1	5.7528	3.546	2,977.7	3,332.3	5.6642
800	5.477	3,365.6	3,803.8	6.2128	4.837	3,328.5	3,763.8	6.1277	4.338	3,292.3	3,726.1	6.0499
900	6.320	3,622.3	4,127.9	6.5018	5.606	3,593.1	4,097.7	6.4253	5.044	3,564.3	4,068.6	6.3553
1,000	7.093	3,857.8	4,425.2	6.7451	6.309	3,833.5	4,401.3	6.6738	5.690	3,809.2	4,378.2	6.6087
1,200	8.534	4,296.4	4,979.1	7.1495	7.617	4,275.5	4,961.0	7.0826	6.889	4,254.7	4,943.6	7.0216

Source: Abridged from J. H. Keenan, F. G. Keyes, P. G. Hill, and J. G. Moore, "Steam Tables," International Edition, John Wiley & Sons, Inc., New York, 1969.

TABLE A-9. Properties of Water-compressed Liquid (Metric Units)

p in bars
t in °C
v in cm³/g
u in joule/g
h in joule/g
s in joule/(g)(°K)

t	$p = 25$ ($t_{sat} = 223.99$)				$p = 50$ ($t_{sat} = 263.99$)				$p = 75$ ($t_{sat} = 290.59$)			
	v	u	h	s	v	u	h	s	v	u	h	s
Sat.	1.1973	959.1	962.1	2.5546	1.2859	1,147.8	1,154.2	2.9202	1.3677	1,282.0	1,292.2	3.1649
0	0.9990	−0.00	2.50	−0.0000	0.9977	0.04	5.04	0.0001	0.9965	0.06	7.54	0.0002
20	1.0006	83.80	86.30	0.2961	0.9995	83.65	88.65	0.2956	0.9984	83.50	90.99	0.2950
40	1.0067	167.25	169.77	0.5715	1.0056	166.95	171.97	0.5705	1.0045	166.64	174.18	0.5696
60	1.0160	250.67	253.21	0.8298	1.0149	250.23	255.30	0.8285	1.0138	249.79	257.40	0.8272
80	1.0280	334.29	336.86	1.0737	1.0268	333.72	338.85	1.0720	1.0256	333.15	340.84	1.0704
100	1.0423	418.24	420.85	1.3050	1.0410	417.52	422.72	1.3030	1.0397	416.81	424.62	1.3011
120	1.0590	502.68	505.33	1.5255	1.0576	501.80	507.09	1.5233	1.0562	500.94	508.86	1.5211
140	1.0784	587.82	590.52	1.7369	1.0768	586.76	592.15	1.7343	1.0752	585.72	593.78	1.7317
160	1.1006	673.90	676.65	1.9404	1.0988	672.62	678.12	1.9375	1.0970	671.37	679.59	1.9346
180	1.1261	761.16	763.97	2.1375	1.1240	759.63	765.25	2.1341	1.1219	758.13	766.55	2.1308
200	1.1555	849.9	852.8	2.3294	1.1530	848.1	853.9	2.3255	1.1505	846.3	854.9	2.3216
220	1.1898	940.7	943.7	2.5174	1.1866	938.4	944.4	2.5128	1.1835	936.2	945.1	2.5083
240					1.2264	1,031.4	1,037.5	2.6979	1.2225	1,028.6	1,037.8	2.6925
260					1.2749	1,127.9	1,134.3	2.8830	1.2696	1,124.4	1,134.0	2.8763
280									1.3288	1,225.4	1,235.4	3.0631

TABLE A-9. Properties of Water-compressed Liquid (Metric Units) *(Continued)*

p in bars
t in °C
v in cm³/g
u in joule/g
h in joule/g
s in joule/(g)(°K)

t	$p = 100$ ($t_{sat} = 311.06$)				$p = 125$ ($t_{sat} = 327.89$)				$p = 150$ ($t_{sat} = 342.24$)			
	v	u	h	s	v	u	h	s	v	u	h	s
Sat.	1.4524	1,393.0	1,407.6	3.3596	1.5466	1,492.2	1,511.5	3.5286	1.6581	1,585.6	1,610.5	3.6848
0	0.9952	0.09	10.04	0.0002	0.9940	0.13	12.56	0.0003	0.9928	0.15	15.05	0.0004
20	0.9972	83.36	93.33	0.2945	0.9961	83.21	95.67	0.2940	0.9950	83.06	97.99	0.2934
40	1.0034	166.35	176.38	0.5686	1.0024	166.05	178.58	0.5676	1.0013	165.76	180.78	0.5666
60	1.0127	249.36	259.49	0.8258	1.0116	248.94	261.58	0.8245	1.0105	248.51	263.67	0.8232
80	1.0245	332.59	342.83	1.0688	1.0233	332.03	344.82	1.0672	1.0222	331.48	346.81	1.0656
100	1.0385	416.12	426.50	1.2992	1.0373	415.42	428.39	1.2973	1.0361	414.74	430.28	1.2955
120	1.0549	500.08	510.64	1.5189	1.0535	499.24	512.41	1.5167	1.0522	498.40	514.19	1.5145
140	1.0737	584.68	595.42	1.7292	1.0722	583.66	597.07	1.7267	1.0707	582.66	598.72	1.7242
160	1.0953	670.13	681.08	1.9317	1.0935	668.91	682.58	1.9288	1.0918	667.71	684.09	1.9260
180	1.1199	756.65	767.84	2.1275	1.1178	755.19	769.16	2.1242	1.1159	753.76	770.50	2.1210
200	1.1480	844.5	856.0	2.3178	1.1456	842.8	857.1	2.3141	1.1433	841.0	858.2	2.3104
220	1.1805	934.1	945.9	2.5039	1.1776	932.0	946.7	2.4995	1.1748	929.9	947.5	2.4953
240	1.2187	1,026.0	1,038.1	2.6872	1.2150	1,023.4	1,038.5	2.6821	1.2114	1,020.8	1,039.0	2.6771
260	1.2645	1,121.1	1,133.7	2.8699	1.2597	1,117.8	1,133.5	2.8637	1.2550	1,114.6	1,133.4	2.8576
280	1.3216	1,220.9	1,234.1	3.0548	1.3148	1,216.6	1,233.0	3.0469	1.3084	1,212.5	1,232.1	3.0393
300	1.3972	1,328.4	1,342.3	3.2469	1.3867	1,322.3	1,339.6	3.2361	1.3770	1,316.6	1,337.3	3.2260
320					1.4895	1,439.9	1,458.5	3.4399	1.4724	1,431.1	1,453.2	3.4247
340									1.6311	1,567.5	1,591.9	3.6546

Sat.	p = 200 (t_{sat} = 365.81)				p = 250				p = 300			
Sat.	2.036	1,785.6	1,826.3	4.0139								
0	0.9904	0.19	20.01	0.0004	0.9880	0.24	24.95	0.0003	0.9856	0.25	29.82	0.0001
20	0.9928	82.77	102.62	0.2923	0.9907	82.47	107.24	0.2911	0.9886	82.17	111.84	0.2899
40	0.9992	165.17	185.16	0.5646	0.9971	164.60	189.52	0.5626	0.9951	164.04	193.89	0.5607
60	1.0084	247.68	267.85	0.8206	1.0063	246.86	272.02	0.8180	1.0042	246.06	276.19	0.8154
80	1.0199	330.40	350.80	1.0624	1.0178	329.34	354.78	1.0592	1.0156	328.30	358.77	1.0561
100	1.0337	413.39	434.06	1.2917	1.0313	412.08	437.85	1.2881	1.0290	410.78	441.66	1.2844
120	1.0496	496.76	517.76	1.5102	1.0470	495.16	521.33	1.5060	1.0445	493.59	524.93	1.5018
140	1.0678	580.69	602.04	1.7193	1.0649	578.76	605.39	1.7145	1.0621	576.88	608.75	1.7098
160	1.0885	665.35	687.12	1.9204	1.0853	663.06	690.19	1.9149	1.0821	660.82	693.28	1.9096
180	1.1120	750.95	773.20	2.1147	1.1083	748.23	775.94	2.1085	1.1047	745.59	778.73	2.1024
200	1.1388	837.7	860.5	2.3031	1.1344	834.5	862.8	2.2961	1.1302	831.4	865.3	2.2893
220	1.1693	925.9	949.3	2.4870	1.1641	922.0	951.1	2.4789	1.1590	918.3	953.1	2.4711
240	1.2046	1,016.0	1,040.0	2.6674	1.1982	1,011.3	1,041.3	2.6580	1.1920	1,006.9	1,042.6	2.6490
260	1.2462	1,108.6	1,133.5	2.8459	1.2380	1,102.8	1,133.8	2.8349	1.2303	1,097.4	1,134.3	2.8243
280	1.2965	1,204.7	1,230.6	3.0248	1.2856	1,197.5	1,229.6	3.0113	1.2755	1,190.7	1,229.0	2.9986
300	1.3596	1,306.1	1,333.3	3.2071	1.3442	1,296.6	1,330.2	3.1900	1.3304	1,287.9	1,327.8	3.1741
320	1.4437	1,415.7	1,444.6	3.3979	1.4200	1,402.4	1,437.9	3.3746	1.3997	1,390.7	1,432.7	3.3539
340	1.5684	1,539.7	1,571.0	3.6075	1.5253	1,518.8	1,557.0	3.5719	1.4920	1,501.7	1,546.5	3.5426
360	1.8226	1,702.8	1,739.3	3.8772	1.6955	1,655.9	1,698.3	3.7986	1.6265	1,626.6	1,675.4	3.7494
380					2.2135	1,879.1	1,934.5	4.1649	1.8691	1,781.4	1,837.5	4.0012

Source: Abridged from J. H. Keenan, F. G. Keyes, P. G. Hill, and J. G. Moore, "Steam Tables," International Edition, John Wiley & Sons, Inc., New York, 1969.

TABLE A-10. Properties of Water-saturated Solid and Vapor (Metric Units)

Temp. °C t	Press. bars p	Specific volume cm³/g		Internal energy joule/g			Enthalpy joule/g			Entropy joule/(g)(°K)		
		Sat. solid v_i	Sat. vapor $v_g \times 10^{-3}$	Sat. solid u_i	Subl. u_{ig}	Sat. vapor u_g	Sat. solid h_i	Subl. h_{ig}	Sat. vapor h_g	Sat. solid s_i	Subl. s_{ig}	Sat. vapor s_g
0.01	0.006113	1.0908	2.061	−333.40	2,708.7	2,375.3	−333.40	2,834.8	2,501.4	−1.221	10.378	9.156
0	0.006108	1.0908	2.063	−333.43	2,708.8	2,375.3	−333.43	2,834.8	2,501.3	−1.221	10.378	9.157
−2	0.005176	1.0904	2.417	−337.62	2,710.2	2,372.6	−337.62	2,835.3	2,497.7	−1.237	10.456	9.219
−4	0.004375	1.0901	2.838	−341.78	2,711.6	2,369.8	−341.78	2,835.7	2,494.0	−1.253	10.536	9.283
−6	0.003689	1.0898	3.342	−345.91	2,712.9	2,367.0	−345.91	2,836.2	2,490.3	−1.268	10.616	9.348
−8	0.003102	1.0894	3.944	−350.02	2,714.2	2,364.2	−350.02	2,836.6	2,486.6	−1.284	10.698	9.414
−10	0.002602	1.0891	4.667	−354.09	2,715.5	2,361.4	−354.09	2,837.0	2,482.9	−1.299	10.781	9.481
−12	0.002176	1.0888	5.537	−358.14	2,716.8	2,358.7	−358.14	2,837.3	2,479.2	−1.315	10.865	9.550
−14	0.001815	1.0884	6.588	−362.15	2,718.0	2,355.9	−362.15	2,837.6	2,475.5	−1.331	10.950	9.619
−16	0.001510	1.0881	7.860	−366.14	2,719.2	2,353.1	−366.14	2,837.9	2,471.8	−1.346	11.036	9.690
−18	0.001252	1.0878	9.405	−370.10	2,720.4	2,350.3	−370.10	2,838.2	2,468.1	−1.362	11.123	9.762
−20	0.001035	1.0874	11.286	−374.03	2,721.6	2,347.5	−374.03	2,838.4	2,464.3	−1.377	11.212	9.835
−22	0.000853	1.0871	13.584	−377.93	2,722.7	2,344.7	−377.93	2,838.6	2,460.6	−1.393	11.302	9.909
−24	0.000701	1.0868	16.401	−381.80	2,723.7	2,342.0	−381.80	3,838.7	2,456.9	−1.408	11.394	9.985
−26	0.000574	1.0864	19.864	−385.64	2,724.8	2,339.2	−385.64	2,838.9	2,453.2	−1.424	11.486	10.062
−28	0.000469	1.0861	24.137	−389.45	2,725.8	2,336.4	−389.45	2,839.0	2,449.5	−1.439	11.580	10.141
−30	0.000381	1.0858	29.43	−393.23	2,726.8	2,333.6	−393.23	2,839.0	2,445.8	−1.455	11.676	10.221
−32	0.000309	1.0854	36.00	−396.98	2,727.8	2,330.8	−396.98	2,839.1	2,442.1	−1.471	11.773	10.303
−34	0.000250	1.0851	44.19	−400.71	2,728.7	2,328.0	−400.71	2,839.1	2,438.4	−1.486	11.872	10.386
−36	0.000201	1.0848	54.44	−404.40	2,729.6	2,325.2	−404.40	2,839.1	2,434.7	−1.501	11.972	10.470
−38	0.000161	1.0844	67.31	−408.06	2,730.5	2,322.4	−408.06	2,839.0	2,430.9	−1.517	12.073	10.556
−40	0.000129	1.0841	83.54	−411.70	2,731.3	2,319.6	−411.70	2,838.9	2,427.2	−1.532	12.176	10.644

TABLE A-11. Properties of Air at Low Pressures

T °R	h Btu/lbm	p_r	u Btu/lbm	v_r	ϕ Btu/(lbm)(°R)
100	23.74	0.003841	16.88	9,643	0.19714
120	28.53	0.007260	20.30	6,123	0.24077
140	33.31	0.012436	23.71	4,170	0.27767
160	38.10	0.019822	27.13	2,990	0.30962
180	42.89	0.02990	30.55	2,230	0.33781
200	47.67	0.04320	33.96	1,714.9	0.36303
220	52.46	0.06026	37.38	1,352.5	0.38584
240	57.25	0.08165	40.80	1,088.8	0.40666
260	62.03	0.10797	44.21	892.0	0.42582
280	66.82	0.13986	47.63	741.6	0.44356
300	71.61	0.17795	51.04	624.5	0.46007
320	76.40	0.22290	54.46	531.8	0.47550
340	81.18	0.27545	57.87	457.2	0.49002
360	85.97	0.3363	61.29	396.6	0.50369
380	90.75	0.4061	64.70	346.6	0.51663
400	95.53	0.4858	68.11	305.0	0.52890
420	100.32	0.5760	71.52	270.1	0.54058
440	105.11	0.6776	74.93	240.6	0.55172
460	109.90	0.7913	78.36	215.33	0.56235
480	114.69	0.9182	81.77	193.65	0.57255
500	119.48	1.0590	85.20	174.90	0.58233
520	124.27	1.2147	88.62	158.58	0.59173
537	128.10	1.3593	91.53	146.34	0.59945
540	129.06	1.3860	92.04	144.32	0.60078
560	133.86	1.5742	95.47	131.78	0.60950
580	138.66	1.7800	98.90	120.70	0.61793
600	143.47	2.005	102.34	110.88	0.62607
620	148.28	2.249	105.78	102.12	0.63395
640	153.09	2.514	109.21	94.30	0.64159
660	157.92	2.801	112.67	87.27	0.64902
680	162.73	3.111	116.12	80.96	0.65621
700	167.56	3.446	119.58	75.25	0.66321
720	172.39	3.806	123.04	70.07	0.67002
740	177.23	4.193	126.51	65.38	0.67665
760	182.08	4.607	129.99	61.10	0.68312
780	186.94	5.051	133.47	57.20	0.68942
800	191.81	5.526	136.97	53.63	0.69558
820	196.69	6.033	140.47	50.35	0.70160
840	201.56	6.573	143.98	47.34	0.70747

TABLE A-11. Properties of Air at Low Pressures (*Continued*)

T °R	h Btu/lbm	p_r	u Btu/lbm	v_r	ϕ Btu/(lbm)(°R)
860	206.46	7.149	147.50	44.57	0.71323
880	211.35	7.761	151.02	42.01	0.71886
900	216.26	8.411	154.57	39.64	0.72438
920	221.18	9.102	158.12	37.44	0.72979
940	226.11	9.834	161.68	35.41	0.73509
960	231.06	10.610	165.26	33.52	0.74030
980	236.02	11.430	168.83	31.76	0.74540
1,000	240.98	12.298	172.43	30.12	0.75042
1,020	245.97	13.215	176.04	28.59	0.75536
1,040	250.95	14.182	179.66	27.17	0.76019
1,060	255.96	15.203	183.29	25.82	0.76496
1,080	260.97	16.278	186.93	24.58	0.76964
1,100	265.99	17.413	190.58	23.40	0.77426
1,120	271.03	18.604	194.25	22.30	0.77880
1,140	276.08	19.858	197.94	21.27	0.78326
1,160	281.14	21.18	201.63	20.293	0.78767
1,180	286.21	22.56	205.33	19.377	0.79201
1,200	291.30	24.01	209.05	18.514	0.79628
1,220	296.41	25.53	212.78	17.700	0.80050
1,240	301.52	27.13	216.53	16.932	0.80466
1,260	306.65	28.80	220.28	16.205	0.80876
1,280	311.79	30.55	224.05	15.518	0.81280
1,300	316.94	32.39	227.83	14.868	0.81680
1,320	322.11	34.31	231.63	14.253	0.82075
1,340	327.29	36.31	235.43	13.670	0.82464
1,360	332.48	38.41	239.25	13.118	0.82848
1,380	337.68	40.59	243.08	12.593	0.83229
1,400	342.90	42.88	246.93	12.095	0.83604
1,420	348.14	45.26	250.79	11.622	0.83975
1,440	353.37	47.75	254.66	11.172	0.84341
1,460	358.63	50.34	258.54	10.743	0.84704
1,480	363.89	53.04	262.44	10.336	0.85062
1,500	369.17	55.86	266.34	9.948	0.85416
1,520	374.47	58.78	270.26	9.578	0.85767
1,540	379.77	61.83	274.20	9.226	0.86113
1,560	385.08	65.00	278.13	8.890	0.86456
1,580	390.40	68.30	282.09	8.569	0.86794
1,600	395.74	71.73	286.06	8.263	0.87130

TABLE A-11. Properties of Air at Low Pressures (*Continued*)

T °R	h Btu/lbm	p_r	u Btu/lbm	v_r	ϕ Btu/(lbm)(°R)
1,620	401.09	75.29	290.04	7.971	0.87462
1,640	406.45	78.99	294.03	7.691	0.87791
1,660	411.82	82.83	298.02	7.424	0.88116
1,680	417.20	86.82	302.04	7.168	0.88439
1,700	422.59	90.95	306.06	6.924	0.88758
1,720	428.00	95.24	310.09	6.690	0.89074
1,740	433.41	99.69	314.13	6.465	0.89387
1,760	438.83	104.30	318.18	6.251	0.89697
1,780	444.26	109.08	322.24	6.045	0.90003
1,800	449.71	114.03	326.32	5.847	0.90308
1,820	455.17	119.16	330.40	5.658	0.90609
1,840	460.63	124.47	334.50	5.476	0.90908
1,860	466.12	129.95	338.61	5.302	0.91203
1,880	471.60	135.64	342.73	5.134	0.91497
1,900	477.09	141.51	346.85	4.974	0.91788
1,920	482.60	147.59	350.98	4.819	0.92076
1,940	488.12	153.87	355.12	4.670	0.92362
1,960	493.64	160.37	359.28	4.527	0.92645
1,980	499.17	167.07	363.43	4.390	0.92926
2,000	504.71	174.00	367.61	4.258	0.93205
2,020	510.26	181.16	371.79	4.130	0.93481
2,040	515.82	188.54	375.98	4.008	0.93756
2,060	521.39	196.16	380.18	3.890	0.94026
2,080	526.97	204.02	384.39	3.777	0.94296
2,100	532.55	212.1	388.60	3.667	0.94564
2,150	546.54	233.5	399.17	3.410	0.95222
2,200	560.59	256.6	409.78	3.176	0.95868
2,250	574.69	281.4	420.46	2.961	0.96501
2,300	588.82	308.1	431.16	2.765	0.97123
2,350	603.00	336.8	441.91	2.585	0.97732
2,400	617.22	367.6	452.70	2.419	0.98331
2,450	631.48	400.5	463.54	2.266	0.98919
2,500	645.78	435.7	474.40	2.125	0.99497
2,550	660.12	473.3	485.31	1.9956	1.00064
2,600	674.49	513.5	496.26	1.8756	1.00623
2,650	688.90	556.3	507.25	1.7646	1.01172
2,700	703.35	601.9	518.26	1.6617	1.01712
2,750	717.83	650.4	529.31	1.5662	1.02244

TABLE A-11.   **Properties of Air at Low Pressures** (*Continued*)

T °R	h Btu/lbm	p_r	u Btu/lbm	v_r	ϕ Btu/(lbm)(°R)
2,800	732.33	702.0	540.40	1.4775	1.02767
2,850	746.88	756.7	551.52	1.3951	1.03282
2,900	761.45	814.8	562.66	1.3184	1.03788
2,950	776.05	876.4	573.84	1.2469	1.04288
3,000	790.68	941.4	585.04	1.1803	1.04779
3,500	938.40	1,829.3	698.48	0.7087	1.09332
4,000	1,088.26	3,280	814.06	0.4518	1.13334
4,500	1,239.86	5,521	931.39	0.3019	1.16905
5,000	1,392.87	8,837	1,050.12	0.20959	1.20129
5,500	1,547.07	13,568	1,170.04	0.15016	1.23068
6,000	1,702.29	20,120	1,291.00	0.11047	1.25769
6,500	1,858.44	28,974	1,412.87	0.08310	1.28268

Source: Abridged from J. H. Keenan and J. Kaye, "Gas Tables," John Wiley & Sons, Inc., New York, 1945.

TABLE A-12.   Properties of Various Gases at Low Pressures

h in Btu/(lbm mole) ϕ in Btu/(lbm mole)($^\circ$R)

Temp. °R	Nitrogen		Oxygen		Water Vapor		Carbon Dioxide		Hydrogen		Carbon Monoxide	
	h	ϕ	h	ϕ	h	ϕ	h	ϕ	h	ϕ	h	ϕ
537	3,729.5	45.755	3,725.1	48.986	4,258.3	45.079	4,030.2	51.032	3,640.3	31.194	3,729.5	47.272
600	4,167.9	46.514	4,168.3	49.762	4,764.7	45.970	4,600.9	52.038	4,075.6	31.959	4,168.0	48.044
700	4,864.9	47.588	4,879.3	50.858	5,575.4	47.219	5,552.0	53.503	4,770.2	33.031	4,866.0	49.120
800	5,564.4	48.522	5,602.0	51.821	6,396.9	48.316	6,552.9	54.839	5,467.1	33.961	5,568.2	50.058
900	6,268.1	49.352	6,337.9	52.688	7,230.9	49.298	7,597.6	56.070	6,165.3	34.784	6,276.4	50.892
1,000	6,977.9	50.099	7,087.5	53.477	8,078.9	50.191	8,682.1	57.212	6,864.5	35.520	6,992.2	51.646
1,100	7,695.0	50.783	7,850.4	54.204	8,942.0	51.013	9,802.6	58.281	7,564.6	36.188	7,716.8	52.337
1,200	8,420.0	51.413	8,625.8	54.879	9,820.4	51.777	10,955.3	59.283	8,265.8	36.798	8,450.8	52.976
1,300	9,153.9	52.001	9,412.9	55.508	10,714.5	52.494	12,136.9	60.229	8,968.7	37.360	9,194.6	53.571
1,400	9,896.9	52.551	10,210.4	56.099	11,624.8	53.168	13,344.7	61.124	9,673.8	37.883	9,948.1	54.129
1,500	10,648.9	53.071	11,017.1	56.656	12,551.4	53.808	14,576.0	61.974	10,381.5	38.372	10,711.1	54.655
1,600	11,409.7	53.561	11,832.5	57.182	13,494.9	54.418	15,829.0	62.783	11,092.5	38.830	11,483.4	55.154
1,700	12,178.9	54.028	12,655.6	57.680	14,455.4	54.999	17,101.4	63.555	11,807.4	39.264	12,264.3	55.628
1,800	12,956.3	54.472	13,485.8	58.155	15,433.0	55.559	18,391.5	64.292	12,526.8	39.675	13,053.2	56.078
1,900	13,741.6	54.896	14,322.1	58.607	16,427.5	56.097	19,697.8	64.999	13,250.9	40.067	13,849.8	56.509
2,000	14,534.4	55.303	15,164.0	59.039	17,439.0	56.617	21,018.7	65.676	13,980.1	40.441	14,653.2	56.922
2,100	15,334.0	55.694	16,010.9	59.451	18,466.9	57.119	22,352.7	66.327	14,714.5	40.799	15,463.2	57.317
2,200	16,139.8	56.068	16,862.6	59.848	19,510.8	57.605	23,699.0	66.953	15,454.4	41.143	16,279.4	57.696
2,300	16,951.2	56.429	17,718.8	60.228	20,570.6	58.077	25,056.3	67.557	16,199.8	41.475	17,101.0	58.062
2,400	17,767.9	56.777	18,579.2	60.594	21,645.7	58.535	26,424.0	68.139	16,950.6	41.794	17,927.4	58.414
2,500	18,589.5	57.112	19,443.4	60.946	22,735.4	58.980	27,801.2	68.702	17,707.3	42.104	18,758.8	58.754
2,600	19,415.8	57.436	20,311.4	61.287	23,839.5	59.414	29,187.1	69.245	18,469.7	42.403	19,594.3	59.081
2,700	20,246.4	57.750	21,182.9	61.616	24,957.2	59.837	30,581.2	69.771	19,237.8	42.692	20,434.0	59.398
2,800	21,081.1	58.053	22,057.8	61.934	26,088.0	60.248	31,982.8	70.282	20,011.8	42.973	21,277.2	59.705
2,900	21,919.5	58.348	22,936.1	62.242	27,231.2	60.650	33,391.5	70.776	20,791.5	43.247	22,123.8	60.002

TABLE A-12. Properties of Various Gases at Low Pressures (*Continued*)

h in Btu/(lbm mole) ϕ in Btu/(lbm mole)(°R)

Temp. °R	Nitrogen h	Nitrogen φ	Oxygen h	Oxygen φ	Water Vapor h	Water Vapor φ	Carbon Dioxide h	Carbon Dioxide φ	Hydrogen h	Hydrogen φ	Carbon Monoxide h	Carbon Monoxide φ
3,000	22,761.5	58.632	23,817.7	62.540	28,386.3	61.043	34,806.6	71.255	21,576.9	43.514	22,973.4	60.290
3,100	23,606.8	58.910	24,702.5	62.831	29,552.8	61.426	36,227.9	71.722	22,367.7	43.773	23,826.0	60.569
3,200	24,455.0	59.179	25,590.5	63.113	30,730.2	61.801	37,654.7	72.175	23,164.1	44.026	24,681.2	60.841
3,300	25,306.0	59.442	26,481.6	63.386	31,918.2	62.167	39,086.7	72.616	23,965.5	44.273	25,539.0	61.105
3,400	26,159.7	59.697	27,375.9	63.654	33,116.0	62.526	40,523.6	73.045	24,771.9	44.513	26,399.3	61.362
3,500	27,015.9	59.944	28,273.3	63.914	34,323.5	62.876	41,965.2	73.462	25,582.9	44.748	27,261.8	61.612
3,600	27,874.4	60.186	29,173.9	64.168	35,540.1	63.221	43,411.0	73.870	26,398.5	44.978	28,126.6	61.855
3,700	28,735.1	60.422	30,077.5	64.415	36,765.4	63.557	44,860.6	74.267	27,218.5	45.203	28,993.5	62.093
3,800	29,597.9	60.652	30,984.1	64.657	37,998.9	63.887	46,314.0	74.655	28,042.8	45.423	29,862.3	62.325
3,900	30,462.8	60.877	31,893.6	64.893	39,240.2	64.210	47,771.0	75.033	28,871.1	45.638	30,732.9	62.551
4,000	31,329.4	61.097	32,806.1	65.123	40,489.1	64.528	49,231.4	75.404	29,703.5	45.849	31,605.2	72.772
4,100	32,198.0	61.310	33,721.6	65.350	41,745.4	64.839	50,695.1	75.765	30,539.8	46.056	32,479.1	62.988
4,200	33,068.1	61.520	34,639.9	65.571	43,008.4	65.144	52,162.0	76.119	31,379.8	46.257	33,354.4	63.198
4,300	33,939.9	61.726	35,561.1	65.788	44,278.0	65.444	53,632.1	76.464	32,223.5	46.456	34,231.2	63.405
4,400	34,813.1	61.927	36,485.0	66.000	45,553.9	65.738	55,105.1	76.803	33,070.9	46.651	35,109.2	63.607
4,500	35,687.8	62.123	37,411.8	66.208	46,835.9	66.028	56,581.0	77.135	33,921.6	46.842	35,988.6	63.805
4,600	36,563.8	62.316	38,341.4	66.413	48,123.6	66.312	58,059.7	77.460	34,775.7	47.030	36,869.3	63.998
4,700	37,441.1	62.504	39,273.6	66.613	49,416.9	66.591	59,541.1	77.779	35,633.0	47.215	37,751.0	64.188
4,800	38,319.5	62.689	40,208.6	66.809	50,715.5	66.866	61,024.9	78.091	36,493.4	47.396	38,633.9	64.374
4,900	39,199.1	62.870	41,146.1	67.003	52,019.0	67.135	62,511.3	78.398	37,356.9	47.574	39,517.8	64.556
5,000	40,079.8	63.049	42,086.3	67.193	53,327.4	67.401	64,000.0	78.698	38,223.3	47.749	40,402.7	64.735
5,100	40,961.6	63.223	43,029.1	67.380	54,640.3	67.662	65,490.9	78.994	39,092.8	47.921	41,288.6	64.910
5,200	41,844.4	63.395	43,974.3	67.562	55,957.4	67.918	66,984.0	79.284	39,965.1	48.090	42,175.5	65.082
5,300	42,728.3	63.563	44,922.2	67.743	57,278.7	68.172	68,479.1	79.569	40,840.2	48.257	43,063.2	65.252

Source: Abridged from J. H. Keenan and J. Kaye, "Gas Tables," John Wiley & Sons, Inc., New York, 1948.

TABLE A-13. Specific Heats of Several Substances

			c_p, cal/(g)($^\circ$C)	c_p, cal/(g mole)($^\circ$C)
Substance	State			
Water	Liquid	0°C	1.00738	
		15°C	0.99976	
		25°C	0.99828	
		50°C	0.99854	
		75°C	1.00143	
		100°C	1.00697	
	Solid	−200°C	0.162	
		−100°C	0.329	
		−60°C	0.392	
		−11°C	0.4861	
		−2.2°C	0.5018	
Aluminum	Solid	100°K	0.115	
		298°K	0.215	
Copper	Solid	100°K	0.061	
		298°K	0.092	
Iron	Solid	100°K	0.052	
		298°K	0.108	
Lead	Solid	100°K	0.028	
		298°K	0.031	
Nickel	Solid	100°K	0.055	
		298°K	0.106	
Silver	Solid	100°K	0.045	
		298°K	0.057	
Tungsten	Solid	100°K	0.021	
		298°K	0.032	
n-Octane	Liquid	25°C		60.7
	Gas	25°C		45.1
Benzene	Liquid	25°C		32.4
	Gas	25°C		19.5
Ethane	Gas	300°K		12.65
	Gas	800°K		25.83
Propane	Gas	300°K		17.66
	Gas	800°K		37.08
n-Butane	Gas	300°K		23.40
	Gas	800°K		48.23

Source: Handbook of Chemistry and Physics, 1972–1973 ed., Chemical Rubber Company, Cleveland, Ohio.

TABLE A-14. Empirical c_p Equations for Various Ideal Gases

$$c_p = \frac{\text{cal}}{(\text{g mole})(^\circ\text{K})} = \frac{\text{Btu}}{(\text{lb mole})(^\circ\text{R})}, \quad \theta = \frac{T(\text{Kelvin})}{100} = \frac{T(\text{Rankine})}{180}$$

Gas		Range °R	Max. Error %
N_2	$c_p = 9.3355 - 122.56\theta^{-1.5} + 256.38\theta^{-2} - 196.08\theta^{-3}$	540–6300	0.43
O_2	$c_p = 8.9465 + 4.8044 \times 10^{-3}\theta^{1.5} - 42.679\theta^{-1.5} + 56.615\theta^{-2}$	540–6300	0.30
H_2	$c_p = 13.505 - 167.96\theta^{-0.75} + 278.44\theta^{-1} - 134.01\theta^{-1.5}$	540–6300	0.60
CO	$c_p = 16.526 - 0.16841\theta^{0.75} - 47.985\theta^{-0.5} + 42.246\theta^{-0.75}$	540–6300	0.42
OH	$c_p = 19.490 - 14.185\theta^{0.25} + 4.1418\theta^{0.75} - 1.0196\theta$	540–6300	0.43
NO	$c_p = 14.169 - 0.40861\theta^{0.5} - 16.877\theta^{-0.5} + 17.899\theta^{-1.5}$	540–6300	0.34
H_2O	$c_p = 34.190 - 43.868\theta^{0.25} + 19.778\theta^{0.5} - 0.88407\theta$	540–6300	0.43
CO_2	$c_p = -0.89286 + 7.2967\theta^{0.5} - 0.98074\theta + 5.7835 \times 10^{-3}\theta^{2}$	540–6300	0.19
NO_2	$c_p = 11.005 + 51.650\theta^{-0.5} - 86.916\theta^{-0.75} + 55.580\theta^{-2}$	540–6300	0.26
CH_4	$c_p = -160.82 + 105.10\theta^{0.25} - 5.9452\theta^{0.75} + 77.408\theta^{-0.5}$	540–3600	0.15
C_2H_4	$c_p = -22.800 + 29.433\theta^{0.5} - 8.5185\theta^{0.75} + 43.683\theta^{-3}$	540–3600	0.07
C_2H_6	$c_p = 1.648 + 4.124\theta - 0.153\theta^2 + 1.74 \times 10^{-3}\theta^3$	540–2700	0.83
C_3H_8	$c_p = -0.966 + 7.279\theta - 0.3755\theta^2 + 7.58 \times 10^{-3}\theta^3$	540–2700	0.40
C_4H_{10}	$c_p = 0.945 + 8.873\theta - 0.438\theta^2 + 8.36 \times 10^{-3}\theta^3$	540–2700	0.54

Source: G. J. Van Wylen and R. E. Sonntag, "Fundamentals of Classical Thermodynamics," John Wiley & Sons, Inc., New York, 1973.

TABLE A-15. Ideal Gas Enthalpy and Absolute Entropy at Atmospheric Pressure for Diatomic Nitrogen (N_2) and Monatomic Nitrogen (N)

Temp. °K	Temp. °R	Nitrogen, Diatomic (N_2) $h_{298} = 0$ cal/g mole $= 0$ Btu/lb mole $M = 28.016$			Nitrogen, Monatomic (N) $h_{298} = 112,965$ cal/g mole $= 203,337$ Btu/lb mole $M = 14.008$		
		$(h - h_{298})$ cal/g mole	s: cal/(g mole)(°K) Btu/(lb mole)(°R)	$(h - h_{537})$ Btu/lb mole	$(h - h_{298})$ cal/g mole	s: cal/(g mole)(°K) Btu/(lb mole)(°R)	$(h - h_{537})$ Btu/lb mole
0	0	-2,072	0	-3,730	-1,481	0	-2,666
100	180	-1,379	38.170	-2,483	-984	31.187	-1,771
200	360	-683	42.992	-1,229	-488	34.631	-878
298	537	0	45.770	0	0	36.614	0
300	540	13	45.813	23	9	36.645	16
400	720	710	47.818	1,278	506	38.074	911
500	900	1,413	49.386	2,543	1,003	39.183	1,805
600	1,080	2,125	50.685	3,825	1,500	40.089	2,700
700	1,260	2,853	51.806	5,135	1,996	40.855	3,593
800	1,440	3,596	52.798	6,473	2,493	41.518	4,487
900	1,620	4,355	53.692	7,839	2,990	42.103	5,382
1,000	1,800	5,129	54.507	9,232	3,487	42.627	6,277
1,100	1,980	5,917	55.258	10,651	3,984	43.100	7,171
1,200	2,160	6,718	55.955	12,092	4,481	43.532	8,066
1,300	2,340	7,529	56.604	13,552	4,977	43.930	8,959
1,400	2,520	8,350	57.212	15,030	5,474	44.298	9,853
1,500	2,700	9,179	57.784	16,522	5,971	44.641	10,748

TABLE A-15. Ideal Gas Enthalpy and Absolute Entropy at Atmospheric Pressure for Diatomic Nitrogen (N_2) and Monatomic Nitrogen (N) (Continued)

Temp. °K	Temp. °R	Nitrogen, Diatomic (N_2) $h_{298} = 0$ cal/g mole $= 0$ Btu/lb mole $M = 28.016$			Nitrogen, Monatomic (N) $h_{298} = 112{,}965$ cal/g mole $= 203{,}337$ Btu/lb mole $M = 14.008$		
		$(h - h_{298})$ cal/g mole	s: cal/(g mole)(°K) Btu/(lb mole)(°R)	$(h - h_{537})$ Btu/lb mole	$(h - h_{298})$ cal/g mole	s: cal/(g mole)(°K) Btu/(lb mole)(°R)	$(h - h_{537})$ Btu/lb mole
1,600	2,880	10,015	58.324	18,027	6,468	44.962	11,642
1,700	3,060	10,858	58.835	19,544	6,965	45.263	12,537
1,800	3,240	11,707	59.320	21,073	7,461	45.547	13,430
1,900	3,420	12,560	59.782	22,608	7,958	45.815	14,324
2,000	3,600	13,418	60.222	24,152	8,455	16.070	15,219
2,100	3,780	14,280	60.642	25,704	8,952	46.313	16,114
2,200	3,960	15,146	61.045	27,263	9,449	46.544	17,008
2,300	4,140	16,015	61.431	28,827	9,946	46.765	17,903
2,400	4,320	16,886	61.802	30,395	10,444	46.977	18,799
2,500	4,500	17,761	62.159	31,970	10,941	47.180	19,694
2,600	4,680	18,638	62.503	33,548	11,439	47.375	20,590
2,700	4,860	19,517	62.835	35,131	11,938	47.563	21,488
2,800	5,040	20,398	63.155	36,716	12,437	47.745	22,387
2,900	5,220	21,280	63.465	38,304	12,936	47.920	23,285
3,000	5,400	22,165	63.765	39,897	13,437	48.090	24,187
3,200	5,760	23,939	64.337	43,090	14,441	48.414	25,994
3,400	6,120	25,719	64.877	46,294	15,451	48.720	27,812

3,600	6,480	27,505	65.387	49,509	16,469	49.011	29,644
3,800	6,840	29,295	65.871	52,731	17,495	49.288	31,491
4,000	7,200	31,089	66.331	55,960	18,531	49.554	33,356
4,200	7,560	32,888	66.770	59,198	19,580	49.810	35,244
4,400	7,920	34,690	67.189	62,442	20,643	50.057	37,157
4,600	8,280	36,496	67.591	65,693	21,721	50.297	39,098
4,800	8,640	38,306	67.976	68,951	22,816	50.530	41,069
5,000	9,000	40,119	68.346	72,214	23,928	50.757	43,070
5,200	9,360	41,935	68.702	75,483	25,059	50.978	45,106
5,400	9,720	43,755	69.045	78,759	26,210	51.195	47,178
5,600	10,180	45,579	69.377	82,042	27,380	51.408	49,284
5,800	10,540	47,406	69.698	85,331	28,570	51.617	51,426
6,000	10,800	49,237	70.008	88,627	29,780	51.822	53,604

Source: Data in Tables A-15 through A-20 are from the JANAF Thermochemical Tables, Second Edition, NSRDS-NBS 37, National Bureau of Standards, Washington, D.C., 1971.

TABLE A-16. Ideal Gas Enthalpy and Absolute Entropy at Atmospheric Pressure for Diatomic Oxygen (O_2) and Monatomic Oxygen (O)

Temp. °K	Temp. °R	Oxygen, Diatomic (O_2) $h_{298} = 0$ cal/g mole $= 0$ Btu/lb mole $M = 32.00$			Oxygen, Monatomic (O) $h_{298} = 59{,}559$ cal/g mole $= 107{,}206$ Btu/lb mole $M = 16.00$		
		$(h - h_{298})$ cal/g mole	s:cal/(g mole)(°K) Btu/(lb mole)(°R)	$(h - h_{537})$ Btu/lb mole	$(h - h_{298})$ cal/g mole	s:cal/(g mole)(°K) Btu/(lb mole)(°R)	$(h - h_{537})$ Btu/lb mole
0	0	-2,075	0	-3,735	-1,608	0	-2,894
100	180	-1,381	41.395	-2,486	-1,080	32.466	-1,944
200	360	-685	46.218	-1,233	-523	36.340	-941
298	537	0	49.004	0	0	38.468	0
300	540	13	49.047	23	10	38.501	18
400	720	724	51.091	1,303	528	39.991	950
500	900	1,455	52.722	2,619	1,038	41.131	1,868
600	1,080	2,210	54.098	3,978	1,544	42.054	2,779
700	1,260	2,988	55.297	5,378	2,048	42.831	3,686
800	1,440	3,786	56.361	6,815	2,550	43.501	4,590
900	1,620	4,600	57.320	8,280	3,052	44.092	5,494
1,000	1,800	5,427	58.192	9,769	3,552	44.619	6,394
1,100	1,980	6,266	58.991	11,279	4,051	45.095	7,292
1,200	2,160	7,114	59.729	12,805	4,551	45.529	8,192
1,300	2,340	7,971	60.415	14,348	5,049	45.928	9,088
1,400	2,520	8,835	61.055	15,903	5,548	46.298	9,986
1,500	2,700	9,706	61.656	17,471	6,046	46.642	10,883
1,600	2,880	10,583	62.222	19,049	6,544	46.963	11,779
1,700	3,060	11,465	62.757	20,637	7,042	47.265	12,676

1,800	3,240	12,354	63.265	22,237	7,540	47.550	13,572
1,900	3,420	13,249	63.749	23,848	8,038	47.819	14,468
2,000	3,600	14,149	64.210	25,468	8,536	48.074	15,365
2,100	3,780	15,054	64.652	27,097	9,034	48.317	16,261
2,200	3,960	15,966	65.076	28,739	9,532	48.549	17,158
2,300	4,140	16,882	65.483	30,388	10,029	48.770	18,052
2,400	4,320	17,804	65.876	32,047	10,527	48.982	18,949
2,500	4,500	18,732	66.254	33,718	11,026	49.185	19,847
2,600	4,680	19,664	66.620	35,395	11,524	49.381	20,743
2,700	4,860	20,602	66.974	37,084	12,023	49.569	21,641
2,800	5,040	21,545	67.317	38,781	12,522	49.751	22,540
2,900	5,220	22,493	67.650	40,487	13,022	49.926	23,440
3,000	5,400	23,446	67.973	42,203	13,522	50.096	24,340
3,200	5,760	25,365	68.592	45,657	14,524	50.419	26,143
3,400	6,120	27,302	69.179	49,144	15,529	50.724	27,952
3,600	6,480	29,254	69.737	52,657	16,537	51.012	29,767
3,800	6,840	31,221	70.269	56,198	17,549	51.285	31,588
4,000	7,200	33,201	70.776	59,762	18,565	51.546	33,417
4,200	7,560	35,193	71.262	63,347	19,586	51.795	35,255
4,400	7,920	37,196	71.728	66,953	20,611	52.033	37,100
4,600	8,280	39,208	72.176	70,574	21,641	52.262	38,954
4,800	8,640	41,229	72.606	74,212	22,676	52.482	40,817
5,000	9,000	43,257	73.019	77,863	23,715	52.695	42,687
5,200	9,360	45,292	73.418	81,526	24,760	52.899	44,568
5,400	9,720	47,332	73.803	85,198	25,809	53.097	46,456
5,600	10,180	49,377	74.175	88,879	26,863	53.289	48,353
5,800	10,540	51,426	74.535	92,567	27,921	53.475	50,258
6,000	10,800	53,479	74.883	96,262	28,984	53.655	52,171

TABLE A-17. Ideal Gas Enthalpy and Absolute Entropy at Atmospheric Pressure for Diatomic Hydrogen (H$_2$) and Monatomic Hydrogen (H)

Temp °K	Temp °R	Hydrogen, Diatomic (H$_2$) $h_{298} = 0$ cal/g mole = 0 Btu/lb mole $M = 2.016$			Hydrogen, Monatomic (H) $h_{298} = 52,100$ cal/g mole = 93,780 Btu/lb mole $M = 1.008$		
		$(h - h_{298})$ cal/g mole	s: cal/(g mole)(°K) Btu/(lb mole)(°R)	$(h - h_{537})$ Btu/lb mole	$(h - h_{298})$ cal/g mole	s: cal/(g mole)(°K) Btu/(lb mole)(°R)	$(h - h_{537})$ Btu/lb mole
0	0	-2,024	0	-3,643	-1,481	0	-2,666
100	180	-1,265	24.387	-2,277	-984	21.965	-1,771
200	360	-662	28.520	-1,192	-488	25.408	-878
298	537	0	31.208	0	0	27.392	0
300	540	13	31.251	23	9	27.423	16
400	720	707	33.247	1,273	506	28.852	911
500	900	1,406	34.806	2,531	1,003	29.961	1,805
600	1,080	2,106	36.082	3,791	1,500	30.867	2,700
700	1,260	2,808	37.165	5,054	1,996	31.632	3,593
800	1,440	3,514	38.107	6,325	2,493	32.296	4,487
900	1,620	4,226	38.946	7,607	2,990	32.881	5,382
1,000	1,800	4,944	39.702	8,899	3,487	33.404	6,277
1,100	1,980	5,670	40.394	10,206	3,984	33.878	7,171
1,200	2,160	6,404	41.033	11,527	4,481	34.310	8,066
1,300	2,340	7,148	41.628	12,866	4,977	34.708	8,959
1,400	2,520	7,902	42.187	14,224	5,474	35.076	9,853
1,500	2,700	8,668	42.716	15,602	5,971	35.419	10,748
1,600	2,880	9,446	43.217	17,003	6,468	35.739	11,642
1,700	3,060	10,233	43.695	18,419	6,965	36.041	12,537

13,430	36.325	7,461	19,854	44.150	11,030	3,240	1,800
14,324	36.593	7,958	21,305	44.586	11,836	3,420	1,900
15,219	36.848	8,455	22,772	45.004	12,651	3,600	2,000
16,114	37.090	8,952	24,255	45.406	13,475	3,780	2,100
17,008	37.322	9,449	25,753	45.793	14,307	3,960	2,200
17,901	37.542	9,945	27,263	46.166	15,146	4,140	2,300
13,796	37.754	10,442	28,787	46.527	15,993	4,320	2,400
19,690	37.957	10,939	30,326	46.875	16,848	4,500	2,500
20,585	38.152	11,436	31,874	47.213	17,708	4,680	2,600
21,479	38.339	11,933	33,435	47.540	18,575	4,860	2,700
22,374	38.520	12,430	35,006	47.857	19,448	5,040	2,800
23,267	38.694	12,926	36,587	48.166	20,326	5,220	2,900
24,161	38.862	13,423	38,178	48.465	21,210	5,400	3,000
25,951	39.183	14,417	41,386	49.040	22,992	5,760	3,200
27,738	39.484	15,410	44,629	49.586	24,794	6,120	3,400
29,527	39.768	16,404	47,909	50.107	26,616	6,400	3,600
31,316	40.037	17,398	51,223	50.605	28,457	6,840	3,800
33,104	40.292	18,391	54,571	51.082	30,317	7,200	4,000
34,893	40.534	19,385	57,949	51.540	32,194	7,560	4,200
36,682	40.765	20,379	61,358	51.980	34,088	7,920	4,400
38,470	40.986	21,372	64,798	52.405	35,999	8,280	4,600
40,259	41.198	22,366	68,267	52.815	37,926	8,640	4,800
42,046	41.400	23,359	71,762	53.211	39,868	9,000	5,000
43,835	41.595	24,353	75,285	53.595	41,825	9,360	5,200
45,625	41.783	25,347	78,835	53.967	43,797	9,720	5,400
47,412	41.963	26,340	82,409	54.328	45,783	10,180	5,600
49,201	42.138	27,334	86,009	54.679	47,783	10,540	5,800
50,990	42.306	28,328	89,633	55.020	49,796	10,800	6,000

TABLE A-18. Ideal Gas Enthalpy and Absolute Entropy at Atmospheric Pressure for Carbon Dioxide (CO_2) and Carbon Monoxide (CO)

Temp °K	Temp °R	Carbon Dioxide (CO_2) $h_{298} = -94{,}054$ cal/g mole $= -169{,}297$ Btu/lb mole $M = 44.011$			Carbon Monoxide (CO) $h_{298} = -26{,}417$ cal/g mole $= -47{,}551$ Btu/lb mole $M = 28.011$		
		$(h - h_{298})$ cal/g mole	s:cal/(g mole)(°K) Btu/(lb mole)(°R)	$(h - h_{537})$ Btu/lb mole	$(h - h_{298})$ cal/g mole	s:cal/(g mole)(°K) Btu/(lb mole)(°R)	$(h - h_{537})$ Btu/lb mole
0	0	-2,238	0	-4,028	-2,072	0	-3,730
100	180	-1,543	42.758	-2,777	-1,379	39.613	-2,483
200	360	-816	47.769	-1,469	-683	44.435	-1,229
298	537	0	51.072	0	0	47.214	0
300	540	16	51.127	29	13	47.257	23
400	720	958	53.830	1,724	711	49.265	1,280
500	900	1,987	56.122	3,577	1,417	50.841	2,551
600	1,080	3,087	58.126	5,557	2,137	52.152	3,847
700	1,260	4,245	59.910	7,641	2,873	53.287	5,171
800	1,440	5,453	61.522	9,815	3,627	54.293	6,529
900	1,620	6,702	62.992	12,064	4,397	55.200	7,915
1,000	1,800	7,984	64.344	14,371	5,183	56.028	9,329
1,100	1,980	9,296	65.594	16,733	5,983	56.790	10,769
1,200	2,160	10,632	66.756	19,138	6,794	57.496	12,229
1,300	2,340	11,988	67.841	21,578	7,616	58.154	13,709
1,400	2,520	13,362	68.859	24,052	8,446	58.769	15,203
1,500	2,700	14,750	69.817	26,550	9,285	59.348	16,713
1,600	2,880	16,152	70.722	29,074	10,130	59.893	18,234
1,700	3,060	17,565	71.578	31,617	10,980	60.409	19,764

1,800	3,240	18,987	72.391	34,177	11,836	60.898	21,305
1,900	3,420	20,418	73.165	36,752	12,697	61.363	22,855
2,000	3,600	21,857	73.903	39,343	13,561	61.807	24,410
2,100	3,780	23,303	74.608	41,945	14,430	62.230	25,974
2,200	3,960	24,755	75.284	44,559	15,301	62.635	27,542
2,300	4,140	26,212	75.931	47,182	16,175	63.024	29,115
2,400	4,320	27,674	76.554	49,813	17,052	63.397	30,694
2,500	4,500	29,141	77.153	52,454	17,931	63.756	32,276
2,600	4,680	30,613	77.730	55,103	18,813	64.102	33,863
2,700	4,860	32,088	78.286	57,758	19,696	64.435	35,453
2,800	5,040	33,567	78.824	60,421	20,582	64.757	37,048
2,900	5,220	35,049	79.344	63,088	21,469	65.069	38,644
3,000	5,400	36,535	79.848	65,763	22,357	65.370	40,243
3,200	5,760	39,515	80.810	71,127	24,139	65.945	43,450
3,400	6,120	42,507	81.717	76,513	25,927	66.487	46,669
3,600	6,480	45,508	82.574	81,914	27,719	66.999	49,894
3,800	6,840	48,518	83.388	87,332	29,516	67.485	53,129
4,000	7,200	51,538	84.162	92,768	31,316	67.946	56,369
4,200	7,560	54,566	84.901	98,219	33,121	68.387	59,618
4,400	7,920	57,601	85.607	103,682	34,930	68.807	62,874
4,600	8,280	60,644	86.284	109,159	36,741	69.210	66,134
4,800	8,640	63,695	86.933	114,651	38,557	69.596	69,403
5,000	9,000	66,753	87.557	120,155	40,375	69.967	72,675
5,200	9,360	69,819	88.158	125,674	42,196	70.325	75,953
5,400	9,720	72,893	88.738	131,207	44,021	70.669	79,238
5,600	10,180	75,976	89.299	136,757	45,849	71.001	82,528
5,800	10,540	79,068	89.841	142,322	47,679	71.322	85,822
6,000	10,800	82,168	90.367	147,902	49,513	71.633	89,123

TABLE A-19. Ideal Gas Enthalpy and Absolute Entropy at Atmospheric Pressure for Water (H_2O) and Hydroxyl (OH)

Temp. °K	Temp. °R	Water (H_2O) $h_{298} = -57,798$ cal/g mole $= -104,036$ Btu/lb mole $M = 18.016$			Hydroxyl (OH) $h_{298} = 9,432$ cal/g mole $= 16,978$ Btu/lb mole $M = 17.008$		
		$(h - h_{298})$ cal/g mole	s: cal/(g mole)(°K) Btu/(lb mole)(°R)	$(h - h_{537})$ Btu/lb mole	$(h - h_{298})$ cal/g mole	s: cal/(g mole)(°K) Btu/(lb mole)(°R)	$(h - h_{537})$ Btu/lb mole
0	0	−2,367	0	−4,261	−2,192	0	−3,946
100	180	−1,581	36.396	−2,846	−1,467	35.726	−2,641
200	360	−784	41.916	−1,411	−711	40.985	−1,280
298	537	0	45.106	0	0	43.880	0
300	540	15	45.155	27	13	43.925	23
400	720	825	47.484	1,485	725	45.974	1,305
500	900	1,654	49.334	2,977	1,432	47.551	2,578
600	1,080	2,509	50.891	4,516	2,137	48.837	3,847
700	1,260	3,390	52.249	6,102	2,845	49.927	5,121
800	1,440	4,300	53.464	7,740	3,556	50.877	6,401
900	1,620	5,240	54.570	9,432	4,275	51.724	7,695
1,000	1,800	6,209	55.592	11,176	5,003	52.491	9,005
1,100	1,980	7,210	56.545	12,978	5,742	53.195	10,336
1,200	2,160	8,240	57.441	14,832	6,491	53.847	11,684
1,300	2,340	9,298	58.288	16,736	7,252	54.455	13,054
1,400	2,520	10,384	59.092	18,691	8,023	55.027	14,441
1,500	2,700	11,495	59.859	20,691	8,805	55.566	15,849
1,600	2,880	12,630	60.591	22,734	9,596	56.077	17,273
1,700	3,060	13,787	61.293	24,817	10,397	56.563	18,715
1,800	3,240	14,964	61.965	26,935	11,207	57.025	20,173

1,900	3,420	16,160	62.612	29,088	12,024	57.467	21,643
2,000	3,600	17,373	63.234	31,271	12,849	57.891	23,128
2,100	3,780	18,602	63.834	33,484	13,681	58.296	24,626
2,200	3,960	19,846	64.412	35,723	14,520	58.686	26,136
2,300	4,140	21,103	64.971	37,985	15,364	59.062	27,655
2,400	4,320	22,372	65.511	40,270	16,214	59.424	29,185
2,500	4,500	23,653	66.034	42,575	17,069	59.773	30,724
2,600	4,680	24,945	66.541	44,901	17,929	60.110	32,272
2,700	4,860	26,246	67.032	47,243	18,794	60.436	33,829
2,800	5,040	27,556	67.508	49,601	19,662	60.752	35,392
2,900	5,220	28,875	67.971	51,975	20,535	61.058	36,963
3,000	5,400	30,201	68.421	54,362	21,411	61.355	38,540
3,200	5,760	32,876	69.284	59,177	23,174	61.924	41,713
3,400	6,120	35,577	70.102	64,039	24,949	62.462	44,908
3,600	6,480	38,300	70.881	68,940	26,735	62.973	48,123
3,800	6,840	41,043	71.622	73,877	28,532	63.458	51,358
4,000	7,200	43,805	72.331	78,849	30,338	63.922	54,608
4,200	7,560	46,583	73.008	83,849	32,153	64.364	57,875
4,400	7,920	49,375	73.658	88,875	33,976	64.788	61,157
4,600	8,280	52,181	74.281	93,926	35,807	65.195	64,453
4,800	8,640	55,000	74.881	99,000	37,644	65.586	67,759
5,000	9,000	57,829	75.459	104,092	39,489	65.963	71,080
5,200	9,360	60,669	76.016	109,204	41,340	66.326	74,412
5,400	9,720	63,520	76.553	114,336	43,197	66.676	77,755
5,600	10,180	66,381	77.074	119,486	45,060	67.015	81,108
5,800	10,540	69,251	77.577	124,652	46,929	67.343	84,472
6,000	10,800	72,131	78.065	129,836	48,803	67.661	87,845

TABLE A-20. Ideal Gas Enthalpy and Absolute Entropy at Atmospheric Pressure for Nitric Oxide (NO) and Nitrogen Dioxide (NO₂)

Temp. °K	Temp. °R	Nitric Oxide (NO) $h_{298} = 21{,}652$ cal/g mole $= 38{,}974$ Btu/lb mole $M = 30.008$			Nitrogen Dioxide (NO₂) $h_{298} = 7{,}910$ cal/g mole $= 14{,}238$ Btu/lb mole $M = 46.008$		
		$(h - h_{298})$ cal/g mole	s: cal/(g mole)(°K) Btu/(lb mole)(°R)	$(h - h_{537})$ Btu/lb mole	$(h - h_{298})$ cal/g mole	s: cal/(g mole)(°K) Btu/(lb mole)(°R)	$(h - h_{537})$ Btu/lb mole
0	0	−2,197	0	−3,955	−2,435	0	−4,383
100	180	−1,451	42.286	−2,612	−1,640	48.387	−2,952
200	360	−705	47.477	−1,269	−835	53.954	−1,503
298	537	0	50.347	0	0	57.343	0
300	540	13	50.392	23	16	57.398	29
400	720	727	52.444	1,309	939	60.046	1,690
500	900	1,448	54.053	2,606	1,936	62.268	3,485
600	1,080	2,186	55.397	3,935	3,001	64.208	5,402
700	1,260	2,942	56.562	5,296	4,123	65.937	7,421
800	1,440	3,716	57.596	6,689	5,291	67.496	9,524
900	1,620	4,507	58.528	8,113	6,496	68.915	11,693
1,000	1,800	5,313	59.377	9,563	7,730	70.215	13,914
1,100	1,980	6,131	60.157	11,036	8,988	71.414	16,178
1,200	2,160	6,960	60.878	12,528	10,265	72.524	18,477
1,300	2,340	7,798	61.548	14,036	11,556	73.558	20,801
1,400	2,520	8,644	62.175	15,559	12,861	74.525	23,150
1,500	2,700	9,496	62.763	17,093	14,176	75.432	25,517
1,600	2,880	10,354	63.317	18,637	15,499	76.286	27,898
1,700	3,060	11,217	63.840	20,191	16,830	77.093	30,294
1,800	3,240	12,084	64.335	21,751	18,167	77.857	32,701

1,900	3,420	12,955	64.806	23,319	19,509	78.583	35,116
2,000	3,600	13,829	65.255	24,892	20,856	79.274	37,541
2,100	3,780	14,706	65.683	26,471	22,207	79.933	39,973
2,200	3,960	15,587	66.092	28,057	23,561	80.563	42,410
2,300	4,140	16,469	66.484	29,644	24,919	81.166	44,854
2,400	4,320	17,354	66.861	31,237	26,279	81.745	47,302
2,500	4,500	18,241	67.223	32,834	27,641	82.301	49,754
2,600	4,680	19,129	67.571	34,432	29,006	82.836	52,211
2,700	4,860	20,020	67.907	36,036	30,373	83.352	54,671
2,800	5,040	20,911	68.232	37,640	31,741	83.850	57,134
2,900	5,220	21,805	68.545	39,249	33,111	84.330	59,600
3,000	5,400	22,700	68.849	40,860	34,482	84.795	62,068
3,200	5,760	24,493	69.427	44,087	37,228	85.681	67,010
3,400	6,120	26,291	69.973	47,324	39,978	86.515	71,960
3,600	6,480	28,094	70.488	50,569	42,731	87.302	76,916
3,800	6,840	29,900	70.976	53,820	45,488	88.047	81,878
4,000	7,200	31,710	71.440	57,078	48,247	88.755	86,845
4,200	7,560	33,523	71.882	60,341	51,008	89.428	91,814
4,400	7,920	35,340	72.305	63,612	53,771	90.071	96,788
4,600	8,280	37,159	72.709	66,886	56,536	90.686	101,765
4,800	8,640	38,982	73.097	70,168	59,302	91.274	106,744
5,000	9,000	40,807	73.470	73,453	62,070	91.839	111,726
5,200	9,360	42,634	73.828	76,741	64,838	92.382	116,708
5,400	9,720	44,465	74.173	80,037	67,608	92.905	121,694
5,600	10,180	46,297	74.507	83,335	70,379	93.408	126,682
5,800	10,540	48,133	74.829	86,639	73,150	93.895	131,670
6,000	10,800	49,970	75.140	89,946	75,922	94.365	136,660

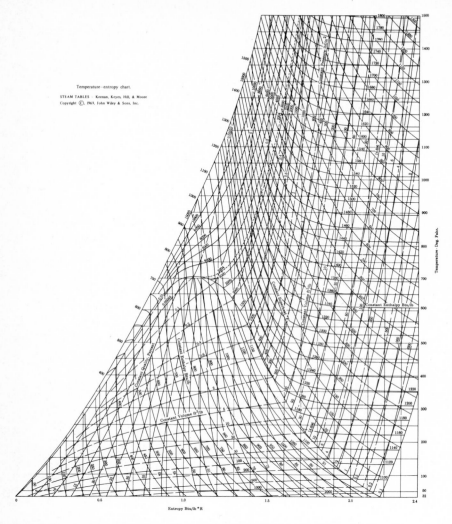

Temperature–entropy chart.

STEAM TABLES · Keenan, Keyes, Hill, & Moore
Copyright © 1969, John Wiley & Sons, Inc.

FIG. A-1. Temperature-entropy diagram for water (English units). [*From J. H. Keenan, F. G. Keyes, P. G. Hill, and J. G. Moore, "Steam Tables (English Units)," John Wiley & Sons, Inc., New York, 1969.*]

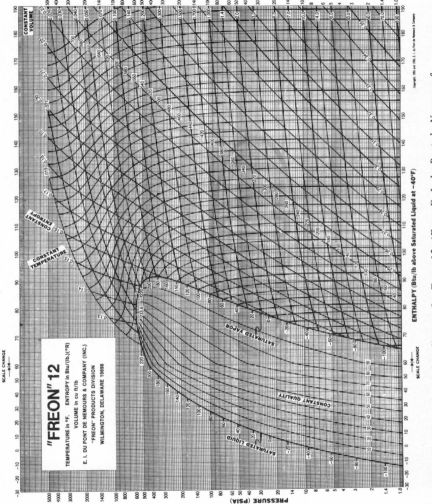

FIG. A-2. Pressure-enthalpy diagram for Freon-12. (*From E. I. du Pont de Nemours & Company.*)

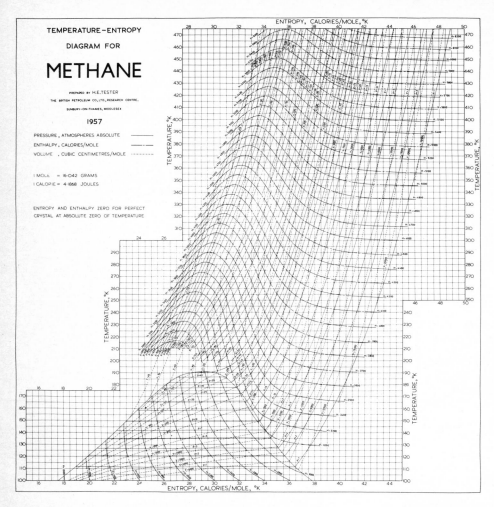

FIG. A-3. Temperature-entropy diagram for methane. [*From F. Din (ed.), "Thermodynamic Functions of Gases," vol. 3, Butterworth & Co., London, England, 1961.*]

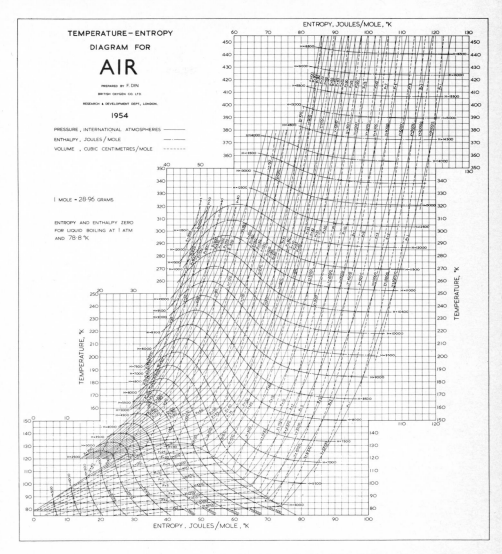

FIG. A-4. Temperature-entropy diagram for air. [*From F. Din (ed.), "Thermodynamic Functions of Gases," vol. 2, Butterworth & Co., London, England, 1956.*]

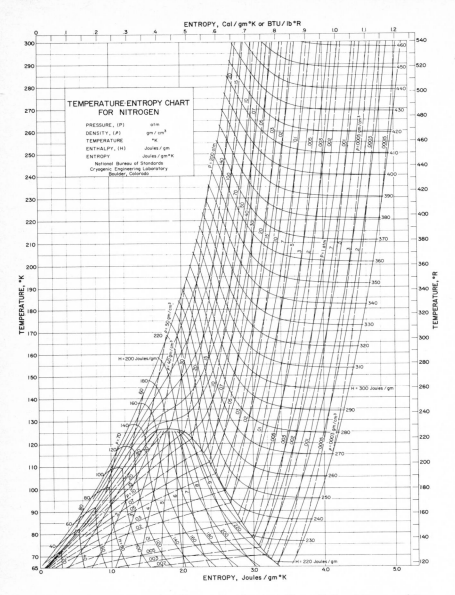

FIG. A-5. Temperature-entropy diagram for nitrogen. (*From National Bureau of Standards, Cryogenic Engineering Laboratory, Boulder, Colo.*)

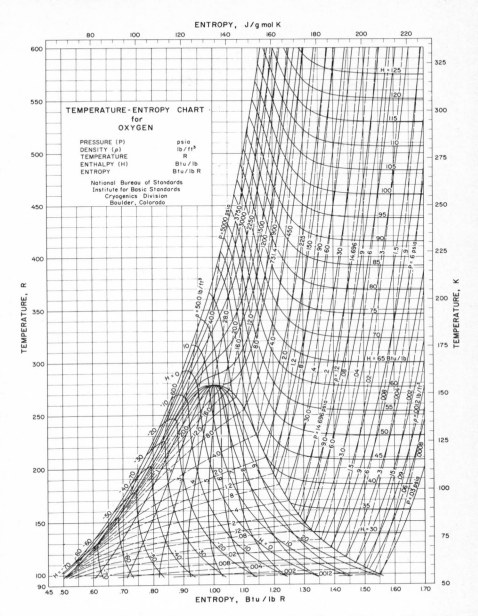

FIG. A-6. Temperature-entropy diagram for oxygen. (*From National Bureau of Standards, Cryogenic Engineering Laboratory, Boulder, Colo.*)

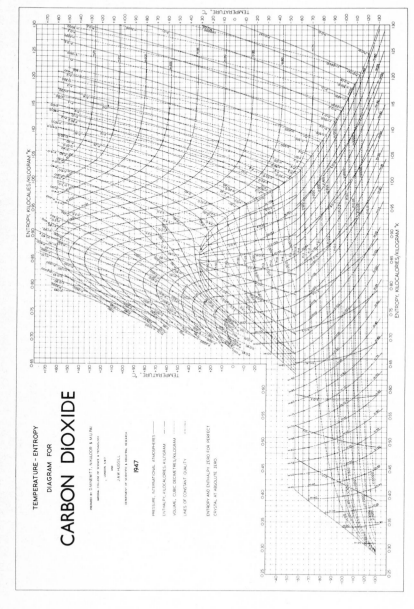

FIG. A-7. Temperature-entropy diagram for carbon dioxide. [*From F. Din (ed.), "Thermodynamic Functions of Gases," vol. 1, Butterworth & Co., London, England, 1956.*]

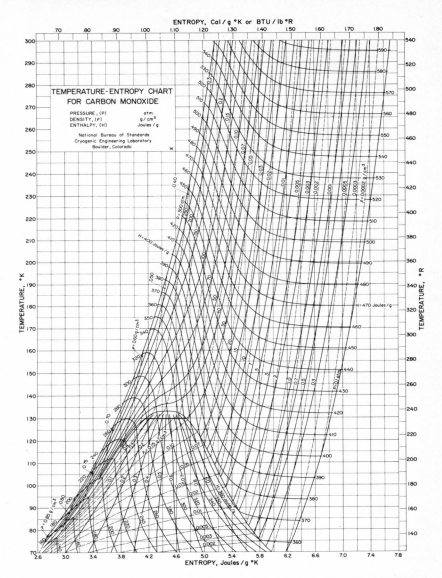

FIG. A-8. Temperature-entropy diagram for carbon monoxide. (*From National Bureau of Standards, Cryogenic Engineering Laboratory, Boulder, Colo.*)

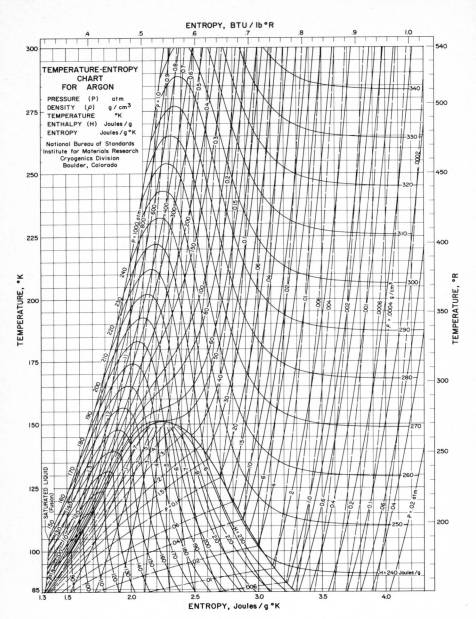

FIG. A-9. Temperature-entropy diagram for argon. (*From National Bureau of Standards, Cryogenic Engineering Laboratory, Boulder, Colo.*)

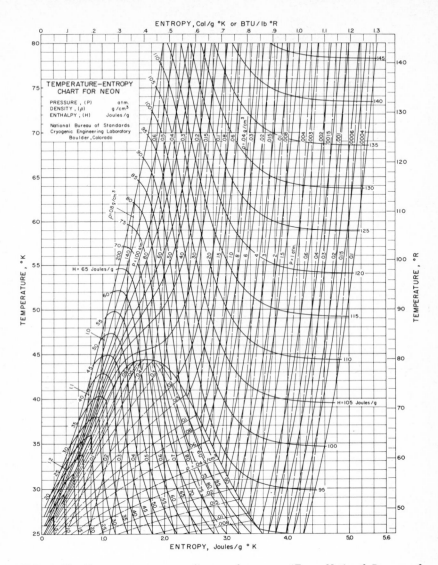

FIG. A-10. Temperature-entropy diagram for neon. (*From National Bureau of Standards, Cryogenic Engineering Laboratory, Boulder, Colo.*)

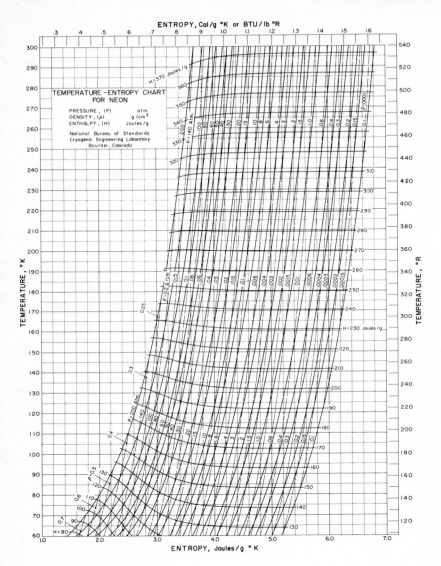

FIG. A-10. (Continued)

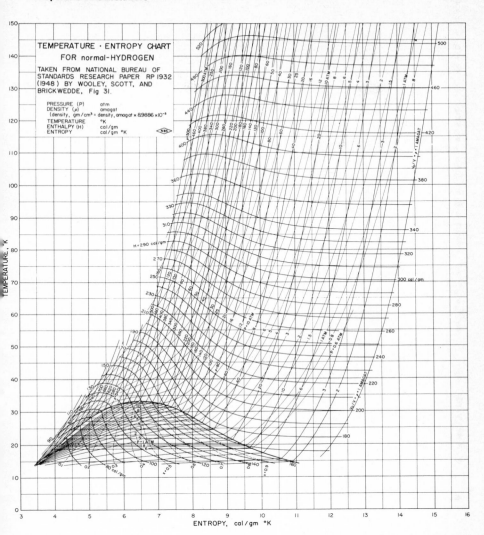

FIG. A-11. Temperature-entropy diagram for normal hydrogen (*From National Bureau of Standards, Washington, D.C.*)

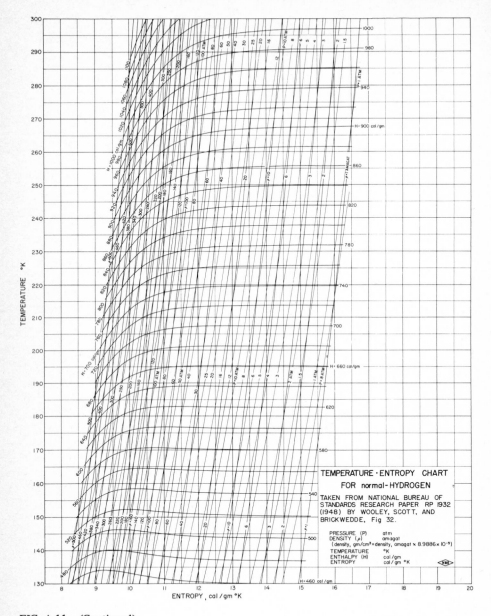

FIG. A-11. (Continued)

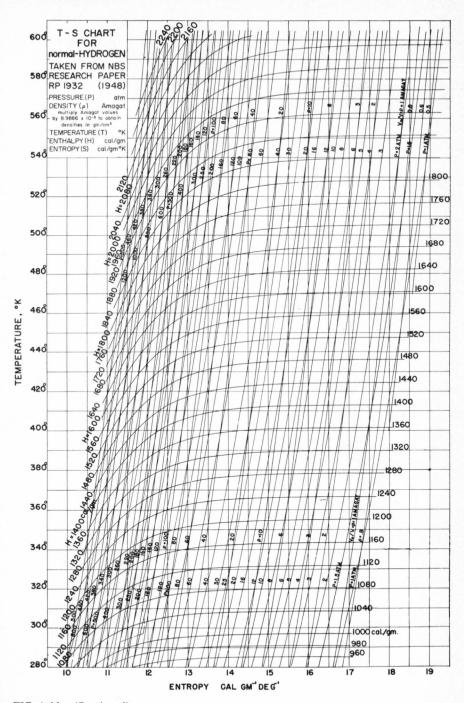

FIG. A-11. (Continued)

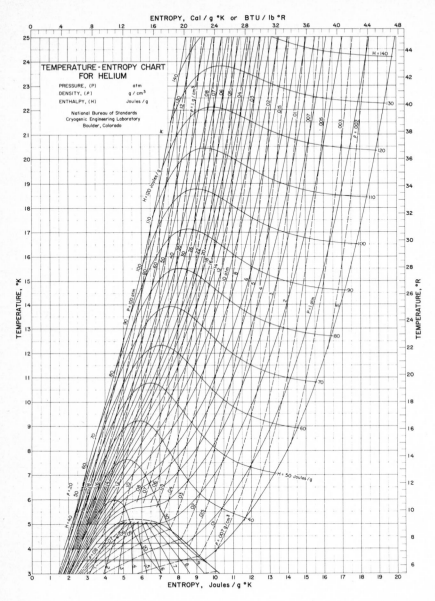

FIG. A-12. Temperature-entropy diagram for helium. (*From National Bureau of Standards, Cryogenic Engineering Laboratory, Boulder, Colo.*)

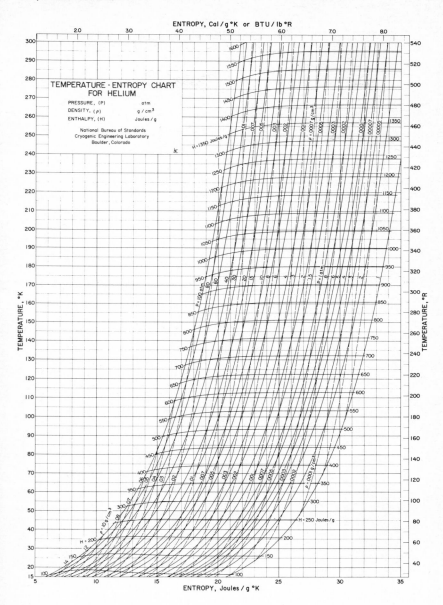

FIG. A-12. (Continued)

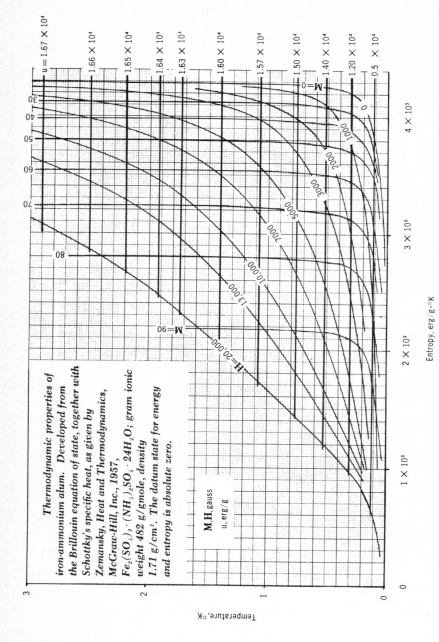

FIG. A-13. Temperature-entropy diagram for iron-ammonium alum. (*From W. C. Reynolds, "Thermodynamics," 2nd ed., McGraw-Hill Book Company, New York, 1968.*)

Thermodynamic properties of iron-ammonium alum. Developed from the Brillouin equation of state, together with Schottky's specific heat, as given by Zemansky, Heat and Thermodynamics, McGraw-Hill, Inc., 1957, $Fe_2(SO_4)_3 \cdot (NH_4)_2SO_4 \cdot 24H_2O$; gram ionic weight 482 g/gmole, density 1.71 g/cm³. The datum state for energy and entropy is absolute zero.

M.H. gauss

u, erg/g

Entropy, erg/g·°K

Temperature, °K

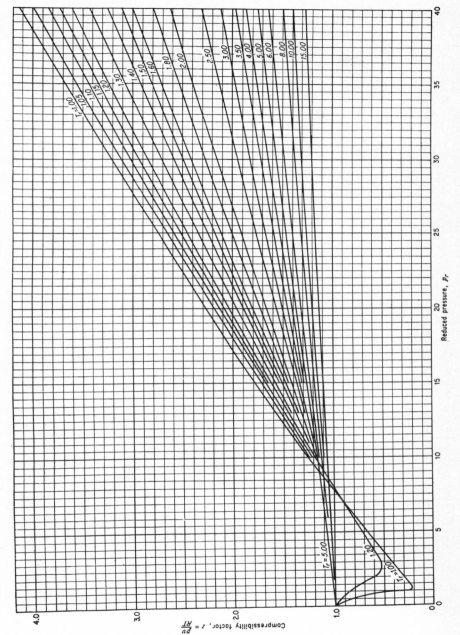

FIG. A-14. Generalized compressibility chart (up to $p_r = 40$). (*After E. F. Obert, "Concepts of Thermodynamics," McGraw-Hill Book Company, New York, 1960.*)

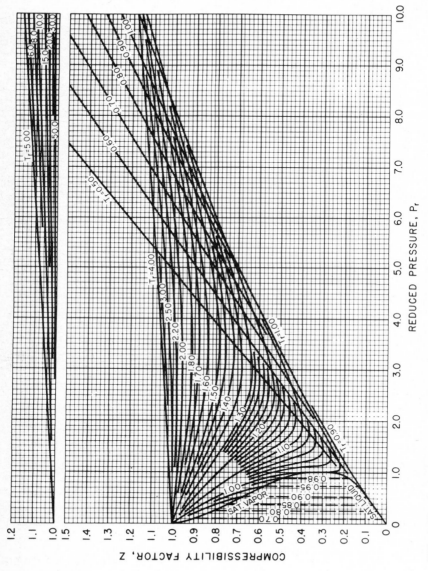

FIG. A-15. Generalized compressibility chart (up to $p_r = 10$) for fluids with a critical compressibility factor of 0.29. [*From J. S. Hsieh, J. Eng. Ind. Trans. ASME, Series B, 88:263 (1966)*.]

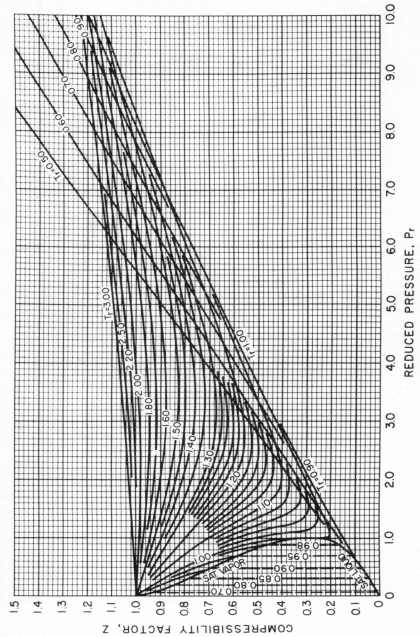

FIG. A-16. Generalized compressibility chart (up to $p_r = 10$) for fluids with a critical compressibility factor of 0.27. [*From J. S. Hsieh, J. Eng. Ind. Trans. ASME, Series B, **88**:263 (1966).*]

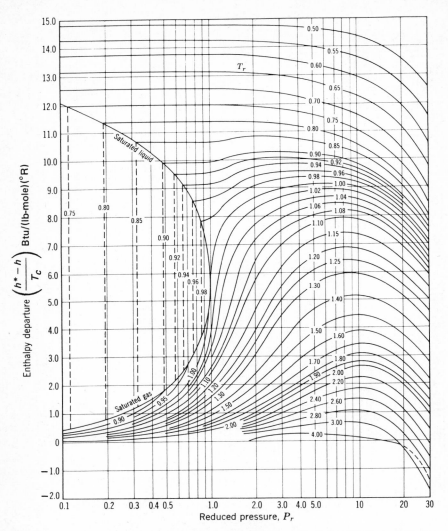

FIG. A-17. Generalized enthalpy correction chart. (*From G. J. Van Wylen and R. E. Sonntag, "Fundamentals of Classical Thermodynamics," 2nd ed., John Wiley & Sons, Inc., New York, 1973.*)

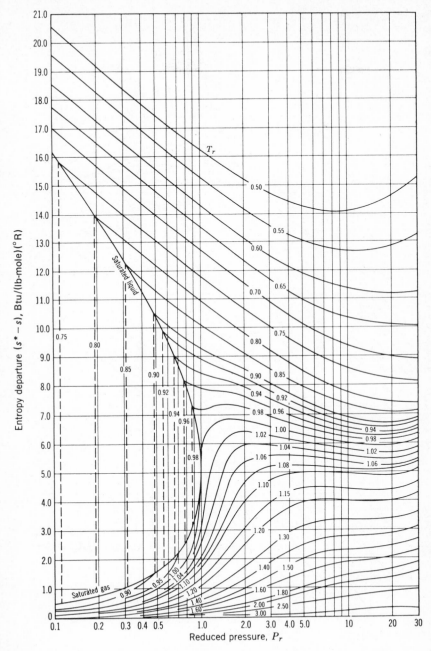

FIG. A-18. Generalized entropy correction chart. (*From G. J. Van Wylen and R. E. Sonntag, "Fundamentals of Classical Thermodynamics," 2nd ed., John Wiley & Sons, Inc., New York, 1973.*)

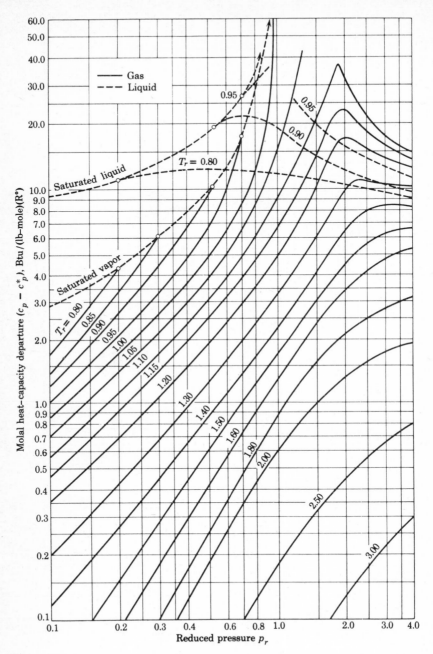

FIG. A-19. Generalized specific heat correction chart. (*From O. A. Hougen, K. M. Watson, and R. A. Ragatz, "Chemical Process Principles, part II, Thermodynamics," John Wiley & Sons, Inc., New York, 1959.*)

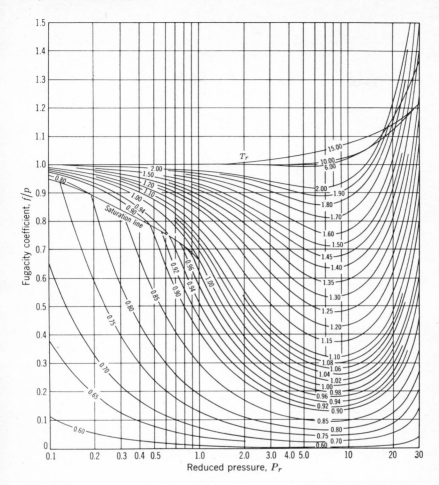

FIG. A-20. Generalized fugacity chart. (*From G. J. Van Wylen and R. E. Sonntag, "Fundamentals of Classical Thermodynamics," 2nd ed., John Wiley & Sons, Inc., New York, 1973.*)

ANSWERS TO
SELECTED PROBLEMS

1-1. (a) 6.93×10^4 joules; (b) 2.33×10^5 joules.

1-2. (a) p (in newton/m^2) $= 2.026 \times 10^6 \; V - 1.013 \times 10^5$;
(b) 7,598 joules; (c) 5,065 joules; 2,533 joules.

1-3. 1,444 ft-lbf

1-4. 1.84×10^4 joules

1-5. $246°F$

1-6. $81°C$

1-7. $151°C$

1-8. 0.103 lbm; 72.3%; -26.1 Btu.

1-9. 2,976 kJ/kg mole; 8.555 kJ/(kg mole)($°$K)

1-10. 808 joules/$°$K

1-11. 236 kW

1-12. 625 joules/g; 1.6 kg/sec.

1-13. 6.59×10^6 joules/sec

1-14. 1.54 hr

1-15. (a) 168,000 ft-lbf; (b) 3.16%.

1-17. 0; 1.364×10^6 joules; 2.781×10^3 joules/$°$K.
3.675×10^5 joules; 4.046×10^5 joules; 6.39×10^2 joules/$°$K.
-2.091×10^5 joules; -1.611×10^6 joules; -3.419×10^3 joules/$°$K.
8.93%.

2-3. (a) $\alpha = T^{-1}$; $\kappa = p^{-1}$

(b) $\alpha = \left(1 + \dfrac{a}{vRT}\right) T^{-1} \left(\dfrac{v}{v-b} - \dfrac{a}{vRT}\right)^{-1}$

$\kappa = p^{-1} \left(\dfrac{v}{v-b} - \dfrac{a}{vRT}\right)^{-1}$

(c) $\alpha = \left(\dfrac{R}{2} + \dfrac{a}{2vT}\right)\left[\dfrac{R^2 T^2}{2p(v-2b)} e^{-a/vRT} - \dfrac{a}{2v}\right]^{-1}$

$\kappa = e^{a/vRT}\left(\dfrac{RT}{v-2b} + \dfrac{a}{2bv}\ln\dfrac{v-2b}{v}\right)^{-1}$

2-4. $\kappa_T - \kappa_s = Tv\alpha^2/c_p$

2-10. 2.36 kJ/(kg)($^\circ$K)

2-11. 9.49°K/atm

2-12. 5,790 joules/g mole

2-13. 0.026 cal/(g)($^\circ$K)

2-14. 82.7 kJ/kg; -78.1 kJ/kg.

2-15. -18.7 Btu;   0;   0.
26.8 Btu;   93.6 Btu;   0.119 Btu/$^\circ$R.
0; -85.5 Btu; -0.119 Btu/$^\circ$R.
8.65%

2-16. -18.77 Btu;   0;   0.
26.75 Btu;   95.23 Btu;   0.1206 Btu/$^\circ$R.
0; -87.23 Btu; -0.1206 Btu/$^\circ$R.
8.40%.

2-17. 0;   3,120 kJ;   3,120 kJ;   7.41 kJ/$^\circ$K.
3,120 kJ;   0; $-3,120$ kJ;   0.
$-2,590$ kJ; $-2,590$ kJ;   0; -7.41 kJ/$^\circ$K.

3-3. (a) 0.271;   (b) 0.375;   (c) 0.333.

3-4. $Z = 1 + \dfrac{1}{v}\left(b - \dfrac{a}{RT^{3/2}}\right) + \dfrac{1}{v^2}\left(b^2 + \dfrac{ab}{RT^{3/2}}\right) + \dfrac{1}{v^3}\left(b^3 - \dfrac{ab^2}{RT^{3/2}}\right) + \cdots$

3-6. $p_r = \dfrac{T_r}{2v_r - 1}\exp\left[2\left(1 - \dfrac{1}{T_r v_r}\right)\right]$

$p_r = (8 - T_r)\exp\left(\dfrac{5}{2} - \dfrac{4}{T_r}\right)$

3-7. (a) 0.62 m^3;   (b) 0.33 m^3.

3-11. 1.851 Btu/(lbm)($^\circ$R);   1,794 Btu/lbm.

3-12. -299 kJ/kg; -316 kJ/kg.

3-13. (a) 198 kW;   47,400 cal/sec;   (b) 233 kW;   62,400 cal/sec.

3-14. (a) 471°K;   90.8 liters; $-691,000$ joules;
(b) 481°K;   84.0 liters; $-664,000$ joules.

3-17. 147.7 kJ/kg; 449 joules/(kg)($^\circ$K).

3-18. 337°K; 94 m/sec.

4-4. 1.550 ft^3/lbm mole; 1.956 ft^3/lbm mole.

4-5. 1.09 ft^3/lbm mole; 1.62 ft^3/lbm mole.

4-6. 82 Btu/lbm mole

4-7. $B = \Sigma(B_i x_i^2)$; $C = \Sigma(C_i x_i^3)$; $D = \Sigma(D_i x_i^4)$; etc.

4-8. (a) 0.168 ft^3/lbm; (b) 0.136 ft^3/lbm; (c) 0.138 ft^3/lbm.

4-10. 1.28 kg

4-11. (a) 196 kg; (b) 292 kg; (c) 312 kg.

4-12. (a) 387 joules/(kg)($^\circ$K); 185 kW;
(b) 352 joules/(kg)($^\circ$K); 180 kW.

4-14. (a) 293 m/sec; (b) 317 m/sec.

5-7. 15,600 atm

5-8. 0.80

5-9. 0.869; 0.531; 0.281; 0.072.
0.956; 0.778; 0.536; 0.178.

5-10. 586; 682; 779; 874; 970; 1,066; 1,162; 1,258; 1354.
0.247; 0.425; 0.558; 0.664; 0.748; 0.816; 0.873; 0.922; 0.964.

5-12. 100.6°C

5-13. -7.41°C

5-14. 8,440 cal/g mole

5-15. $\Delta T = \dfrac{RT_0^2}{h_{1fg}^0}\left(1 - \dfrac{k_2}{p_1^0}\right)x_2$

6-1. $\dfrac{T}{T_0} = \exp\left[\dfrac{\beta(\epsilon - \epsilon_0)}{c_\epsilon}\right]$

6-2. 50.75 newtons

6-3. (a) 0.176 kN; (b) 0.0841 kN.

6-4. -0.0317 joule; 0.359 joule; 0.391 joule.

6-5. -0.121°C

6-10. (b) -0.718°C/m; 2.19°C/m.

6-12. -162 joules/kg; 270 joules/kg.

6-14. 15.2 atm

6-15. $W = -8\pi\gamma R^2 + \frac{4}{3}\pi R^3 p_{atm}$

7-1. (a) 800 amp/m; (b) 1.005×10^{-3} weber/m^2; (c) 2.01 weber/m^2; 1.60×10^6 amp/m.

7-6. (b) $dU = T\,dS + V\mathbf{E}\,d\mathbf{P}$; $dH = T\,dS - V\mathbf{P}\,d\mathbf{E}$; $dA = -S\,dT + V\mathbf{E}\,d\mathbf{P}$; $dG = -S\,dT - V\mathbf{P}\,d\mathbf{E}$.

7-9. (b) $dU = T\,dS - p\,dV + \mathbf{E}\,d\mathbf{P}'$
$dH = T\,dS + V\,dp - \mathbf{P}'\,d\mathbf{E}$
$dA = -S\,dT - p\,dV + \mathbf{E}\,d\mathbf{P}'$
$dG = -S\,dT + V\,dp - \mathbf{P}'\,d\mathbf{E}$

7-11. 12.2 kJ

7-13. 0.633 atm; 7.41×10^{-4} g/cm^3; 0.123 atm; 0.510 atm.

7-14. 53 km

7-15. (a) 33.2%; 66.8%; (b) 0%; 100%.

8-2. 27/4

8-4. 11.4 kW

8-5. 250 atm; 475 kJ/kg.

8-6. 0.079 kg liquefied/kg gas compressed; 1.48 kW-hr/kg liquefied; 171°K.

8-7. 0.115 kg liquefied/kg gas compressed; 1.02 kW-hr/kg liquefied; 267°K; 166°K.

8-8. 0.154 kg liquefied/kg gas compressed; 2,520 kJ/kg liquefied; 4,450 kJ/kg liquefied.

8-12. 1.87 (cm^3)(°K)/g ion; 23.5×10^{-6} (m^3)(°K)/g ion

8-13. (a) 12.3 joules; 12.3 joules; 1.23 joules/°K; (b) 1.52°K.

9-5. 0.148 joule; -0.352 joule.

9-8. 2.748×10^4 amp/m; -1.385×10^4 amp/(m)(°K).

9-10. 207.4

9-11. 1.87×10^6 amp/m; 3.85×10^3 amp/m.

9-13. 0.9998; 4,999; 52.1 m/sec.

9-14. $\dfrac{\rho}{\rho_n} = 1 + \dfrac{0.4(2.476T - 2.955)}{(1.238T^2 - 2.955T + 1.847)^2}$

10-4. 48.41 cal/(g mole)(°K)

10-5. 62.58 cal/(g mole)(°K)

10-6. 36.31 cal/(g mole)($^\circ$K)

10-7. 36.53 cal/(g mole)($^\circ$K)

10-8. 434 cal/g mole

10-9. 1.459 cal/(g mole)($^\circ$K)

11-1. 10.03, 3.82, 86.15%;   130°F.

11-2. (a) 4.65 ft^3/ft^3 (e) 5.32 lbm/lbm
         (b) 12.5, 4.7, 82.8%   (f) 128°F
         (c) 6.32 lbm/lbm   (g) 136°F
         (d) 0.57 lbm/lbm

11-3. (a) 90.39, 9.61%;   (b) 22.08 kg/kg;   (c) 162%.

11-4. $-$ 493 kcal

11-5. 962°K

11-7. (a) $\log_{10} K(T) = -5.023$;   (b) $\log_{10} K(T) = -5.018$.

11-8. $\log_{10} K(T) = 0.859$

11-10. $K(T) = 0.05281$

11-11. 0.624

11-12. (a) 0.358,   0.463,   0.179;
          (b) 0.254,   0.619,   0.127.

11-13. 0.08732,   0.00725,   0.18616,   0.00298,   0.00512,
0.71117; $-$ 6,438 cal/g mole.

11-14. (b) 1.06 volts;   92%.

11-15. (b) 1.02 volts;   100%.

INDEX

INDEX